T0310134

Managing Performance in Construction

Managing Performance in Construction

Leonhard E. Bernold
Simaan M. AbouRizk

JOHN WILEY & SONS, INC.

This book is printed on acid-free paper. ♾

Published by John Wiley & Sons, Inc., Hoboken, New Jersey
Published simultaneously in Canada

Library of Congress Cataloging-in-Publication Data:

Bernold, Leonhard E. (Leonhard Emil), 1952-
Managing performance in construction / Leonhard E. Bernold and Simaan M. AbouRizk.
p. cm.
Includes index.
ISBN 978-0-470-17164-6 (cloth)
1. Construction industry–Personnel management. 2. Labor productivity. I. AbouRizk, S. M. II. Title.
HD9715.A2B483 2010
338.4'76240285–dc22

2009045968

Printed in the United States of America

10 9 8 7 6 5 4 3 2 1

We would like to dedicate this book
to our families for their continued love and support.
Our gratitude goes to our wives Marilyn and Marleine and
to our five beautiful daughters Elizabeth, Sarah, Hala,
Jenna and Deema

Contents

Preface

I hear and I forget. I see and I remember. I do and I understand.

——Confucius

Addition for construction managers: I think and I innovate.

In 2005, we were offered the opportunity to write this book to present the state-of-practice and knowledge related to the wide range of performance aspects in the construction field. From the outset we felt that a book of this sort should engage the students on many different levels in order to provide a rich set of learning experiences that support one another. For example, we include process simulation software, to provide an opportunity for students and practitioners to gain hands-on experience with modeling and learn how to rapidly evaluate different construction methods and supply-chain configurations. Other important features in the book are pedagogical support of the teaching and learning process, and a companion website, with a wide selection of additional material and supports.

1.1 TAPESTRY OF LEARNING

Deep learning is a complex process, whereby the teacher is both the enabler and facilitator. Einstein once said that one does not understand something until one can explain it to one's grandmother. Obviously, he had more in mind than just parroting what was presented in class. Why would someone's grandmother even be interested in hearing about theories or mathematical formulas? The answer is, only if it is relevant to her and builds on what she already knows. Furthermore, it must be translated into language that she understands.

A textbook does not have to be dry and sterile. This one engages the reader in a conversation, with the intent of stimulating interaction with the teacher. It presents small problems leading the student to follow the thoughts of an experienced engineer, in the same way an apprentice looks over the shoulder of his or her master. Equally important, each reader is recognized as having unique learning

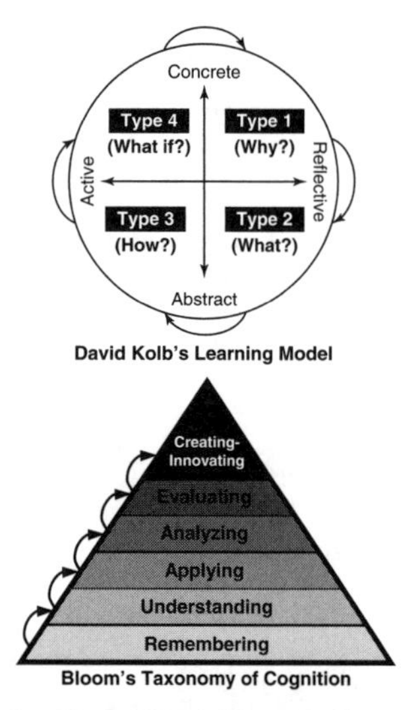

Figure P.1 David Kolb's Learning Model (top); Bloom's Taxonomy of Cognition (bottom)

strengths, as illustrated by David Kolb's learning model, shown at the top of Figure P.1. Various methods and techniques are employed in each chapter to enable all readers to apply their individual learning strengths.

What is the feeling one gets when one really understands new material (before explaining it to grandmother)? One understands the theory behind it, knows how to apply it to solve problems, knows why it is important, and, finally, has an instinctive feeling what would happen if critical variables were changed. In fact, one has visited all four quadrants of Kolb's circle, while scaling Bloom's learning pyramid (bottom of Figure P.1), beginning with memorizing the facts and ending up being able to apply the new knowledge creatively. The teacher is considered the coach, who leads the students through the process of acquiring knowledge via a personal and guided inquiry into the subject. In particular, students will be inspired to tap into their creative and innovative capabilities, and find many opportunities to practice and, ideally, earn well-deserved credits toward their final grades.

The vast majority of engineering students receive very little education related to the historical accomplishments that led to the construction technologies they see almost every day. To fill this vacuum, we add short sidebars to the main text when appropriate in order to provide interesting stories and to encourage an extended inquiry. In the same vein, we "assign" homework and short classroom exercises, which require students to collect, analyze, and evaluate data as additional means for building bridges to the new knowledge they are acquiring. Students can assess how well they understand what they have learned, and detect where their weaknesses are. These activities ensure that students are in control of the learning process, and take an active role. Finally, this book offers a unique opportunity for the teacher to foster reflective learning through journaling

There is no doubt that the most prominent engineer "journaler" lived and worked in the sixteenth century. Who does not admire Leonardo da Vinci's sketches of lifting cranes and flying machines, and the text and calculations that accompany them explaining how and why the proposed system would work.

Generic Structure of a Class Journal

A. Executive Summary

B. Table of Contents

C. Weekly Journal Entries. Address the following five topics:

1. Weekly review:
   What happened in class?
   What did we have to do?
2. Personal discussion. Reflections about each main subject of the week:
   Why are the new principles and methods important?
   How does the new material relate to what I already know?
   What are real-world problems that I can solve now?
   What helped me most to understand?
3. Special journaling questions relating to the chapter (assigned by the professor)
4. New terms: the meaning of 10 new (to me) terms in construction
5. My favorite sketch/figure/picture that communicates most effectively.

D. Supplementary Materials:
   Personal observations
   Evaluations
   Poems
   Newspaper articles
   PowerPoint presentations
   Websites
   Other

Many of the skills that students practice while journaling are among those suggested by the Accreditation Board for Engineering and Technology (ABET): communication, analytic skills, information literacy, lifelong learning, knowledge of contemporary issues, holistic approach to engineering problems, and reflective thinking about professional issues. One of the most satisfying experiences for a teacher is to observe students struggling during a semester, then see them present their journals at its end with pride and a smile, and say, "This was the toughest thing I did in school, but it will help this class experience guide my future professional life."

1.2 BOOK CONTENTS

The material in *Managing Productivity in Construction* is presented using two integrated media: the traditional textbook and the supporting website, the latter which offers additional supplementary materials for students to explore, if they wish.

Some of the chapters contain features called header problem and/or worked-out problem, positioned at "pressure points." Both features are nontraditional, in that they are embedded in real-world situations that can occur during a construction project. Rather than providing straightforward solutions, they guide readers through problem-solving exercises, requiring them to make assumptions and collect and analyze data; the problems conclude with a discussion of the results, adding a qualitative component and final recommendations to make to a "virtual boss."

Each chapter ends with a chapter review and bibliography. The chapter review includes three components: (1) journal questions, (2) traditional homework problems, and (3) open-ended problem. The journal questions are tailored to each chapter and can be selected by the teacher to amplify the generic outline of a journal.

Managing Performance in Construction covers a wide spectrum of topics in 11 chapters, beginning with the concept of productivity measurements and process modeling. Here is a quick overview of the chapters:

Chapter 1, "Indicators of an Industry in Transition," identifies some of the early indicators of an industry in transition. A major change in the way we design, plan, and even control construction projects will be caused by the industry-wide adoption of the Building Information Modeling (BIM) system. This thrust will be accompanied by the further growth of e-construction management, which will eventually digitally link construction equipment in real time to a project's website. Another major impact on the industry is the drive toward sustainable and ecoefficient construction.

Chapter 2, "Productivity in the Spotlight," describes the many different ways productivity is measured. It details the concept of process productivity in construction, while highlighting the causes of nonproductive work. This chapter also presents the seven types of *muda*, Japanese for "waste," or "wasteful activity," a very effective concept first introduced by Taiichi Ohno, Toyota's chief engineer. In short, a plan has little meaning if it is not accompanied by a method to control the process to meet it. Thus, the chapter introduces scientific methods to measure how well construction is progressing on the most critical level: the production process. Finally, it offers various methods useful in identifying the factors leading to high, as well as poor, productivity.

Chapter 3, "Cornerstones of Efficient Site Operation," begins by quoting the thirty-fourth president of the United States, Dwight D. Eisenhower: "Plans are nothing; planning is everything." This chapter strives to live up to this premise, one that has been uttered by many others.

The disappearance of the medieval Master-Builder who constructed those gigantic cathedrals in Europe caused a long list of problems for construction professionals, key of which was the loss of construction expertise during the design phase. This chapter reviews the emergence of constructability as a critical issue of securing efficiency during construction. It goes on to present the Work Breakdown Structure (WBS) and Organizational Breakdown Structure (OBS) as key elements of integrating the various participants in a project. Finally, Chapter 3 works through several examples that demonstrate how construction processes are tightly linked to the efficiency of the supply chain. Real-world construction examples also are used, to illustrate the importance of planning and implementing highly efficient procedures.

Chapter 4, "Introduction to Simulation and Its Use in Modeling Production Systems," takes as its premise that as the twenty-first century advances, there will be no physical labs, that most experimentation will take place using computer simulation. Experimenting in the real world is often too expensive and impractical, especially in construction. In this chapter, we introduce readers to a new generation of computer simulation systems that facilitate experimenting with complex construction operations with relative ease. Specifically, we introduce Simphony and its simulation templates and show how they can be used in modeling construction systems in general and construction operations in particular. Once a model is built, we can then experiment with it for the purpose of understanding its production structure, its bottlenecks, and the means of improving its productivity.

Chapter 5, "A Case Study: Applying Simulation in Tunnel Construction," assesses, models, and simulates a real-life project. The purpose here is to give the reader as near as possible an actual experience of building large-scale models of complex systems. To that end, we introduce a project that we developed during a consulting assignment, involving the installation of a large water main inside a tunnel over a stretch of approximately 500 meters. The operation we model includes construction of the shafts and the tunnel and the installation of the pipe inside it. We simplify the modeling strategy in this chapter to suit the broad range of readers, then on the website provide the full models so that the advanced reader can further investigate their features.

Chapter 6, "Competencies That Drive the Company," examines the present thinking around the competencies successful workers and managers need to develop. The knowledge and skills necessary today have gone far beyond what was thought sufficient just 30 years ago. The chapter describes how core competences depend, as never before, on the personal and continued initiative of the individual, one who makes lifelong learning a key element of his or her development.

A second section in chapter 6 presents modern methods for assessing the performance of a manager and explains how to turn the evaluation into a tool of constant improvement.

Chapter 7, "Productivity in a Healthy and Safe Work Environment," discusses the many physical and the emotional strains that impact both on-site workers and managers in their offices. Ulcers and insomnia are as destructive to the performance of workers as accidents and overexertion on the construction site. Along with the more traditionally known detrimental effects of working in the open environment, such as extremes in temperature and humidity, the health risks of noise and vibration from tools and equipment will also be reviewed. The mechanisms leading to the epidemic of back injuries in the industry also are presented.

The chapter closes with a look at the increasingly destructive forces of the numerous on-the-job stressors—harassment for example, which, when allowed to smolder, costs companies large sums of money. In some cases, targeted individuals and their families are bullied into an abyss from which many never return. A major part of this interesting topic is dedicated to skills training and the learning curve, brought to life via problems and exercises.

Chapter 8, "The Complexity of Human Motivation," addresses the question of what motivates people to do certain things, a topic of study for over 3,000 years. One of the first who formulized needs-based motivation of humans was the American psychologist Abraham Maslow, closely followed by Yale business professor Victor Vroom and the psychologist Frederick Hertzberg. The theories of each of these men are discussed, and some are applied to solve a problem faced by individuals. Modern efforts to improve job satisfaction, such as job enrichment, are also appraised.

Chapter 9, "Performance Factors of Leaders and Teams," considers the question: "Is a manager also a leader?" Warren Bennis, an American pioneer in leadership studies, once wrote: "The manager asks how and when; the leader asks what and why!" This would imply the answer to the question is no. This chapter looks at the differences between the manager and the leader before delving into some of the known leadership models. The need for leaders to possess emotional intelligence is emphasized; then the focus shifts to the behavior patterns of groups, since a group leader must have special competencies to be successful. Finally, a Worked-Out Problem highlights some of the key issues that commonly emerge when groups are brought together to address a complex issue.

Chapter 10, "Communication: The Nerve System of Construction," is dedicated to examining communication from several different perspectives. Miscommunication has started wars, and lack of effective communication has been cited in many sources as the single most important barrier to overcome to improve productivity in construction.

The chapter begins with an overview of the long and fascinating history of the blueprint, which can be traced back more than 4,000 years. Next, the theory of two-way communication applied to the use of the Internet in construction is presented. This is followed by a discussion of the special needs of construction, and their relationships to different established methods and technologies. The chapter ends with a review of the remarkable opportunities offered today by agent-based communication, balanced by a look at the perils of face-to-face group discussions, addressed using theory and examples.

Chapter 11, "Performance Management," concludes *Managing Performance in Construction* by pulling together all the main topics discussed in the book. It makes the case for moving from a productivity-only focus to measuring business success to assessing performance. Here, performance function is defined as consisting of two factors: productivity and effectiveness. It emphasizes that the short-term focus on the output of the operation neglects the fact that the business consists of many other components, which need to be aligned to the long-term business objects. Thus, the performance of the purchasing department running the supply chain is as critical to the success of the company as the financial department securing critical loans at "effective" conditions. The Balanced Scorecard method is used to demonstrate how to utilize a companywide performance assessment as a basis for continuous improvement. The chapter ends with a top-to-bottom review of supply-chain management as one of the keys to improving future construction performance.

Chapter 1

Indicators of an Industry in Transition

Adam Smith (1723–1790) was once hailing the great opportunities offered by the division of labor. This mechanistic view of work has since been matched with views that stress the needs and goals of humans within a social organization. In many ways, the human relations movement has brought laborers and management closer together. Many of the expected changes in the future will bring employees even closer together, through so far un-mined communication technologies and electronics. Many construction equipment companies around the world are already putting the "smart" communication infrastructure into their equipment that will radically change the way they manage construction projects in the future. For the first time ever, construction managers will know at every second how well the project is performing; more, they will able to supervise the site from anywhere in the world.

1.1 BREAKING INTEROPERABILITY BARRIERS

The Building Information Modeling (BIM) system, a parametric and object-oriented approach to drawing buildings, is being embraced by architects and design engineers. BIM is more than detailed 3D modeling: It attaches functions and features to each object, to identify it as a unique element in relationship to its surroundings. For example, the function attached to a door might be its ability to swing around the hinge point; and its feature would include the make, weight, color, and so on. BIM allows designers and builders to identify all attributes of a building element, such as windows, HVAC ducts or elevators, and makes it easy to conduct interference and constructability checks.

The potential savings in time—to design, distribute the drawings electronically, and switch smoothly between different software packages (e.g., structural analysis)—has captured the attention of many change agents inside large Architectural-Engineering-Construction (AEC) companies. Even more, the structural steel suppliers in the United States have been on the forefront in taking advantage of creating interoperable software systems. Tests demonstrated how last-minute design changes can be transmitted electronically from a design software directly to the fabricators shop equipment. This ability is unprecedented in the construction industry and gives designers and fabricators many opportunities to reduce costs, while ensuring that project objectives are met. This is just the beginning; we can expect more radical changes in the future in the way we bid, supply, and generally manage construction projects. These changes will encourage the industry to adopt an ontology that will become commonplace, not only between architects and construction engineers in the United States, but those around the world. The education of architects and engineers could even be organized around one of the primary strengths of BIM: collaboration. In fact, its support of cross-organizational and global collaboration among project participants is expected to become a major benefit for the industry.

1.2 CONSTRUCTION BECOMES SUSTAINABLE

According to the World Business Council for Sustainable Development (WBCSD), critical aspects of ecoefficiency are:

- Reduction in the material intensity of goods or services
- Reduction in the energy intensity of goods or services
- Less dispersion of toxic materials
- Improved recyclability
- Maximum use of renewable resources
- Greater durability of products
- Increased service intensity of goods and services

Like BIM, ecoefficiency is changing the way we construct, as we seek to produce sustainable, ecofriendly structures. Sustainable buildings are achieved through integrated building design, an approach to design and construction that involves the participation of owners, contractors, suppliers, building users, and design professionals. Notably, the U.S. Green Building Council (USGBC) has, to date, certified more than 20,000 projects using its Leadership in Energy and Environmental Design (LEED) rating system, introduced in 1998. The nonprofit organization promotes buildings that are environmentally responsible, profitable, and healthy places to live and work. LEED awards points for relative efficiency improvements in the use of energy, water, and materials. But LEED covers only the final building, so a special standard has since been added: LEED for New Construction and Major Renovation (LEED-NC). It focuses on safeguarding and conserving water, energy, and the atmosphere, while eliminating the creation of material waste. Green building points can be earned, for example, by recycling old building materials or breaking up concrete on-site to create a permeable ground for rainwater to pass through. Replacing cement with fly-ash, or reducing the amount of cement, also earns points for producing less CO_2.

Low-Impact Development (LID) is also generating a lot of interest, as it addresses one of the main problems facing the United States, the dry-pumping of aquifers, combined with evaporation of large amounts of water. This innovative stormwater management approach uses rainfall as a major resource, rather than a nuisance. It creates opportunities for the rainwater to infiltrate the local ground, to be filtered and stored, instead of being piped away.

Maintaining a sustainable earth means ensuring that the next generation will take over a world in much the same state and with the same amount of resources as in the previous generation. Two building materials that meet this goal extremely well are wood and bamboo. Using wood as a building material will resurrect an old and almost-lost craft, lead to innovation in wood structural design, and reduce the cost of timber by creating global economic incentives to plant different kinds of trees for economic benefits.

Construction and demolition (C&D) waste contributes 25 percent of the total waste stream in the United States, and many landfills have already reached capacity. Instituting stricter new regulations, raising public awareness, and opening new landfills are not always options, or will become expensive to sustain. Therefore, reducing and recycling C&D waste has become paramount. One new concept calls for mining old landfills to recover and recycle decades of C&D waste.

These developments are additional signs that construction will continue to change.

1.3 E-CONSTRUCTION MANAGEMENT

E-business, e-collaboration, e-communication, and wireless communication have made major inroads into many industries. E-business refers to electronic or Internet-based commerce, supply-chain management, customer relations, and companywide communications. The resulting benefits are faster interactions, better prices and delivery terms, and better support from suppliers, who are able to reduce their administrative burden and human error. Although construction is lagging behind in seeing the increased efficiencies and savings that are being reaped by users in other industries, this will undoubtedly change. The benefits of a matured technology that is able to survive the rough construction environment is something the construction cannot afford to pass up.

Similarly, Internet-based project collaboration services have seen steady growth. They offer functions such as team directories, file sharing, document archiving, task management, project scheduling, and current budget status, all accessible by authorized personnel 24/7. These services will further develop to include virtual, live project-site meetings. An inspector on-site will be able to show a problem with his wireless mobile network camera as engineers and suppliers, all connected to the Internet, all review the video and the relevant drawings simultaneously. Internet phone services will provide the audio, to

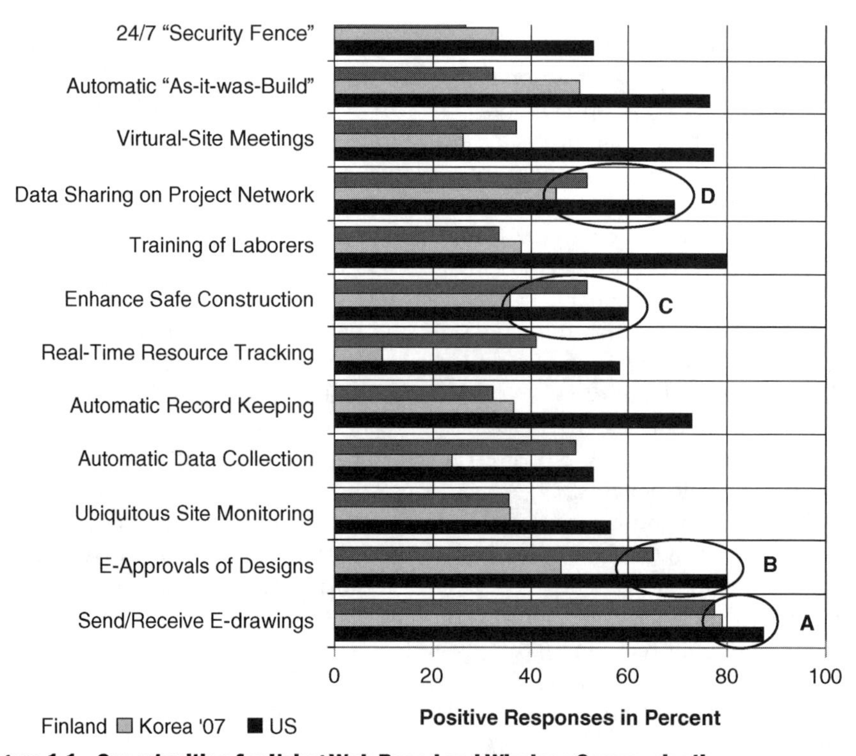

Figure 1.1 Opportunities for Using Web-Based and Wireless Communication

enable brainstorming on a shared whiteboard. These services are all about saving significant amount of time solving a real problem, eliminating weeks of waiting for an answer to a Request for Information (RFI).

Another indicator of future changes was a survey made in 2007 of 300 construction managers from three countries—the United States, South Korea, and Finland—about their beliefs regarding where Web-based wireless communication could provide important benefits. While there were significant differences in their responses, as shown in Figure 1.1, there were also some interesting commonalities.

In general, the 100 American managers, located in different regions of the country, were more optimistic about the use of Web-based communications than their counterparts in Finland and South Korea. Indicated by the labels A, B, C, and D are areas where one or two country representatives have the highest expectations. We point out, however, that A, B, and D are those areas where these services are already being heavily used, and thus are most familiar to the respondents. We already discussed how BIM will facilitate the rapid growth of those same areas. New areas include construction safety as automated sensor systems protect the workers from exposure to hazardous environments. Finally, the American managers thought that Web cameras could change the way laborers are trained, provide automatic "as-it-was-built" updates, and support virtual site meetings.

1.4 LINKING UP TO SMART CONSTRUCTION EQUIPMENT

Change is progressing in the way construction professionals manage, maintain, and protect their equipment. Combining "smart" agent-based software, global positioning systems (GPS), and wireless satellites and local communication systems, vast networks have been created to track and receive information from sensors built into equipment.

Geofencing is one very useful application of what is referred to as *telematics*. With this application, a contractor is able to install a virtual fence around a construction site, which is being monitored by GPS and local positioning tools. Every time a person, vehicle, piece of equipment, or anything else equipped with a hidden or open tag crosses the virtual fence, the incident is automatically recorded, or an alarm notification is sent to an "agent"—such as the local police station. Wireless sensors or tags, such as smart radio frequency identification (RFID) tags, can be integrated into the system to collect data or to send a warning. Thus, for example, equipment rental agencies will be able to automatically track how their units are being treated, and protect them from being operated by an untrained or unauthorized person.

A related change has begun in the way operators run dozers, graders, or scrapers. After equipment manufacturers switched to electronic controls and differential GPS or laser-based systems, real-time 3-D position data of cutting blades are generated electronically. These sophisticated equipment still require human operators, but this, too, will change over time. Having the ability to send commands to hydraulic actuators electronically opens the door for full automation. In other words, for suitable tasks, such as directing the blade of a dozer or grader, the human operator is being replaced with a "smart" and fast-reacting automatic controller. A computer calculates the discrepancy ($\Delta Z(x,y)$) to the desired contour and sends signals to the hydraulic cylinders, which move the blade into the desired position, while the human operator focuses on steering the equipment safely and quickly.

How Global Positioning Systems Work

A global positioning system (GPS) is made up of three parts: (1) 24 satellites, which orbit the Earth every 12 hours on almost the same track; (2) control and monitoring stations on Earth; and (3) GPS receivers owned by users. GPS satellites broadcast signals from space that are picked up and identified by GPS receivers. Each GPS receiver then provides a three-dimensional location (latitude, longitude, and altitude), plus the time. Due to ionospheric and tropospheric distortions of the signals, however, errors may be introduced into the calculation of coordinates of the target receiver. To correct most of these inaccuracies, differential GPS uses the location of a known reference station to constantly recalibrate the coordinates.

Source: www.gps.gov

1.5 HIGHLY SUCCESSFUL ENGINEERING MANAGERS

Managers in engineering and construction spend, on average, 60 percent of their time on managerial tasks and 35 percent on engineering-related work. Further, the majority of the managerial time is spent addressing human resources issues, which requires what is referred to as "soft leadership skills" or "emotional intelligence." Ever since David McGregor published

The Human Side of Enterprise in 1960, in which he introduced his Theory X and Y, managers and employees have started to move away from X toward Y. This change was accelerated by research results that showed that successful managers depend 85 percent on emotional intelligence and 15 percent on technical skills and knowledge. It was also found that 80 percent of managers who fail do so because they lack soft leadership skills. This parallels observations by successful engineer managers, who state that they use only 5 percent of what they learned in college, but had to acquire 85 percent of what they really needed on their own—a large part of which was learning to work with people.

Organizations wishing to systematically improve their performance must expand their knowledge base. In this regard, change is underway in the way knowledge gain is fostered. The main driver of this development is the highly successful 360-degree appraisal, or feedback method, for managers ("360 degrees" refers to the full circle that surrounds an individual working inside a company). It is used to identify, early, key skills that are lacking in an aspiring manager. When an appraisal is done, feedback is solicited organizationwide—from subordinates, peers, and supervisors—and includes a self-assessment. The results are used to direct skills training and continuing education.

Soft skills that are typically needing development in today's engineers are:

- Synergistic thinking
- Interpersonal communication
- People development—coaching
- Leadership
- Motivation
- Workplace conflict resolution
- Teamwork collaboration
- Change management
- Stress management

A Story of Whats and Whys

Michelle Brink couldn't believe her eyes as she looked at her computer screen, which displayed the results of several what-if runs she had made with her new process simulation software. She was looking at a possible productivity increase between 20 to 30 percent in laying block and brick, compared to the present method. And the necessary change to achieve it was so minor.

Michelle received a BS in civil engineering three years ago, and has been working for Prime Value Builder, Inc. (PVB) a major developer of single-family homes and town houses. For the first year she worked in scheduling, and then spent a year and a half in estimating, where she received excellent 360-degree performance evaluations. Aspiring to become a construction manager, she was promoted to assistant to Jerry Foresight, a reliable "old hand" in the company.

PVB's human resources department had been following her career and decided to manage her further growth in the company. First, she was recommended for a continuing education course, called Kaizen management, offered by a major management consultant company. Michelle had never heard of Kaizen management, and so had no clue what to expect. The HR people cautioned her that the course would be tough but that afterward she would be well informed and well prepared to advance her career.

From the first minute she was intrigued as the course first covered the history of management theories, concepts she had heard about but never fully understood. But what most fascinated her was the period following World War II with the rapid introduction, in quick succession, of new management concepts such as the Work Breakdown Structure (WBS) and the Critical Path Method of scheduling, or the Gantt chart. But she was most impressed by what the Japanese car company Toyota had implemented to improve the quality of its cars while at the same time increasing productivity and reducing costs. What was ironic was that the foundation for the "miracle" Japanese management system had been laid by an American, by the name of W. Edward Deming. He had taught them the basics and they listened while he was dismissed by US companies for many years.

The previous day, the class had covered the concept of Just-in-Time (JIT), which Michelle thought was less interesting than what she learned in the second half of the class, which was dedicated to Just-in-Sequence (JIS) delivery. The key to JIS, a very simple concept, was not just that deliveries were coming to the plant on time but that they also were organized in a first-in-first-out (FIFO) manner, meaning that the parts that were needed first were in front or on top of the containers or pallets. This was in direct opposition to the method used to store luggage in airplanes, which operate on a first-in-last-out (FILO) basis.

Later they were introduced to computer simulations of processes. The class was shown how to change delivery times, to account for rush hour, switching JIS off and on, and so on. Then they could watch on the screen as the effects of the changes took place, in terms of production time and productivity.

What struck Michelle was the realization that when ordering brick and block, for example, she currently worried only about "getting it there" before the subcontractor arrived. But by ordering with JIS in mind, that is telling the supplier to put it right where it was needed on the construction site—the blocks could be delivered inside the footprint next to the foundations, or the bricks along the outside wall. Of course, this must be timed so that the bricks

(*continued*)

and blocks were delivered after the foundations had been poured, at which point Michelle realized that a major problem would be communicating the correct locations to the supplier—or, more importantly, to the truck driver delivering the bricks and blocks. She also realized she would have to factor in that their supplier used trucks with long-reach hydraulic crane booms.

To solve these problems, she thought back to a case study they had worked on in class. The delivery trucks in the case study used *telematics*. Once at the correct address, the driver simply accessed a website that directed him, via a live video feed, from the entrance of the site to his destination, using direction arrows overlaid on the screen. The driver was able tosee exactly where to turn and which door to back up to. In fact, the door opened just as he pulled up, and the forklift driver was already there, ready to unload. They were expecting him, since the telematics system had announced when he was entering town.

But Michelle did not anticipate that the supplier was using telematics and PVB did not have a wireless camera on-site. So she considered alternatives she might use. How about colored stakes? The day before delivery, she would put up the stakes at the locations where she needed the operator to unload cubes of bricks or blocks. She would also calculate how many were needed, and space the cubes appropriately. And, she thought, "If I tell them at the plant to attach coordinating colored paper to the cubes, there is no way the driver will get confused. This has to work!"

Michelle also remembered another lesson emphasized in the Kaizen class: performance measurements as the necessary basis for managing improvements. This motivated her to design a time-and-motion study, to compare performance at a traditionally organized site versus a JIS site. Probably, she thought, she would be able to enlist two of the summer interns to conduct the study, after she trained them how to do it. What an exciting learning experience would this be for them while, at the same time, establishing valuable data to the company.

CHAPTER REVIEW

Journaling Questions

1. Draw up a list of all the concepts and ideas that you do not understand from reading Michelle's story.
2. How will the drive toward sustainable construction affect the way we manage projects?
3. Sketch the communication network for a virtual site meeting to solve a problem detected by a quality control inspector. It should involve the structural engineer, the steel manufacturer, and the architect, each in his or

her respective offices in New York, Detroit, and Baltimore, and the steel erector foreman and inspector, who are both on-site in Nashville.

BIBLIOGRAPHY

AEC3, *Information Delivery Manual*, Thatcham, UK, October 2009. www.aec3.com/2/2_04.htm.

Alsamsam, I., L. Lemay, and M. G. Van Geem. Sustainable High-Performance Concrete Buildings. *J. Structures Congress 2008: Crossing Borders*, ASCE, 2008.

Anderson, L. L., Jr., R. D. Douglass, and B. C. Kaub. Successful Partnering Program on a Megaproject. *J. Leadership and Management in Engineering*, July 2006.

Antony, J., R. Narain, and R. C. Yadav. Productivity Gains from Flexible Manufacturing—Experiences from India. *Inter. J. Productivity and Performance Mgmt*, vol. 53, no. 2, 2004.

Arslan, G., and S. Kivrak. E-Business Transformation Stages for Construction Companies. *Proc. Computing in Engineering*, ASCE, 2007.

Baker, D. The New Economy Does Not Lurk in the Statistical Discrepancy. *Challenge*, vol. 41, no. 4, 1998.

Beaudreau, B. C. Engineering and Economic Growth. *Structural Change and Economic Dynamics*, vol. 16, no. 2, 2005.

Bernold, L. E. Automatic As-Built Generation with Utility Trenchers. *J. Constr. Eng. and Mgmt.*, vol. 131, no. 6, 2005.

— Spatial Integration in Construction. *J. Constr. Eng. and Mgmt.*, vol. 128, no. 5, 2002.

Bosworth, D., S. Massini, and M. Nakayama. Quality Change and Productivity Improvement in the Japanese Economy. *Japan and the World Economy*, vol. 17, no. 1, 2005.

Buchanan, J. Let Us Understand Adam Smith. *J. History of Economic Thought*, vol. 30, no. 1, 2008.

Coleman, G. S., and J. W. Jun. *Interoperability and the Construction Process.* American Institute of Steel Construction, Inc. (AISC), Chicago, IL, 2004; www.aisc.org.

Crane, A. *Rethinking Construction: 2002.* Rethinking Construction Ltd, London, UK, June 2002.

Denzer, A. S., and K. E. Hedges. From CAD to BIM: Educational Strategies for the Coming Paradigm Shift. *Proc. Building Integration Solutions*, Architectural Eng. Institute, ASCE, 2008.

Dickson, M., and R. Harris. Timber Engineered for C21 Architecture. *Proc. Structures Congress 2008: Crossing Borders*, ASCE, 2008.

Earles, A., D. Rapp, J. Clary, and J. Lopitz. Breaking Down the Barriers to Low-Impact Development in Colorado. *Proc. World Environmental and Water Resources Congress 2009*, Great Rivers, CO, ASCE, 2009.

East, E. W. An Overview of the U.S. National Building Information Model Standard, *Proc. Computing in Engineering*, ASCE, 2007.

Egan, J. *Rethinking Construction.* Report of the Construction Task Force to the Deputy Prime Minister, Department of Trade and Industry, London, 1998.

FIATECH. *Capital Projects Technology Roadmapping Initiative, V2.0.* Austin, TX, 2001; www.fiatech.org.

Froese, T., and S. Staub-French. A Unified Approach to Project Management. *Proc. Info. Technology Conf. 2003*, ASCE, 2003.

Goodrum, P. M., C. T. Haas, and R. W. Glover. The Divergence in Aggregate and Activity Estimates of U.S. Construction Productivity. *Construction Management and Economics*, vol. 20, no. 5, 2002.

Gwartney, J. D., R. G. Holcombe, and R. A. Lawson. Economic Freedom, Institutional Quality, and Cross-Country Differences in Income and Growth. *Cato Journal*, vol. 24, no. 3, 2004.

Halfawy, M., and T. Froese. Building Integrated Architecture/Engineering/Construction Systems Using Smart Objects: Methodology and Implementation. *J. Comp. in Civil Engineering*, April 2005.

Hanna, A., and K. T. Sullivan. Impact of Overtime on Construction Labor Productivity. *Cost Engineering*, vol. 46, no. 4, 2004.

Hedges, K. E., and A. S.Denzer. Visualizing Energy: How BIM Influences Design Choices. *Proc. of the 2007 Int. Design Engineering Technical Conf. and Computers*, ASME, 2007.

Hersh, A., and C. Weller. Does Manufacturing Matter? *Challenge*, vol. 46, no. 2, 2003.

International Standards Organization (ISO). *Building Construction—Organization of Information about Construction Works, Part 3: Framework for Object-Oriented Information.* ISO/FDIS 12006-3, 2007; www.iso.org/iso/en/CatalogueDetail.

Issa, R. R., D. Fukai, and M. O. Danso-Amoako. Evaluation of Computer Anatomic Modeling for Analyzing Pre-Construction Problems. *Proc. Construction Research Congress 2003*, ASCE, 2003.

Johnson, D. Selection and Implementation of Web-based Project Management and Technical Collaboration Systems for Port Development Use. *Proc. Ports Conf. 2004*, ASCE, 2004.

Jorgenson, D. W., M. S. Ho, and K. Stiroh. U. S. Economy: Will the Productivity Resurgence Continue? *Business Economics*, January 2006.

Kepke, J., K. Wright, and W. Williamson. Water Resource Management Challenge: Botany Case Study in Sydney, Australia. *Proc. World Environmental and Water Resources Congress 2008*, ASCE, 2008.

Kuran, M. S., and T. Tugcu. A Survey on Emerging Broadband Wireless Access Technologies. *J. Computer Networks*, vol. 51, no. 11, 2007.

Liberatore, M. J., A. Hatchuel, B. Weil, and A. C. Stylianou. An Organizational Change Perspective on the Value of Modeling. *European J. Operational Research*, vol. 125, no. 1, 2000.

McCormick, M., C. Swan, D. Matson, D. M. Gute, and J. Durant. Expanding the College Classroom: Developing Engineering Skills through International Service-Learning Projects. *Proc. World Environmental and Water Resources Congress*, ASCE, 2008.

McLeod, R. Human Factors Assessment Model Validation Study. Health and Safety Executive, Sudbury, Suffolk, UK, 2004.

Metcalfe, J. S., J. Foster, and R. Ramlogan. Adaptive Economic Growth. *Cambridge J. of Economics*, vol. 30, no. 1, 2006.

Michael, L. *Modernising Construction*. Report by the Comptroller and Auditor General, National Audit Office, London, December 2000.

Mirsky, R., and A. D. Songer. Beyond Sustainability: The Contractor's Role in Regenerative System Design. *Proc. Constr. Res. Congress*, April 5 7, Seattle, WA, 2009.

Pheng, L. S. Managing Building Projects in Ancient China: A Comparison With Modern-Day Project Management Principles and Practices. *J. Management History*, vol. 13, no. 2, 2007.

Powell, C. Who Did What: Division of Labour among Construction-Related Firms. *Proc. of the First Int. Congress on Construction History*, Madrid, Spain, January 20–24, 2003.

Rapp, R. R. Evolution, Not Revolution. *J. Management in Engineering*, July, 2001.

Rojas, E. M., and P. Aramvareekul. Labor Productivity Drivers and Opportunities in the Construction Industry. *J. Mgmt. in Engineering*, vol. 19, no. 2, 2003.

ShahramTaj, S., and L. Berro. Application of Constrained Management and Lean Manufacturing in Developing Best Practices for Productivity Improvement in an Auto-Assembly Plant. *Int. J. Productivity and Performance Mgmt.*, vol. 55, no. 3/4, 2006.

Smith, L., J. Smith, and H. Li. The Propagation of Rework Benchmark Metrics for Construction. *International Journal of Quality & Reliability Management*, vol. 16, no. 7, 1999.

Steindel, C., and K. J. Stiroh. Productivity: What Is It, and Why Do We Care About It? *Business Economics*, October 2001.

Van der Zee, D. J. Modeling Control in Manufacturing Simulation. *Proc. of the 2003 Winter Simulation Conference*, New Orleans, LA, 2003.

Williams, T., L. Bernold, and H. Lu. Adoption Patterns of Advanced Information Technologies in the Construction Industries of the United States and Korea. *J. Constr. Engrg. and Mgmt.*, vol. 133, no. 10, 2007.

Winsor, T. A.,Genre and Activity Systems: The Role of Documentation in Maintaining and Changing Engineering Activity Systems, written communication, Sage Publications, 2003; www.sagepublications.com.

Zollinger, W. R., and T. T. Calvey. Is Noncritical Progress Critical? *AACE International Transactions*, CDR. 18, AACE, 2004.

Chapter 2

Productivity in the Spotlight

Productivity isn't everything, but in the long run it is almost everything. A country's ability to improve its standard of living over time depends almost entirely on its ability to raise its output per worker.
—Paul Krugman, *The Age of Diminishing Expectations* (1994)

Productivity is a ratio of output over input that indicates the efficiency of a productive system, be it a three-person crew building a brick wall, the masonry industry of a country, or an entire nation producing many different products. While the first relates to the efficiency of laborers producing specific goods, the latter relates to the economic performance of a country.

2.1 MEASURING NATIONAL PRODUCTIVITY

As Paul Krugman states, productivity growth within a nation or region is a crucial requirement in raising the living standard of its citizens. Productivity growth means that more value is added by productive enterprises, so that more income (profit) is available for distribution. Labor productivity is typically measured as a ratio of output per labor hour and the most widely known measure of productivity is gross domestic product (GDP) per hour worked.

Gross Domestic Product

GDP is equal to the total expenditures for all final goods and services produced within a country during a given year.

GDP = private consumption
+ gross investment
+ government spending
+ (exports − imports)

National productivity measures are often used to compare the economic performance that nations need, to generate economic growth. Figure 2.1 presents such a comparison from 2007 measured in U.S. dollars. National productivity improvements depend on many factors, which interact with each other and may be heavily dominated by one industry sector, especially in small countries. The technological and managerial aptitudes of a country, both fostered by a highly educated

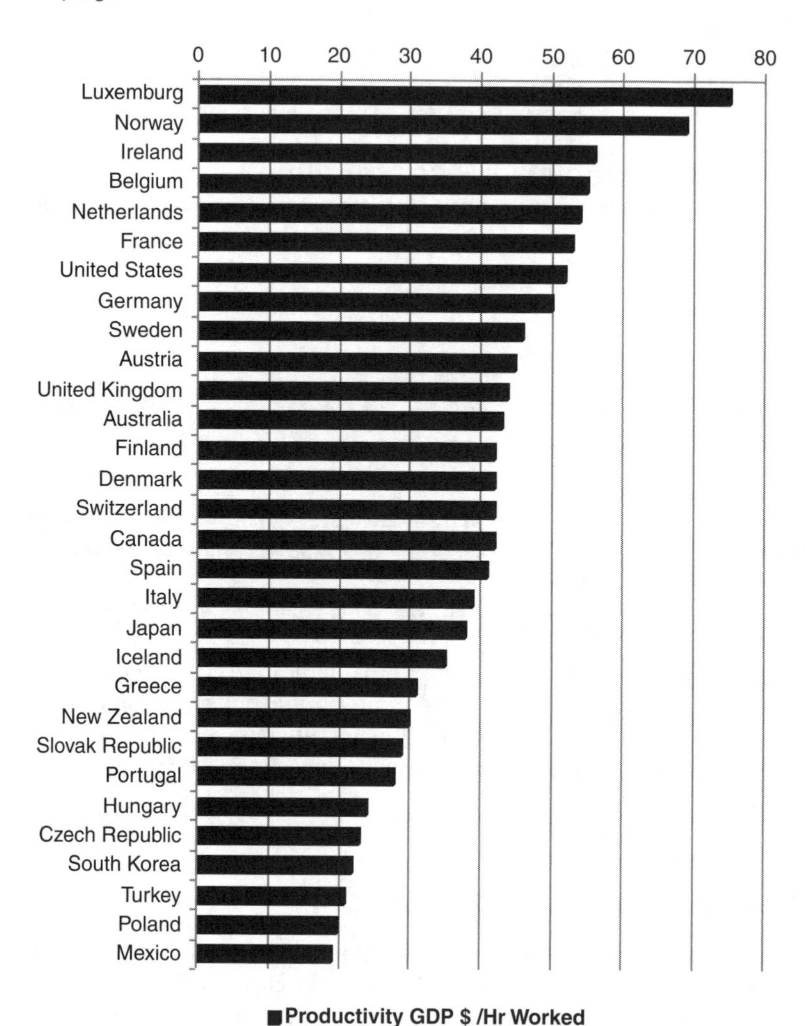

Figure 2.1 Productivity in Gross Domestic Product (GDP) per Hour Worked, Measured in U.S. Dollars

workforce, are critical to fostering constant improvements that drive increases in national productivity. Over time, other factors, such as innovation, the development of human resources through continued education, and incentives from stronger industry sectors, promote the increase in productivity. Ultimately, governmental policies, institutional as well as cultural factors determine a nation's success in improving productivity.

The U.S. Bureau of Labor Statistics (BLS) produces two different types of productivity data. As defined on the BLS website, (www.bls.gov/bls/productivity.htm): "Productivity and related cost measures are designed for use in economic analysis and public and private policy planning. The data are used to forecast and analyze changes in prices, wages, and technology."

The BLS cites two primary types of productivity statistics:

- ***Resource productivity*** measures *output per hour of one resource.*

  Example 1:

Labor Based Productivity = Number of Value-Add Units Produced ÷ Number of Laborers Utilized × Hours Worked

Example 2:

Equipment Based Productivity = Number of Value-Add Units Produced ÷ Number of Equipment Utilized × Hours Worked

- ***Multifactor productivity*** measures *output per unit of combined inputs,* which consist of labor and capital and, in some cases, intermediate inputs such as fuel.

  Example 3:

Cost Based Productivity = Number of Value-Add Units Produced ÷ Cost of All Resources ÷ Hour × Hours Used

Instead of using only one input value as a divisor, the costs of all resources are added up.

By definition, the output can vary widely. It may be an entire building, only the steel structure, or only one floor. As we focus our attention on the construction process that produces all the elements of a building, consuming costly resources, it is necessary to understand the mechanism of a productive process. While there are many other definitions of productivity, it is important to recognize that there are layers around the core process of transforming an input into an output using a set of resources. Figure 2.2 illustrates four levels for defining productivity in construction.

2.2 BASIC RELATIONSHIPS AFFECTING PRODUCTIVITY

Business success in a capitalist system depends not only on a high quality product but also on a constant increase in productivity to stay competitive. This rule certainly applies to construction. Thus, construction companies constantly review their methods and technologies with the goal of finding ways to generate greater value-added outputs with fewer resources. Figure 2.3 depicts several strategies for improving productivity through changes that can take place on a construction site.

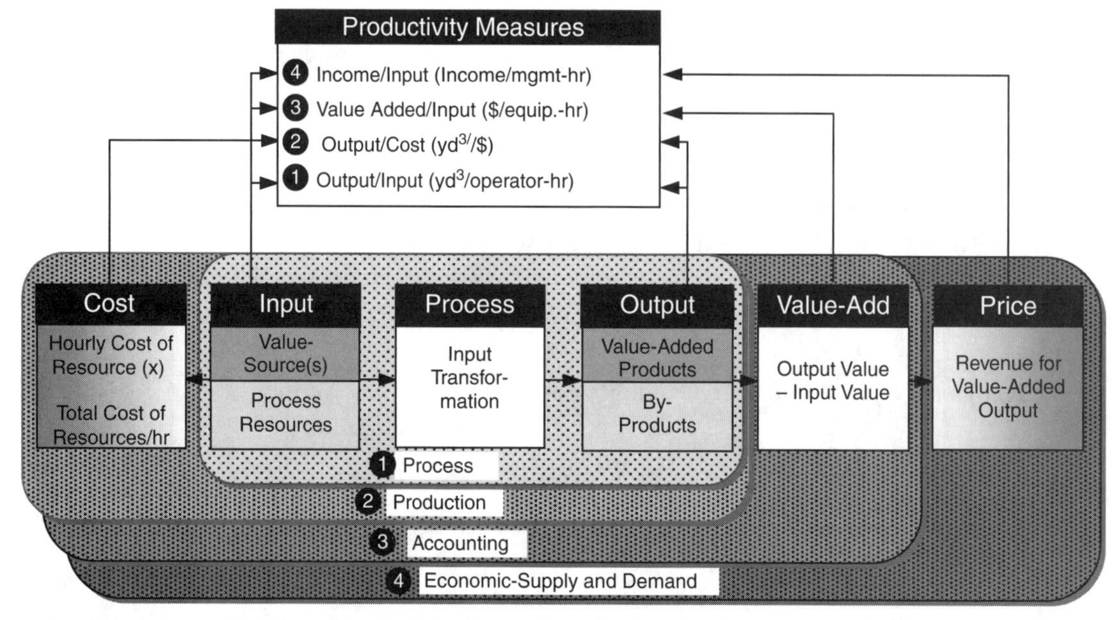

Figure 2.2 A Layered Display of Productivity Calculations for Construction

In the figure, for simplification, productivities appear as slopes that pass through the origin of a two-axis plot, with the inputs on the x-axis and the outputs along the y-axis. Both A and B represent labor productivity, with B at a steeper slope to indicate a higher level of productivity. Changes 1 and 3, both the consequence of an increase in efficiency, start at an assumed initial productivity P_I. Change1 results in a reduction in inputs needed for the same number of outputs, while change 3 indicates a drastic increase in outputs produced with the same amount of resources. While change 2 also results in a reduction of inputs, it really is an example of substituting resources but not necessarily of improved overall productivity.

Line D plots the productivity of equipment that can't function without some labor hours, at a minimum its operator. If we assume that B represents production without the use of equipment, change 2 characterizes a situation where labor hours are substituted with equipment hours, resulting in a further increase in labor productivity. Similarly, change 4 is the consequence of substituting bulk material, such as concrete, with prefabricated elements, such as precast panels. As depicted in Figure 2.3, this second type of substitution results in lower bulk material and, at the same time, lower labor hours. In a final tally, some of those labor hours will not totally disappear; they will be taken up by the precast plant producing the panels.

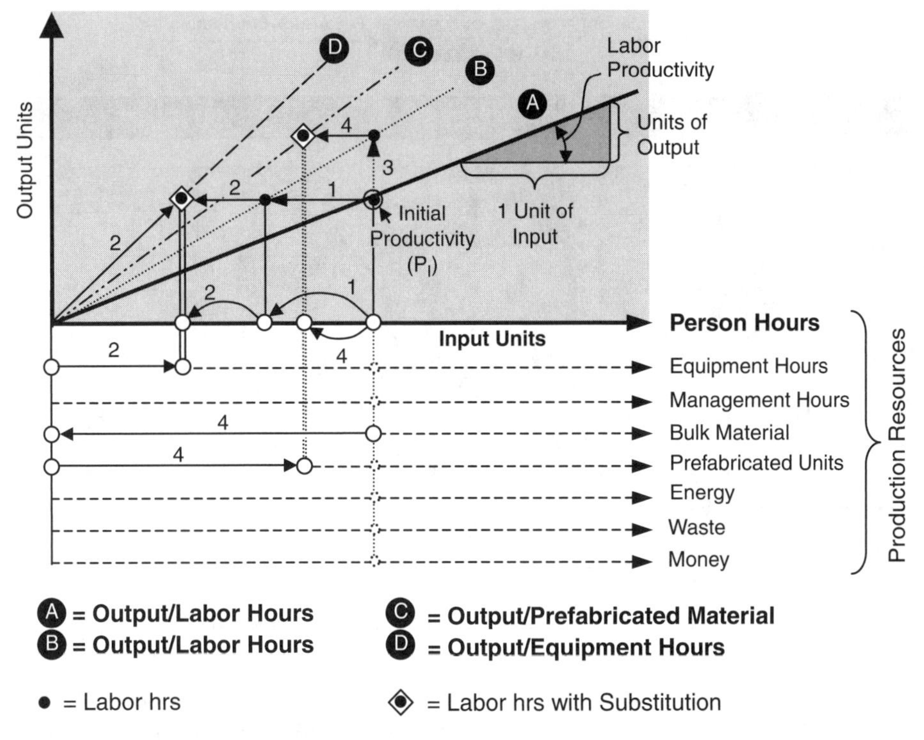

Change 1 = Improved productivity leads to lower labor hours for the same output
Change 2 = Substitution of some labor hours with equipment hours
Change 3 = Improved productivity leads to more increased output for same input
Change 4 = Substitution of bulk with prefabricated elements requires less labor hours

Figure 2.3 Plotting Changes in Productivity and the Effects of Resource Substitution

2.3 FACTORS RELATED TO PROCESS PRODUCTIVITY

So far, we have taken a mainly mechanistic view of productivity, without considering all those factors that enter the function as variables and coefficients. As we all know, a multitude of factors impact our personal efficiency during the day, as students, employees, parents, and so on. Similarly, productivity in construction is affected by a myriad of issues, including the quality of tools, availability of up-to-date information, and the detail of planning. Figure 2.4 offers a model that encompasses the most critical factors.

2.3.1 Necessary Work Resources

A key function of management is to allocate scarce resources in such a way that the business objectives can be reached in the most efficient manner. One can only admire the management skills of the Egyptian pyramid builders who organized more

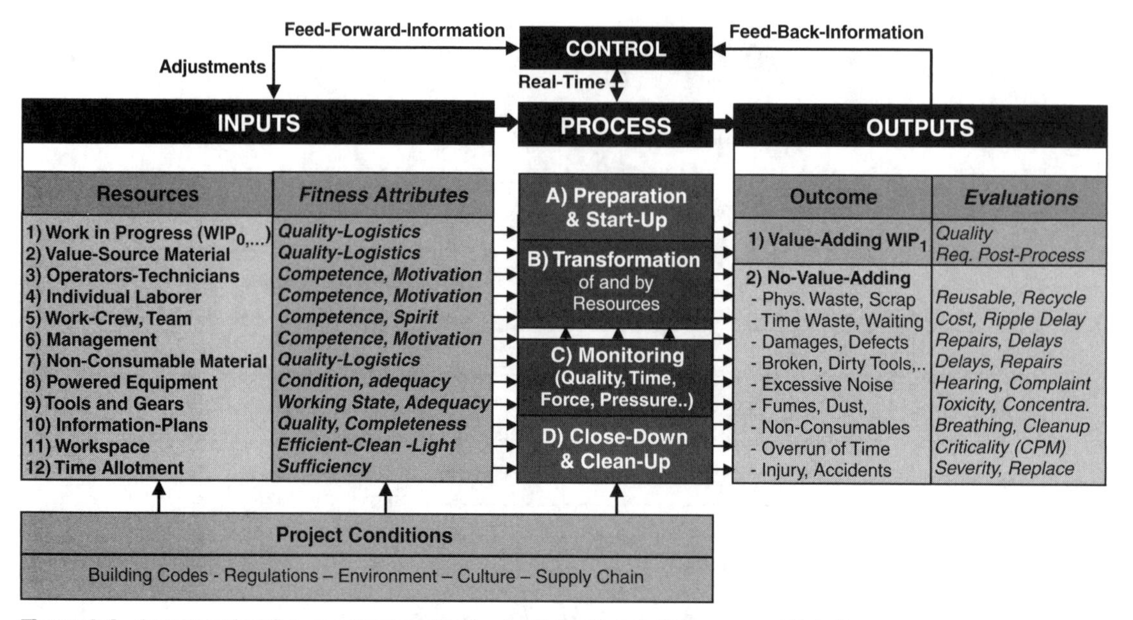

Figure 2.4 Construction Process Model with Productivity-Related Factors and Relations

than 10,000 workers and other resources in such a way that large stone blocks were cut precisely, transported over long distances, and assembled with a high level of accuracy. All that was done without large equipment, computers, or telephone.

Figure 2.4 lists 12 resource groups that are commonly deployed today to create value-added products, including management. Humans are considered a company's most critical asset because they possess the skills and the knowledge that is vital to turning ideas and visions into drawings, and drawings into complex physical structures that provide the desired functions.

It is important to recognize the changing focus among various resource allocation techniques used during project scheduling versus those used during process management. The Project Management Institute's (PMI) Project Management Body of Knowledge (PMBOK) focuses on five higher-level phases: (1) initiating, (2) planning, (3) executing, (4) controlling and monitoring, and (5) closing. While the goal of *project* management is to establish an efficient supply chain, assure sufficient cash flow to acquire necessary resources, and match human resources to project needs. It is the goal of construction *process* management to orchestrate a transformative Input-Output (I/O) system able to construct a value-added product with the least amount of waste, on time, and at a cost budgeted by the project management. Many have compared this complex task to a juggling act.

What is not immediately apparent here is the existence of a third dimension: technology. Often, the technology called for is dictated by the work to be done (e.g., excavate a trench) and the cost of deploying a key resource. For example, in a developed country, a large excavator may be easily rented for a short period of time, whereas in a developing country the low cost of labor makes the use of smaller, simpler machinery, combined with hand-operated carts, a cheaper and more dependable option.

As indicated in Figure 2.4, in order to construct a value-added building segment, a set of resources needs to be made available. The role of the construction manager is, first, to determine the specifics of each resource—for example, the size of a truck crane—that will be capable of performing the required operation safely and effectively. Second, the needed resources have to be procured and possibly assembled by a specific date and time, to carry out the operation as planned. At this time, the condition and quality of the resources that were pulled together will dictate the efficiency of the work and, with it, the productivity of construction.

2.3.2 Job Fitness

Let's assume that an old HVAC unit on a flat roof of a highrise building needs to be replaced. Simply renting a truck crane that will reach the roof to hoist the old and the new HVAC can lead to a "disastrous" outcome. For example, a critical resource not only needs to fit the situational conditions of a specific job but also match the constraints of other resources. Suppose a truck crane has just arrived when the operator recognizes the surrounding make it impossible to install the crane very close to the building. After measuring, it is found that the required 15 meter reach from the roof's edge necessitates a flexible jib; but the crane has only a fixed jib. Further, the weight of the HVAC unit would exceed the safe lifting capacity of the crane, given the desired reach. Even using a bigger crane would require that the unit be disassembled into two halves of about equal weight.

The second column in Figure 2.4, "Fitness Attributes," aligns with the 12 resources in column 1 and presents examples of terms describing attributes such as quality, competence, condition, and completeness. In later chapters of this book we will revisit this list many times and come to appreciate its importance to productivity.

It is important to be aware of an issue that is equally important: the need for synergy between resources. In other words, it is not enough that the resources fit the work to be done; they also must match each other. Take for example, a first-class piece of new equipment that is operated by an average operator who has never worked with this type of machine. In addition, the condition of the job does not precisely match the equipment. Because of these two mismatches, this top-level production equipment will be able to function only at an average or below-average productivity level.

2.3.3 Where the Rubber Meets the Road: The Process

Preparation and Start-up

The focal point of this phase is the effort required to deploy all the resources so that the process can begin. Following the process plan, tools and gears (e.g.,

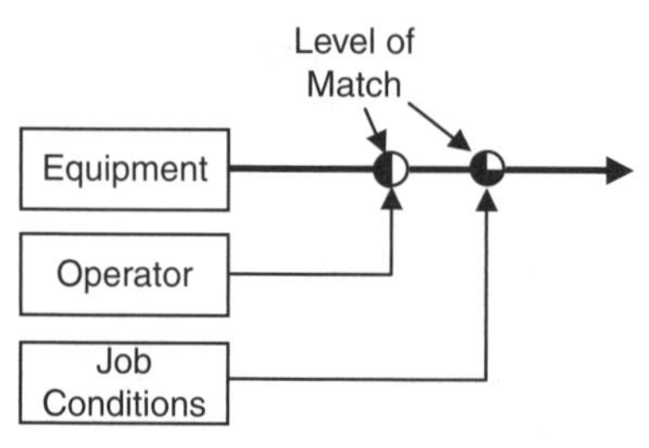

Figure 2.5 Moderate Resource Synergy

concrete vibrators and power generators) have to be moved to the workfront. Laborers must be instructed about the planned progress of the work; truck drivers need direction for real-time delivery; and so on. Finally, the construction manager must inspect the preparations to eliminate any miscommunications. Once this is complete, the start-up process may begin.

Similar to the way that a dry river slowly fills up after a rainstorm, the first "trickles" of material, or work in progress (WIP), that reach the work site will turn into a continuous flow, which, if planned properly, will keep all resources productively involved.

Transformation

This stage may be compared to the *Paris to Dakar Rally*, an annual off-road automobile race, in which participants speed along highways, climb sand dunes, cross mud streams, and battle camel grass, rocks, and sandy desserts. Many things can go wrong bringing a race car to a halt or severely slow its progress. A loose stone could hit the windshield; a sharp stone might slice a tire; or a dust particle could get into the fuel injector and block the fuel supply—to name just a few. Experienced racers are prepared to either avoid those perils, through preventive devices, or to fix a problem on the fly. Let's find out how we can apply such "winning" strategies to construction.

Figure 2.6 adds a new dimension to the input/output (I/O) concept that we have been discussing: namely the repetitive process cycle. The world, humans, and animals—all experience many cycles that move things perpetually forward. For example, the yearly cycle of spring, summer, fall, and winter, invoke an agricultural cycle to produce the food necessary for humans to survive. Similarly, the great variety in size and weight of the pieces needed to construct a steel structure make it impossible to install everything in one crane lift; thus, a crane must cycle back and forth between the pickup and assembly locations. As Figure 2.6 illustrates, the continuous nature of the process requires that all the needed resources be available without interruption. It is apparent that there exist many opportunities for the process to be disrupted, such as changing conditions, human error, or lack of information.

Manufacturing was the first industry that recognized the cumulative impact of small interruptions in production lines—especially car manufacturing, which

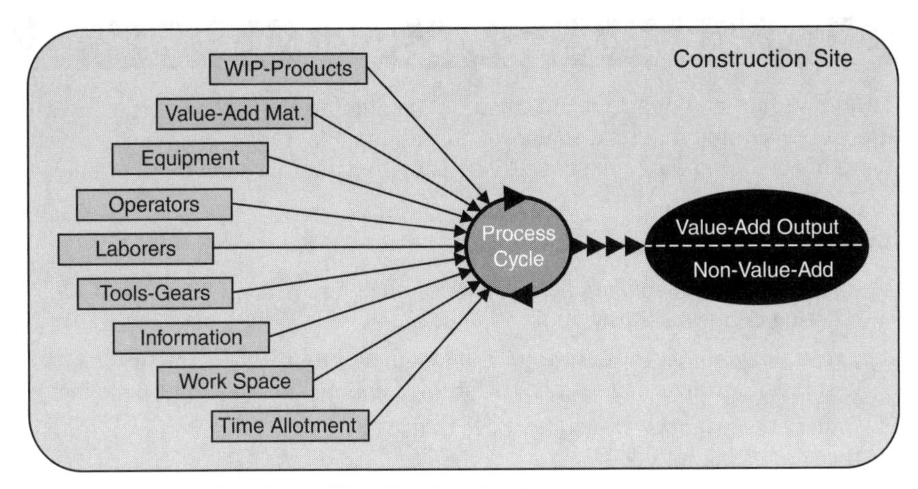

Figure 2.6 Flow Model of Repetitive Construction Process

experienced major growth after World War II. Japan, following its devastation at the end of the war, was driven to become a major player on the global market by exporting high-quality cars produced by a workforce eager to restart their lives. One name in particular became well known all over the world: Taiichi Ohno, chief engineer at Toyota. Ohno devised a production system that later was called *lean manufacturing*, Built on the concept of seven *muda* (Japanese for "waste") he developed a strategy to reduce or eliminate the effects of these seven muda, thereby improving productivity and quality.

Given that construction methods and circumstances differ significantly between various product manufacturing processes, the principles of Lean Manufacturing had to be translated for use in the construction industry. Consider, for example, that overproduction of a product, such as cars, is not an issue in construction. However, digging a pit or trench too deep at a project site could be considered a case of "overproduction", as it will require corrective backfilling. Another difference in construction is that, with some exceptions like precasting or rebar fabrication, it does not have fixed workstations, such as automated milling or welding machines. Similar to a Navy ship at sea, the construction workfront and its tools, equipment, and labor are constantly moving. Thus, the transportation of materials and parts has to constantly update its delivery plan to eliminate shipments to the wrong location. This makes the staging of process material an ongoing topic in regard to avoiding muda/ waste. The same holds true for inefficient processing. The constant changing of project conditions—including worksite location, weather and soil conditions—adds significant complexity to the process of managing resources to fit changing situations.

The construction industry has one of the highest incidence of workplace accidents. Many times, defects in tools and gears (e.g., slings and ropes) not only lead to damages of machinery or WIP, they also lead to injuries and even deaths, caused by falling loads, electrocution, broken ladders, or toppling equipment.

Taiichi Ohno—February 1912-May 1990

Toyota's chief engineer, Taiichi Ohno, is considered the father of the Toyota Production System (TPS), also known as lean manufacturing. He defined seven types of waste (*muda*) to describe those activities that add cost but no value.

The Seven Muda

1. *Overproduction:* Products are manufactured before needed ''just in case,'' creating costly inventory.
2. *Excess inventory:* Any raw material, work in progress (WIP), or finished good that is not moving requires capital, storage space, and additional handling.
3. *Unnecessary motions:* Workers have to use extra time and effort to perform, thus becoming fatigued.
4. *Idle waiting:* People or equipment waiting for other resources to be supplied wastes valuable time.
5. *Unnecessary transportation:* Poor work-floor layouts require extra handling of parts and materials, increasing risk of damages and nonproductive time.
6. *Inefficient processing:* Lack of adequate machinery, sufficient training, and up-to-date standards.
7. *Defects, breakdowns, injuries:* Missed quality targets and equipment malfunctions result in rework, generate scrap, and slow down the entire operation, unless a costly spare inventory is maintained. Accidents and injuries not only disrupt the workflow but may cause pain and suffering for the employees.

Table 2.1 presents examples for each of the seven muda in construction including potential causes. Add at least one of your own examples to each muda box. We also encourage you to fill in remedies that come to mind at this point, in the rightmost column. We will pick up this issue again, later in the book.

Monitoring

Visual supervision of the construction process, to ensure its execution according to plan, is increasingly supported by electronic technologies. Some of the emerging methods are built into the machinery, such as slurry wall drills, while others can be attached to equipment, such as a GPS or data logger. The goals of each system vary, but the general objective of electronic monitoring is to improve productivity mostly through the prevention of poor quality, idleness, and accidents. Figure 2.7 illustrates three simple examples of electronic sensors that are being deployed to monitor operations in construction.

Table 2.1 Ohno's Seven Construction Muda

Muda	Examples	Possible Causes	Remedies
1. Overproduction	• Digging trench too deep • Casting too many pipes • Concrete • Overdrilling shaft	• Surveying error • Mistake in design • Seasonal needs • Miscalculation	
2. Excess Inventory	• Rebar • Precast elements • Formwork • Steel • Equipment	• Seasonal needs • Price fluctuations • Business cycles • Buy versus rent • Tax advantages	
3. Unnecessary Motions	• Hand-lifting mortar • Hand-digging dense soil • Carrying heavy rebar • Chipping concrete • Pushing cart in mud	• Lack of proper equipment • Lack of appropriate tools • Material too heavy • No housekeeping • Insufficient training	
4. Idle Waiting	• Crane unloading slow • Trucks waiting to be loaded • Asphalt paver waiting to be resupplied • Concrete crew waiting to for the next bucket • Drill rig operator waiting for air compressor	• Lack of efficient tools and gears • Traffic fluctuations • Lack of supply trucks • Natural part of technology • Lack of preparation	
5. Unnecessary Transportation	• Carrying bricks to mason • Lifting sheetrock pieces • Restaging rebar • Storage of precast on-site, instead of direct installation	• Lack of capital • Lack of planning • Delay in construction • Change in plans • Unavailable resource	
6. Inefficient Processing	• Repair of chipped, cracked concrete • Retightening of bolts • Grouting of cast-in-place concrete piles • Equipment not fit to job	• Insufficient protection • Lack of experience • Faulty measurement • Careless drilling • Lack of planning • Insufficient information	
7. Defects, Breakdowns, Injuries	• Vibrator breakdown • Accident on-site • Damage of buried utility • Collapse of crane • Missing inserts • Misaligned joints	• Preinspection not done • Ignoring safety rules • Lack of preventive maintenance • Lack of quality control • No subsurface engineering • Poor operator training	

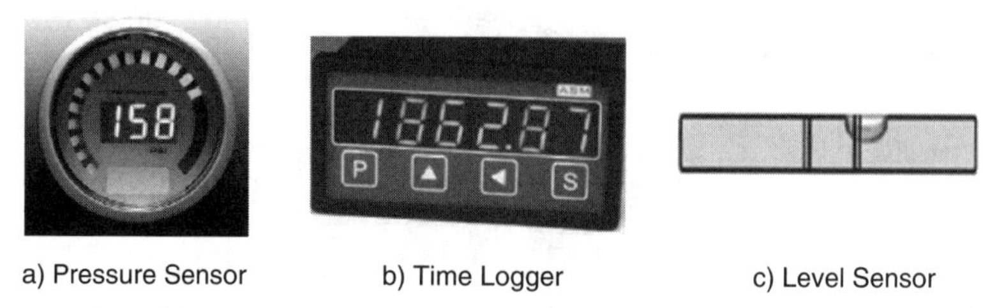

a) Pressure Sensor b) Time Logger c) Level Sensor

Figure 2.7 Process Monitoring Devices Used in Construction

Closedown and Cleanup

Analogous to the start-up phase, the closedown/cleanup process does not terminate abruptly. Step by step, the tasks comprising the process "empty out" of work, until the last part or piece reaches the workfront. Laborers at the end of the supply chain, one by one, clean their "workstations" and return to storage the tools and gear they no longer need. Often at this stage, measures are needed to protect the new value-added outputs from environmental effects, such as rain, wind, or freezing. For example, fresh concrete or sloped earth have to be covered; or wall forms need to be braced against severe wind, and concrete pumps must be emptied before the concrete starts to set inside the pipes.

Again and again, studies show that good housekeeping on a construction site is the key to a safe and highly productive operation. As a consequence, this last phase of the process plays a critical role in achieving the required quality and, equally important, for the performance level of the subsequent processes.

2.3.4 Value-Added and No-Value-Added Outputs

To illustrate how value-added WIPs and by-products evolve into a final product, Figure 2.8 presents the floor and wall systems of a common platform for a single-family house built in the United States.

Three value-added process outputs can be recognized here:

A) Foundation, with footing and beam support
B) First floor with box sill, beam, joists, and plywood subfloor
C) Walls made of soleplate, studs, and top plate.

Each process depends on the completion of a preceding operation, a sequence that can be easily drawn as an arrow diagram, as shown in Figure 2.9.

Given that each process depends on the completion of the previous one, the actual duration of each becomes important. If one of the processes gets delayed due to poor productivity, the subsequent *one(s)* will have to be postponed. For example, a delay in the carrying out process A causes it to take longer than planned. As a direct consequence, the concrete curing starts later by the length of

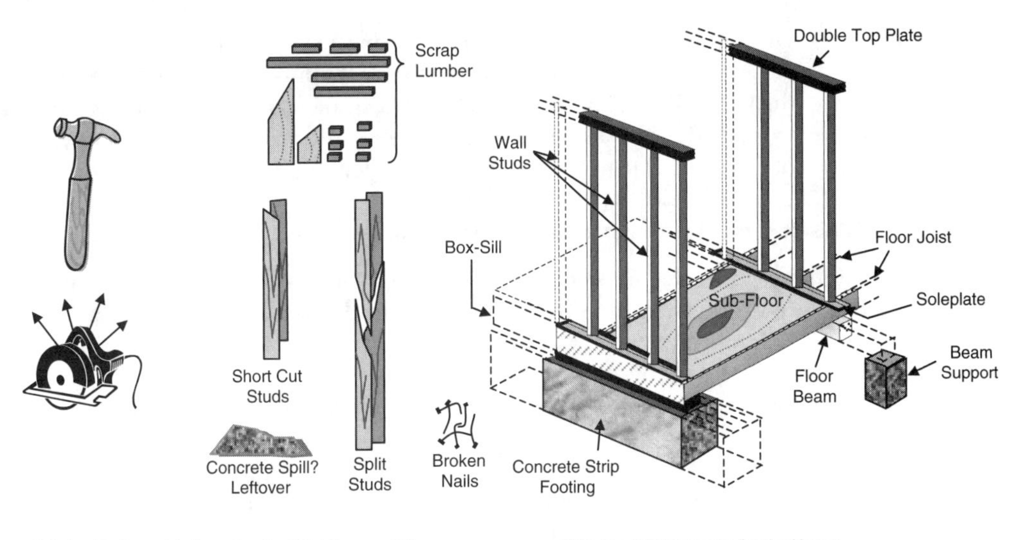

Figure 2.8 Process Outputs for Building Structure

the delay; as a result, the start time of process B is also moved back. Even a shortening of the time to construct the base floor is not enough to catch up to the planned start time of process C. As floor construction and framing processes will continue, the small delays in framing the walls will add up to days, rather than hours. The example highlights another interesting effect: some process tasks, such as concrete curing, may only require one basic resource, time.

The types of muda produced by the three processes A), B) and C) are varied, and caused by different problems:

- Several tools used by framers, such as circular saw and nail guns, are extremely loud, and are dangerous if not used with proper safety protection (e.g., glasses).
- Scrap pieces of wood result when, for example, the studs for the walls are too long, or the plywood sizes don't fit exactly to the floor's dimensions.

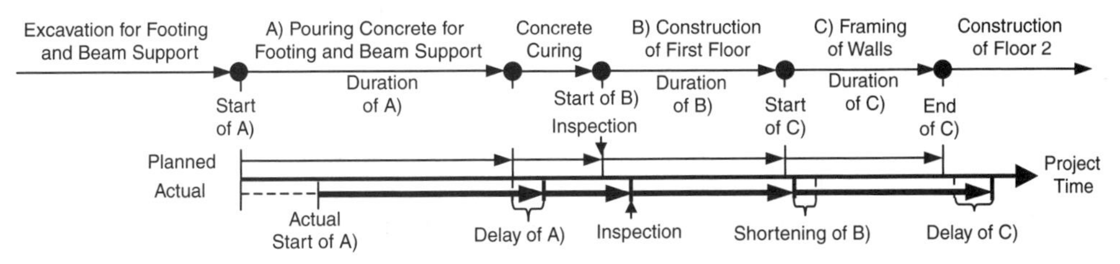

Figure 2.9 Critical Sequencing for Value-Added Processes

- Other physical waste occurs when the framer makes an error by cutting a piece too short, or when a stack of lumber contains pieces that are split through. Bent nails must either be pulled back out or hammered into the wood and then be supplemented by a second full-length nail. An oversupply of concrete, due to incorrect volume calculations, is dumped on the ground.

2.3.5 Control Mechanisms for Keeping Processes on Target

Highway traffic acts like a production process in that the inputs—cars, buses, and trucks of various sizes, and their drivers' qualifications—enter the road at a start-up ramp. Several rules and regulations apply, with the sole goal of ensuring that the inputs will exit the highway safely, when the "value-added" location is reached. Most countries enforce a maximum speed for vehicles, which is controlled by each driver or operator, either by means of an accelerator foot pedal or by setting an automatic speed controller. As everyone who has ever traveled in a car knows those official speed limits are treated by many drivers more as recommendations than laws, which, when broken, can have consequences.

To address this, South Korea installed speed control cameras at conspicuous overhead cross-beam across the highways. Approximately 1 kilometer before the location of the next camera the audio of the GPS system announces how much above the speed limit the car is traveling. Every car, bus, or truck slows down to reach the speed limit exactly at the location of the camera. Of course, most pick up speed immediately afterward.

What can we learn from this observation? Naturally, the central theme is the control of a process (driving on the highway) to meet a desired outcome (reaching the correct exit), with and without providing information (warnings from the GPS). Let's go back in history to learn about the origin of the first process control system.

James Watt is considered the first to build a successful proportional feedback control mechanism, for the steam engines being manufactured by his company in England in the late 1780s. He knew that if he could control the amount of steam that was entering the engines, it would be possible to keep the pistons at a constant speed.

Figure 2.10 sketches Watt's invention, the fly-ball governor (i.e., speed limiter). As the engine speeds up, a connected gear increases its rate of rotation, from $\omega_1 = 30$ RPM to $\omega_2 = 60$ RPM, which in turn rotates a vertical spindle faster and faster, thus increasing the centrifugal force from F_1 to F_2. This increase pushes the steel balls of mass, m, outward, which is translated into a rotational motion by the lever and center pin. Then, via a mechanical two-pin link and an additional lever, this motion eventually activates the throttle valve built into the pipe

James Watt (1736–1819)

James Watt was a Scottish inventor and mechanical engineer, who became an enthusiastic hands-on inventor. Similar to Isaac Newton, he became a gifted craftsman and was capable of performing systematic scientific measurements that quantified the improvements he came up with. Together with his business partner, Matthew Boulton, Watts installed many steam engines to pump water out of mines. But it was his centrifugal governor or speed controller, that opened the door for steam engines to become the fundamental device to power the Industrial Revolution.

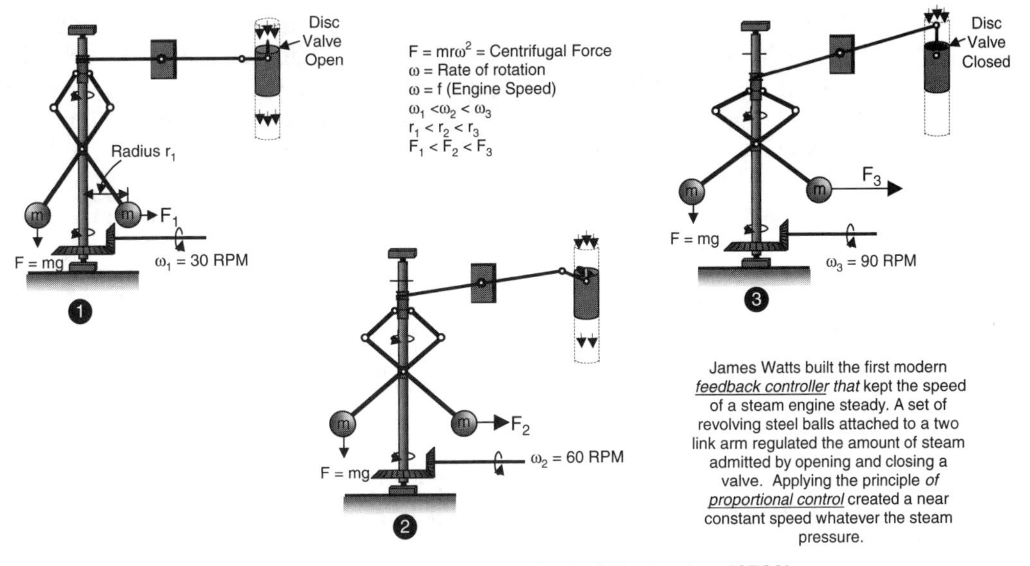

Figure 2.10 Principles of the First Proportional Process Control Mechanism (1789)

that supplies steam to the engine. As shown, the valve is able to change from wide open in 1, leaving the steam to pass freely through the pipe, to totally closed in 3.

The amount of steam that is allowed to enter the cylinder is continuously regulated by the rotating disc inside the valve, called the butterfly valve. If, for example, the desired speed is 60 RPM, the control motion shown in Figure 2.11 will be automatically executed in a proportional manner.

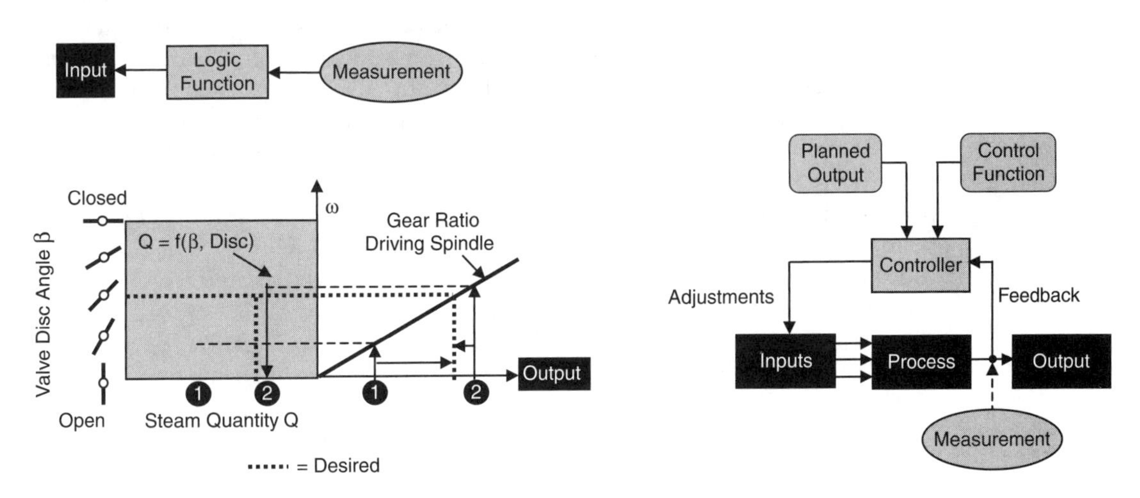

Figure 2.11 Modeling the Main Features of Process Control

When planning construction, it is essential to have a performance target in mind, whether it is the production rate of a backhoe excavator or the framing crew, or the time each process task should take. Rarely, however, is the target met, for doing so depends on the accurate prediction of *all* the conditions and behavior of *all* the input factors. Clearly, this is impossible. So what is the construction manager to do? Should he just "let it happen" and "back schedule" after the project is finished? Or can he use the principles of Watt's proportional feedback control?

Principles of Proportional Feedback Control

Watt's controller builds on four main capabilities that are generic to all process controls:

1. **Capability 1:** Measurement of one or several output parameters
2. **Capability 2:** Real-time feedback of output measurement data to the controller
3. **Capability 3:** Generation of directives to adjust the present input flow(s) so that the present output will shift toward a desired or goal status
4. **Capability 4:** Modification of input flow parameter(s) based on control logic.

Figure 2.11 shows two models that illustrate schematically how the four capabilities are integrated to control the speed of an engine. We'll apply these principles to the house framing process shown previously in Figure 2.8, where the pouring of concrete for the foundation is delayed by one or two days. Keep in mind, the time for concrete curing cannot be shortened, and inspection of the foundation requires open access to observe the condition; that means construction of the first floor by the carpentry crew has to be delayed, too, and so on down the line.

Capability 1: Two of the earliest measurement needs arose from trade and the ownership of land. Historically, one of the reasons for establishing the weight and distance measurement system was to prevent fraud in trading goods and selling land. Today, a variety of measuring devices and electronic sensors are available that offer an easy way to turn physical quantities into values that are recognizable by humans or electronic devices. For example, temperature can be measured either by using a mercury-filled tube that indicates the rise in temperature, or a thermocouple that changes the position of a needle according to the changes in voltage. Information technologies have also been developed data-acquisition boards with multiple channels that convert analog voltage into digital readings, called analog-to-digital conversion. This advancement enabled a computer to monitor the values of a wide range of sensors simultaneously.

In the past, the vast of number of process output parameters and measurements depended largely on the observation capabilities of humans. Today, a wide range of electronic sensors supply readings with a precision and speed impossible to match by human sensory capabilities. Take the requirement to measure the performance of pouring concrete for the footings of a single-family home. What should be measured: the seconds, minutes, hours, days, or weeks it takes to complete the pour? Or the mm^3, cm^3, or m^3 of concrete that has been poured by a particular time? Or is it the exact time that the last batch of concrete was poured? Today, it is possible to assess any of these; but the question is what would provide the most value or the most valuable information for the purpose of controlling the overall construction? One important issue to consider is the entire time frame to pour foundation concrete for a single-family structure, which may last, at a maximum, four hours—assuming concrete delivered by trucks and placed via chutes or pumps.

A second issue to consider in selecting the appropriate means of measurement is its value to the overall control of the processes. In the case of controlling the building of a house, gaining detailed knowledge of how much concrete is poured every minute is not very valuable. Much more important would be to know the quality of concrete, the start and end times of unloading a concrete truck, and the start and end times of the entire operation. This information is essential in passing the quality tests, detecting time delays, and the required curing time. In other words, the most important measurements for the case at hand are the characteristics of the concrete and the completion times. The first data set can be taken from the delivery certification and from proper sensors or on-site tests (slump test), the second from the times when concrete trucks arrive and leave the site, and the third from the time when the last vibrator is turned off. All three values are simple to attain and to input to a controller, automatically or by hand.

Capability 2: Real-time feedback of output data to the controlling "brain" requires a communication medium that connects both ends. Today's electronic or wireless communication channels provide very effective means to bridge the gap. And the continued move away from paper-based to electronic communication offers more opportunities to tap into traditional information flows. One case is concrete delivery by truck: Each concrete batch arriving on-site is accompanied by a delivery receipt/bill, providing time, volume, truck number, batch mix, and so on, which is handed to the site manager for payment. This information will in the future be in electronic form either as a radio frequency ID (RFID) tag or part of a telematics system accompanying each vehicle. Thus, while the electronic delivery receipt provides necessary information about the concrete delivery history, time of process completion is easily determined from the departure time of the last truck.

In summary, real-time feedback from output to controllers, virtually anywhere in the world, not only is feasible but also is extremely reliable and cost-efficient. One note of caution—and this is addressed further in Chapter 10 on communication—the contractor has to study the need for security and validate the reliability of the data. A wide range of expert services are available to help develop a safe, company-controlled network.

Capability 3: The controller's "brain" is capable of interpreting the feedback data and determining what to do to correct errors, defined as the discrepancies between goal and present state. Figure 2.11 shows a polynomial function relating the speed of an engine with an ideal valve opening. Today, this task is handled by an electronic programmable logic controller (PLC) or a proportional integral derivative (PID) controller, which can easily be reprogrammed. On/off or discrete controllers are simple non-proportional machines useful where the input can be either on or off (e.g., light), or where it allows only a limited set of control options (e.g., stop, slow, fast, very fast).

Regardless of the type of control method used, the first step always is to define the desired outcome to be used as a setpoint. In the case of the single-house construction, the objective might be straightforward: to stay on schedule.

When will the "brain" conclude that the project is falling behind schedule? According to Figure 2.9, this occurs by the time concrete pouring begins, as it starts later than planned.

The second phase in adjusting the input, after goal setting and error detection, involves defining a logic as the basis for decision making. It is apparent that concrete pouring is not a process that can be sped up proportionally; neither can concrete curing. Also, an on/off logic would not be helpful. What might be useful is a discrete way of making decisions based on some simple IF-THEN rules or rule trees. For example:

IF Actual Start Time *Project-Process*—Planned Start Time *Project-Process* ⇒ 1 day, THEN Take Action

Assuming that the concrete pour is delayed at least one day (e.g., due to bad weather), the controller needs to know what action to take. Again, IF-THEN logic seems to be most appropriate:

Take Action

IF *Project-Process* = Concrete Pouring, THEN . . .
IF *Project-Process* = Construction First Floor, THEN . . .
IF *Project-Process* = Framing First Floor, THEN . . .

What should be done at each decision point is a function of the available options and the strategy that management is willing to pursue. For example, one company might decide that the manager should receive an email reminding him/her about the delay of the process. Thus, the Take Action rule would state:

IF *Project-Process* = Concrete Pouring, THEN Send Email to Manager, *Project-Process*

Another strategy might be more direct, initiating actions without waiting. Some possible steps would be to require admixtures or additives so that the concrete achieves higher compressive strength early, or the use of a concrete pump to speed up the pouring process. As both of these actions would cost additional money, management might choose measures that do not involve the present process but rather future activities. For example, typically, carpenters are not allowed to start work that covers the concrete foundation until it is inspected; but might they be able to work next to the foundation and prefabricate sections that match the foundation exactly? Such prefabricated modules could be lifted into place with a small crane. Changing from sequential to parallel processes, work on the first floor could begin in on-site prefabrication mode while the concrete is curing. Such a strategy would result in following rule:

IF *Project-Process* = Concrete Pouring, THEN Start Construction First Floor as On-Site Prefab on Planned Start

Capability 4: Finally, a successful controller has to be accompanied by a mechanism capable of manipulating those input parameters that will modify the output toward a goal state. James Watt used a lever to open or close a valve built into the steam pipe, to either slow down or ramp up the engine. Modern speed controllers in cars work much the same way.

Construction management is still not automated to the point that equipment will take commands directly from controllers—although it is getting closer. Depending on the sophistication level of a company's IT system, an automated process controller will be instructed to print standard reports, send electronic emails, electronically create and distribute job assignment recommendations, and more.

The Feed-Forward Data Stream

Before closing the discussion of the construction process control model, the second information channel to the controller needs to be mentioned: the *feed-forward data stream*. This stream originates at the input, rather than the output side of the process. It is the basis for implementing proactive process management with the goal to prevent problems the might occur, by eliminating problematic inputs (e.g., drawings that are not up to date) or by stopping the process to avoid an impending hazard (e.g., approach of a storm).

2.4 TAXONOMY OF WORK TIME

Work time is not always spent on value-added tasks. In fact, much work is spent on necessary but no-value-adding tasks, such as preparing of work site, cleaning non-consumable materials (e.g., concrete formwork), or waiting for a truck to arrive with the gravel for a road bed. Of course, the less time is spent on no-value-adding work, the higher its productivity.

Construction is considered a "one-off" industry, in that it never produces two buildings that are exactly the same; nevertheless, its production functions are related to the applied technology and the skill of the laborers. For example, building the shoring for an elevated concrete slab does not vary, whether it is done for the third or the twenty-third floor. On the other hand, the time needed to assemble it will be a function of the shoring system being employed and how familiar the laborers are with putting it together (i.e., where they are on the learning curve). In addition, various adverse conditions may emerge, or delay-causing events may occur at random.

With this premise in mind, the total time spent on a process, such as installing the shoring for a concrete floor, can be broadly classified into *basic work time* and *added time* (caused by negative impacts) or into *efficient*, *inefficient*, and *wasted* time. Figure 2.12 illustrates a framework useful in dissecting the work time of a laborer, operator, or crew.

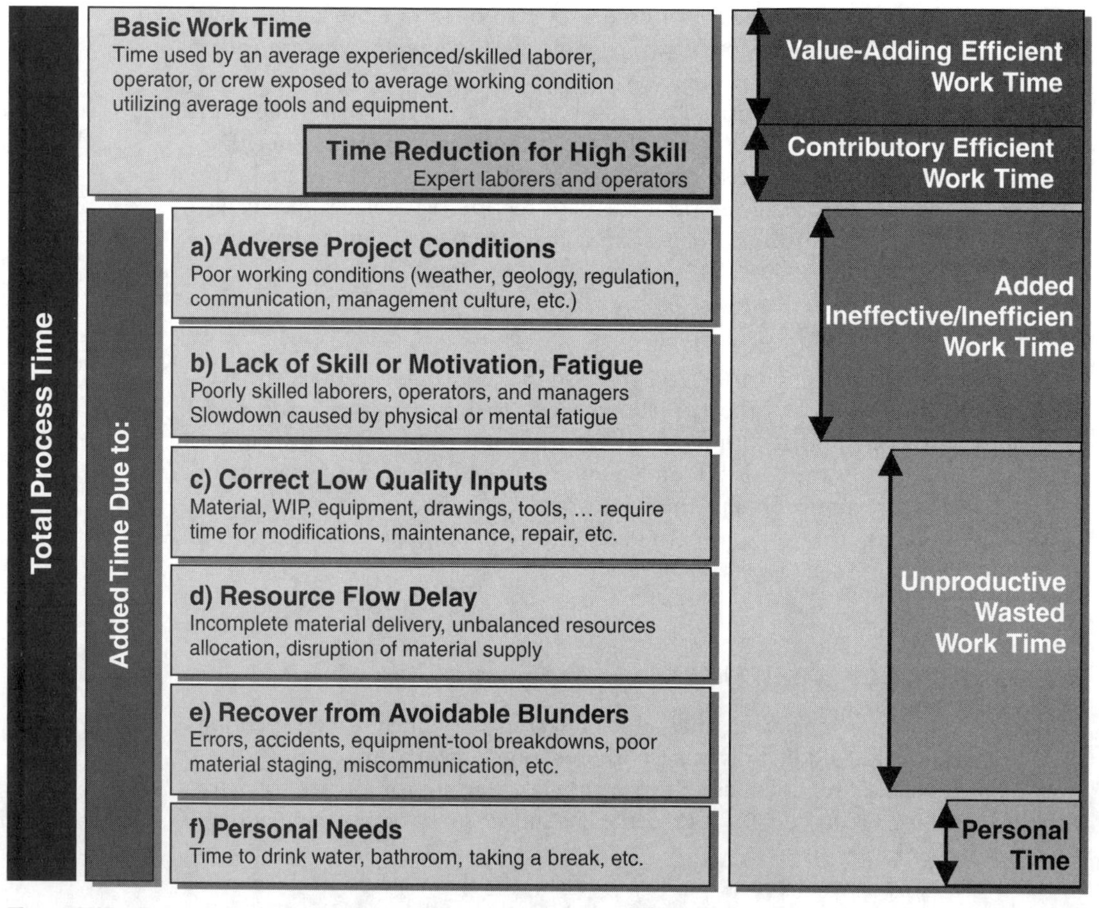

Figure 2.12 Framework for Categorizing Labor Hours Spent in Construction

Worked-Out Example Problem 2.1: Productivity of Walt Holz's Framing Crew

Mr. Walt Holz, president of Production Framers, Inc., feels that some of the crews he employs are not as productive as they could be. He consulted with us to randomly pick one of his construction sites and evaluate the performance of its framing crew. In particular, he wants to know how much time they are spending on inefficient and unproductive work.

Background and Assumptions

Holtz's concern focuses squarely on the productivity of his framing crews. It is interesting to note that he does not believe it is not a lack of technology or the effect of difficult work conditions are the cause of unsatisfactory performance by his crews.

Preparing the Site Observations

Table 2.2 itemizes the tasks of the carpenters on the job, and categorizes the time they spend on each.

Table 2.2 Defining Carpentry Work

Work Time		Task Content
Value-Added	1.	Cutting wall studs, laying out bottom/top plates and wall studs, assembling walls and corner posts, laying out window/door openings, nailing on sheathing, raising walls into place, and bracing walls
Contributory	2.	Carrying (includes picking up and unloading) wall studs, plates, sheathing, and other lumber pieces within the staging area; and obtaining tools needed for the operation
	3.	Measuring and marking the locations of the walls and their window/door openings on the subfloor, marking top and bottom plates, inspecting the straightness of wall studs before cutting or assembling them
	4.	Loading nail guns, and checking the quality of work being performed
	5.	Receiving/giving instructions and reading drawings containing instructions communicated to or by supervisors and among crew members; studying drawings and planning works by foreman
Ineffective	6.	Walking empty-handed to and from and within the work area
	7.	Carrying (including picking up and unloading) materials or tools more than 35 feet from the work station
	8.	Searching for material in or outside the staging area
Unproductive	9.	Waiting for materials, tools, instructions, prerequisite work, other crafts, etc.

(*continued*)

Table 2.2 (*Continued*)

Work Time	Task Content
	10. Correcting an error
Personal	11. Talking about nonwork-related issues while not actively working
	12. Taking a break or being idle for no apparent reasons
	13. Not observable—failure to observe worker who is assigned to a specific work location

Site Layout and Work Assignment

Figure 2.13 illustrates the various areas at the site location where the carpentry crew will be working.

Data Collection and Analysis

The initial survey of the residential building construction at 37 Bentley Court reveals that the crew consists of one carpenter who is measuring and cutting, two framers who are nailing the walls, and two helpers.

The data was collected using the statistical method called *work sampling*, which will be discussed shortly. In brief, each worker was observed 180 times during the day, at random, noting what the person was doing; this information was classified by checking the appropriate task box, 1 to 13, in Table 2.2. The totals were entered into the form shown in Table 2.3, followed by the calculation of percentages, as shown.

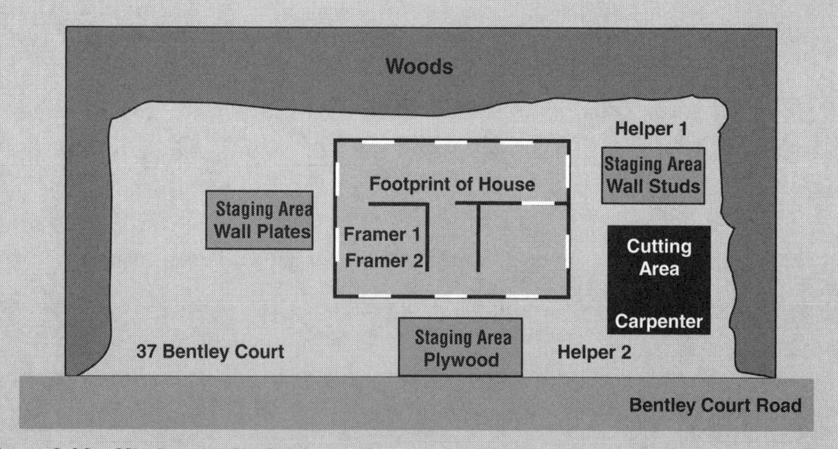

Figure 2.13 Site Layout for Carpentry Crew

Table 2.3 Work Sampling Summary Sheet after 180 Observations Rated According to the Task List

Project	Production Framers	Date	June 20, 2009
Day	Tuesday	Weather	Sunny
Crew activity	Wall framing	Temperature	Morning: 75-82°F
Humidity	65%		Afternoon: 82-89°F
Observer	Jing Zhang		

	Carpenter		Framer 1		Framer 2		Helper 1		Helper 2	
	Times Observed	Percent	Times Observed	Percent	Times Observed	Percent	Times Observed	Percent	Times Observed	Percent
1.	37	20.6	84	46.7	69	38.3		0.0	58	32.2
2.	28	15.6	15	8.3	14	7.8	2	1.1	18	10.0
3.	40	22.2	32	17.8	23	12.8		0.0	6	3.3
4.	2	1.1	1	0.6	3	1.7		0.0	6	3.3
5.	3	1.7	13	7.2	15	8.3		0.0	2	1.1
6.	16	8.9	4	2.2	7	3.9	54	30.0	4	2.2
7.	2	1.1		0.0	2	1.1	106	58.9	12	6.7
8.	12	6.7	6	3.3	9	5.0		0.0	18	10.0
9.	21	11.7	9	5.0	6	3.3		0.0	7	3.9
10.	6	3.3		0.0		0.0		0.0		0.0
11.	5	2.8	5	2.8	2	1.1	4	2.2	7	3.9
12.	8	4.4	11	6.1	25	13.9	14	7.8	42	23.3
13.					5	2.8				
	180	100%	180	100%	180	100%	180	100%	180	100%

Table 2.4 and Figure 2.14 present the field data organized in a form that is easy for Mr. Holz to understand. The graph presents a quick synopsis of the situation, and the table supplies detailed data points in a matrix format.

Results

The graph highlights three key issues that should interest Mr. Holz:

a. **Helper 1** shows no value-added work. Almost all of his time is unproductive.

b. **Carpenter** measuring and cutting the lumber spends a large percentage of time on inefficient and unproductive work. At the same time he takes the least amount of personal time.

c. **Helper 2** spends a large amount of his time on unproductive tasks and personal time.

Table 2.4 Distribution of Work Time Use by Each Worker

Work Time	Tasks	Carpenter	Framer 1	Framer 2	Helper 1	Helper 2
		Percent of Total Time (Percent)				
Value-Added	1	20.6	46.7	38.3	0.0	32.2
Contributory	2, 3, 4, 5	40.6	33.9	30.6	1.1	17.8
Ineffective	6, 7, 8	16.7	5.5	10.0	88.9	18.9
Unproductive	9, 10	15	5	3.3	0.0	3.9
Personal	11, 12, 13	7.2	8.9	17.8	10.0	27.2

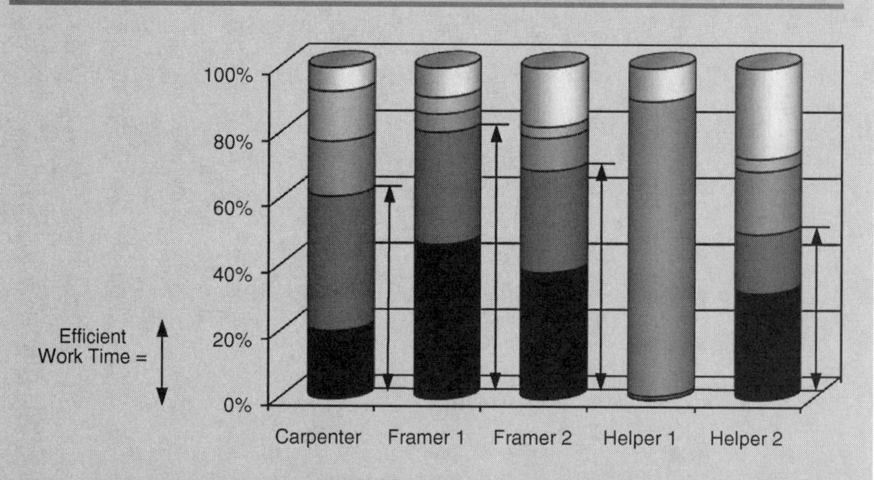

Figure 2.14 Graphical Display of Time Use by Each Worker

Discussion of Results

The data confirm Mr. Holz's feeling that at least one of his crews is not performing with high efficiency. From the site layout and a detailed analysis of the task ratings, it seems that there are several areas that, if redesigned, would bring substantial improvements. After we present the data to Mr. Holz, he might ask us to elaborate on what lessons could be learned from this random observation of his construction site.

2.5 GAUGING CONSTRUCTION PROCESS EFFICIENCY

How can a construction manager know how well his or her crews are performing? One way, of course, is to review the monthly cost reports at the end of the month. Unfortunately this time delay is not very helpful to a manager who is trying to steer the operation in order to maximize productivity. In any event, we have learned that costs are not necessarily clear indicators of process productivity—in particular, not the amount of wasted time.

The Egyptians were the first to advance the art of taking engineering measurements. Just imagine the shape of the pyramids if they had not found a way to create a foundation that was square and perfectly level, or to align the sides in a perfect slope to the top, or ensure that large blocks of rock cut miles away would fit together like puzzle pieces. The construction manager has the same need; he must be able to assess the process quantitatively and accurately, in order to determine in real time how the efficiency of the crews compares to the plan.

How should a construction manager measure the productivity of the many crews on a bridge project as depicted in Fig. 2.15? What is needed is a standardized, uniform method and scale, similar to that used to measure kilograms, pounds, meters, feet, liters, or pints. But before using a new measuring device or instrument, such as a pan balance scale, it needs to be calibrated, by comparing its readings against a known value. For example, the accuracy of a pan balance scale can be validated by putting masses of exactly the same weight on either side of the balance.

Figure 2.15 Incheon Bridge

Example VA Output Measures

- Length of driven precast piles
- Length of drilled piles
- Volume of compacted sub-base
- Length of concrete road poured
- Area of brick wall finished
- Number of precast panels hung
- Length of earth-anchor drilled
- Area of driven sheet- piles
- Area of formwork completed
- Tons of rebar placed
- Volume of concrete placed on slab
- Days to launch a bridge section
- Distance slipping an elevator shaft
- Tons of structural steel installed

The scale passes the test if the pointer on the beam falls within an allowed error band. Of course, the smaller the allowed error, the more precise the scale has to be.

What could we use to measure productivity in construction that is uniform and can be used by anyone after a short training? First we need to agree on a base measure that can be standardized. We learned earlier that a variety of different productivity definitions can be used for different applications, but they all depend on a basic ratio: output divided by input. Clearly, construction does not lend itself to defining a standard for value-added (VA) output as it is as varied as sand on the beach.

The input side of the construction process (see the model in Fig. 2.4) comprises a long list of different inputs yet offers the opportunity to select from quantifiable measures that are common to virtually all processes today—time and resource units engaged or absorbed. And as all construction still requires the involvement of workers and, usually, equipment, two common input standards for measurements can be defined:

$$\textbf{Labor-hours} = \text{process time} \times \text{the number of laborers}$$
$$\textbf{Equipment-hours} = \text{process time} \times \text{the number of equipment.}$$

The attentive reader will note that this is the base concept used by the U.S. Bureau of Labor Statistics. Can these two measures really help a construction manager control the production "engine" in real time, similar to Watt's steam control device?

It is apparent that one key element is missing: the control function that links the measure to a corrective action. In other words, what will a manager do when, by 11:30 AM, he or she learns that the labor hours are 25 percent higher than planned? The availability of real-time output data would be extremely helpful, but this is generally hard to come by. Even if both measures, labor hours and output, were available, the manager would still be without any substantial information about how efficient the crew is working, because efficiency depends on using the hours effectively. Only with data about the muda generated, especially time wasted while waiting for missing resources, will the manager be able to understand not only about efficiency but most importantly, where and what could be done to improve even a smooth working operation.

2.5.1 Scientific Measurements of Individuals at Work

Frederick W. Taylor (1856–1915), called the father of scientific management, is considered by many to be the most famous management pioneer of all time.

Taylor rose from a common laborer to chief engineer in six years, and completed a home study course to earn a degree in mechanical engineering in 1883.

He became interested in improving the productivity of a single laborer by optimizing the work setup, including the tools (e.g., shovels) he was using. Taylor began a scientific study to determine what workers ought to be able to produce, which opened the door to scientific management. Taylor used time studies to break down tasks into elementary movements, and designed complementary piece-rate incentive systems.

In 1911, Taylor published *Principles of Scientific Management* (Harper and Brothers) in which he defined a method aimed at determining the single best way to do a job. Central to the determination was the breakdown of an activity into smaller tasks, each of which could then be measured and assessed for efficiency.

According to Taylor, time studies should follow a specified process. First, there are three steps involved in planning:

Step 1: Divide a laborer's cyclic work into smaller tasks, or subtasks, that are executed repeatedly.

Step 2: Decide how many repetitions of the task to record.

Step 3: Prepare a form that prompts you to record all the pertinent information (e.g., date, temperature, humidity).

The next steps can be executed by measuring the task durations, either by observing the laborer directly or viewing video recordings, and then computing the collected data:

Step 4: Use a stopwatch to time and record task durations for as many repetitions decided in step 2 (the number of observations).

Step 5: Using the recorded data of repeated task durations, compute averages of observed time.

$$\text{Average observed time} = (\text{Sum of the times recorded to perform each work element}) \div \text{Number of observations}$$

Step 6: Assess the person being observed in terms of how much his or her performance differed from an average work pace—by assigning a factor to normalize the observed times. If the laborer was apparently working faster than a normal sustainable performance, the resulting rating factor would be less than 1.0 (e.g., 0.9). (It goes without saying that assigning a meaningful performance rating factor takes a lot of experience.)

Step 7: Calculate the normal times of each element or subtask.

$$\text{Normal time} = (\text{Average observed time}) \times (\text{Performance rating factor})$$

Step 8: Sum up all the normal times for each element to develop the total normal time for the task.

Step 9: Special conditions that existed during the observed activity should be accounted for in order to calculate a standard time that includes factors such as personal time for breaks, etc. Furthermore, extremes of temperature will cause the laborer to perform less efficiently than when the temperature is a ideal 70 degrees Farenheit. This step must account for all these atypical situations by providing allowances:

A) Personal allowance 5%
B) Basic fatigue allowance 4%
C) Standing allowance. 2%
D) Awkward (bending) 2%
E) Very awkward (lying, stretching) 7%
F) Weight lifted (pounds):

    20 . . . . 3%
    40 . . . . 9%
    60 . . . . 17%

G) Poor light:

    Well below recommended 2%
    Highly inadequate 5%

H) Atmospheric conditions (heat and humidity) 0–10%
I) Mental strain:

    Complex work 4%
    Very complex 8%
    Tedious 2%
    Very tedious 5%

Allowance factor = the sum of all individual allowances from the difference factors A–I

Step 10: Compute the standard time:

$$\text{Standard time} = \text{Total normal time} \div 1 - \text{Allowance factor}$$

It is important to point out here that Taylor's scientific management principles were focused primarily on factory-type work, characterized by stationary workstations, more or less consistent material flows, and lots of repetition. Construction work, in contrast, is much less regular, and is marked by constantly changing work locations, tool variations, and in-process material flows that are dependent on the completion of other activities. Therefore, in the following sections, we describe approaches that allow us to modify Taylor's basic time study concept to make it suitable for construction work.

2.5.2 Measuring Value-Added Work

There are two main concepts involved in measuring laborers and machines at work: (1) continuous time study, and (2) work sampling. The first is based on data that represent the time it takes to complete the sequenced actions by a person or a machine continuously; the second relies on statistics derived from data collected at random intervals during the process. Naturally, the two require a determination of what should be measured and analyzed in order to arrive at information that tells us something about efficiency.

Section 2.4 presented a taxonomy for breaking down work time into five clusters, separating the value-added from contributory and ineffective time. In the worked-out problem defined by Walt Holz, we also learned that process-specific tasks, such as the framing-carpentry tasks, can be classified according to work clusters (e.g., measuring and marking is contributory work). Finally, the results of the site observations were given as percentages of total time that a worker spent doing a certain task, or doing nothing.

We can establish that the measurement scheme is, in fact, a matrix composed of five work-time clusters on one axis, and observation counts and time spent in each cluster. Figure 2.16 presents schematically the framework of the measurement matrix we can use.

To help flush out the differences, we'll look at an example of how the method and the outcome of the two concepts, continuous time and work sampling, compare. Figure 2.17 offers such a comparison, focusing on two workers involved in building a concrete formwork.

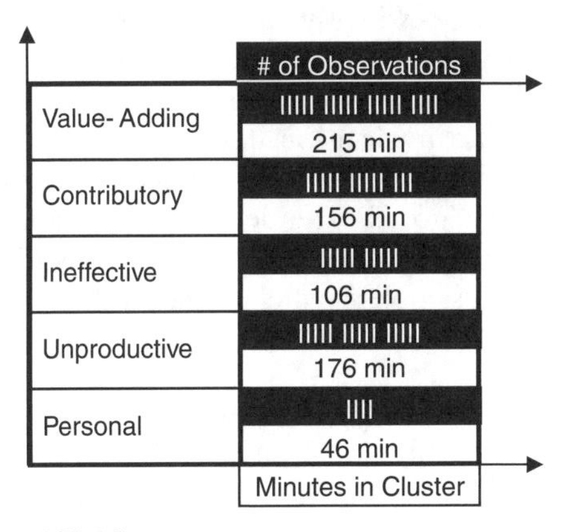

Figure 2.16 Measurement Matrix

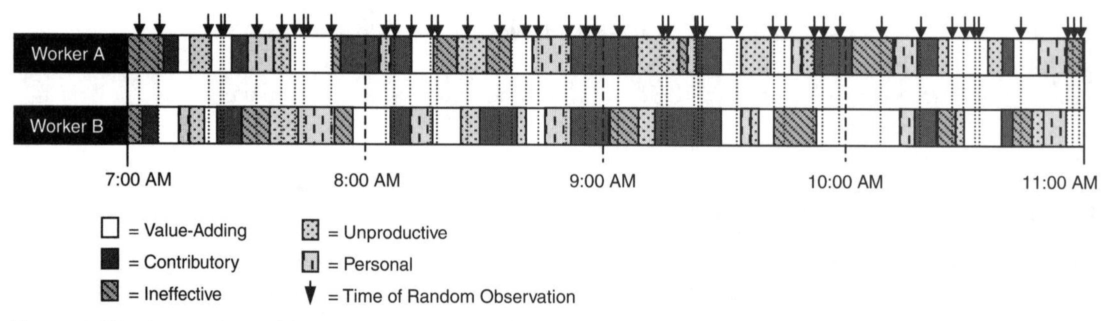

Figure 2.17 Comparison of Continuous Time Study and Work Sampling for a Four-Hour Work Period

The bases for the comparison are two detailed work-time records for a four-hour period, starting at 7:00 AM. Each "stream" has been broken into variable--size time blocks representing the five work-time classes we defined earlier. For example, worker A begins at 7:00 AM, and for a short time does ineffective work, followed by period of contributory and value-added work. Worker B also starts by doing ineffective work, immediately followed by contributory and value-added work, then some personal time. The streams are translations of continuous work into categorized time blocks that require a human translator, who has to make decisions, which are not always straightforward. For example, when is work ineffective and when is it unproductive?

Everybody who has tried to do this kind of "translation" for a long period of time for two workers simultaneously can attest to how tiring it becomes. For that reason, it is recommended to videotape the work, with the clock running. Modern video systems are digitized and can be processed using computer software, a vast improvement over the time-lapse cameras of the 1980s. Soon, however, this technology too will become obsolete, to be replaced by automatic and wireless data-acquisition systems that will provide much faster and easier access to crucial data for measuring process efficiency.

A much different process is involved in using work sampling to establish a measurement. The down arrows indicate the random times when the observer has to assess into which of the five classes of work the observed person belongs at that precise moment. Although this decision has to be made only at random times, rather than continuously, the categorization still requires a clear understanding of "what belongs in which category." The decision is especially tricky when, at the very moment, the worker switches from one to another work segment. This highlights the need for a "work sampler" who is sufficiently trained, a well-developed plan, and consistency. In the next section we discuss in greater detail how to apply this technique.

Tables 2.5 and 2.6 list the results of the two methods applied to measuring the work efficiency of the two workers, A and B.

The difference in measurement between the two methods surfaces in the first column for each worker. Whereas the continuous time study utilizes the sum of

Table 2.5 Result of Four-Hour Continuous Time-Study

Work Time Cluster	Worker A		Worker B	
	Min.	%	Min.	%
Value-Added	60.2	25	71.5	30
Contributory	57.5	24	60.3	25
Ineffective	38.3	16	45.8	19
Unproductive	45.6	19	28.8	12
Personal	38.4	16	33.6	14
Total	**240.0**	**100**	**240.0**	**100**

Table 2.6 Result of Incomplete Work Sampling

Work Time Cluster	Worker A		Worker B	
	Times Observed	Percent	Times Observed	Percent
Value-Added	14	21	22	34
Contributory	16	25	13	22
Ineffective	12	19	11	17
Unproductive	10	16	10	15
Personal	12	19	8	12
Total	**64**	**100**	**64**	**100**

the minutes for each of the five work-time groups, the work sampling method makes use of the cumulated number of observations in each. Following the underlying principle, the totals reflect either the total observation time—4 hours = 240 minutes—or the total number of random observations—64 in 4 hours.

The key to the comparison is the difference in the final result of the work efficiency measurement: how much time each worker spent productively. Let's focus on worker B. According to the continuous time study, he worked 30 percent on value-added tasks, versus 34 percent using the work sampling method. Taken together, value-added and contributory work, the comparison is very close: 55 percent versus 56 percent. The closeness of these numbers must be considered a coincidence, as the number of random samples is small, statistically speaking. On the other hand, it might point to the difficulty of the observers to distinguish consistently between contributory and value-added work, a difference that disappears when these two numbers are added up.

At this juncture, it's important to point out that the work study would not be finished at this point, as the result represents only 50 percent of one day. It would be necessary to observe the afternoon, as well as other randomly picked days of the week when the two are involved in the same kind of construction process. To that end, the next two sections offer more details for implementing each of the two methods.

2.5.3 The Continuous Time Study

The essential measurement tool of time-studies was formerly the stopwatch, but it has been replaced by computer software for digital video tracking, integrated into a bank of electronic controls. The goal, however, is the same as it has been since the inception of this work measurement method: to develop time records for the various tasks comprising a process. What was once done by a group of people positioned at appropriate places on a construction site, where they would conduct direct, real-time observations with stopwatch, pencil, and paper, can now be done cost-efficiently by one person in an office, which may be nowhere near the actual site.

The enabling technologies include digital cameras, digital video, and remotely accessible, controllable, and programmable Internet cameras. Not only is it unnecessary for the observer to be present on-site, but the data capture can be done remotely in a real-time processing mode, or recorded automatically for processing later. All that is needed is one or more network cameras, which may be wireless, and a computer with accompanying software in the office. Benefits of remote video time studies include:

1) Travel time to install a video camera on-site is eliminated.
2) Real-time analysis can be combined with delayed analysis of recorded data.
3) Cameras can be used for other managerial purposes such as visual safety inspection,
4) Video records can be stored as documents, in case of litigations.
5) Playbacks allow analysis of multiple processes by the same person.
6) Custom-made software can be created or purchased, to automate data analysis.
7) The time study can be conducted from almost anyplace with access to the Internet.
8) Data recording directly into spreadsheets allows rapid processing and documentation.
9) Worker training can be drastically improved, thanks to the availability of captured video.
10) No file compression from handheld video cameras to an efficient Internet format is necessary.

The following highlights the main steps in performing a time study of an average construction process, using one framer from Walt Holz's crew as an example.

Step 1: Study of the process to be observed in order to list and describe the various tasks that constitute a work cycle. Most work on construction is cyclic at the process level, meaning that a series of tasks are sequenced and performed repeatedly, like a framer working on a single-family house.

The model in Figure 2.18 recognizes two common exceptions that are important to be recognized.

First, not every task is being part of every cycle. For example, it is not necessary for him to study the drawings for every set of lumber he nails as a wall is commonly made of several studs of the same length. In addition, the distance between two

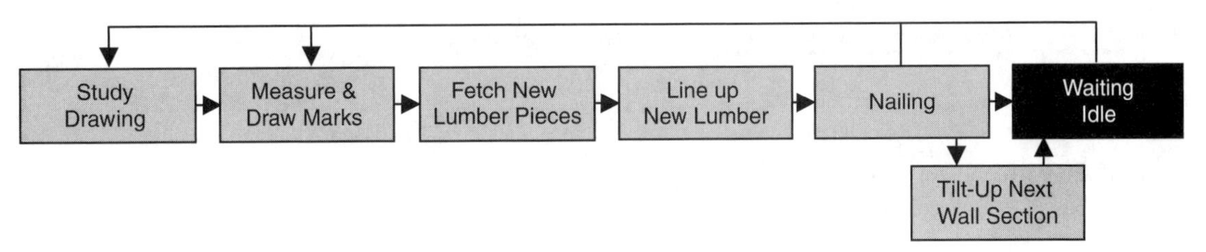

Figure 2.18 Task Sequence of a Framer's Work Cycle

studs is commonly 16 inches. Similarly, he does not wait idle at the end of every cycle.

Second, one task might need to be executed only after several previous cycles have been completed. The example is the tilt-up of a completed wall section that occurs only after of the required studs have been nailed to the top-header and the sole plate.

Finally, this step of the process needs to review the purpose of the time-study as the level of detail has to fit its intent. If the level is too detailed, recording effort will be expended to generate useless information. Should the level be too general, the study will not provide the desired results. For example, **Waiting Idle,** could be split into more specific time consuming but unproductive work segments such as: a) waiting for **framer B,** b) waiting for next lumber from cutting, c) resting, d) drinking wat**er, etc.**

Step 2: A data recoding scheme must be established, starting with a framework for the efficient entry of data into the computer. For example, the software used for the time-study will require a list of the tasks involved in order to prepare a user-friendly computer interface. For the purpose of explaining the concept, a standard format is used and presented in Table 2.7. The header includes the relevant data about the overall process, including the name of the observer or recorder. The second row of information lists all the tasks that could possibly be observed at the desired level, according to Step 1. Finally, space for recording times during each cycle, as well as summary values, must be included.

Step 3: Carefully plan for the positioning of either a recorder, video recorder, or Internet camera. Naturally, of prime importance is line of sight to what will be recorded. For example, focusing a camera on what will happen first will not capture later work. Similarly, selecting a too-narrow angle of view will cut out what happens outside the frame. The planning has to anticipate any view-obstructing sections of the building that will be erected during the period of observation.

Step 4: Inform workers of the need for and purpose of a signed consent form before filming an individual worker. It needs to be made clear that the purpose of the study is not to find out the personal habits of a worker, but rather

Table 2.7 Time Study Results for 31 Minutes of a Framer's Performance

Continuous Time Study

Project:	Production Framers	Date:	June 20, 2009
Day:	Tuesday	Weather:	Sunny
Crew Activity:	Wall Framing	Temperature:	Morning: 75–82°F Afternoon: 82–89°F
Humidity:	65%		
Recorder:	Jing Zhang	Study Object:	Jack Singleton, framer

Duration of Working Tasks (Seconds)

Cycle	Start (Sec)	Study Drawing	Measure and Mark	Fetch Lumber	Line up Lumber	Nailing	Wall Tilt-up	Waiting	Total Time (Sec)
1	0	27	62		4	18		46	157
2	157		51		7	23		22	103
3	260			12	6	34		38	90
4	350		17	16	22	57			112
5	462	30	41			46		17	134
6	596			11	9	21		31	72
7	668				10	20		43	73
8	741			18	12	21		17	68
9	809	16			17	34			67
10	876			17	12	30		28	87
11	963		28		24	61		20	133
12	1096	31		10	19	41	45		146
13	1242					40			40
14	1282		34	11	8	21		33	107
15	1389		26	9	7	33		19	94
16	1483	24		10	9	21			64
17	1547		52	14	11	19		25	121
18	1668	18			13	22		10	63
19	1731		23	10	17	31			81
20	1812			12	11	27		12	62
Tot Time at Tasks (sec)		146	334	150	218	620	45	361	1874
Average Task Time (sec)		24.3	37.1	12.5	12.1	31.0	45.0	25.8	
Percent at Task		**7.8%**	**17.8%**	**8.0%**	**11.6%**	**33.1%**	**2.4%**	**19.3%**	**100.0%**

Contributory = 33.6% (Study Drawing, Measure and Mark, Fetch Lumber)

Value-Added = 47.1% (Line up Lumber, Nailing, Wall Tilt-up)

to find ways to improve the overall operation by cutting out non-productive work time caused by, for example, the set-up of the operation around the worker. If possible, the worker should be invited to participate in the analysis of the video. As we will learn later in this book, such inclusive collaboration can result in improved outcomes, as the laborers typically know more than anyone about the problems that will become apparent upon viewing the video. Unfortunately, asking a laborer at the beginning of the study will not provide the same insights, as workers are rarely aware of inefficiencies around them.

Step 5: At this point in the process, everything should be in place to run a trial. It is extremely unlikely that everything will work as planned the first time, especially for novices of this method. Thus, it is wise to plan at least two site visits.

Step 6: Now everything should be ready to execute a successful video and time recording session. Here's how Table 2.7 was created: For each of the 20 cycles, lasting a total of 1,874 seconds, or 31 minutes, the time increments related to each task performed by Jack Singleton were recorded in a spreadsheet. As discussed earlier, there are various ways to produce a spreadsheet like this, including hands-on entry into forms or using various software that are available on the market.

Step 7: The final step in the process is a discussion of the time-study outcome. What can be reliably concluded from the data? Once more, the 20 cycles used in the example are not sufficient to draw definite conclusions, but they serve to explain the essentials of the continuous time study method.

What stands out immediately is the 47.1 percent for value-added work and the 19.3 percent waiting time, which appear consistently at the end of almost every cycle. Note next the 33.6 percent of time the laborer spends on contributory work, with measuring and marking taking up the majority, at 17.8 percent.

Final recommendations depend on the goal of the study. If the objective was to identify opportunities to improve the crew's productivity, then the cause of the 19.3 percent waiting time should be studied further. This would be the ideal time to invite Singleton to review the video in order to get his input as to what is happening that is not necessarily apparent on the video.

Methods for identifying and eliminating the key causes of waiting time will be discussed later in the book.

2.5.4 Work Sampling

As just noted, statistical work sampling is an alternative to studying the work time of a laborer or an operator. Steps 1 to 5 are, however, exactly the same as in the continuous time study—although the data-recording software will use a slightly different program and the data-recording format will be different, as will be explained in the following.

Step 6: Valid work sampling necessitates a sound sampling design, which must consider the nature of the process and the purpose of the measurement. Mathematically, the number of observations required is calculated using the following:

$$N = k^2 \times p(1 - p) \div s^2$$

where:

N = number of needed observations
k = number of standard deviations for a given confidence limit
s = absolute limit of error
p = chance that observation elements might not be visible (decimal)

For sampling construction operations, there is general agreement that a confidence limit of 95 percent and a limit of error *s* of plus/minus 5 percent are sufficient for producing sound results. The value of *p* is a factor used to reduce the number of observations needed for processes, wherein each cycle is almost equivalent to all the others, such as work on a conveyor. In such a situation, *p* would be a low decimal (e.g., 0.05, or 5 percent). However, as demonstrated by the framing example, some tasks are left out of some cycles while others are only executed from time to time. It is only natural that in order to increase the opportunity to observe every task, the number of observations has to be higher. For the minimum *p* value of 0.5, a confidence level $k = 95\%$ and $s = \pm 5\%$ the minimum number of random observations = 384.

After a establishing a statistically sound number for the measurements, we need to ensure the proper execution of the observations. The first key factor is *consistency*, the other *comprehensiveness*. Construction processes commonly last at least one eight-hour shift, so a comprehensive sample should cover the entire eight hours. But if performances tend to vary on different days of the week (Monday being the "headache day"), an extension of the study to include other days should be considered. At this point, the requirement for consistency comes to the fore. If the goal is to achieve an overall measurement of a crew or one laborer, the sample should cover the entire week, with some of the 384 observations carried out randomly every day. The need for consistency, however, would require that the members of the crew itself not change, and that they work on the same job under the same conditions.

Unlike in manufacturing, such conditions are hard to achieve. Thus, it is recommended to pick a comprehensive observation "window," to capture all the common cycles during an entire day. In order to study the effect of changed conditions—say, on a different day of a week—the work sampling can be repeated, leading to a doubling of the observations and thus to opportunities to detect discrepancies that are normally not apparent.

A final remark should be made concerning assessing an entire crew as compared to an individual. If the objective of a study is to measure a crew consisting of 5 workers, all observable at the same time, the minimum observations of 384 can be divided by 5, as at each random time 5 data points are established. Thus, the number of minimum random observations for a 5-member crew is 77, resulting in an overall evaluation of the crew as a whole. On the other hand, the final results for one crew member alone would be statistically unsound in these circumstances, since only 77 observations would be made, not 384.

Step 7: Table 2.8 provides the result of 18 random observations of two framers working for Production Framers during a two-hour time block starting at 8:00 A.M. For each observation a random number has to be chosen, either from a table or a random number generator, such as the function RAND() in Excel, which provides numbers between 0 and 1. After sorting the resulting random numbers from smallest to highest, they can be translated into minutes. This can be done by aligning the highest number, or 1, as with the total duration (e.g. 2 hours, or 120 minutes), and the lowest number with the first observation (e.g., 8:00 A.M.). Thus, each increment between two random numbers, starting with the lowest, represents a fraction of the total time, (e.g., $0.183687 - 0.058742 = 0.124945 \times 120 = 14.99$), is rounded to the next minute, and added to the last observation (e.g., 8:00 A.M. + 15 minutes).

Step 8: There are several ways to execute the observation, assuming an internet camera is installed at the site. Software is available to capture images at either preprogrammed or random times. Otherwise, the entire two hours has to be watched in real time or stored for quick analysis at a later date. Whatever method is chosen, the data recorded may be of different types, as shown in Table 2.8. The main data elements to be captured are the five main work clusters:

VA = Value-Added
CO = Contributory
IN = Ineffective
UN = Unproductive
PE = Personal

At the time of observation, the recorder makes a quick—and sometimes biased—assessment of each framer's work in terms of the five predefined options. This is recorded with a checkmark, or when using a computer, inputting a number. This could be the end of an observation. But as the example illustrates,

Table 2.8 Work Sampling Recording Form for Two Framers

WORK-SAMPLING

Project:	Production Framers	**Date:**	Jun-20-2009
Day:	Tuesday	**Weather:**	Sunny
Crew Activity:	Wall Framing	**Temp:**	Morning: 75-82 degree F
Humidity:	65%	**Study Objects:**	Jack Singleton, framer
Recorder:	Jing Zhang		Vince Perry, framer

# Obs	Random Number	Time of Obser.	Framer 1					Framer 2				
			VA	CO	IN	UN	PE	VA	CO	IN	UN	PE
1.	0.058742	8:00					√ 11		√ 3			
2.	0.183687	:15					√ 11		√ 3			
3.	0.452753	:47		√ 5					√ 2			
4.	0.464919	:48		√ 5				√ 1				
5.	0.470881	:49		√ 2				√ 1				
6.	0.479587	:50	√ 1					√ 1				
7.	0.499287	:52	√ 1					√ 1				
8.	0.554698	:59	√ 1						√ 3			
9.	0.564838	9:00		√ 3					√ 3			
10.	0.619949	:07		√ 3					√ 2			
11.	0.63546	:09		√ 3				√ 1				
12.	0.663062	:12		√ 3				√ 1				
13.	0.692963	:16		√ 3				√ 1				
14.	0.801708	:29		√ 4					√ 5			
15.	0.825561	:32		√ 4					√ 5			
16.	0.849609	:35		√ 4					√ 5			
17.	0.893628	:40	√ 1						√ 3			
18.	0.928605	:44	√ 1						√ 3			
	Tot Observation		5	11	0	0	2	7	11	0	0	0
	Percent (Tot/18)		28	61	0	0	11	39	61	0	0	0
	Data from a Larger Sample		46.7	33.9	5.5	5.0	8.9	38.3	30.6	10.0	3.3	17.8

this examination can be made more interesting by adding a second layer of detail, similar to the continuous time study example discussed previously. This is accomplished by attaching a second marker, a label, that points to one of the 13 work tasks that the entire framing crew can be expected to engage in. This flexible way of recording can be done efficiently with computer software.

Step 9: The final analysis begins with the summing up of the total checkmarks in each category and then dividing by the total number of observations, which results in the percentage of time a framer was observed doing a task that fit the respective class. For example, framer 1 spent 28 percent of his time doing value-added work and 61 percent doing contributory work.

The bottom row in Table 2.8 highlights most prominently the importance of sample size. During the two hours of sampling, neither of the framers, for example, did any unproductive or ineffective work. Also, framer 2 did not take a break. Only a sufficient sample size, a consistent process, and a comprehensive consideration of all important factors will result in meaningful outcomes.

2.5.5 Sensor-Based Work Measurement

Since the late 1990s, major equipment manufacturers have switched to using electronic controls for their equipment. They have abandoned mechanical methods for opening and closing a hydraulic valve, for example, and replaced them with the ubiquitous "joysticks" that change the voltage input at the valve and thus direct the hydraulic fluid. This switch to electronics changed the control scheme of the equipment. Because the electronics required a totally new "infrastructure" of protected cables, switches, hubs, processors, etc. to be built into the equipment, adding new sensory capabilities was made easy. In turn, previously unavailable data and information became suddenly accessible for little money.

Equipment manufacturers that want to offer new services to their customers must focus on keeping large engines in good condition, as contractors are hit hard when an engine breaks down. Thus, sensors that collect data regarding engine health from, for example the engine oil, and the software that analyzes such data have become a mainstay of preventive maintenance. It demonstrates that electronics does indeed have a place in construction, and can provide cost-saving information to a contractor. Not surprisingly, this has led to the development of myriad applications that have slowly penetrated an industry that has been very reluctant to test new technologies on low bid projects where everything has to work as planned.

Similar to a framing crew, construction equipment works on a cyclical basis. A tower crane on a building site goes through repetitive cycles of picking up, lifting, placing, and lowering; and an asphalt delivery truck goes back and forth from the asphalt plant to the asphalt spreader. Because of this repetitive nature, involved equipment equipped with the correct sensor devices could become direct sources of information about their efficiency, without having to resort to work measurement techniques such as work sampling. Let's study how this could work for an asphalt delivery truck, starting with a model of the delivery cycle.

Figure 2.19 indicates the two main locations of the truck cycle: the plant and the spreading operation. Marked with A through F are the points at which a truck begins a certain task. They can be represented by time and location, and both can be sensed automatically. For example, a GPS is able to recognize when the truck

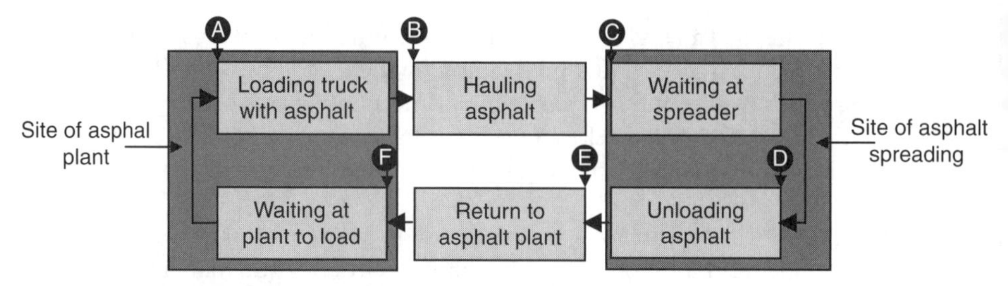

Figure 2.19 Task Sequence of an Asphalt Delivery Truck

> **Digital Data Loggers**
>
> Digital data loggers are self-contained, battery-powered data-collection devices that have drastically expanded the capabilities of the old card recorders. Equipped with a multichannel input device, digital converter, microprocessor, and large memory, data loggers can collect data once every second from a large number of electronic sensors, including voltage, counters, truck speed, location, temperature, humidity, wind speed, and weight. Data loggers typically use real-time clocks to time-stamp the collected data. The processors can execute many functions, such as calculating statistical averages.

leaves the plant to start hauling, labeled in the figure as B. The start time of unloading, D, could be triggered by the raising of the truck bed containing the asphalt for refilling the spreader-hopper. In other words, sensors are able to generate the raw data to be transmitted wirelessly to a central processing unit.

IT companies have made large investments in various telematics applications, so that they can offer these services to the equipment industry. One such application is theft protection for trucks and other expensive equipment. By means of small sensor that sends signals as soon as a piece of equipment is moved, local police can be quickly alerted to a theft in progress. Other systems will broadcast the location of the equipment as it is being moved. Telematics systems, in combination with data loggers, are excellent technologies for managing an entire equipment fleet; equally important, they improve the safety and the productivity of construction operations.

There is no doubt that Internet cameras, data loggers, and telematics systems will drastically improve the timeliness and efficiency of feedback and feed-forward data communications for the construction industry; however, the question about what to do with all the data still has not been fully addressed. It's true that we can record the percentage of time a laborer is working on a value-added task, or how long a truck is waiting to be loaded at the excavator; but what can the manager do with that information at 10:30 in the morning, or even at the end of the day? James Watt would point out very quickly that what is missing is the controller "brain," the capability to turn output data into commands to regulate the input, and thus achieving an improvement of the output. We know that our declared goal is to maximize process productivity; we also know we can assess the productive and unproductive work time for two key input factors, labor and equipment, we are now ready to close the loop.

Equipment Telematics

Telematics (a contraction of "telecommunications" and "informatics") encompasses electronic, electromechanical, and electromagnetic devices that operate in conjunction with microprocessors and radio transceivers to provide a large number of functions, among them:

- Real-time mapping (GPS)
- Data logging
- Automatic wireless messaging
- Remote locating
- Geofencing

With geofencing, the equipment owner can create a virtual fenced-in area on a map, such as a loading zone of an earthmoving operation. The telematics system can be programmed to send a message, in various forms, whenever equipment crosses the fence. Data loggers integrated into the system offer a reliable technology for monitoring the operator, reconstructing accidents, recording engine health, and more.

Some of the data can be stored; or it can be sent automatically using a global system for mobile communication (GSM), the standard used by most mobile phone companies worldwide. Data can also be sent using the less expensive general packet radio service (GPRS), a packet-oriented mobile data service available to users of GSM.

2.6 IDENTIFYING CRITICAL IMPACT FACTORS

As we learned from studying James Watt's first proportional control mechanism, we need a logic function that relates an input to an output. He used the centrifugal force created by a spinning three-link mechanism, which raised and lowered a lever depending on the speed of the engine. Today, as just described, we have electronic controllers that take electronic input, process it, and, if a program is built in, react appropriately by initiating desired actions all on their own. We also learned earlier that the logic functions can be polynomial, discrete on/off mechanisms, or even IF-THEN rule trees.

This last section of the chapter will address the underlying concepts that will allow us to model and build functioning control systems for construction.

2.6.1 Understanding the Cause of Inefficiency

Understanding a phenomenon means to have the ability to predict its behavior. What makes it predictable is the recognition of cause-and-effect relationships that control the behavior of the phenomenon. Take for example scurvy, an illness that killed many sailors before the nineteenth century. Nobody understood what caused it and nobody had a predictable remedy until 1753, when Doctor James

Lind published the result of his experiments that identified citrus fruits as a clear remedy that effectively cured scurvy. Although the understanding of *why* it happened took much longer, recognizing the cause-effect relationship and providing sailors with citrus fruits had the predictable effect—nobody got scurvy anymore.

Similarly, the result of observing the framing process of the crew from Production Framers, Inc. gives us the opportunity to study how to apply cause/effect thinking to develop an understanding of behavior.

Recall from the list of 13 framing tasks shown in Table 2.2 that several were clearly nonproductive: Four examples are:

> **Task 7: Carrying** (including picking up and unloading) materials-tools *more than 35 feet* from work
>
> **Task 8: Searching** for material in or outside the staging area
>
> **Task 9: Waiting** for materials, tools, instructions, etc.
>
> **Task 10: Correcting an error**, such as a wrong measurement

Note that task 7 differs from task 8, although both involve carrying: the difference is in the distances that things are carried. Task 8 involves carrying inside the work area, considered a "circle" with a radius of 35 feet. Why is number 2 contributory and number 7 wasteful and ineffective? Why is searching for a tool or material considered ineffective? Before addressing these questions, let's quickly review how cause-and-effect relationships can be visualized using circles and arrows. Anyone familiar with Critical Path Method (CPM) scheduling will recognize that directed arrows are equivalent to the sequencing relationships of project activities. The effect of completing one activity in a finish-start relationship (the cause) is that the following activity is able to begin. Similarly, the cause-and-effect relationship between citrus juice and scurvy can be formulated as follows: Drinking lemon juice (= cause) prevents scurvy (= effect).

Hence, we now have a simple but elegant method to model ineffectiveness: by considering the task at hand as being affected by one or several causal relationships (see Figure 2.20). What will help in this effort is to quickly revisit the list of seven muda wastes, to remind ourselves that not all actions that take time and effort are productive—walking around searching for a missing tool comes to mind. Figure 2.21 shows a multilevel cause-and-effect diagram for tracking down a lost hammer

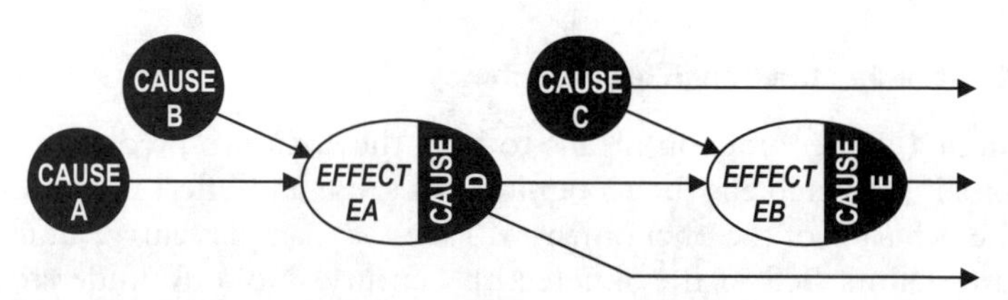

Figure 2.20 Principle of Cause-And-Effect Modeling

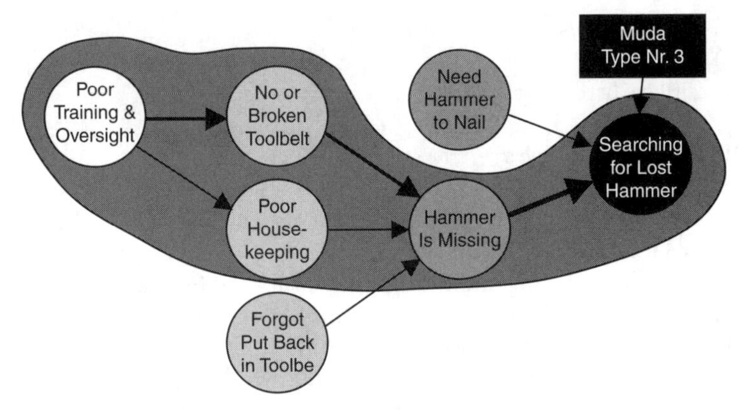

Figure 2.21 Cause-and-Effect for Muda Waste in Framing

needed for nailing. Why is searching for a tool a muda? It's a muda—type 3, to be exact—because unnecessary time has been spent searching for a tool that should have been readily accessible, given that carpenters are rarely seen without their toolbelts around their waist, loaded with everything they could possibly needed at any time (e.g., different-size nails in pouches, cordless devices, and, most certainly, a hammer, which also commonly serves as a crowbar).

The diagram in Figure 2.21 indicates that there are two obvious causes for the search: a need to nail and a missing hammer.

A more interesting question arises from investigating: **Why is the tool missing?** Let's assume two immediate possibilities: (1) the framer does not wear a toolbelt or the one he has is broken, or (2) he forgot to put the hammer back into the belt when he used it last. The model assumes that these two are needed but not sufficient to lose a hammer. What has to go along is poor housekeeping during framing, such as scrap, leftovers, trash, cables, drawings, and so on strewn all over the area, making it extremely easy for a lost tool to "hide" from view. **Why does the framer not wear a proper belt?** and **Why is there poor housekeeping?** As shown, the answer to both questions is modeled as poor training and oversight. How can the answers to the two "Why's?" be helpful in regulating the productivity of framing? Before finding the answer, let's apply the **3 Why's?** to a second case, task **Number 9:** Framer waits for cut lumber pieces.

Some Muda Waste Revisited

3. *Unnecessary motions:* Workers use more time and effort to perform, thus becoming fatigued.

4. *Idle waiting:* Workers or equipment must wait for other resources to be supplied.

5. *Unnecessary transport:* Poor work floor layouts require extra handling of parts and materials, increasing risk of damages and nonproductive effort.

6. *Inefficient processing:* Lack of adequate machinery, sufficient training, and up-to-date standards.

7. *Defects, breakdowns, injuries:* Missed quality targets and equipment malfunctions result in rework, cause scrap, and so on.

1. *Why is waiting for cut lumber a muda?*

Waiting idly is listed as muda type 4.

2. *Why is the framer waiting for cut lumber?*

To answer this question we need to revisit the site layout and the job assignment at Walt Holz's construction site, 37 Bentley Court. Two framers, Jack and Vince, are working on nailing and tilting-up walls for the first floor. One carpenter, Fred Mercy, is responsible for cutting the lumber and plywood that the framers will need next. Helper 1 is supposed to carry lumber to the cutting area, while helper 2 is assigned to assist the cutting carpenter in pulling and carrying cut pieces to the framers. In other words, the framers do not cut lumber pieces. Instead they depend on the pieces being supplied by the cutting carpenter or helper 2. Now the answer in this case becomes straightforward. There are two causes behind the framer's wait for lumber: (1) He finished everything he could do productively with the resources he had available, and (2) the lumber needed for the next sequence has not been supplied.

3. *Why haven't the needed pieces of lumber been supplied to the framer?*

There is no question that the possible reasons for this are almost limitless. At this point we will focus on inherent problems, rather than "fluke" rationales. The apparent direct cause is the pace at which new pieces are cut and/or the pace at which cut lumber pieces are carried to the framers, or a combination of the two. The pace measurement is, of course, the production rate, defined as pieces per hour. As one depends on the other, the smaller of the two will dictate the supply rate.

We can conclude, then, that there is a twofold cause behind the framer having to wait: (1) The rate of cutting lumber pieces is slower than the rate at which the framer can frame new pieces, and (2) the rate of carrying cut pieces from the cutting area to the framer is slower than the rate at which the framer can frame new pieces. This leads to a fourth, two-part, question:

4a. *Why is the cutting rate slower than the framing rate?*

Again, there is a large number of possible answers, ranging from poor-quality equipment (the saw) to hard-to-read drawings to extensive time needed to mark and make perfect cuts, or time needed to do noncutting related work. Whatever the reason, the possible answers to this question are: (1) the carpenter spends too much time per cut, or (2) the carpenter does not spend enough time cutting.

4b. *Why is the rate of carrying pieces from the cutting area slower than the framing rate?*

Answering this question must be done independently of answering question 4a. So, assuming that the cutting rate is equal to or faster than the framing rate, what might cause the slow supply rate? Is it the distance between the cutting area and the framer's location? Is it a slow-moving laborer, who has to climb over obstacles to reach the framer? Or is it that no one is readily available to pick up the cut pieces and transport them to the framer?

According to the job responsibilities, helper 2 was assigned to assist the cutting carpenter. Should the carpenter be fully occupied with cutting the lumber, and helper 2 with supplying it? In other words, the causes here are all three of those presented as possible: distance, obstacles, and unavailability.

2.6.2 Functions for Corrective Interventions

Assume the Walt Holz happens to check the Internet camera set up at the project site from his office and sees one of his framers searching for his hammer. Helper 2 is nowhere to be seen. At the same time, the second framer, together with helper 1, is tearing apart a wall section that was already nailed and erected. Since this is a type 7 muda—spending time fixing an error—Mr. Holtz is considering an intervention. What should he do to eliminate this form of waste?

Anyone familiar with framing could come up with at least one recommendation for Mr. Holz; but in order to develop a well-founded response, it is necessary to review all the options and pick the one that is most likely to be successful. Cause-and-effect modeling, used to create a list of possible causes of undesired effects, is also useful in anticipating the effect of corrective actions taken as the final step in controlling the process.

Figure 2.22 lists four possible corrective actions: (1) hire a third helper, (2) add a foreman to the crew, (3) supply a new toolbelt to the framer, and (4) offer a free training course to the crew. As indicated, there are two types of effect: *immediate* and *delayed*. Adding a helper or a foreman would have to wait until tomorrow morning, whereas bringing a toolbelt to the framer, if in fact he needs one, should make the crew more effective by eliminating the observed muda—searching, waiting, and errors. On the other hand, none of those actions will guarantee higher productivity by the crew (output/labor hours). For one, the elimination of time spent searching might just lead to more waiting time. Likewise, the increase in the number of laborers has to be balanced against a decrease in time needed to produce the same output. If the additions will not have the necessary countereffect on the time, productivity might drop even further.

Option 4, on the other hand, would create a more long term effect. By organizing a series of training sessions right away (the first one tomorrow) there might also be near-term benefits to the present job by cutting out inefficiencies. But it probably would cost some money, since paid time must be made available, and the top carpenter in the company might have to be brought in. The true long-term benefits, should the crew stay with the company, is that they will be able to take the methods they have learned for achieving high value-adding output to new

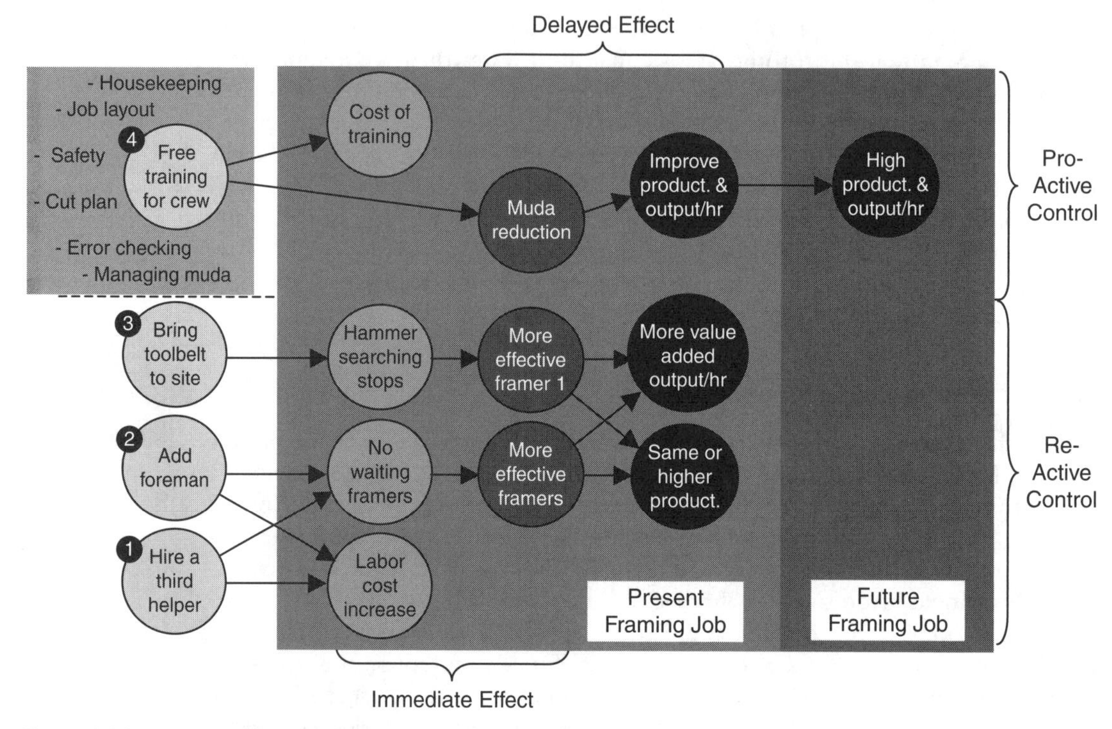

Figure 2.22 Cause-and-Effect Model for Possible Control Actions

jobs, as illustrated in Figure 2.22. Thus, option 4 can be considered as a proactive control strategy, rather than reactive to the present situation.

The example has hopefully made one point very clear, that there are usually many corrective actions from which a manager may choose. However, his or her choice should align with the company's strategies and objectives. It goes without saying that it also needs to be based on a solid understanding of the probabilities of specific effects and the measures needed to make them successful. This leads us to the last topic in this chapter, the identification of the most important factors that promise success.

2.6.3 Critical Success Factors

If we expand our small framing example to a large construction site, with several hundred workers and numerous pieces of equipment moving about, the number of inefficiencies will skyrocket, and with it the number of possible interventions. The ensuing complexity involved in taking corrective actions reactively while a process is already in progress, favors pro-active control activities, including training and detailed planning. Proactive "regulation" of process productivity does not inhibit real-time control, and has been found extremely effective in achieving high productivity. However, the sheer number of factors that can be "fine-tuned" before the start of the process is excessive. One way to deal with this dilemma is to apply the 80/20 rule, which allows the manager to concentrate his or her efforts on those factors that promise to be most effective in achieving the desired outcome.

The process of separating the few vital factors from the majority trivial ones is referred to a *critical success factor analysis.* The following demonstrates the basic mechanism, with the help of experienced managers in the field of home building. (The use of experts is only necessary if one does not possess the necessary expertise to select and evaluate factors from several options.) While such a prioritization can be accomplished in many different ways, we will use Likert scale questionnaires, because they offer a straightforward tool. Table 2.9 shows a simple example survey used to query 10 experienced managers in home construction. The 20 factors listed had been put together based on a literature review and the personal experiences of those who compiled the questionnaire. Note that randomness of the listed items is necessary in order to avoid biases of the designers creeping in (e.g., by putting the least favored at the end).

Each completed questionnaire was processed by assigning a numeric value to each response, such as 1 for "None" or a 5 for "Major"—which was then entered into a spreadsheet for easy averaging and sorting. Table 2.10 gives the result from the 10 managers. Keep in mind that due to the small sample size, this is by no means representative of the entire industry.

The resulting ranked list of the top seven factors contains many we have already discussed. The almost unanimous "winner" is planning and control of site operation. This first place is most interesting in that the availability and skill of labor only made it to third place. Even the importance of good communication ended up in second place.

This is only the first step in identifying those managerial activities that most effectively impact the productivity of site operation. The second step is to ask the whys, whats, and hows associated with each of the seven highest-ranking factors. The listed items are still too general to provide specifics for a new manager, who will ask: "What are the critical success factors for each of the seven identified factors?" Indeed, each factor is only the top of an entire new hierarchy of more specific factors that can be investigated for their importance. This is actually a feature of nature, as well as human-made systems and organizations.

Assertions Based on the 80/20 Rule

- 80 percent of a result is achieved by 20 percent of the team members.
- 20 percent of an individual's effort will generate 80 percent of his or her output.
- 2 difficult tasks completed are worth as much as 8 easy ones.
- 20 percent of a company's customers bring in 80 percent of the total value.
- In any process, a few factors are vital; but most are trivial.

Rensis Likert (1903–1981)

Rensis Likert was an American educator and organizational psychologist who developed the Likert scale for questionnaires, the most widely used scale in survey research. Respondents are asked to specify their level of agreement to a statement. The format of a typical four-level Likert item is:

1. Strongly disagree
2. Disagree
3. Agree
4. Strongly agree

Data from Likert scales are commonly reduced to a nominal level.

Table 2.9 Survey Questionnaire: Critical Success Factors in Building Construction

Item	Factor Description	Influence on Process Productivity				
		None	Minor	Some	Big	Major
1	Work space					
2	Communication/coordination					
3	Weather conditions					
4	Management competence					
5	Relations with community					
6	Work methods (sequence, technology, etc.)					
7	Planning and control of site operations					
8	Site conditions					
9	Inputs from previous process (WIPs)					
10	Material staging and site transportation					
11	Material purchase and delivery					
12	Change orders					
13	Labor's skill and motivation					
14	Quality of drawings/specifications					
15	Safety program					
16	Working environment (noise, lighting, etc.)					
17	Owner's requirements					
18	Government regulations					
19	Waste disposal and recycling					
20	Equipment and tools					
21						
22						
23						

Note to respondents: The table lists 20 factors that may or may not impact the productivity of building construction. Please consider how important each item is in achieving high efficiency on the site. Express your opinion by checking the appropriate column. Do not hesitate to add factors that you feel are missing from the list, in the blank spaces provided.

Table 2.10 Critical Success Factors in Building Construction, Based on 10 Respondents

Rank	Average Point Value	Factor Description
1	4.9	Planning and control of site operations
2	4.7	Communication/coordination
3	4.6	Labor (availability, skill, motivation, etc.)
4	4.4	Equipment and tools (appropriateness, quality, etc.)
5	4.4	Working methods (sequences, technology, etc.)
6	4.0	Site conditions
7	3.9	Material delivery, staging, and site transportation

Instead of pursuing the further breakdown of the seven factors in the example, we offer yet another tool, in the form of a fish skeleton, that is helpful in representing success factors in detailed layers. Figure 2.23 presents a two-layered fishbone diagram, organized according to cause/effect relationships as they affect construction productivity.

The fishbone diagram in Figure 2.23 presents higher-level critical success factors organized into six main groups. It is apparent that the seven factors ranked most critical by the 10 building construction managers will fit nicely into the structure. Nevertheless, the diagram requires more work, possibly by a team of experts, who can break each factor down further. The main contribution of fishbone diagrams is, in fact, to foster such brainstorming, a process we will discuss later in the book. It is also apparent, that the outcome of such efforts is closely linked to the work the company specializes in. To that end, the next section offers a set of tools

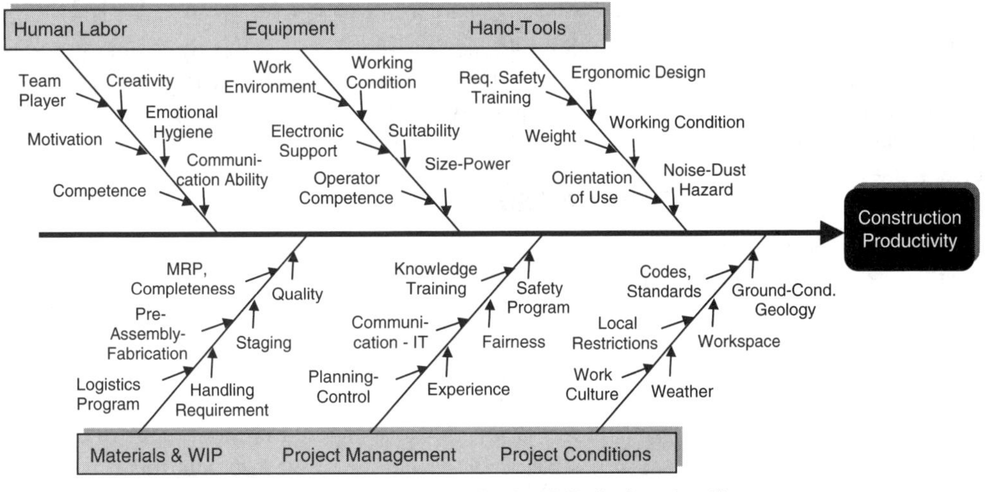

Figure 2.23 Two-Layer Fishbone Diagram Related to Productivity in Construction

Kaoru Ishikawa (1915–1989)

Japanese chemical engineering professor Kaoru Ishikawa introduced the concept of quality circles in 1960. He is best known for pioneering the fishbone, or cause-and-effect diagram, as a tool for modeling and working on complex issues in teams.

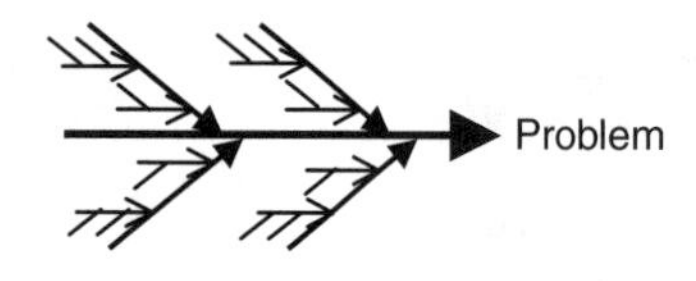

Here's how to create your own fishbone diagram:

1. Define the main problem or goal.
2. Draw the backbone and first-level spines.
3. Brainstorm the main factor groups related to the problem or goal.
4. Define a key word to connect each factor to spines.
5. List HOW-WHY factors causing the problem, or list HOW-WHY factors contributing to the defined goal.
6. Attach answers to second-level spines.
7. Repeat brainstorming for the third level, and so on.

that might be helpful to those interested in generating critical success factors for a specific business. It also adds yet another dimension to the approach we have been using so far.

2.6.4 Bipolar Success Factor Analysis

Should success factors related to productivity be equally important in achieving high productivity as in avoiding low productivity? Let's take the case of the framers who are waiting an extraordinary amount of time for the next delivery of lumber pieces. Adding one helper will reduce their waiting time, and thus cut out this muda/waste. On the other hand, this addition will not boost productivity, as it just remedies an obvious problem. Improving the productivity of a crew that is performing poorly is really a two-step effort:

1. Eliminate poor productivity to an expected level.
2. Advance expected productivity to a higher level.

These are two totally different managerial missions. This concept is at the foundation of the forms presented in Tables 2.11 and 2.12, created for use in establishing prioritized success factors to avoid poor, and achieve high, productivity for a construction business.

Step 1: Rank according to the importance in achieving **HIGH** productivity

Step 2: Distribute numeric weights to each factor according to its rank (e.g., Rank 1= Weight 36%).

Step 3: Cover what you just entered in Table 2.11.

Step 4: Repeat Steps 1 and 2, this time ranking the factor groups for being most responsible for causing **LOW** productivity.

SUMMARY

Step 1: Copy the results from Table 2.11 into column a of Table 2.15 and those from Table 2.12 into column a of Table 2.16

Step 2: Add the check marks in a significance column (0–3) of a factor group in Tables 2.13 and 2.14 and multiply the sum with the associated numeric value of the Likert scale (0–3). Enter the result into columns c, d, e, and f of Table 2.15 and 2.16 respectively.

Table 2.11 Weighting of Factor Groups Leading to HIGH Productivity

Factor Groups	Rank	Weight
Human labor		
Project management		
Materials supply		
Equipment		
Hand tools		
Project conditions		
		$\sum = 100\%$

Table 2.12 Weighting of Factor Groups Leading to LOW Productivity

Factor Groups	Rank	Weight
Human labor		
Project management		
Materials supply		
Equipment		
Hand Tools		
Project Conditions		
		$\sum = 100\%$

Table 2.13 Critical Success Factors for Achieving High Productivity

Step 1: Add at least one factor at the end of each group list that you think should be included because of its relevance to reaching high or finish up with low productivity

Importance of Individual Factors to Achieving High Productivity	Significance			
	0	1	2	3
A: Project Management				
1. Design drawings				
2. Scheduling/project planning				
3. Daily Planning				
4. Daily report reviews				
5. Communication				
6. Information technology				
7. Experience in construction work				
8. Fair Treatment of laborers				
9. Safety program				
10. Support for continuing education				
11. Training programs for laborers				

(*continued*)

Table 2.14 Critical Factors Leading to Low Productivity

Step 2: For each factor check one Significance (0–3) in contributing to achieving high and low productivity Likert scales: 0 = Irrelevant; 1 = Of minor importance; 2 = Of substantial importance; 3 = Key

Importance of Individual Factors to Achieving Low Productivity	Significance			
	0	1	2	3
A: Project Management				
1. Design drawings				
2. Scheduling/project planning				
3. Daily planning				
4. Daily report reviews				
5. Communication				
6. Information technology				
7. Experience in construction work				
8. Fair treatment of laborers				
9. Safety program				
10. Support for continuing education				
11. Training programs for laborers				

(*continued*)

Table 2.13 (*Continued*)

12. Supporting culture of change				
13. Collaboration/sharing responsibilities				
14.				
Total checks A				
B: Human Labor				
1. Competence (skills, knowledge)				
2. Communication ability				
3. Experience				
4. Motivation to perform				
5. Emotional health				
6. Creativity				
7. Team player				
8.				
Total checks B				
C: Equipment				
1. Working conditions				
2. Suitability				
3. Operator competence				
4. Size/Power				
5. Electronic support				
6. Work environment				
7.				
Total checks C				
D: Hand Tools				
1. Working conditions				
2. Ergonomic design				
3. Noise/dust hazard				
4. Use Orientation				
5. Weight and size				

Table 2.14 (*Continued*)

12. Supporting culture of change				
13. Collaboration/sharing responsibilities				
14.				
Total checks A				
B: Human Labor				
1. Competence (skills, knowledge)				
2. Communication ability				
3. Experience				
4. Motivation to perform				
5. Emotional health				
6. Creativity				
7. Team player				
8.				
Total checks B				
C: Equipment				
1. Working Conditions				
2. Suitability				
3. Operator competence				
4. Size/power				
5. Electronic support				
6. Work environment				
7.				
Total checks C				
D: Hand Tools				
1. Working conditions				
2. Ergonomic design				
3. Noise/dust hazard				
4. User competence and skill				
5. Weight and size				

6. Required safety training				
7.				
Total checks D				
E: Materials and WIP				
1. Quality				
2. Staging				
3. Handling requirement				
4. Logistics program				
5. Preassembly/fabrication				
6. MRP, completeness				
7.				
Total checks E				
F: Project Conditions				
1. Weather				
2. Work space				
3. Ground condition/geology				
4. Work culture				
5. Local restrictions				
6. Codes and standards				
7.				
Total checks F				

6. Required safety training				
7.				
Total checks D				
E: Materials and WIP				
1. Quality				
2. Staging				
3. Handling requirement				
4. Logistics program				
5. Preassembly fabrication				
6. MRP, completeness				
7.				
Total checks E				
F: Project Conditions				
1. Weather				
2. Work space				
3. Ground condition/geology				
4. Work culture				
5. Local restrictions				
6. Codes and standards				
7.				
Total checks F				

Step 4: Add the weighted points (c + d + e + f) for each factor group and enter the result into column g of Tables 2.15 and 2.16 respectively.

Step 5: Divide the totals in column g with the number of factors listed in b to calculate the average significance of one individual factor per group (column h).

Step 6: Multiply the average point values in column h with the corresponding weights in column a to obtain the weighted average for each factor group (column i).

Table 2.15 Calculating Weighted Averages for High Productivity

Factor Groups	Weighted Success Factor Groups for High Productivity								
	A	B	c	d	e	f	g	H	I
	Weight	Number of Factors	Total Points from Table 2.13				Added Points	Average Points	Weighted Average
	Table 2.11	Table 2.13	0	1	2	3	c + d + e + f	G ÷ b	h × a
A: Project Management									
B: Human Labor									
C: Equipment									
D: Hand Tools									
E: Materials and WIP									
F: Project Conditions									
$\sum =$	100%								

Table 2.16 Calculating Weighted Averages for Low Productivity

Factor Groups	Weighted Factor Groups Responsible for Low Productivity								
	a	B	C	d	e	f	g	H	i
	Weight	Number of Factors	Total Points from Table 2.14				Added Points	Average Points	Weighted Average
	Table 2.12	Table 2.14	0	1	2	3	c + d + e + f	g ÷ b	h × a
A: Project Management									
B: Human Labor									
C: Equipment									
D: Hand Tools									
E: Materials and WIP									
F: Project Conditions									
$\sum =$	100%								

CHAPTER REVIEW

Journaling Questions

1. The process model separates between value-added and nonvalue-added by-products. Apply this to concrete placement for an elevated slab. Draw and discuss how process inputs—for example, the completed formwork—

become part of the process, then emerge either as value-added or nonvalue-added outputs.

2. An important factor that dictates productivity of a process is job fitness. Provide a personal example of this factor that involves you or another person. What happened?
3. Assume you are watching a large asphalt spreading operation that lasts an entire day. You pass it several times during the day and notice that there are always about five full trucks idling, waiting to unload. The spreader sometimes stops working, and the operator is calling to one of the many workers, who then has to walk over to the operator. Only after that worker does something does the spreading continue. What are the causes of muda you can identify? What control system, including feedback sensors and a control logic function, would you recommend to reduce the amount of "muda" that is being created right now?
4. Continuing from question 3, assume the asphalt paving contractor likes your recommendations. However, he first would like to see some sound data about the inefficiency. He has another day on the same job tomorrow, and invites you to do produce this data for him. Sketch out the study. What will you do tomorrow, and what format will you use? If you plan to use forms, for example, specify what the forms will comprise and what will be gained from their use. Be specific.

Traditional Homework

1. An undergraduate student is doing a project for a rebar placing company but needs some help. Together with a foreman, she made a list of 15 tasks that the crew is involved in during the day, including time off for personal business. She asks you to help her categorize the list into the five main work classes. These categories will help her to analyze the data she will collect later using the work sampling method.
   a. Carrying rebar, wire cutters, tie wires, lumber pieces for temporary holding of rebar
   b. Receiving/giving instructions and reading drawings involving instructions communicated to or by supervisors and among crew members. (Note: Casual talk is not considered instruction.)
   c. Staging rebar with crane (first time)
   d. Rehandling/moving rebar with crane (second time)
   e. Placing rebar, placing bar supports/separators, aligning, spacing, tying rebar
   f. Measuring or marking bar location, cutting bars with torch, and so on
   g. Walking empty-handed in the work area

h. Searching for rebar in or outside the staging area
i. Obtaining and transporting tools and rebar outside the staging area
j. Waiting for tools, materials, instructions, crane delivery, and so on
k. Correcting an error
l. Idling for no apparent reason
m. Talking while not actively working
n. Not observable—worker not visible

2. Refer to Table 2.17, which shows the result of a time study. Calculate the percent of the time a carpenter assigned to cut the lumber to the correct size is actually working, and classify that time according to these categories: (a) value-added, (b) contributory, (c) ineffective, (c) unproductive, and (d) personal. Use the portion of a continuous time study presented in the table.

Table 2.17 Time Study of a Framing Process

Cycle	Duration of Work Tasks (sec)					
	Fetch Lumber	Measure Lumber	Make Cut	Deliver Piece(s) to Farmers	Return to Cut Area	Waiting/ Other
1	8	16	10	12	5	20
2	6	12	12	10	6	
3	2	20	20	12	8	23
4	5	20	13	11	5	32
5	6	11	8	10	7	
6	3	10	38	8	6	15
7	3	17	14	9	6	10
8	7	13	14	10	5	35
9	5	17	16	11	6	52
10	3	29	17	13	10	67
11	5	15	14	8	8	13
12	4	15	19	10	7	80
13	10	13	17	12	9	6
14	5	14	13	10	7	
15	4	11	8	13	6	19

16	6	37	16	11	7	
17	8	20	14	14	8	33
18	5	13	20	12	5	
19	11	16	15	9	6	28
20	3	14	9	12	7	

3. Let's assume that the times in the table of homework 2 are in fact minutes and not seconds and the process be an earth-anchor-drill operation with following tasks:

Align Drill Steel	Drill to Depth	Remove Steel-Insert Cap	Measure Next Location	Reset Drill Rig	Waiting-Other

Using the random numbers that were established for the sampling of Jack Singleton and Vince Perry, calculate the result of sampling the earth-anchor-drill operation with the above continuous time-study data.

4. Organize an interview with an expert in an area of construction to fill out the surveys on bipolar success/failure factor analysis. Use the data to identify the weighted critical-success-factors. Explain in a few sentences what you found.

Open-Ended Problems

1. The fishbone diagram in Figure 2.23 indicates that Electronic Support impacts the contribution of Equipment to Construction Productivity. Draw a separate fishbone diagram to show second, third, and forth layers, with cause-and-effect relationships indicating how electronics influence the productivity of an earthmoving fleet, consisting of trucks, an excavator, a bulldozer, and a compactor.
2. Assume that you are trying to come up with a control scheme for two drink-water reservoirs, dam 1 and dam 2, that also serve as flood prevention schemes. A heavy rainstorm is approaching the valley, expected to arrive at the upper ridge in about the hours, at the middle of dam 1 in four hours, and at dam 2 in five hours. The storm is predicted to be over in one hour, but the surface water is expected to continue to be high for at least six hours. Currently, the upper dam, dam 1, is half-empty, but it will not be able to hold all the rain that is anticipated when keeping the gates closed. The lower dam, dam 2, is almost full.

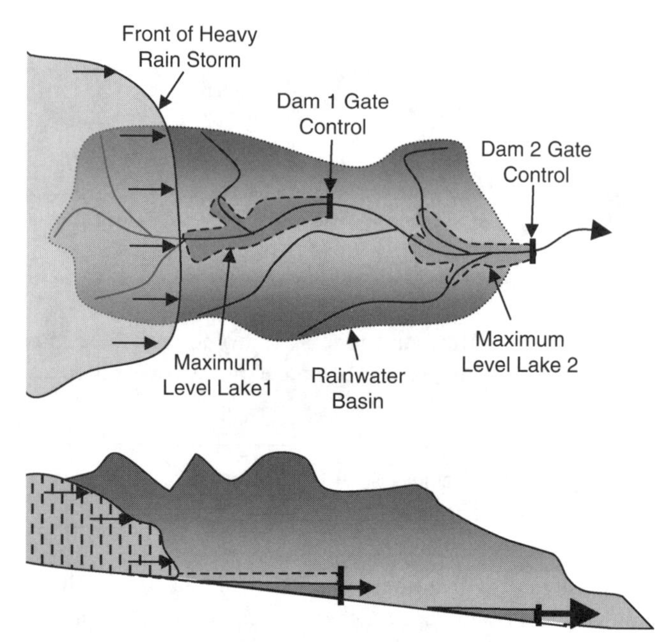

Figure 2.24 Top and Cross-Section View of Dammed-up Valley

a. What should your feedback and feed-forward information channels report to the controller?

b. Which control function(s) would minimize the risk of flooding beyond dam 2?

c. What mechanism would you install to achieve a real-time control system that encompasses dams 1 and 2?

d. How would this system differ from Watt's first proportional control system for steam engines?

BIBLIOGRAPHY

Anthony, F., T. Y. Chou, and S. Ghosh. Training for Design of Experiments. *Work Study*, vol. 52, no. 7, 2003.

Baines, A. Work Measurement—The Basic Principles Revisited. *Work Study*, vol. 44, no. 7, 1995.

Barttlet, J. E., J. W. Kotrlik, and C. C. Higgins. Organizational Research: Determining Appropriate Sample Size in Survey Research. *J. Info. Tech., Learning, and Perf.*, vol. 19, no. 1, 2001.

Bititci, U. S., T. Turner, and C. Begemann. Dynamics of Performance Measurement Systems. *Int. J. Oper. Prod. Mgmt.*, vol. 20, no. 6, 2000.

Brown, S. The Role of Work Study in TQM. *The TQM Magazine*, vol. 6, no. 3, 1994.

Chan, Y. E. IT Value: The Great Divide Between Qualitative and Quantitative and Individual and Organizational Measures, *J. Mgmt. Info. Systems*, vol. 16, no. 4, 2000.

Cohen, Y., B. Bidanda, and R. E. Billo. Accelerating the Generation of Work Measurement Standards through Automatic Speech Recognition: A Laboratory Study. *Int. J. Prod. Res.*, vol. 36, no. 10, 1998.

Deffenbaugh, R. L. Total Quality Managements at Construction Jobsites, *J. Mgmt. Engrg.*, vol. 9, no. 4, 1993.

Dunlop, P., and S. D. Smith. Planning, Estimation and Productivity in the Lean Concrete Pour. *Engineering, Constr. Arch. Mgmt.*, vol. 11, no. 1, 2004.

El-Diraby, T. E. Web-Services Environment for Collaborative Management of Product Life-Cycle Costs. *J. of Constr. Eng. and Mgmt.*, vol. 132, no. 3, 2006.

Elnekave, M., and I. Gilad. Rapid Video-Cased Analysis System for Advanced Work Measurement. *Int. J. Prod. Res.*, vol. 44, no. 2, 2006.

Ericksen, J., and L. Dyer. *Toward a Strategic Human Resource Management Model of High Reliability Organization Performance.* Center for Advanced Human Resource Studies, WP04-02, Cornell University, Ithaca, NY, March 2004. http://catalog.library.cornell.edu.

Goodrum, P. M., C. T. Haas, and R. W. Glover. The Divergence in Aggregate and Activity Estimates of U.S. Construction Productivity. *Constr. Mgmt. Econ.*, vol. 20, 2002.

Gulezian, T., and F. Samelian. Baseline Determination in Construction Labor Productivity-Loss Claims. *J. Mgmt. Eng.*, vol. 19, No. 4, 2003.

Gunesoglu, S., and B. Meric. The Analysis of Personal and Delay Allowances Using Work Sampling Technique in the Sewing Room of a Clothing Manufacturer. *Int. J. of Clothing Sci. and Tech*, vol. 19, no. 2, 2007.

Hildreth, J., M. Vorster, and J. Martinez. Reduction of Short-Interval GPS Data for Construction Operations Analysis. *J. Constr. Eng. Mgmt.*, vol. 131, no. 8, 2005.

Ho, C., and E. S. Pape, E.S., Continuous Observation Work Sampling and Its Verification. *Work Study*, vol. 50, no. 1, 2001.

JenkinsJ. L., and D. L. Orth. Mechanical and General Construction Productivity Results. *Cost Eng.*, vol. 46/ No. 3, 2004.

Kestenbaum, M. I., and R. L. Straight. Procurement Performance: Measuring Quality. Effectiveness, and Efficiency. *J. Public Productivity & Management Review*, vol. 19, no. 12, 1995.

Lofthouse, T. The Taguchi Loss Function. *Work Study*, vol. 48, no. 6, 999.

Martinez-Costa, M., and A. Martinez-Lorente. A Triple Analysis of ISO 9000 Effects on Company Performance. *Intl. J. Prod. Perform. Mgmt.*, vol. 56(5/6), 2007.

Miller, M. E., M. K. James, C. D. Langefeld, M. A. Espeland, J. A. Freedman, D. K. Martin, and D. M. Smith. Some Techniques for the Analysis of Work Sampling Data. *Statistics in Medicine*, vol. 15, 1996.

Miyata, E. S., H. M. Steinhilb, and S. A. Winsauer. *Using Work Sampling to Analyze Logging Operations.* St. Paul, MN: Forest Service, U.S. Dept. of Agriculture, 1981.

Mohanty, R. P., and O. P. Yadav. Linking the Quality and Productivity Movements. *Work Study*, vol. 43, no. 8, 1994.

Motwani, J., A. Kumar, and M. Novakoski. Measuring Construction Productivity: A Practical Approach. *Work Study*, vol. 44 No. 8, 1995.

NewtonL. A., and J. Christian. Impact of Quality on Building Costs. *J. Infra. Syst.*, vol. 12, no. 4, 2006.

Ovararin, N., and C. Popescu. Field Factors Affecting Masonry Productivity. *AACE International Transactions*, EST.09, 2001.

Peer, S., An Improved Systematic Activity Sampling Technique for Work Study, *Construction Management and Economics*, 4, 1986.

Phusavat, K., P. Jaiwong, S. Sujitwanich, and K. Kanchana. When to Measure Productivity: Lessons from Manufacturing and Supplier-Selection Strategies. *Ind. Mgmt. & Data Syst.*, vol. 109 No. 3, 2009.

Salaheldi, S. I. Critical Success Factors for TQM Implementation and Their Impact on Performance of SMEs. *Intl. J. Prod. Perform. Mgmt.*, vol. 58, no. 3, 2009.

Scarnati, J. T., and B. F. Scarnati. Empowerment: The Key to Quality. *The TQM Magazine*, vol. 14, no. 2, 2002.

Shin, D., R. A. Wysk, and L. Rothrock. An Investigation of a Human Material Handler on Part Flow in Automated Manufacturing Systems. *IEEE Transactions on Systems, Man, and Cybernetics—Part A: Systems and Humans*, vol. 36, no. 1, 2006.

Stewart, M. G., Modeling Human Performance in Reinforced Concrete Beam Construction, *J. Construction Eng. and Mgmt.*, vol. 119, no. 1, 1993.

Thomas, R., and I. ZavrskiI. Construction Baseline Productivity: Theory and Practice. *J. Constr. Eng. and Mgmt.*, vol. 125, no. 5, 1999.

Tommelein, I. D., and M. Weissenberger. More Just-in-Time: Location of Buffers in Structural Steel Supply and Construction Process. *Proc. 7th Ann. Conf. Intl. Group for Lean Constr. Proceedings*, IGLC-7, University of California, Berkeley, CA, July 1999

Van Goubergen, D., and F. Vancauwenberghe. Using Time Studies for Quantifying Waste and Improvement Opportunities in Work Methods, *Proc. 2007 Ind. Eng. Res. Conf.*, May 19–23, Nashville, TN, 2007.

Wyrick, D. A., and C. Eseonu. *Integration of Automated Vehicle System Data Acquisition into Fleet Management.* St. Paul, MN: Minnesota Department of Transportation, March 2008.

Chapter 3

Cornerstones of Efficient Site Operation

Plans are of little importance, but planning is essential.

—Winston Churchill

Plans are nothing; planning is everything.

—Dwight D. Eisenhower

In the previous chapter, we learned that planning and control of on-site construction are considered two of the most important success factors in achieving high productivity. When should planning for high productivity of the construction process begin? What is the subject or the goal of an effective planning effort? In order to provide answers to these questions it is helpful to quickly review the phases of a project and what tools are used to link them together.

3.1 RETURN OF THE MASTER-BUILDER

Four hundred years ago, the Master-Builder was the sole person in charge of an extremely flat project organization of gigantic building projects, even by today's standards. He was the architect, engineer, and contractor all in one. As the person in charge, he had the responsibility for designing, planning, and construction. One of the last successful projects put together by a single Master-Builder is the Suez Canal, built between 1859 and 1869 by the Frenchman Ferdinand de Lesseps. However, even with forced labor provided by the Egyptian ruler, the final cost was double the estimate. What saved it from financial ruin was its immediate success—which had not been assured—providing a significant income stream from ships paying tolls.

The financial risks of such large projects became clear when Lesseps led a second canal project, the Panama Canal, which became an engineering and financial disaster for the French. During the same period, the building of the Eiffel Tower marked the emergence of steel as a new economical building material. This development, along with the rise of steam-powered construction equipment, lead to the "death" of the Master Builder, and the rise of specialized design, fabrication, and construction companies.

The stress of leading some of these risky projects led to actual deaths of the lone "man in charge", such as the fatal heart attack of the Swiss engineer Louis Favre (1825–1879) who did not live to see the completion of his masterpiece. For eight years he not only led the work but was also responsible for securing additional financing from all over Europe for the nine-mile double-track train tunnel through the Gotthard Massif in Switzerland. He died just days before a long-overdue breakthrough in the middle of the mountain proved his calculations accurate.

Win-Win/Nonzero-Sum Game

In a *zero-sum* game, one player loses what the other wins. In a win-win, or nonzero-sum, situation, the parties' *aggregate* gains and losses can be less than or more than zero. John Nash showed mathematically that if the "players" would collaborate, instead of playing the zero-sum game, their overall gains would increase.

Planning

Planning is the creative process of developing a plan to express a desired goal, and the path to achieve it. The resulting plans consist of sketches, diagrams, checklists, drawings, maps, physical models, charts, and more, represent probabilistic anticipation of future developments that normally do not become reality. The true benefit of planning lies in the thinking through, brain-storming, documentation, and discussion of all possible outcomes. Having a plan will not guarantee success, but having no plan will certainly guarantee failure.

3.1.1 Separating Design from Construction Expertise

The separation of functional specialties, responsibilities, and risk was paralleled by the separation of labor that argued in favor of increasing productivity through specialization, which in theory allowed everybody to do what they did best. Eventually this backfired, as people became dissatisfied and bored with their work, developed repetitive motion syndromes, and so on. Similarly, in construction, architects, civil engineers, mechanical engineers, electrical engineers, and contractors all became narrow specialists, a development that was synergized by the move to the "lowest responsible bid," which over time opened a large and antagonistic chasm between design and construction, between engineers and contractors, and especially between general contractors and subcontractors. In fact, it gave rise to the "zero-sum game" culture, which assumes that if one side wins the other side loses proportionately. Mathematicians, such as Nobel Prize-winner John Nash, eventually proved that this is not the best strategy for "players" to improve their situation.

Figure 3.1 presents the design-bid-build delivery scheme commonly used in construction, while also noting that others are also used, such as the design-build (Master-Builder) scenario.

The two dashed curves in Figure 3.1 make apparent the first of two inherent dichotomies between the two main phases—planning and design, and construction—separated by the opening bid. They illustrate in qualitative terms the relationships of decisions made during design with the efficiency and cost of the construction process that transforms the designs into reality. After the bid opening mistakes made during the design stage can not be corrected without additional cost.

Since the contractor estimated the bid-price based on the engineers' design modifications can only be implemented by issuing official change-orders. Because the contractor has to change the original plans the additional cost will have to be added to the agreed bid price which leads to the "famous" cost overruns.

The second problem is represented by the money bag, which is being filled up by the owner based on the progress of construction. As shown, the general contractors, the subcontractors, and the suppliers all get paid from the same bag, which is basically linked to the bid prices that all had previously submitted. If they all insist on playing a zero-sum game, the inevitable change orders will open the "gates" to play for more of the "pie" to the detriment of other players. Should the contractors and suppliers instead adopt a win-win/nonzero-sum strategy, everybody will come out ahead without creating the adversarial relationships that commonly develop.

Designing

Designing is the creative process of developing scaled and detailed representations of structures, mechanisms, systems, and components that are needed to implement aspects of a plan. Building designs normally involve the iterative work of architects, a variety of engineers, contractors, users, and scientists to come up with a final blueprint for turning a plan into reality.

3.1.2 The Rise of Constructability Expertise

It became apparent in the 1970's that the large owners, private and governmental, were the ones who suffered the most from this "separation of disciplines."

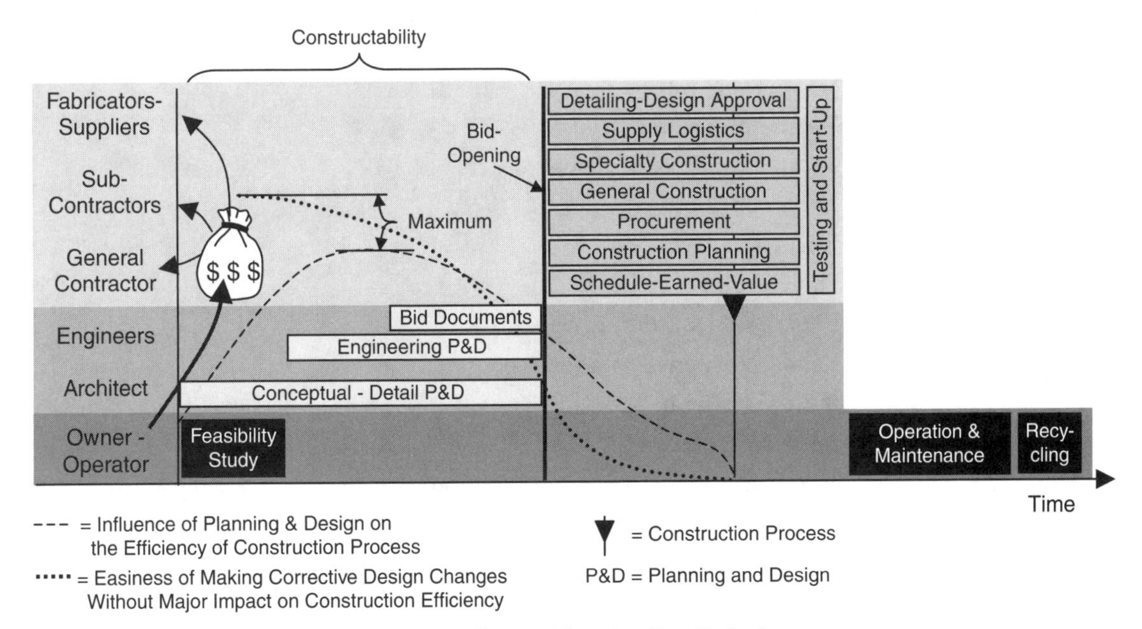

Figure 3.1 Schematic of Design-Bid-Build Delivery of Construction Projects

Not surprisingly, this led to a long series of expensive litigations in which mainly the lawyers benefited. Recognition of this led to a two-pronged response: (1) a return to the Master-Builder concept, with negotiated or design-build contracts; and (2) the rise of constructability as a new service program for advising designers about construction-related problems raised by their designs.

From the perspective of construction productivity, inserting constructability thinking into the design process is nothing more than the early elimination of predictable muda waste by experts, who otherwise would have to deal with it once the bid process is complete. Constructability thinking can be considered a reverse supply of expertise so that up-stream work does not create "canals" and "dams" that would make the downstream construction extremely difficult. With this in mind, it is interesting to review some of the goals and objectives that have been used to define *constructability*:

- The extent to which a design facilitates efficient use of construction resources, and enhances ease and safety of construction on-site.
- The integration of construction knowledge and experience in the planning, design, procurement, and construction phases.
- Service provided by knowledgeable, experienced construction personnel who are part of a project team.

Some of the benefits of using constructability programs are mentioned in the literature, not all of which have been proven. They include:

1. Savings in construction costs
2. Shorter project duration
3. Reduced number of change-orders
4. Improved safety
5. Reduction in rework
6. Higher team morale
7. Fewer problems during start-up
8. Reduced material inventories
9. Built-in flexibility
10. Increased goodwill
11. Smoother, pro-active communication
12. Quicker problem solving
13. Fewer damages to materials and equipment
14. Higher incidence of reuse and recycling

3.1.3 The Emergence of Intelligent Building Data Models

The second relevant movement, which is just taking hold but is expected to grow in its importance in coming years, is the Building Information Modeling (BIM) concept that is slowly replacing the 2- and 3-dimensional CAD (computer-aided-drafting) with an object-oriented approach. Instead of lines, boxes or symbols, the designer now works with the walls, doors, hinges, or electrical conduits that are associated with rooms and floors of a building. Walls can be made of concrete, blocks, steel or wooden frames. The software is able to automatically maintain databases on all items, each with its attributes relevant to the different engineering requirements and rules about procurement or construction. Of course, all the project members will be able to take advantage of instant access to the most updated design, and procurement can be done paperlessly via the Internet.

It is evident that BIM is also a program that will re-establish one of the strong points of the Master-Builder that had been lost: the intimate familiarity of the builder with the many aspects of his designs. BIM in fact equips the design objects with markers that are readily understood by the many engineering software in use.. Specialty applications are already emerging. One example is the Masonry Information Management System (MIMS), which automates the search for information similar to the way a human purchase agent uses a database to get price quotes, sorting the responses, getting credit information or past records on suppliers, and executing the final transaction for purchase. A second example is Structural Information Modeling (SIM), which supports multiple structural analysis models (i.e., foundation, steel frame, curtain wall), case studies, versioning and bi-directional interfaces to exchange data between models.

Finally, the Mechanical, Electrical, and Plumbing System (MEPS) models the interactive dynamic systems of a building, simulating the environment as well as the behavior and impact of the various systems on the building itself and providing information about the fabrication processes that will be necessary. By interfacing with scheduling, it can detect potential "disasters", such as a shipment for a large pump set for a date after the mechanical room is completed, leaving only a tiny door for access.

3.2 PLANNING THE SUPPLY, AND RESUPPLY, OF THE CONSTRUCTION PROCESS

The human capacity to plan future actions based on a set of assumptions about what might happen on a project, and to communicate that plan to others, was essential for the survival of early hunters. Similarly, nomads had to plan when to move to their winter quarters; and early settlers had to plan their planting and harvest times. Military leaders from ancient Egypt, China, and Peru were excellent planners of strategic alliances, resupply logistics, and the optimal use of the newest technologies. The Inca maintained well-stocked warehouses containing everything

Just-in-Time (JIT) and Order (JITO)

The concepts of JIT and JITO define the delivery of resources to the point of consumption on an as-needed basis. Henry Ford organized the supply of raw material and parts so that they fed directly into the car production process, eliminating the need for costly inventories. Everything was shipped not only on time (JIT) but also in the order it was needed (JITO). The philosophy behind both is simple: inventory is waste. The goal is to have the right material, at the right time, at the right place, and in the exact amount. The concept works well if the supply chain is uninterrupted, but creates significant problems if it breaks down.

Production Logistics

The military, in ancient times, was the first to recognize the need to plan for the resupply of its advancing forces, with fresh troops, food, ammunition, and strategic replacement parts. Logistics became known as the art and science of planning and carrying out the movement and maintenance of military forces. Production logistics today involves managing the flow of materials, information, and other resources, including energy and people, between the point of origin and its transformation into value-added outputs.

a moving army needed, whereas the horse-bound hordes of Genghis Kahn depended on the food they "found" wherever they went. In contrast, maintaining the efficiency of construction processes has more in common with the smooth functioning of a navy ship that is constantly on the move.

The emergence of constructability as an effective approach to avoiding costly design decisions being made upstream during the construction phase teaches us that the construction process needs to be considered as an element of a larger system that impacts performance before it even begins. In other words, decisions are made in the design or procurement phase that will influence the efficiency of the construction process. This, of course, was the premise for the introduction of constructability programs in the planning and design phases, discussed earlier. However, this does not mean that there are not other similarly negative effects unrelated to planning and design decisions. Recall that Chapter 2 introduced the concept of muda/waste as related to the work in progress, such as a craftsman waiting for resources, or errors that must be corrected. Keep in mind, too, that the cost of construction includes more than paying for labor, equipment, and materials. One source of nonvalue-adding costs is inventories of material stockpiled on the site or, conversely, the unavailability of materials when needed. Also recall from Chapter 2 that the concept of just-in-time (JIT) material delivery in manufacturing was a feature of the early Toyota production system (TPS). It eliminated the muda caused by maintaining inventories, whose costs include not only the financial and operating costs for the space but also the labor/equipment costs for storage and retrieval, for damages during handling, and more.

3.2.1 Modeling the Construction Input Supply Chain

In Chapter 2 we modeled construction as an I/O process and investigated how to plan and control the system in order to produce the most value-added output and the least amount of muda/waste. So far, however, we have said nothing about the source for all the inputs. It is not only material that are needed. Also necessary are equipment and tools, and operators trained to use them; information, in the form of drawings and charts; and even a clean workspace, one that is properly drained, lighted, and ventilated. During winter, tents, heaters, and blankets should be ready to be deployed. If only one resource is not operable or available, the entire process cannot begin. If the number and/or quality of resources do not meet the plan, the efficiency of the entire process will suffer.

In Roman times, keeping the army moving, fed, clothed, and armed was the responsibility of the *Logistikas*. Today, logistics engineering has taken over the important role of ensuring the reliable and efficient supply of the right resources, at the right price, at the right time, in the right amount, and delivered to the right place, which include the purchasing, packaging, transport, warehousing, and distribution of raw or bulk materials, work-in-progress (WIP), and final products.

There is an old saying that "information is power." This gives the impression that the person within an organization who hoards information will become the most powerful. But if everyone applies this same logic, it will lead to the slow crippling of the system and its eventual collapse. Information is only power if it is shared and distributed so that informed decisions can be made and the need for changes in the environment can be identified and responded to. The information revolution spurred by the widespread distribution and use of computers and electronic communications technologies brought with it the need to organize, store, "package," find, and distribute large quantities of data that was relevant, accurate, and timely. From this revolution was born a new field of study: *information logistics*.

Since we will discuss communication in the construction industry in a later chapter we will spend very little time on the importance of information, or the devastating effect it can have if it is unavailable. The core characteristics of information is that it has a sender, one or more recipients, and contains a message, which may raise a question, share facts, or request an action or response by the recipient (s). The possible formats of information are many, but are concentrated in paper and electronic systems. BIM is an example of an effort to standardize data and information on the physical models of buildings, from "cradle-to-grave"—that is, from planning to decommissioning. In effect, a building is put together first in the form of data and information, before the contractor turns it, element by element, into a physical replication of the plan. After it is complete, the BIM model will be updated with "as-it-was-built" and "as-built" data, for the actual system rarely turns out exactly as it was planned. The result, then, are two building systems: one on "paper" and one in steel and concrete.

The efficient transfer from the virtual BIM building to creating the physical objects begins with information, namely the "letter to proceed" that is sent to the general contractor. Preceding any physical action are the contracts, approvals, and agreements among the various parties, followed by the first set of engineering drawings that slowly permeate the expanding information system of the organization. This information instructs the subcontractors where, when, what, and how; orders materials; confirms what has been said and done; reports quality problems; requests payment; and so on. The heart and hub of this massive undertaking is the construction process, whereby an idea represented in designs becomes real. The foundation of success of this undertaking has been laid much earlier—we call it "upstream" or up the "supply chain." Figure 3.2 diagrams the heart of the

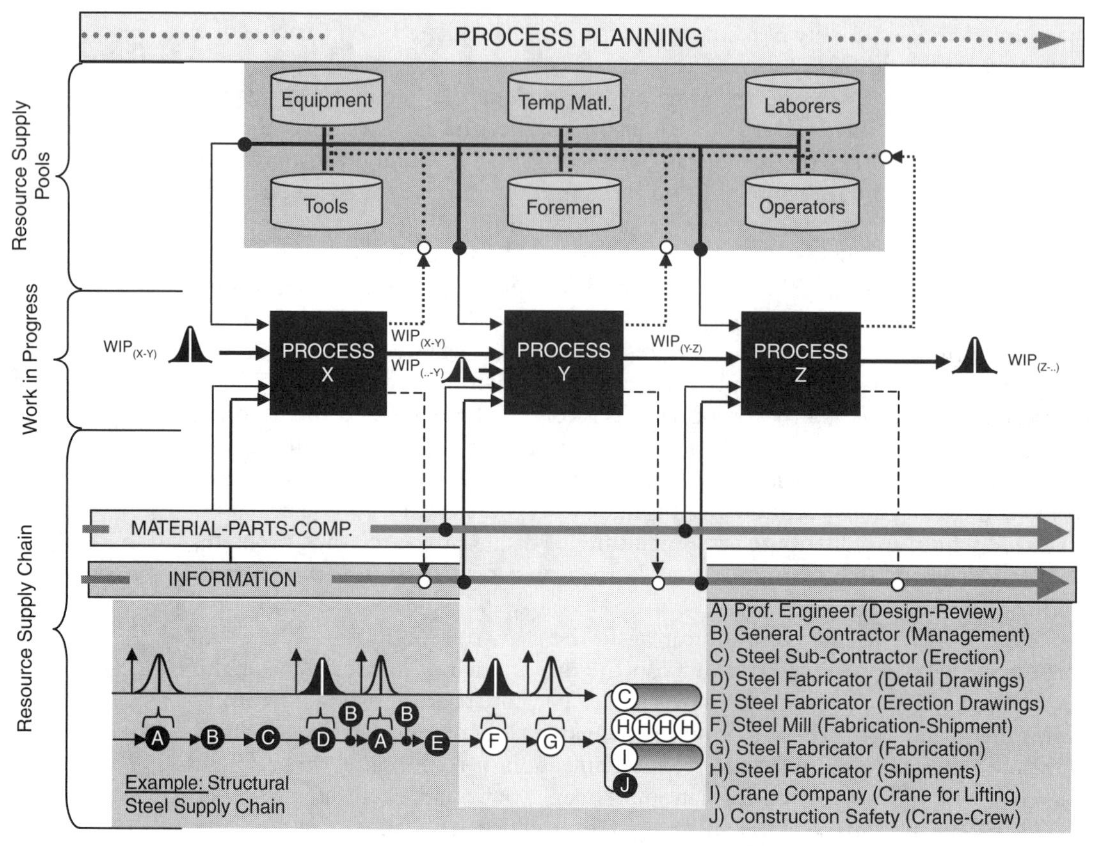

Figure 3.2 Diverse Resource "Streams" Supplying the Construction Processes

construction process, and indicates how the different "arteries" join to transform value-adding resources into a building.

The model rests on three basic origins of resources: (1) pools, (2) WIPs, and (3) the supply chain. The pools represent resources that are in direct control of the project management, such as laborers; the WIPs, as well as the resources linked to the supply chain, are subject to the performance of external systems not directly linked to the organization of the project. The example of the structural steel supply chain illustrates the flow of design information from the engineer to the steel detailer and back, for approval, from there to where the virtual designs become reality in the fabrication shop, until they are shipped, in truckloads, and assembled on-site by skilled erectors and a crane operator according to drawings supplied by the fabricator.

An important difference between the resource streams is highlighted by the bell curves in the figure, which indicate a normal distribution of the time during which the remote "processors" develop the required new information that feeds the stream. Program Evaluation and Review Technique (PERT), a computerized project planning tool, has the capability to model this probabilistic feature of any productive system, whereby it is possible to define a shortest, longest, and most likely duration of an activity. From that, PERT develops various scenarios to forecast the most probable time of occurrence. As shown, the ripple effects of the nondeterministic flow of information, materials, and hardware also reaches the WIPs, as the possible completion times vary around a moving mean.

It is important to recognize, and appreciate, the difficult problems faced by a construction manager responsible for achieving the highest level of process efficiency on the construction site. In addition to a constantly changing environment on the construction site are the unforseeable events that occur upstream on the supply chain that are out of his/her control. It is apparent, that effective management of the resulting complexity requires tools and techniques that not only address this reality but also support the objective of construction management—the minimization of muda waste.

In the remainder of this chapter we will establish a framework in which to model the "world" around the construction process before returning to the question of how to design a supply chain in support of highly effective operations on the construction site.

3.3 TOP-DOWN FRAMEWORKS FOR MANAGING PROJECTS

The building of Hoover Dam (1931–1936), an enormous project even by today's standards, required that the work of a great number of engineers, a six-contractor joint venture, and sub-contractors all be coordinated down to the day. The main planning tool at that time was the Gantt chart, which provided a visual but vastly inadequate method for coordinating the many stakeholders. It was only through an incredible effort by all involved, especially the construction workers, that the project was completed two years ahead of schedule. World War II further showed the need for a different approach to planning and control for the next generation of large projects, such as nuclear power plants or submarines.

The computer revolution beginning in the 1940s in Britain and the United States enabled the implementation of mathematical approaches to developing and updating large and complex Gantt charts. Computers made it possible to program mathematical algorithms, which could then be executed up to a million times within a very short period of time. One such algorithm, called the Critical Path Method (CPM) was developed in the 1950s by the DuPont Corporation, together

Henry Laurence Gantt (1861–1919)

American mechanical engineer and management consultant Henry Gantt worked with Frederick Taylor on projects related to scientific management. His contribution was a graphical chart onto which managers could input all their planned work tasks for projects that extended over several days. By notating progress in different colors, they could quickly recognize when something was going wrong. This chart was later popularized under the name Gantt Chart and is still used today.

with the Remington Rand Corporation. At about the same time, Booz-Allen & Hamilton Corporation, on behalf of the United States Navy in conjunction with the Lockheed Corporation, developed the aforementioned PERT. The construction industry was eager to adopt these methods, as they provided an efficient solution to a real problem: how to plan the construction of large nuclear facilities.

3.3.1 Organizational Structures That Facilitate Integration

The nucleus of a CPM or PERT is the *activity* or *deliverable,* constituting both a physical element (e.g., a basement) of a project and a time needed to construct. The time, referred to as activity duration, is defined by a beginning and an end date when all the necessary resources need to be available. In order to work with computers and their programming codes, the necessity emerged in the industry to organize construction projects into formats that could be much more efficiently handled by increasingly sophisticated computers.

In 1962, the National Aeronautics and Space Administration (NASA) and, in 1968, the U.S. Department of Defense (DoD), issued "Work Breakdown Structures for Defense Material Items." The purpose of the WBS was to define and group deliverables so that an entire project could be organized and categorized according to structured codes that could be understood by a computer program. The logic behind this goal was to enable the DoD to pursue the benefits of computers along many different avenues. Figure 3.3 presents a partial example of a WBS for a segmental bridge project constructed of prefabricated segments that are erected and posttensioned span by span with the assistance of a launching gantry.

The WBS for the bridge project shows three levels—A, B, and C—breaking the main structural bridge elements into smaller and smaller building blocks. As shown, the superstructure includes the bridge girder, organized in spans, expansion joints, special diaphragm elements on top of the piers, and other elements. The component at the bottom of the WBS is commonly referred to as a *work package* (WP), which is further broken down into activities as a basis for planning. Thus, an *activity* defines a specific element of the project requiring one or more resources to complete. The projected durations of the activities are sequenced and used by the CPM or PERT algorithms to calculate the best- and worst-case scenarios, as well as the most likely. The logic for sequencing is based on the chosen method of construction, predicted weather conditions (e.g., rain, snow, or ice), the availability of resources, and other factors. Since each activity requires resources, such as gantry crane, trucks, laborers, foremen, concrete, the unit costs of each can be multiplied by the quantity needed and then tallied into total expected costs.

Where do the main resources come from? As Figure 3.3 points out, the participating organizations, presented in the Organizational Breakdown Structure (OBS), supply the equipment—manpower, materials, expertise, and so on—and all will submit bills for their services. By integrating the vertical WBS with the

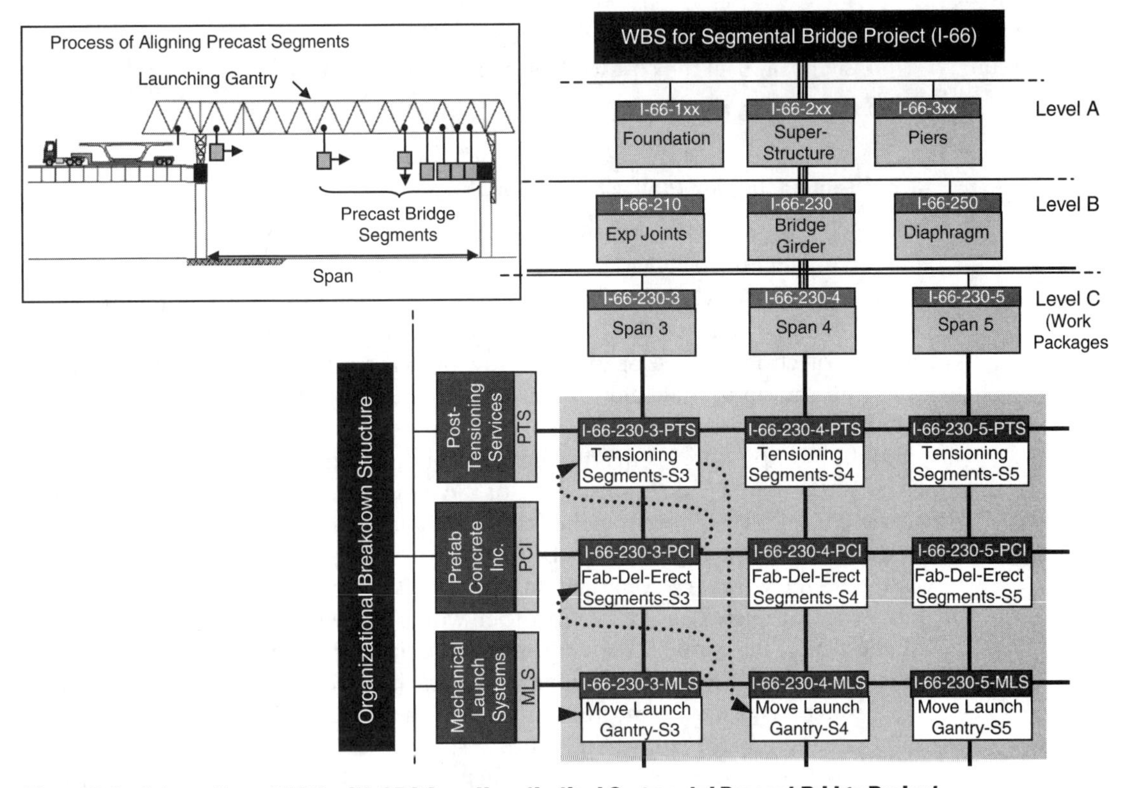

Figure 3.3 Integration of WBS with OBS for a Hypothetical Segmental Precast Bridge Project

horizontal OBS, a matrix is created, with the intersection points identifying which organizational unit works on which WP. Hence, such intersections can be assigned a cost code, which can then be broken down further into cost accounts (e.g., 100.20 for material—concrete). An example in Figure 3.3 is I-66-230-4-PCI, a code that has been concatenated from:

Project Name:	I-66
Bridge Superstructure:	2xx
Bridge Girder:	30
Bridge Span Code:	4
Cost Code:	PCI (Contractor: Prefab Concrete, Inc.)

The beauty of such a structure becomes apparent when identifying and comparing costs—for example, for the spans, including the foundations, piers, and so on. In such a situation, a computer program, into which all the cost information has been entered according to the code, could be queried as follows:

List and sum all the cost for I-66-xxx-1-xxx

List and sum all the cost for I-66-xxx-2-xxx

List and sum all the cost for . . .

3.3.2 Sequencing the Plan, Start to Finish

The final aspect portrayed in Figure 3.3 is its link to scheduling, an effort to sequence the identified activities, based on the WBS, so as to identify when and how many resources have to be available over time. Let's use the few relationships shown here to understand the importance of scheduling to the management of an entire project. Figure 3.4 opens a window onto the master Gantt chart, which displays the line-up of the main, or master, activities related to constructing the segmental bridge girder for span 3. The horizontal axis charts time, which for simplification purposes is defined here as nondimensional time units (TU). Indicated by the bar in TU 9, tensioning of the precast segments in span 2 is finished at the end of TU 9, allowing the launching gantry to be disconnected from the suspended precast segments and moved to span 3. This activity, called Move Launch Gantry—S3, can only start at the beginning of TU 10, after the tensioning has been completed in TU 9. In other words, moving the gantry to S3 is sequenced after tensioning S2, referred to as the *finish-start relationship* between the preceding and the subsequent activity. The graphical equivalent of this relationship is a dashed arrow line connecting the end of the preceding with the beginning of the subsequent bar, as shown in Figure 3.4.

A crucial point in time emerges when the launching gantry is in place across span 3. Now the precast concrete bridge girder elements can be connected to cables or rods suspended from the gantry and lifted off a vehicle that supplies them. For this to happen, however, the concrete segments must have been fabricated, cured, and transported to the gantry crane. The Gantt chart reveals that the

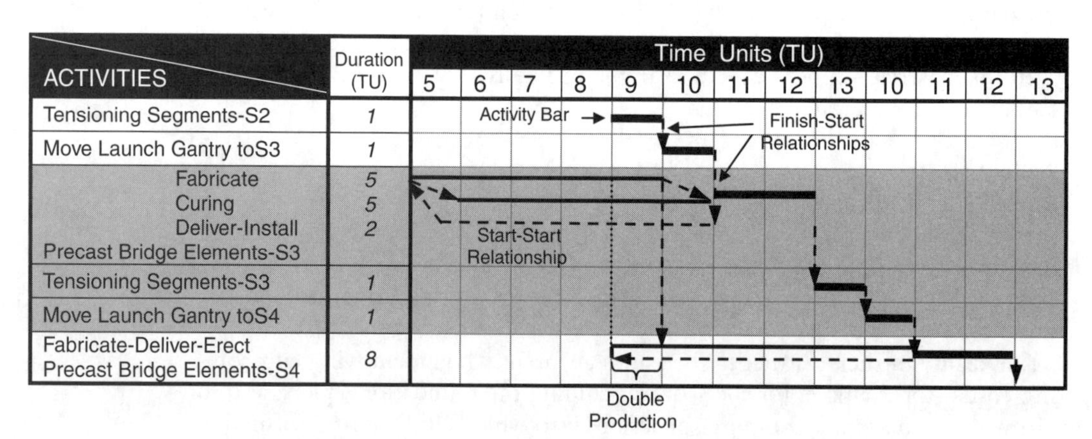

Figure 3.4 Master Gantt Chart of Main Activities to Construct Precast Bridge Girder Span 3

duration for precasting all the necessary segments will take 5 TUs, followed in the curing process by 1 TU. As a consequence, in order to be able to ship the first segments starting with TU 11, the production process has to start at the beginning of TU 5. This connection is called a *start-start relationship* between fabrication and delivery/installation, with a time value of 6 TUs.

One important feature of the Gantt chart is its capability to highlight conflicts between different activities, such as fabricating segments for spans 3 and 4. Delivery and erection for span 4 is ready to begin at TU 11, requiring that fabrication start at TU 9, when the precasting plant is still working on a segment for span 3, which is scheduled to be completed at the end of TU 9. A simple solution would be to shift the beginning of the installation by one TU. Another solution would be to double the production of the plant by introducing a second shift. Still another possibility might be to shorten the curing time. Note that the value of the Gantt chart is not to find solutions to the problem but to highlight problems that need to be addressed.

Indeed, as this example highlights, the function of the WBS or activities is to provide a framework for managing large projects in a structured manner, to include estimating, scheduling, and cost accounting. However, a WBS does not provide measures that relate to productivity, quality, or safety of the operations needed to produce the related Work Packages. To understand and model those relationships, we need to move beyond the static bars shown in Figure 3.4, into an entirely new world, a world of dusty roads, tools that break down at the worst moment, incomplete shipments of material, buried gas lines where none should be, rain and snow, cranes that need to be shared among different crews, and poorly trained equipment operators. Far from being deterministic, as a bar in the Gantt chart might suggest, the real world behind the bar is full of unpredictable events, a dynamic "battleground" where "foot soldiers" are suddenly left without supplies, and vehicles get stuck in the mud. This is the field of construction at the operational or process level, where equipment and laborers have to interact with each other in an ever-changing environment. This is where "the rubber meets the road," the place where money is made and money is lost. This is where we will focus our attention for the remainder of this chapter.

3.4 BOTTOM-UP QUANTITATIVE PLANNING

In Chapter 2 we learned about the repetitive or cyclical nature of construction processes, a fact that made it convenient to apply the sampling method to assess the work time distribution of the framing crew. We were also introduced to the concept of cause-and-effect relationships, with which we were able to identify problems and lay out a series of steps that we could take to produce remedies. What we have not been able to do so far is to plan the process and predict its productivity. Different from Gantt chart activities, the timescale of processes is in minutes, even seconds. And, as we will soon observe, the duration of the cyclical tasks

The Construction Process . . .

. . . transforms *inputs* into *outputs* consisting of *value-adding* and *non-value-adding by-products*. The *conversion* is performed by interdependent *work tasks* each requiring a set of *resources* and operating within *constraining boundaries* and *conditions*. Construction processes are inherently *cyclic* and *dynamic in nature*. Construction processes can be represented as *input-process-output* (IPO) models *hierarchically* integrated into larger production systems.

is tightly connected to the quality of the resources and the characteristics of the job and the environment.

We'll start with a clear definition of what a process is, before going through some applications step by step.

3.4.1 Defining the Process Model

A graphical representation of this definition will take some time to go through but doing so will make it much easier to understand. To that end, Figure 3.5 builds on what we have covered already, by modeling the problem of supplying aggregates to a precast plant.

The simple loading and hauling process depicted in the figure provides the opportunity to introduce a set of modeling elements that need to be understood. Table 3.1 lists some of more important ones and discusses their meanings as applied to the aggregate supply process.

3.4.2 Computing Process Production

The repetitive nature of the construction process, whereby various resources cycle through a series of tasks, each taking a certain amount of time, allows us to calculate the **total time** it takes to produce a certain amount of value-added outputs (e.g., hours or days) or its **production rate** (e.g., cuyd/hr) or its **productivity** (e.g., cuyd/labor-hr or cuyd/equipment-hr). As previously, we'll begin by creating a model that sequences all the time-consuming tasks (see Figure 3.6).

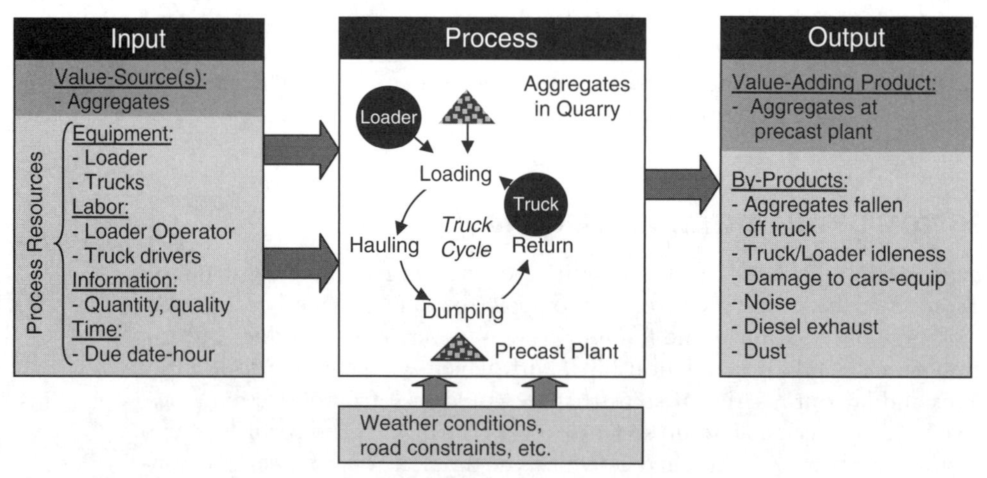

Figure 3.5 Example IPO Model for Supplying Aggregates to a Precast Plant

Table 3.1 Input-Process-Output Model: Terminology

IPO Terms	Definitions
Construction Process	Using loaders and trucks, aggregates at a rock quarry are loaded and transported to a precast plant, where they are dumped into storage bins (from where conveyors move it to concrete mixers).
Inputs	*Value-source(s)*: These inputs are transformed into desired products that contribute to the construction project. In our case, the precast company needs aggregates to make concrete to cast the bridge girder segments at the plant. Thus, the aggregates stored in large piles in the quarry comprise the source material that will be transported to the plant by trucks. *Process resources*: Without one of these resources, the construction process would stop. Listed in the Input box in Figure 3.4 are *equipment* (loader and trucks), *labor* (operators), *information* (material specifications and quantities), and *time*. For example, if a loader is not available, the trucks can't be loaded, resulting in a stoppage. Similarly, if no information about the needed aggregate mix is available, the loader operator will not know what to put into the trucks.
Work Tasks	The model in Figure 3.4 shows four interdependent work tasks: (1) loading, (2) hauling, (3) dumping, and (4) returning. They are interdependent, in that one can't start until the previous one is completed. A work task requires time to be executed. Durations for loading, hauling, dumping, and returning depend on a diverse set of functions, such as the size of the loader relative to the trucks, distance between the quarry and precast plant, average speed of truck, traffic situation, skill of the operators, and others.
Cycles	For a sizable job, the hauling trucks will drive back and forth continuously on the same road. This is called a *cycle*—in this case, a truck cycle. The loader is also cycling, between truck and the aggregate heap, to fill the bucket. This is called the *loader cycle*.
Dynamic	Dynamic is the opposite of static, fixed, or deterministic. For example, the hauling duration will fluctuate based on traffic conditions, traffic lights, and so on. Similarly, the time for loading one truck will vary slightly as a result of variations in operator controls.
Conversion	Aggregate material stored at the quarry is transformed into aggregate that is ready to be batched into a concrete mixer at the precast plant. At this time, it is still aggregate, but its value has increased dramatically, as it is now accessible without delay for producing concrete on demand.
Conditions	Most construction processes take place exposed to the environment, where rain, snow, ice, wind, geology, and other related factors can cause dramatic effects on them. Most critically, snow or ice could shut down the aggregate transport operation.
Constraints	Constraints originate from state and federal laws (e.g., maximum load to be hauled on public roads), local codes and ordinances, as well as material and building standards issued by professional groups such as the American Concrete Institute (ACI) or the American Society for Testing and Materials (ASTM).
Outputs	*Value-added products*: The difference between the prices one is willing to pay for an input as compared to an output is the value added during the process. In this case, the precast company is willing to pay more for the aggregate in its storage bin than when it is stored at the rock quarry. *Nonvalue-added by-products*: From a productivity point of view, the most important by-product is muda, or waste, such as the time wasted when the loader is idling, waiting for a truck to be loaded. Similarly, truck drivers and operators that have to wait to be loaded at the quarry are not productive, but nevertheless have to be paid. By-products also can impact the environment. In this case, aggregates could spill from the truck and strike cars; cause noise; emit CO_2 fumes; or raise dust.

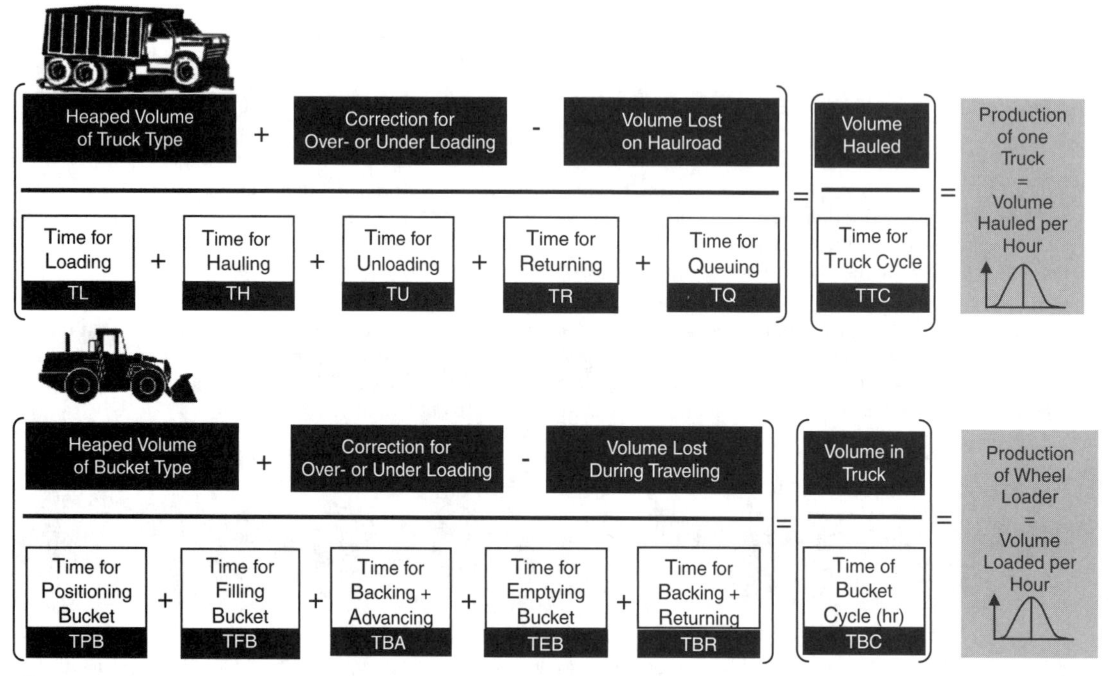

Figure 3.6 Cycle Times and Quantities Handled by Truck and Loader

Reviewing the formulas makes it clear that it is now necessary to have more detailed information about the equipment, in particular the truck and loader bucket volumes, as well as the speed of the truck on the road. Worked-Out Problem 3.1 goes through a case where these values are given.

Worked-Out Problem 3.1: Calculating Production Rate and Total Time to Haul Aggregates

We have been charged with calculating the total time it takes to haul aggregates from the stone quarry at Wheeler Road to the precast plant owned by International Casting Company (ICC), which is 20 miles from the quarry. ICC has ordered 770 yd^3 (589 m^3) to refill its storage capacity.

The skill of the loader operator is presumed to be good, resulting in the following mean durations:

TPB = 13 seconds, **TFB** = 25 seconds, **TBA** = 30 seconds, **TEB** = 7 seconds, **TBR** = 25 seconds.

The heaped volume of the loader bucket is 3.5 yd^3 (2.7 m^3), and it usually loses 4 percent to spills. Two-axle dump trucks will be used, each capable of hauling about 17 yd^3 (13 m^3) of aggregates when fully loaded. The average speed when full, factoring in all red lights, is approximately 35 mph

(56 km/hr); when empty, the average speed increases to 45 mph (72 km/hr). The time spent driving inside the plant, spotting and unloading is approximately 9 minutes.

Assumptions

There are several assumptions that we need to make:

1. The volume of loaded aggregates will not cause the truck to exceed the acceptable gross vehicle weight (GVW).
2. For a first calculation, the work task durations are considered fixed averages, which is true only in theory.
3. There is no difference in the average truck speed between rush hour and off-peak traffic.
4. The time the truck spends in the queue waiting to be loaded, TQ, is 0.

Process Model

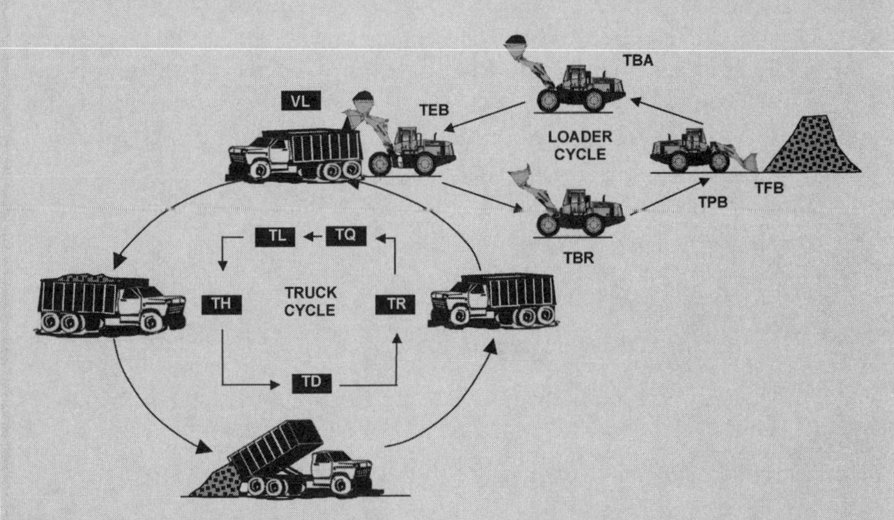

Figure 3.7 Modeling the Truck Cycle

Calculations

Loading Time to Fill One Truck (Five Loading Cycles):

Cycle time to load one bucket into the truck is calculated as follows:

$$TPB + TFB + TBA + TEB + TBR = 13\text{ sec} + 25\text{ sec} + 30\text{ sec} + 7\text{ sec} + 25\text{ sec}$$
$$= 100\text{ sec}$$

(*continued*)

(*continued*)

Volume loaded at each loader cycle is:

$$0.96 \times 3.5\,\text{yd}^3(\text{spillage } 4\%) = 3.36\,\text{yd}^3$$

Capacity of one truck = $17.0\,\text{yd}^3$

Number of required loading cycles is calculated as:

$$17\,\text{yd}^3 \div 3.36\,\text{yd}^3 = 5.06(\text{rounding down}) = 5\text{ cycles}$$

Total loading time with 5 cycles is:

$$\mathbf{5 \times 100\,sec = 500\,sec = 8.3\,min}$$

Truck Cycle Time

The formula of cycle time is as follows (see Figure 3.7):

$$\text{TL} + \text{TH} + \text{TD} + \text{TR} + \text{TQ}$$

where:

TL (Time for Loading) = 8.3 minutes

TH (Time for Hauling) = Distance/Speed Full = 20 miles ÷ 35 mph = 34.3 min

TU (Time for Unloading) = 9.0 min

TR (Time for Return) = Distance/Speed Empty = 20 miles ÷ 45 mph = 26.7 min

TQ (Time in Queue) = 0.0 min

TTC (Total Truck Cycle time) = 78.3 min

Production Rates for Truck and Loader

$$\text{VL (Volume Loaded) in } 8.3\text{ min} = 5\text{ cycles} \times 3.36\,\text{yd}^3 = 16.8\,\text{yd}^3$$

$$\text{Loader production} \div \text{hr} = 60\,\text{min/hr} \times (16.8\,\text{yd}^3 \div 8.3\,\text{min}) = 121.0\,\text{yd}^3/\text{hr}$$

$$\text{Volume hauled by truck during each cycle} = 16.8\,\text{yd}^3$$

$$\text{Truck production/hour} = 60\,\text{min/hr} \times (16.8\,\text{yd}^3/78.3\,\text{min}) = 12.9\,\text{yd}^3/\text{hr}$$

Discussion

While the loader is able to load $121\text{yd}^3/\text{hr}$, a single truck can haul only $12.9\,\text{yd}^3/\text{hr}$. Because these two vehicles have to work together, and depend on each other, the slower of the two will dictate the outcome, will cause the bottleneck. By adding a second truck, the production will double (2 ×

12.9 yd^3/hr = 25.8 yd^3/hr). On the other hand, the loader will be idle for fewer minutes per hour (60 min – 2 × 8.3 min) = 43.4 min. Obviously, the hourly cost will go up, too. Thus, we should find out how many trucks to use to achieve the most efficient earthmoving operation.

This planning method is called *balance point analysis*, as we need to balance the two production resources.

Balance Point Analysis of Aggregate Hauling Process

Figure 3.8 plots the production rate of the entire fleet (the loader plus trucks) against the number of trucks being used to haul the aggregate. The horizontal line at 121 yd^3/hr indicates that only one loader will be available; this is its maximum or bottleneck production rate, if there were always a truck available to load.

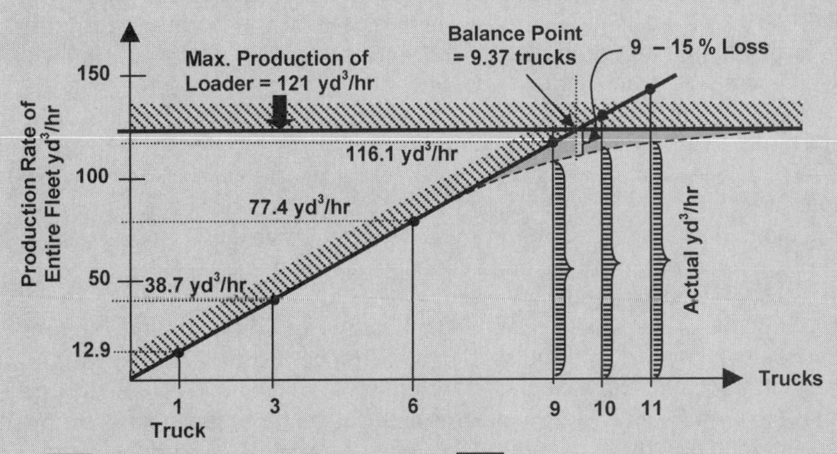

Figure 3.8 Balance Point Analysis for Aggregate Hauling

The sloped line through the origin represents the production of an increasing number of trucks, starting with one truck that is able to haul 12.9 yd^3/hr. If three trucks are operated, that number goes up the 38.7 yd^3/hr; and with nine trucks, 116.1yd^3/hr. The graph illustrates the situation with 10 trucks (129 yd^3/hr), which theoretically at least, causes the sloped line to cross the horizontal line—the maximum rate of 121 yd^3/hr at which the loader is able to fill the trucks. The intersection is the point at which the loader becomes the bottleneck, when the operator will work continuously. This theoretical point is called the *balance point*, since, here, the production of 9.37 trucks balances that of the loader. To the left of the balance point,

(*continued*)

(*continued*)
the number of trucks defines the maximum production rate of the fleet; to the right of the point, it's the loader.

There are two main issues associated with the balance point: issue 1: It is impossible to operate 9.37 trucks; and issue 2: the *bunching effec*t reduces the theoretical production value.

Issue 1: We have to decide whether to run a number of trucks that is either below or above the balance point. If the number is rounded down, the loader will have to wait, and thus be idle some of the time. If, however, more trucks are operating, trucks will have to be idle while the loader is filling the previous truck. The choice depends on the cost difference, which will be discussed shortly.

Issue 2: Bunching is a common phenomenon in everyday traffic. Cars bunch up behind a car ahead that is going slower, for any number of reasons (e.g., the driver is talking on the phone, or a car stalls as a traffic light turns green). Traffic light synchronization, or lack thereof, can also cause bunching. These are random events and cannot be predicted but, can add a significant amount of minutes to the total cycle time of the truck delivering aggregates.

As the following scenario demonstrates, a slow truck will also slow down subsequent trucks. Let's compare two trucks that cycle at a mean travel time: **Scenario 1**, in which the first truck is slowed down and the truck behind it drives at an average speed; and **Scenario 2**, in which the time for unloading at precast plant stays the same as before (9 min.)

The focus of our study is the loading conditions at the quarry. The first truck is driven by Gus and the second by Terrance. Figure 3.9 presents the time distribution for the haul and the return trips, which have the mean durations of 34.3 minutes and 26.7 minutes, respectively. The probability for driving faster or slower is reduced according to the bell curve of a normal distribution. In addition, since an empty truck is quicker to pick up speed, the deviations from the mean return duration are much smaller. The labels (1) and (2) point to the haul and return durations that Gus experienced in scenario 2, whereas Terrance consistently ends up with a mean duration.

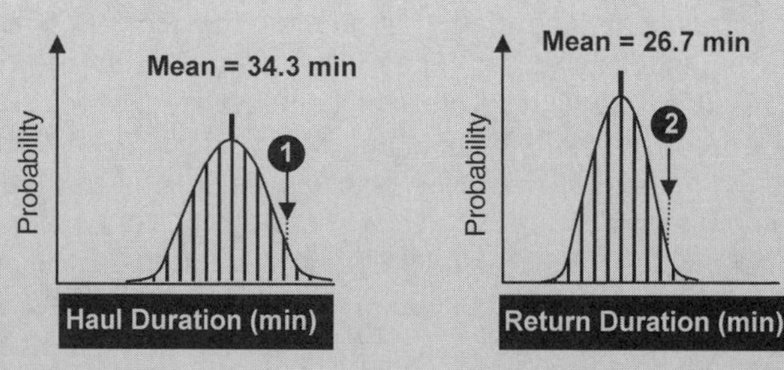

Figure 3.9 Distributed Travel Times for Truck on Haul and Return Trips

Figures 3.10 (a) and (b) exhibit the bar charts (i.e., Gantt charts) for the first truck cycles for Gus and Terrance. In scenario 1, everything runs like clockwork. As soon as Gus returns, the loader is ready and fills the dump truck in 8.3 minutes, at which time Gus takes off again just as Terrance is ready to be loaded for the second time.

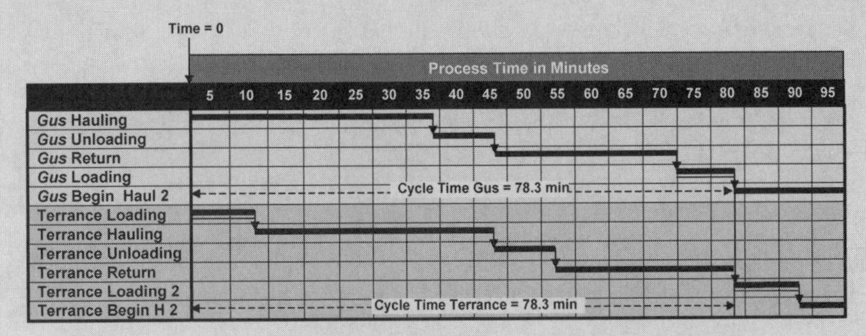

a) Truck Cycles for Gus and Terrance with Mean Durations: Scenario1

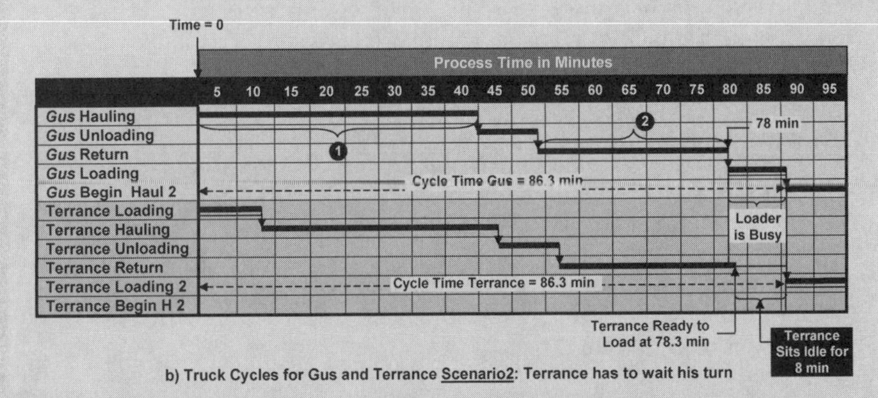

b) Truck Cycles for Gus and Terrance Scenario2: Terrance has to wait his turn

Figure 3.10 Bunching Effect of Cycle Times Longer Than the Mean

In scenario 2, the durations of both Gus's haul and return trips are longer than the mean, adding up to an 8-minute delay in returning from the plant. As the second bar chart indicates, Gus returns to the quarry after 78 minutes and is loaded right away. Terrance returns to the quarry slightly behind Gus, at 78.3 minutes. At this time, the loader is servicing Gus's truck, so Terrance has to sit idle until 86.3 minutes, when Gus's truck is full and moving out. Thus, instead of beginning the second cycle at 78.3 minutes, as in scenario 1, Terrance's second cycle starts at 86.3 minutes, even though his haul and return times were at the mean. Had he come one minute earlier, he might

(*continued*)

(*continued*)
have bypassed Gus and been loaded first. But then Gus would have had to wait instead.

The difference in production is calculated as:

$$12.9\,yd^3/hr - (60\,min/hr \times (16.8\,yd^3/86.3\,min)) = 1.2\,yd^3/hr\ (9.3\%)$$

Terrance's experience is called bunching, since he came right behind Gus, who had a "bad" run. So even though he achieved an average cycle time, he was delayed at the loader, and had to wait in line. The bunching effect is the most severe at the balance point, producing a production loss of 9 to 15 percent, depending on the situation

TIME TO HAUL 770YD^3

The loader with a skilled operator will load a maximum, but only if he works a full hour—60 minutes. This is impossible to sustain, as humans do have to go to the bathroom; they are interrupted by someone with a question, and so on. To account for this reality, it is common to introduce an *operating factor,* which determines how many minutes a laborer or operator will actually be working. In our case, let's assume an operating factor for the loader of 50 minutes, or 0.83. As a result, the maximum production rate of 121 yd^3/hr has to be reduced by 17 percent, to a corrected value of 100.8 yd^3/hr. Similarly, a truck driver will stop to get something to drink, take a break, and so on, so we'll assign an operating factor of 0.83, thus reducing his production rate from 12.9 yd^3/hr to 10.7 yd^3/hr.

Because we reduce both production values by the same 17 percent, the number of trucks at the balance point will stay at 9.37. Lets assume that the bunching effect at 9 trucks is 8 percent, 9 percent at 10 trucks, and 6 percent at 11 trucks. Table 3.2 shows the differences in production between 9 and 10 trucks.

Table 3.2 Comparison of Production Rates with Different Number of Trucks

Number of Trucks	Operating Factor	Adjusted Production (yd^3/hr)	Bunching Effect	Actual Production (yd^3/hr)	Daily Production (yd^3/day)
9	0.83	$9 \times 0.83 \times 12.9 = 96.3$	0.08	$0.92 \times 96.3 = 88.6$	$8 \times 88.6 = 708.8$
10	0.83	$0.83 \times 121 = 100.8$	0.09	$0.91 \times 100.8 = 91.7$	$8 \times 91.7 = 733.6$
11	0.83	$0.83 \times 121 = 100.8$	0.06	$0.94 \times 100.8 = 94.7$	$8 \times 94.7 = 758.0$

RECOMMENDATION

Although the loader would be able to load the required 770 yd^3 in 7.6 hours, the bunching effect reduces the actual fleet production rate substantially. As Table 3.2 shows, utilizing 10 or even 11 trucks would not make it possible to haul the required amount. While 10 trucks would be 36.4 yd^3 short, at 11 trucks it would come to 12 yd^3. Nevertheless, since one truck holds 17yd^3, it should not be difficult toward the end of the day to reassign one or two trucks from another job-site to make one run to achieve the goal; it is not necessary to add another truck for the entire day. In the same vein, a second loader might be available for 15 minutes or so, to load the extra truck(s). In summary, it should be possible to haul 770yd^3 in one day.

Still, the question about how many trucks to actually use during the day should be decided based on an economic analysis of the unit cost: $/$yd^3$ aggregates loaded and hauled.

3.4.3 Optimizing the Process

A business can survive only if it uses its resources in the most productive and cost-efficient manner. Operating at levels where the resource-based and cost-based productivities are the highest will be critical for financial survival in the long run.

The simplest way to explain how to use productivity calculations to assess the efficiency of the construction process is to use another example. Here we'll follow the recommendation from Worked-out Problem 3.1: production rate and total time.

The balance point analysis in the aggregate supply problem demonstrated that using 10 or 11 trucks, with some help from one or two additional trucks at the end of the day, would be sufficient to haul the 770 yd^3 of aggregates in an eight-hour day. But which option is the better for the company?

Finding an answer to the question is not simple, as multiple factors have to be considered. The safest and most direct approach is to calculate the productivities for the key variables the number of trucks. Figure 3.11 contains three graphs to illustrate the relationships between production, productivity, and unit cost for transporting the value-added aggregates. The analysis considers one loader at a daily cost of $800; it is further assumed that the loader operator costs $28.00/hour, and a truck driver costs $18.00/hour. Two of the graphs present the cost-related data for four different truck costs: (a) $150/day, (b) $200/day, (c) $250/day, and (d) $300/day.

In Graph 1, the solid line extends the production rates for the number of trucks studied in the worked-out example. As shown, 7 to 13 trucks are included with 13, resulting in a 1 percent bunching effect, approaching the maximum production of the loader, 100.8 yd^3/hr. The dotted line in the graph presents the resource-based productivity, yd^3/operator hour, as a function of the size of the

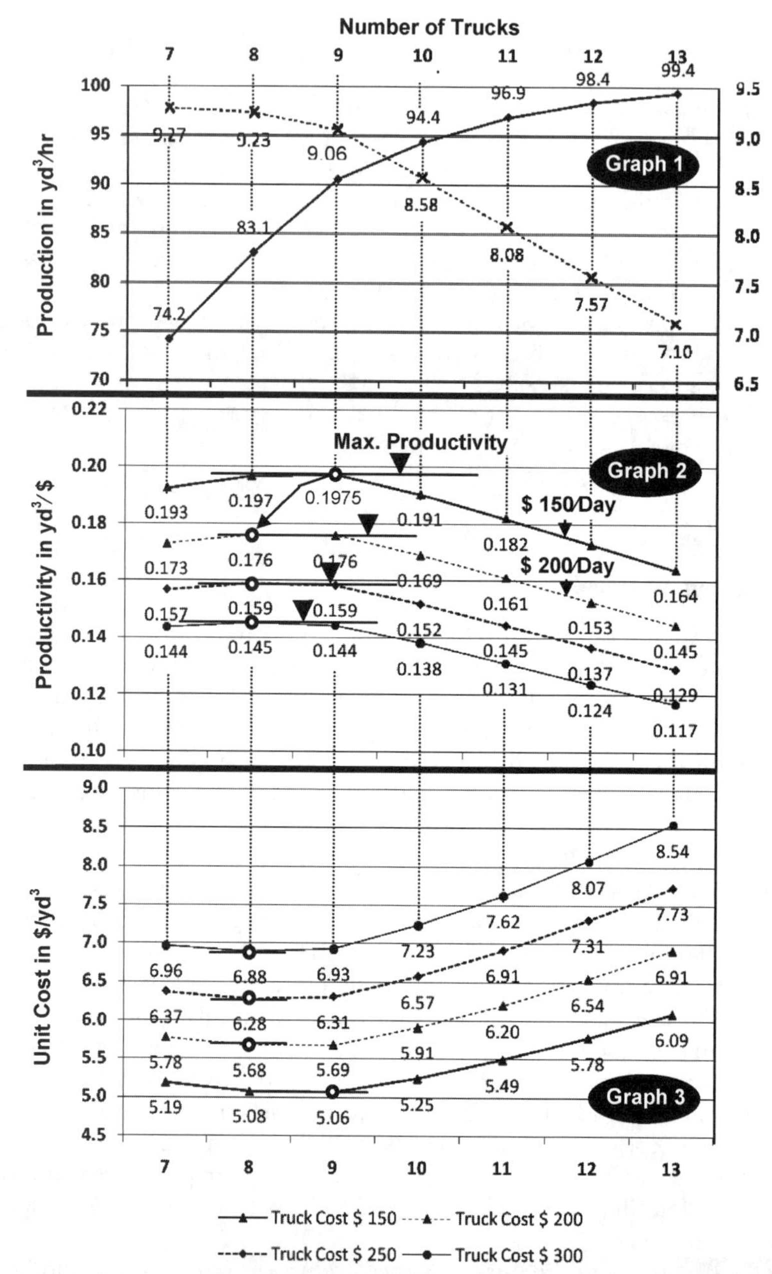

Figure 3.11 Productivity and Unit Cost Analyses

truck fleet. It is apparent that the number of operators is always one person more than the number of trucks, because of the loader. The bunching effect results in a severe decline to productivity, starting with 9 trucks (bunching = 6 percent) and continues due to the bottleneck production rate of the loader. Above 10, adding one truck slightly increases production, from 94.4 to 96.9 yd^3/hr, or 2.5 percent, but diminishes productivity by approximately 0.5 yd^3/hour, or 6 percent.

In summary, Graph 1 demonstrates clearly that production might still increase while productivity would shrink rapidly.

Graph 2 presents the cost-based productivity for four different truck costs. The solid line on the top plots the results for $150/day trucks. The concave shape of the curve tells us that there exists an absolute maximum productivity, identified at that point with a horizontal tangent. As shown, that point is reached with 9 trucks, where $1.00 would be sufficient to haul 0.1975 yd^3. When the daily cost for one truck reaches $200, two important changes occur: (1) The 33 percent cost increase for one truck reduces the amount of aggregates that can be hauled per $1.00 by 6 percent, from 0.1975 to 0.176; and (2) the horizontal tangent to the productivity curve now meets the curve where it intersects with 8 trucks.

The first change is easy to understand; the second requires some explanation. The balance point analysis showed that 9.37 trucks were necessary to match the production rate of the loader. We also learned that this theoretical production value needs to be adjusted downward to account for the bunching effect. Since we cannot use 9.37 trucks, it is necessary to choose between 9 and 10, or any other number. Graph 1 shows that production drops when switching from 9 to 8 trucks from 90.5 to 83.1 yd^3/hr, respectively, while operator productivity increases from 9.06 to 9.23 yd^3/operator-hr.

What is of primary importance is the cost difference between the single loader with one operator and multiple trucks with their drivers. Of course, a critical role is played by the bunching factor. Let's examine in detail the changes between the $150/day and $200/day trucks (See Table 3.3).

The arrows point to the differences arising from the 33 percent increase in truck cost. By adding a ninth truck and a tenth truck, production rates increases first by 8 percent (to 108 percent), and second by 5 percent (to 113 percent). However, total cost for the entire fleet consisting of loader and trucks increases at a higher rate of 9 and 8 percent, respectively for the cheaper trucks but 10 and 10 percent for the $200/day trucks. Although this represents a minor difference, it impacts the optimal amount of trucks embodied by the lowest unit cost. Assuming that the two trucks have the same capacity and horsepower, the $50 higher cost for the $200/day trucks, compared to the fixed $800 for the loader, made the cost of idling the loader less expensive than adding a ninth truck. Thus, using 8 trucks results with $ 5.67 in the lowest unit cost. The same does not hold true for the cheaper truck. Here, the cost of idling the loader is higher than the added cost of the ninth truck, which makes the 9 truck option with $ 5.07/yd^3 the optimal solution.

This analysis demonstrates the importance of the relative cost differences between the main resources. In addition, we learn that the optimum solution is not

Table 3.3 Detailed Analysis of Production, Productivity, and Cost around the Balance Point

Trucks	Oper Factor	Bunching	Production		Tot. Equipment Cost DTC = \$150 or *(\$200)/Day)* + Loader	Cost of Operators	Fleet Cost /hr				Unit Cost (\$150/ truck)		*Unit Cost (\$200/truck)*	
			yd^3/hr	%	\$/hr	(\$/hr)	%	\$/hr	*(\$/hr)*	%	yd^3/\$	\$/yd^3	*(yd^3/\$)*	*(\$/yd^3)*
8	0.83	0.03	83.1	100	(8 × DTC + 800)/8 = 250 *(300)*	8 × 18 + 28 = 172	100	422	*(472)*	112	0.197	5.08	*0.176*	*5.67*
9	0.83	0.04	90.5	108	(9 × DTC + 800)/8 = 269 *(325)*	9 × 18 + 28 = 190	109	459	*(515)*	122	0.198	5.07	*0.175*	*5.69*
10	0.83	0.07	93.7	113	(10 × DTC + 800)/8 = 287 *(350)*	10 × 18 + 2 8 = 208	117	495	*(558)*	132	0.189	5.28	*0.168*	*5.95*

DTC = Daily Truck Cost; italic is used for \$200/day truck.

readily apparent from looking only at the data derived from the balance point analysis.

Graph 3 compares the dimensions on the y-axis of graphs 2 and 3, making it obvious that they are inverse. Thus, it is expected that the concave curve will become convex. Indeed, the maximum cost-based productivity finds its inverse in the minimum unit cost of 5.07 $\$/yd^3$ for 9 trucks costing \$150/day. By adding a tenth truck, the unit cost rises to 5.28 $\$/yd^3$, while fleet production goes up from 90.5 yd^3/hr (724 yd^3/day) to 93.7 yd^3/hr (750 yd^3/day). It is evident that bunching costs money, incurred by the idling trucks and the loader. Production continues to increase, yes, but at a higher and higher cost per added unit.

3.5 PROCESS SYNCHRONIZATION IN THE SUPPLY CHAIN

So far, we have presented that construction processes as "stand-alone," operating independent of anything else—with the exception of traffic that can slow down trucks. In practice, however, most processes depend on others; they supply critical inputs to a downstream operation, without which it could not function. Particularly in manufacturing, managing the supply channels for materials, parts, components, and even information has become increasingly important, for their sources have become global and customers are accustomed to buying products specifically tailored to their needs and desires. A good example is the personal computer, which is composed of parts and components whose origins are from all over the world. What makes it possible for a customer to special-order a combination of parts that come from many locations yet can be assembled and delivered within a week? The answer is supply-chain management, which integrates all the parts suppliers into one collaborative network, capable of delivering within days all the necessary parts, tested, ready for assembly, testing of the final product, and "short-time" shipment to the customer.

Such a production system requires open and real-time communication channels, matched by agile responses from the links in the supply chain. While construction processes can't be compared to a production scheme suited to consumer products that can easily be shipped by trucks or airplane; nevertheless, some of the basic principles that underpin this extremely efficient system of production can be applied to construction work.

Supply Chain in Construction

The *supply chain* is an integrated and collaborative system of suppliers, service companies and processes through which products pass, gaining value at each link. *Supply-chain management* includes all the planning and integration of activities involved in the sourcing, procurement, and logistics of construction materials; mechanical or electrical systems; parts, prefabricated work-in-process (WIP), and information.

The *green supply chain* incorporates sustainable efforts such as preventive maintenance of transportation equipment, efficient lighting, paperless communications, and minimizing the carbon footprint.

3.5.1 Costly Two-Way Ripples in the Supply Chain

What bonds construction IPOs to one another and to off-site work are the WIPs, pooled resources that have to be shared (e.g., a tower crane), and the supply streams consisting of

a) Crushed Stone Bulk Storage in Rock Quarry

b) Concrete Supply Truck

c) Transitional Storage of Stone

Figure 3.12 Views of the Concrete Supply Chain

information, materials, parts, components, and so on—all necessary inputs. Recall that the automobile industry eliminated one muda/waste, the cost of storing materials and parts, by adopting the JIT concept, which has as its goal to supply the right material, at the right time, to the right place, and in the right amount. That said, it has been recognized that maintaining strategic "safety" inventories can provide critical advantages in atypical situations that interrupt the supply chain. Unfortunately, creating such inventories in the construction industry is prohibited by the perishable characteristics of the materials (e.g., concrete), the lack of adequate space (e.g., a project site in a crowded downtown area), and the need to storage materials where they are protected against bad weather.

The many barriers to applying the principles of JIT in construction can result in troubles in the supply chain that will ripple all the way through, to the last process in the chain, and cause costly disruptions and production losses. We'll use concrete to study the ripple effect of supply-chain problems. Figure 3.12 presents three different material channels linking several IPOs.

The heart of the concrete supply chain is the mixing process, where aggregates, cement, water, and additives are combined according to a project-specific recipe. Two main channels deliver the bulk cement and the cleaned aggregates via a transportation system. (Note: Because of the high cost of transportation, cement factories and the rock quarries are, whenever possible, built locally, with trucks as the preferred method of transport.) Cement and dump trucks can be treated as transitional mobile storage containers that are filled at one end and emptied at the opposite end of the truck cycle. Both the rock quarry and the concrete plant agree on a strategic storage capacity, which allows them to stabilize their operations while serving the fluctuating yearly cycle (e.g., high demand during summer), or to take advantage of lower bulk prices. Of course, the variable storage costs are the incentive for maintaining it at an economically optimal size.

Special trucks with drum mixers are commonly used to deliver fresh concrete to the construction site. Because significant time can elapse between batching and unloading—for example, during rush hour—the concrete must be prevented from setting or hardening, by turning the drum slowly during transit and idling. Delaying curing is further assured by adding chemicals.

Three main methods are used for unloading: (a) gravitational flow directed by a shoot, (b) filling crane buckets, and (c) discharge into the hopper of a concrete pump. A truck-mounted concrete pump pushes the concrete through boom-mounted pipes and a hose trunk to the specified location. A crew of 8 to 10 laborers is needed to guide the hose nozzle, spreading and vibrating the concrete, followed by screeding and, finally, floating and finishing the surface.

The concrete trucks have to adjust their discharge rate to the pumping rate, which in turn is tied to the speed at which the concrete crew is able to place it. The two controlling constraints are, one, the maximum amount of concrete allowed to pile up on the formwork, restricted by the load capacity of the formwork, and, two, the short amount of time before a "resting" concrete starts to harden. Thus, the drum of the truck represents an important transitional storage capacity. After the drum has been emptied, it will be moved to a cleaning location to remove the cement adhering to the inside of the drum and prepare it for the return to the plant for reloading.

In order to understand the result of the following quantitative analysis, it's important to appreciate two important features of the concrete supply chain using trucks and a pump. First, these two resources—and along with them the dual processes of concrete delivery and concrete pumping are deeply constrained by the production rate of the concrete crew, and the traffic situation faced by the truck driver. Second, the two resources are tightly linked to each other: the pump can't perform without concrete from the drum, and the truck can't unload at a higher rate than the pump.

As we will see, these two features makes planning difficult, and control mechanisms complex. On the other hand, if the pump has to wait for a longer period of time, a lot of muda/waste accumulates, as laborers become idle and the pump pipes have to be emptied to avoid the concrete hardening inside.

The model of the concrete supply chain shown in Figure 3.13 focuses on the supply streams for the key resources, aggregates, cement, and concrete along the chain and the inflows from resource pools. Also depicted are three material storage states: bulk material, mobile truck based, and zero storage; the model also traces the sequenced WIP "stream" that has to be set in motion before the concrete can be placed. Finally, each IPO is modeled as a cyclical system, similar to a truck cycle for an earthmoving operation.

3.5.2 Just-in-Sequence Material Supply

Section 3.2 presented the concept of Just-in-Time (JIT) delivery of resources as a means to cut the costs associated with managing large inventories. Henry Ford

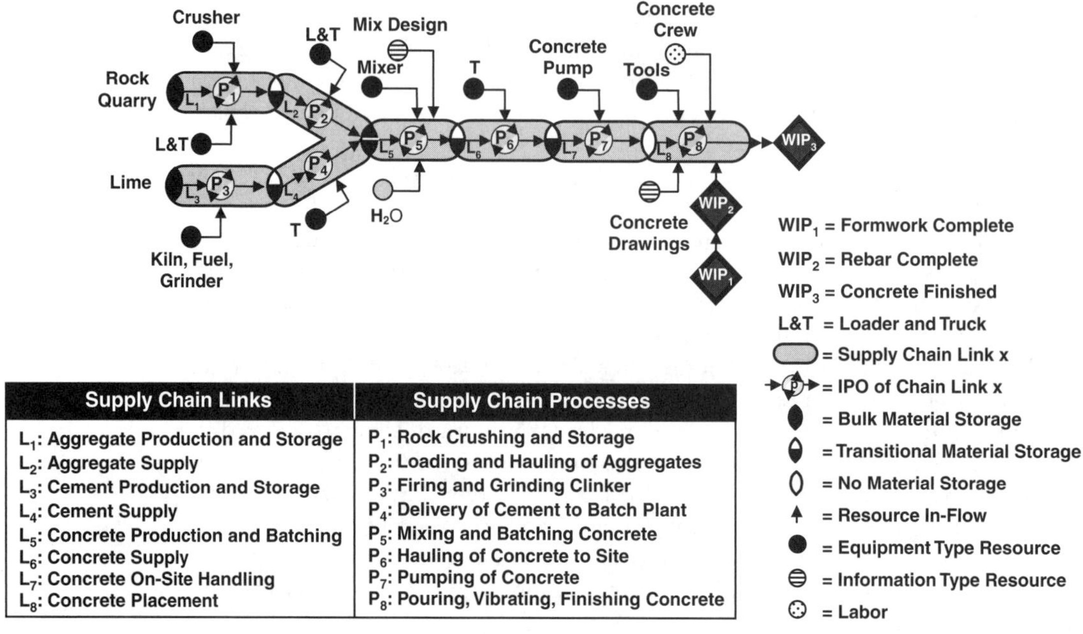

Supply Chain Links	Supply Chain Processes
L_1: Aggregate Production and Storage	P_1: Rock Crushing and Storage
L_2: Aggregate Supply	P_2: Loading and Hauling of Aggregates
L_3: Cement Production and Storage	P_3: Firing and Grinding Clinker
L_4: Cement Supply	P_4: Delivery of Cement to Batch Plant
L_5: Concrete Production and Batching	P_5: Mixing and Batching Concrete
L_6: Concrete Supply	P_6: Hauling of Concrete to Site
L_7: Concrete On-Site Handling	P_7: Pumping of Concrete
L_8: Concrete Placement	P_8: Pouring, Vibrating, Finishing Concrete

Figure 3.13 Model of Concrete Supply Chain Processes and Resource Flows

Worked-Out Problem 3.2: Upstream and Downstream Supply Ripples

John Franklin is the project manager of a 20-story concrete building in San Francisco. He is planning a new concrete pour for tomorrow that involves the pouring 81 yd^3 of concrete walls for the elevator shaft and 270 yd^3 onto an elevated slab. He expects his concrete crew to be able to pour 0.75 yd^3/min into the walls and 1.5 yd^3/min onto the slabs at a *crew operating factor* of 0.80. The rented concrete pump, which is scheduled to arrive at 5:45 AM can pump 2.0 $yd^{3/}$min. Installation should not take longer than 20 to 30 minutes.

Crew Operating Factor

The *crew operating factor* is the percentage of time a crew is fully operational, minus the time needed to set up the job, move to another location, address minor malfunctions, take short breaks, receive new instructions, and so on.

Mr. Franklin was extremely dissatisfied last week when they had a similar pour for the level below. He thinks the problem was caused by a mismanaged supply of concrete from the plant, partially due to a change in the daily traffic pattern after Labor Day. He acquired the traffic record for last Wednesday,

September 10, for the area where the concrete trucks circulated, as he expects a repeat of last-week's traffic pattern since the conditions will be similar. The weather is also predicted to be similar.

Mr. Franklin also suspects that the dispatcher at the concrete plant moved trucks from his job during the day, rather than keep cycling them between plant and site. He had a record of truck arrival and return times, between 6:00 A.M. and 1:00 P.M., drawn up from the truck delivery receipts (Table 3.4). In order to avoid a similar disaster tomorrow, Mr. Franklin requests a recommendation for how to handle the situation.

Table 3.4 Truck Load Arrival/Departure Record

Truck	Drum Vol.	Load Time		Arrival Time		Depart Time	
Code	yd^3	Hr	Min	Hr	Min	Hr	Min
M34	10	6	5	6	26	6	49
M21	10	6	15	6	41	7	17
S3	8	6	25	6	50	7	37
M3	10	6	35	7	4	7	59
M34	10	6	58	7	34	8	21
M21	10	7	8	7	38	8	42
T4	8	8	22	8	47	9	6
M34	10	9	0	9	22	9	43
S3	8	9	35	9	55	10	14
M34	10	10	21	10	39	10	54
M21	10	10	46	11	4	11	19
S3	8	10	51	11	10	11	24
M11	10	11	13	11	34	11	50
M34	10	11	36	11	56	12	11
M21	10	11	49	12	11	12	26
S3	8	12	4	12	26	12	51
M34	10	12	47	13	8	13	36
M21	10	12	57	13	17	13	49

(continued)

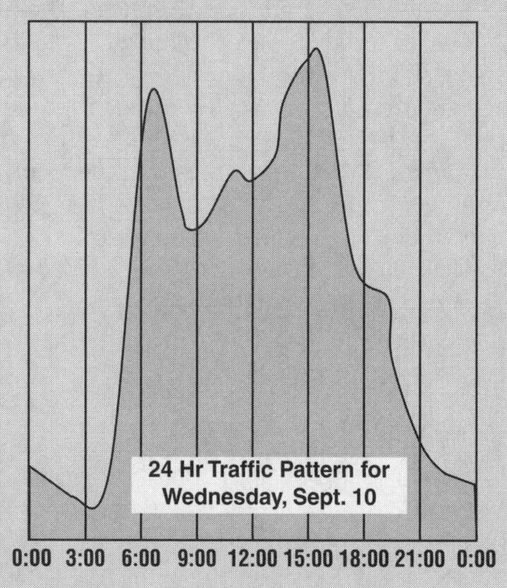

Figure 3.14 Traffic Pattern

Assumptions

The concrete mixing plant is an independent business that tries to optimize the allocation of its trucks while keeping the mixing machinery operating at full capacity. A truck waiting at a construction site to be unloaded at the pump is a muda/waste for the firm, one that will ripple back to the plant, in that the truck is less available to be batched. On the contractor's side, as Mr. Franklin mentioned, is the cost of an idle pump and crew, who will have to be paid even while they wait for a concrete truck to arrive. Therefore, we assume that:

The average time each truck needs for maneuvering is three minutes, and for cleaning and departing the site, five minutes. The pumps will be ready at 6:15 A.M at the latest.

- The traffic count pattern applies to all the possible truck routes.

Data Analysis

As a first step, we should study the data from the truck records in detail. Table 3.4 lists the haul duration for each truck, along with interarrival times (minutes between two trucks arriving on-site). It also shows the ideal pump rates for the walls and the slab, with the switchover occurring at around 10:00 A.M. There are two pump durations for the ideal 0.75 yd^3/min—13.3 and 10.7 minutes—reflecting the two different trucks capacities of 10 yd^3 and 8 yd^3.

Table 3.5 Truck Movement Durations

Haul Dur Min	Arrival TGap Min	Pump Rate yd^3/min	Pump Dur Min	Spot&Clean Min
21		0.75	13	8
26	15	0.75	13	8
25	9	0.75	11	8
29	14	0.75	13	8
36	30	0.75	13	8
30	4	0.75	13	8
25	69	0.75	11	8
22	35	0.75	13	8
20	33	0.75	11	8
18	44	1.5	7	8
18	25	1.5	7	8
19	6	1.5	5	8
21	24	1.5	7	8
20	22	1.5	7	8
22	15	1.5	7	8
22	15	1.5	5	8
21	42	1.5	7	8
20	9	1.5	7	8

More illustrative are the plots in Figure 3.15, with the time at the x-axis. Clearly visible are three problem spots. First, as expected, the haul time increases between 6:00 and 7:00 A.M. Second, and more disturbing, is the truck time on-site, the period between arrival and departure. Ideally, the time should consist of 8 + 13.3 (10.7) minutes, for pumping, spotting, and cleaning. However, those durations climb to over 60 minutes at around 7:20 A.M., before dropping back to 20 minutes. Problem spot 3 is equally serious, as it points to the fact that the trucks stopped coming, incurring a time gap between two trucks of 70 minutes following the on-site waiting period, which peaked at 7:20 A.M. It seems that the dispatcher at the concrete plant did

(*continued*)

(*continued*)

respond to the long time the trucks had to wait at the site by reducing the number of trucks he sent to the site. The immediate effect was a drastic reduction in on-site time for trucks staying fewer than 20 minutes, until 12:00 P.M. We would expect that the long on-site waiting time and the long interarrival times of trucks would be directly linked to the performance of the crew and, with it, of the pump. We need to keep in mind that it is impossible to ''inventory'' fresh concrete after it has been discharged from the slowly turning drum of the truck.

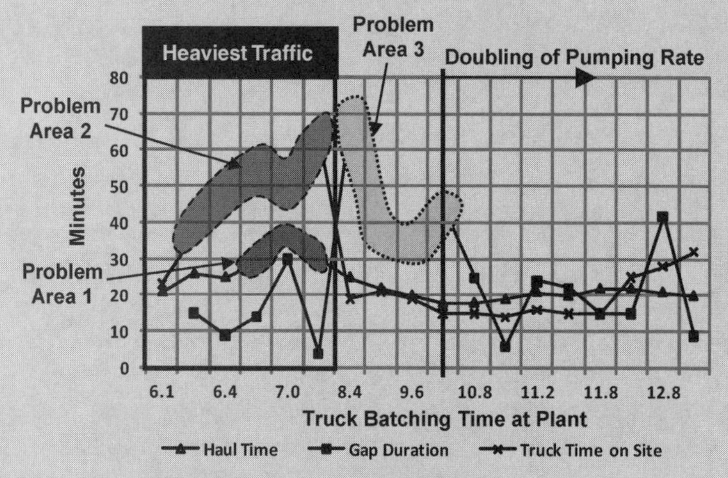

Figure 3.15 Changing Truck Times On-Site and Off-Site

An eye-catching feature is the alignment of the period of time the pump was idle and the time gap between truck arrivals, which starts at 9:00 A.M. In the early morning, the pump never had to wait for a truck; to the contrary, the trucks had to wait for the pump. This changed when, as depicted in Figure 3.15, the dispatcher sent fewer than the two and three trucks per half hour. Between 7:30 and 8:30 A.M., only one truck was sent.

A second important aspect is the alignment of pump delays, most probably due to problems faced by the concreting crew and the length of time trucks had to wait to discharge the concrete. Naturally, when the pump had to stop, the truck was unable to continue unloading its concrete into the hopper of the pump. While a 10 yd^3 truck should take 13 minutes to empty, it took up to 15 minutes longer in the early morning hours. After 8:00 A.M., those interruptions ceased, but the damage was done, at which time the dispatcher drastically cut back on the number of trucks he sent. When that happened, the pump lay idle, and with it the crew, giving it ample time to fix any problem it faced.

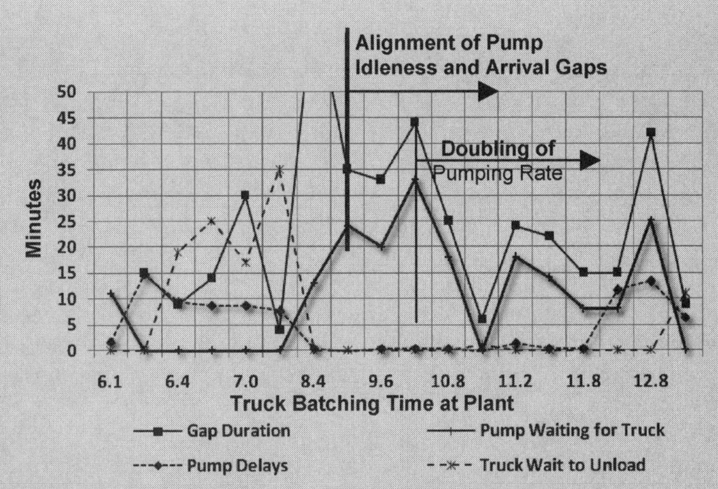

Figure 3.16 Changing Waiting Times at the Pump

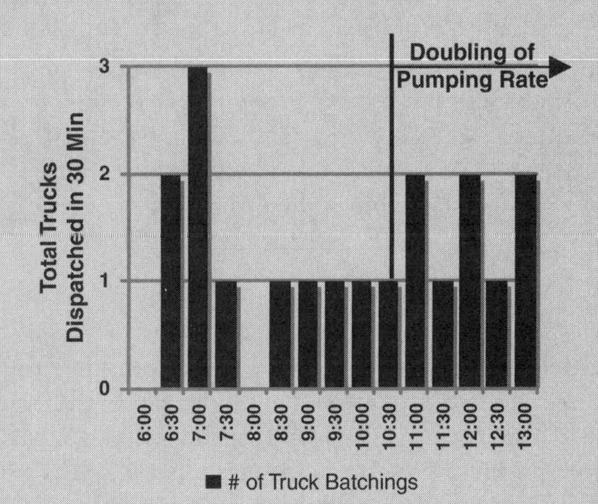

Figure 3.17 Batching Pattern

Discussion and Recommendation

After the first truck arrived at 6:26 A.M, by 1:30 P.M. only 160 yd^3 had been pumped, which computes into an average pumping rate of 0.38 yd^3/min. At 9:30 A.M., 66 yd^3 had been pumped, resulting in a rate of 0.37 yd^3/min, instead of the 0.75 yd^3/min that had been planned. The dynamics of the concrete supply chain changed dramatically at around 8:30 A.M. after a series of concrete trucks were unable to discharge their loads for 50 minutes, a typical upstream ripple effect caused by a downstream delay. The cause appears to have been severe problems faced by the crew, necessitating stoppages at

(continued)

(continued)
the pump lasting 10 to 15 minutes per truck. At the same time, the dispatcher at the concrete plant kept the trucks coming, assuming that the pump rate would be 0.75 yd^3/min or higher. When that did not occur, trucks were reassigned (e.g., M^3), in order to cut back on truck waiting time.

The second phase started when the problems of the crew seem to have become less severe and the pump operated without interruptions. At this time, the interarrival time of the trucks had increased to the point at which the pump was sitting idle, a typical downstream effect caused by a slowing of the supply. It is apparent that the concrete hauling process was not in line with the concrete pumping and placing processes.

Based on the analysis of the available data, we can make two recommendations:

Recommendation 1: Improve the planning and preparations for the concrete crew. Find out what went wrong last week and brainstorm with the crew, today or early tomorrow morning, over what needs to be done to avoid it. Mr. Franklin might be the best person to lead such a meeting.

Recommendation 2: Open a direct and real-time communication with the dispatcher. Both sides will gain from aligning the two supply-chain links, pumping and trucking. This is a perfect example of a win-win situation, whereby cooperating both parties will benefit from reducing their muda: idling trucks and waiting pump and crew.

synchronized the delivery of components by prescribing not only time but also order. Today, the concept is being expanded to also cover the supply of parts, under a program called Just-in-Sequence (JIS). In the automobile industry, each vehicle needs the same group of prefabricated elements, as well as a predefined set of bolts, screws, and nuts. Under JIS, such items are prepackaged in small containers and in quantities defined by the bill-of-material, before being transferred to larger holding carts or trailers, and delivered in synchrony with the assembly line. Prepackaging in this way minimized errors and confusion, especially when the car model on the assembly line changed frequently. The time to refill and pick materials from bins also was eliminated.

The construction industry, with its wide range of different parts that have to be assembled at specific locations, is ideally suited to benefit from JIS. To examine how JIS works in construction, let's look at an example involving reinforcing bars, more commonly referred to as rebar.

Concrete is commonly used to construct elevated slabs with beams and joists, some of which are post-tensioned. Figure 3.18 is a schematic view of a shoring, formwork, and rebar assembly used in an elevated slab with beams.

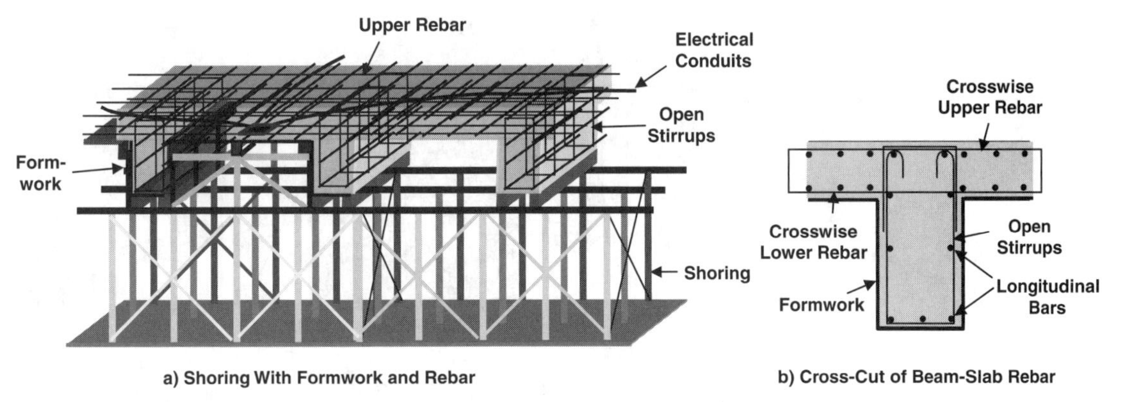

Figure 3.18 Cutaway of Formwork and Rebar for Slab with Beam

On the face of it, this design looks fairly simple, but the irregular dimensions and joists that run perpendicular to the beam combine to create a large puzzle, which features long and heavy straight bars that are very difficult to handle. Likewise, the assembly of the rebar for the beam, which has to be tied in with the lower and upper slab rebar, is not as simple as it looks. Figure 3.18 (b) shows one open stirrup in a cross section of a beam with hooks at both ends. A second open stirrup, with straight ends, is inserted upside down and tied in order to close the open end on the top. The question is, how to get the longitudinal bars inside, tying them to the stirrups while crossing the slab bars? As this represents the last link in the supply chain, it's the step that would benefit most from good supply-chain management—and in this case, Just-in-Sequence delivery. For this reason, we need to look upstream, at what is influencing the productivity of rebar placement on-site.

In the United States, a detailer working for a fabricator develops placement drawings and rebar lists, according to the engineer's design. Figure 3.19 presents two streams: (a) material and (b) information. The source of the reinforcing bars is deformed steel in coils or straight lengths (2), which stream toward the placement on the formwork (10), for casting in concrete as a slab. The information stream starts during the design process (1) and continues even after the encasement of the rebar with concrete, as bills for shipped rebar have to be paid and the as-builts need to be established. The stream consists of many substreams that emerge, among them: (a) blueprints, (b) electronic data for shear line/rebar bender, (c) printed bundle tags, (d) loading list/bill-of-lading, and finally (e) an (electronic) bill for the shipped rebar (9). When a new shipment is called for, the bar lists are sent to the shop floor, where the bars are pulled from the material stock (2), cut and bent (3), assembled into bundles according to prescribed shapes (4), to be laid out on the shop floor. As soon as a shipment is complete, the bundles are loaded onto a trailer meeting its weight limitations, for which bills-ofl-ading or material lists are created (5). When needed on the construction sites, the

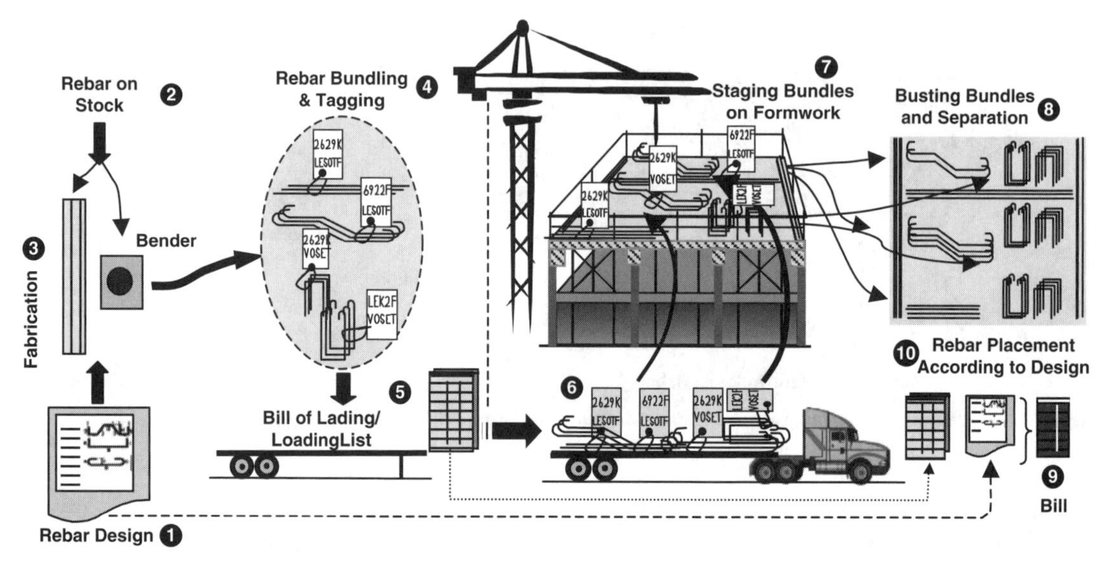

Figure 3.19 Traditional Rebar Supply Chain

tractor will pull the trailers to the site (6), where they will be unloaded by crane onto the empty formwork (7). Subsequently, a rebar laying crew will "bust" the bundles (8) and place the bars (10) according to the rebar design (1).

Funded with a grant from the National Science Foundation (NSF) one of the authors led a research project designed to study whether a change in the bundling and staging process would help boost the productivity of even a hard-working crew. A multistory building was found that allowed a team of researchers to observe and measure the rebar placement on each floor.

First, the team evaluated the availability of space and crew movement; quantity, size, and shape of rebar; crane operation; and other related factors. Second, the bundles of bars were looked at in detail to assess what happened when the bundles were busted, to determine where the bars end up. That is when the effect of the non-sequencing of the bars was most visible. It was observed that from a bundle of 12 bars, 4 had to be carried to location A, 4 to B, and 8 to C.

Problem Area 1: Misalignment of Supply and Construction Process

Bundling rebar that has the same shape neglects the sequence of the placement process, which commonly starts at a corner, end, or beam that is picked by the foreman. Instead of having the rebar available in the sequence they are needed, as in the automobile industry, the bars lay in bundles across the formwork, which have to be searched to find the right size.

A second observation was of the repositioning (rehandling) of the rebar that had not been used as progress was being made.

Problem Area 2: Misalignment of Staging and Construction Process

Because the bundles were created according to size and shape, it was impossible not to have to move some rebar to make space for the crosswise bottom bars.

As a result of this observation, three changes in the supply chain were initiated. First, the concept of *master bundles*, made up several microbundles was implemented. Instead of making one large bundle, the shear line operator was asked to create several microbundles (e.g., instead of 1 large bundle with 12 bars, make 3 bundles with 4). This caused no problem since the mechanisms at the shear line allow up to six bundles at the same time. Other bundles were similarly broken up and added to the first one, establishing a master bundle consisting of a series of microbundles of different sizes but all straight or similarly bent. Similarly, the bundles of 50 stirrups were broken up according to need on the process.

Second, the staging area on the formwork was planned and organized according to the placement strategy provided by the foreman. The primary goal was to avoid any rehandling. This was accomplished by spreading the master bundles in such as way so they were all used up by the time the location was needed. Equally important was that the long bars were all staged nearby, and parallel to their final placement location.

Third, a very effective means of communication was found: the use of color tags. Master bundles were labeled with color tags that matched those on their staging areas. A primary purpose of color-coding the bundles was to facilitate rapid staging by easy recognition/identification and matching of colors. Having a color-coded staging scheme in his hand enabled the foreman to easily direct the crane operator to the appropriate staging area, where crew members quickly released the crane hook.

Table 3.6 presents the result of the work sample analysis for the two key floors. Clearly, the effect was dramatic. Direct work for the third floors, which had accounted for 27.8 percent of the crew's time, jumped to 54.9 percent when the rebar was delivered JIS. As expected, a major drop was experienced in the time spent searching for the next set of rebar: almost 10 percent, from 3.9 to 4.2 percent. Another significant drop was in the time spent measuring rebar that had lost its tag, to verify its length.

Implementing JIS in rebar supply perfectly demonstrates the principle of the Nash Equilibrium. The two companies, the rebar fabricator and the erector, traditionally work at such an equilibrium. While bundling differently and applying colored tags in place of the white paper tags would add a small amount of work in the

Rebar Facts

"All protruding reinforcing steel, onto and into which employees could fall, shall be guarded to eliminate the hazard of impalement."

—*OSHA 1926.701(b)*

Deadly Rebar

"An employee pulling a concrete trunk hose along the form fell two stories and hit his head on steel rebar, which punctured his brain."

(http://www.osha.gov/Publications/100most/100most.html)

Each reinforcing bar has a series of markings: The top letter or symbol identifies the producing mill and deformation pattern. Next comes the bar size. The third symbol designates the manufacturing material —either "S" for carbon-steel (ASTM A615) or "W" for low-alloy steel (ASTM A706).

Table 3.6 Work Comparison

Categorv of Work	Traditional %	Using JIS %
Value-Adding Work	27.8	54.9
Carrying	11.1	10.2
Communication	6.9	5.8
Staging	0.7	0.9
Measuring	7.9	3.9
Walking empty	9.9	7.9
Searching rebar	13.9	4.2
Leave staging area	1.2	0.2
Waiting	3.5	1.4
Correcting	0.5	0.2
Idle	5.6	3.7
Private communicat.	4.9	4.6
Not observable	6.2	2.1
	100	100

shop, the gain to the erector would be significant. But because both were committed to their best position, this gain could not be reaped. To motivate the two companies to collaborate would take *value gain-sharing*. Say that both agree to collaborate and that the fabricator would get 30 percent of the gain made by the erector; then everyone would benefit, given that the increased cost by fabricator was smaller than the 30 percent return gain.

CHAPTER REVIEW

Journaling Questions

1. The increasing size, complexity, and risk involved in construction projects in the nineteenth century caused the demise of the Master-Builder. What was the effect on the constructability of designs? How did the industry react?
2. The proliferation of computer software in the 1990s, along with increased adoption of electronic communication, surfaced the very costly problem of interoperability. Why is this a problem, and what is being done to solve it?

3. The WBS organizes a project top-down while IPO planning approaches the project bottom-up. Why do they come from opposite directions? How can they link up?
4. The design of the supply chain can have major impacts on the productivity of the construction process. Summarize the main problem areas discussed and apply them to another situation in construction that you know (don't hesitate to talk to a construction manager in the field, if you do have not enough experience).

Traditional Homework

1. Section 3.5.2 explained in detail the supply chain for rebar. Transfer that description to the following a supply-chain model by:
   a. Attaching the missing labels
   b. Completing the tables
   c. Correcting obvious errors in the presented model

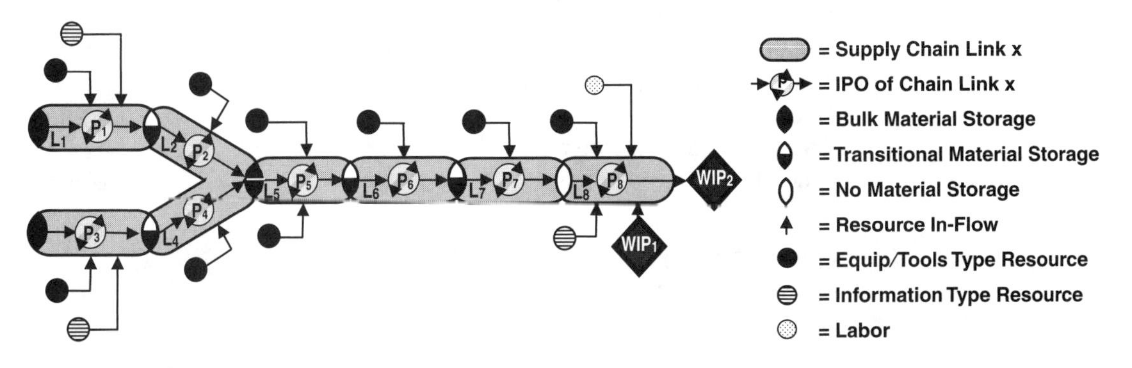

Supply-Chain Links	Supply-Chain Processes
L_1:	P_1:
L_2:	P_2
L_3:	P_3:
L_4:	P_4:
L_5:	P_5:
L_6:	P_6:
L_7:	P_7:
L_8:	P_8:

2. What is the balance point for concrete placement using a pump with the following resources and under these conditions?
   a. Capacity of concrete pump = .33 m^3/min
   b. Batching one 5 m^3 concrete truck = 5 minutes

c. Hauling time for truck = 15 minutes

d. Cleaning truck = 5 min

e. Return time for truck to batch plant = 10 minutes

3. A contractor is excavating a caisson set in dense material. The process employs a small tower crane to lower and lift a container that is filled by a small excavator at the bottom of the caisson. After it is full, the crane hoists it up and swings it over to two large hoppers, where 10 yd^3 trucks remove the spoils from the hopper. The density of the material at the bottom of the caisson requires the use of a dozer-ripper that pushes the loosened material toward the backhoe for loading into the container. The average duration for each task in the process are given in the table here:

TASK	Duration (min)
Ripping 2 yd^3 dense material by small dozer-ripper	3
Loading 4 yd^3 into container by small backhoe	4
Crane hoisting and swinging container to hopper	1.5
Emptying container into hopper	0.5
Crane lowering container to bottom of caisson	1.5
Filling 10 yd^3 truck from hopper	1.5
Hauling soil by 10 yd^3 truck	30.0
Spotting and emptying truck	3.0
Truck returning to hopper	20.0

a. How many cycles can you identify?

b. Draw the IPO diagram for the entire operation.

c. How many trucks are needed to keep up with the excavation?

d. What two simple modifications would you recommend to boost productivity under the present circumstances (all the major equipment must remain the same, because it would be too expensive to change).

Open-Ended Question

1. This chapter used several cases to show how important real-time information is to controlling the construction process. Chapter 2 explained that telematics provides many opportunities to measure productivity of fleets, totally automatically and in real time. Design a telematics system that could be used by the concrete batch plant to immediately reassign trucks before they arrive on the site so that they don't have to wait at the pump. What decision would the dispatcher have to make, in light of the information that you would provide him or her in real time? How would you include the concrete truck in the system (i.e., what would you monitor) and what would be the benefits?

BIBLIOGRAPHY

Arditi, D., A. Elhassan, and Y. C. Toklu. Constructability Analysis in the Design Firm. *J. Constr. Eng. Mgmt.*, vol. 128, no. 2, 2002.

Bernold, L. E. Developing a Resource Supply Chain Planning System for Construction Projects. Discussion in *J. Constr. Eng. Mgmt.*, vol. 132, 4, 2006.

Bernold, L. E., and M. Salim. Placement-Oriented Design and Delivery of Concrete Reinforcement. *J. Constr. Eng. and Mgmt.*, vol. 119, no. 2, 1993.

Caridi, M., and S. Cavalieri. Multi-Agent Systems in Production Planning and Control: An Overview. *Prod. Planning Contr.*, vol. 15, no. 2, 2004.

Chong, W. K., and S. P. Low. Latent Building Defects: Causes and Design; Strategies to Prevent Them. *J. Perform. Constr. Facilities*, vol. 20, no. 3, 2006.

Christopher, M., and T. Towill. An Integrated Model for the Design of Agile Supply Chains. *Inter. J. Phys. Distr. Log. Mgmt*, vol. 31(4), 2001.

Cuthbertson, R., and W. Piotrowicz. Supply Chain Best Practices—Identification and Categorization of Measures and Benefits. *Inter. J. Prod. Perform. Mgmt*, vol. 57, no. 5, 2008.

Davis, P. R. A Relationship Approach to Construction Supply Chains. *Ind. Mgmt. Data Systems*, vol. 108, no. 3, 2008.

Dunston, P. S., and C. E. Williamson. Incorporating Maintainability in the Constructability Review Process. *J. Mgmt. Eng.*, vol. 15, no. 5, 1999.

Figueiredo, J. M., and J. Sá da Costa.A Two-Level Hierarchical Control Strategy Applied to an Intelligent House. IEEE International Workshop Intel. Signal Process., Faro, Portugal, September 2005.

Fisher, D. J., S. D. Anderson, and S. P. Rahman. Integrating Constructability Tools into the Constructability Review Process. *J. Constr. Eng. Mgmt.*, vol. 126(2), 2000.

Forslund, H., P. Jonsson, and S. Mattsson. Order-to-Delivery Process Performance in Delivery Scheduling Environments. *Inter. J. Prod. Perform. Mgmt.*, vol. 58(1), 2009.

Gallaher, M. P., A. C. O'Connor, J. L. Dettbarn, and L. T. Gilday. *Cost Analysis of Inadequate Interoperability in the U.S. Capital Facilities Industry.* Gaithersburg, MD: National Institute of Standards and Technology, NIST GCR 04-867, 2004.

Gavilan, R. M., and L. E. Bernold. Source Evaluation of Solid Waste in Building Construction. *J. Constr. Eng. and Mgmt.*, ASCE, vol. 120, no. 3, 1994.

Goodrum, P., A. Smith, B. Slaughter, and F. Kari. Case Study and Statistical Analysis of Utility Conflicts on Construction Roadway Projects and Best Practices in Their Avoidance. *J. Urban Planning Develop.*, vol. 134, no. 2, 2008.

Graham, L. D., S. D. Smith, and P. Dunlop. Lognormal Distribution Provides an Optimum Representation of the Concrete Delivery and Placement Process. *J. Constr. Eng. Mgmt.*, vol. 131, no. 2, 2005.

Gujarati, D. N. The Economics of the Davis-Bacon Act. *Journal of Business*, vol. 40(3), 1967. www.jstor.org/stable/2351753.

Hallowell, M., and T. M. Toole. Contemporary Design-Bid-Build Model. *J. Constr. Eng. Mgmt.*, vol. 135, no. 6, 2009.

Harper, D. G., and L. E. Bernold. Success of Supplier Alliances for Capital Projects. *J. Constr. Eng. and Mgmt.*, vol. 131, no. 9, 2005.

Ingemansson, A., T. Ylipa, and G. S. Bolmsjo. Reducing Bottle-Necks in a Manufacturing System with Automatic Data Collection and Discrete-Event Simulation. *J. Manu. Tech. Mgmt*, vol. 16, no. 6, 2005.

Jenkins, J., and D. L. Orth. Productivity Improvement Through Work Sampling. *Cost Engineering*, vol. 46, no. 3, 2004.

Klotz, L., M. Horman, H. H. Bi, and J. Bechtel. The Impact of Process Mapping on Transparency. *Inter. J. Prod. Perform. Mgmt.*, vol. 57, no. 8, 2008.

Kuprenas, J. A., and A. S. Fakhouri. A Crew Balance Case Study: Improving Construction Productivity. *CM eJournal*, Construction Mgmt. Assoc. of America, 2001.

Lam, P. T., A. P. Chan, F. K. Wong, and F. W. Wong. Constructability Rankings of Construction Systems Based on the Analytical Hierarchy Process. *J. Arch. Engrg.*, vol. 13, no. 1, 2007.

Lee, J., S. J. Lorenc, and L. E. Bernold. A Comparative Performance Evaluation of Tele-Operated Pipe Laying. *J. Constr. Eng. and Mgmt.*, vol. 129, no. 1, 2003.

Lindholm, A., and P. Suomala. Learning by Costing: Sharpening Cost Image Through Life Cycle Costing? *Inter. J. Prod. Perform. Mgmt.*, vol. 56, no. 8, 2007.

Liu, M., and D. M. Frangopol. Multiobjective Maintenance Planning Optimization for Deteriorating Bridges Considering Condition. Safety, and Life-Cycle Cost, *J. Struct. Engrg.*, vol. 131, no. 5, 2005.

Menches, C. L., and A. S. Hanna. Quantitative Measurement of Successful Performance from the Project Manager's Perspective. *J. Constr. Engrg. Mgmt.*, vol. 132, no. 12, 2006.

Motwani, J. Measuring Critical Factors of TQM. *Measuring Business Excellence*, vol. 5, no. 2, 2001.

National Electronics Manufacturing Initiative (NEMI). *In Search of the Perfect Bill of Materials (BoM)* (white paper), Herndon, VA, March 2002.

Nenadal, J. Process Performance Measurement in Manufacturing Organizations. *Int. J. Prod. and Perform. Mgmt.* vol. 57 no. 6, 2008.

Newton, L. A., and J. Christian. Impact of Quality on Building Costs. *J. Infra. Systems*, vol. 12, no. 4, 2006.

Nyhuis, P., and M. Vogel, M. Adaptation of Logistic Operating Curves to One-Piece Flow Processes. *Inter. J. Prod. Perform. Mgmt.*, vol. 55, no. 3/4, 2006.

Palaneeswaran, E., M. Kumaraswamy, and S. T. Ng. Formulating a Framework for Relationally Integrated Construction Supply Chains. *J. Constr. Res.*, vol. 4, no. 2, 2003.

Pocock, J. B., S. T. Kuennen, J. Gambatese, and J. Rauschkolb. Constructability State of Practice Report. *J. Constr. Engrg. Mgmt.*, vol. 132, no. 4, 2006.

Project Management Institute (PMI). *A Guide to the Project Management Body of Knowledge (PMBOK® Guide)*, Project Management Institute, Newtown Square, PA, 2000.

Pulaski, M. H., M. J. Horman, and D. R. Rile. Constructability Practices to Manage Sustainable Building Knowledge. *J. Arch. Eng.*, vol. 12, no. 2, 2006.

Pulaski, M. H., and M. J. Horman. Continuous Value Enhancement Process. *J. Constr. Eng. Mgmt.*, vol. 131, no. 12, 2005.

Qiping, S., and L. Guiwen. Critical Success Factors for Value Management Studies in Construction. *J. Constr. Eng. Mgmt.*, vol. 129, no. 5, 2003.

Ruhnke, J., and C. J. Schexnayder. Description of Tilt-Up Concrete Wall Construction. *Pract. Per. Struct. Design Constr.*, vol. 7, no. 3, 2002.

Sacks, R., C. M. Eastman, and G. Lee. Process Model Perspectives on Management and Engineering Procedures in the Precast/Prestressed Concrete Industry. *J. Constr. Eng. Mgmt.*, vol. 130, no.2, 2004.

Saiz, J. J., A. O. Bas, and R. R. Rodriguez. Performance Measurement System for Enterprise Networks. *Inter. J. Prod. Perform. Mgmt.*, vol. 56, no. 4, 2007.

Salim, M., and L. E. Bernold. Effects of Design-Integrated Process Planning on Productivity in Rebar Placement. *J. Constr. Eng. and Mgmt., ASCE*, 120, no. 4, 1994.

Shi, J. J., and D. W. Halpin. Enterprise Resource Planning for Construction Business Management. *J. Constr. Eng. Mgmt.*, vol. 129, no. 2, 2003.

Song, Y., and D. K. Chua. Modeling of Functional Construction Requirements for Constructability Analysis. *J. Constr. Eng. Mgmt.*, vol. 132, no. 12, 2006.

Storey, J., C. Emberson, and D. Reade. The Barriers to Customer Responsive Supply Chain Management. *Int. J. Operating Prod. Mgmt.*, vol. 25, no. 3, 2005.

Tam, C. M., T. Tong, and Y. W. Wong. Selection of Concrete Pump Using the Superiority and Inferiority Ranking Method. *J. Constr. Eng. Mgmt.*, vol. 130, no. 6, December 1, 2004.

Tanskanen, K., and J. Holmstrom. Vendor-Managed-Inventory (VMI) in Construction. *Inter. J. Prod. Perform. Mgmt.*, vol. 58, no. 1, 2009.

Thomas, R. Schedule Acceleration, Work Flow, and Labor Productivity. *J. Constr. Eng. Mgmt.*, vol. 126, no. 4, 2000.

U.S. Department of Defense. *Extension to: A Guide to the Project Management Body of Knowledge (PMBOK® Guide), Version 1.0.* Fort Belvoir, VA: Defense Acquisition University Press, 2003.

———. *Handbook Work Breakdown Structure*, MIL-HDBK-881, January 1998.

U.S. Department of Defense-Systems Management College. Systems Engineering Fundamentals. Fort Belvoir, VA: Defense Acquisition University, January 2001.

U.S. Department of Energy. *Project Management Practices: Work Breakdown Structure.* Office of Management, Budget and Evaluation, Rev. E, June 2003.

Winn, M. T.The Benefits of Work Breakdown Structures. *Contract Mgmt.*, May 2007.

Yi, Z., A. Yong, W. and W. Weinong. Modeling and Analyzing of Workflow Authorization Management. *J. Net. Systems Mgmt*, vol. 12, no. 4, 2004.

Zhang, J., D. L. Eastham, and L. E. Bernold. Waste-Based Management in Building Construction. *J. Constr. Eng. and Mgmt.,* ASCE, vol. 131, no. 4, 2005.

Chapter 4

Introduction to Simulation and Its Use in Modeling Production Systems

Simulation is "the mathematical representation of the interaction of real-world objects."

—*Free Online Encyclopedia*
(http://encyclopedia2.thefreedictionary.com/simulation)

Simulation is the process of building digital models of real-world systems and experimenting with them on a computer.

—Paul A. Fishwick

Production systems such as those discussed in this book lend themselves to simulation modeling. If you think of a production system, you normally would visualize the item being produced and the resources (people, equipment, etc.) producing it. The production process can, therefore, be analyzed by tracing or modeling how the resources are engaged during the production process. We can generalize and simplify this to say that we can model and analyze a production process by trying to describe which resources are involved in manufacturing a product, and the activities they follow in the process.

Simulation is a great medium to help us describe and analyze production systems. Within the vast realm of simulation there are many tools, such as 3D visualization, which lets us represent a product in CAD and navigate through it. And four-dimensional modeling lets us visualize how the product is being built in three dimensions, by following each step in assembling the 3D components. And so on.

In this chapter, we will focus on one branch of simulation called *process interaction discrete event simulation*. This branch enables us to model a production system, with its resources, activities and overall processing logic, in order to analyze it.

Worked-Out Problem 4.1: Analyzing Cycle Times

The most simplistic form of analyzing production systems is to model and analyze their cycle times. If we define a production in the form

$$\text{Production} = \frac{\text{Units produced}}{\text{Time}}$$

then we can measure what has been produced and the time it takes to produce it, to calculate production. Normal measures of production in construction include units/hour (e.g. m^3/hr), units per shift (e.g., linear meters in a tunnel per eight-hour shift), units per day (e.g., lane-km of highway per day), and so on.

In many construction production systems we identify cycles, to make it easier to measure the time in the production equation. Halpin[1] (1976) showed that by modeling cycles followed by resources, we can effectively model production systems. A cycle is simply what it takes a given resource or group of resources to complete the production of one unit of the product, as shown in Figure 4.1. In this figure, the excavation process is composed of two activities or tasks, namely: load, back-cycle (travel to dump, dump, and return). Collectively, those four tasks make up the cycle time of a truck.

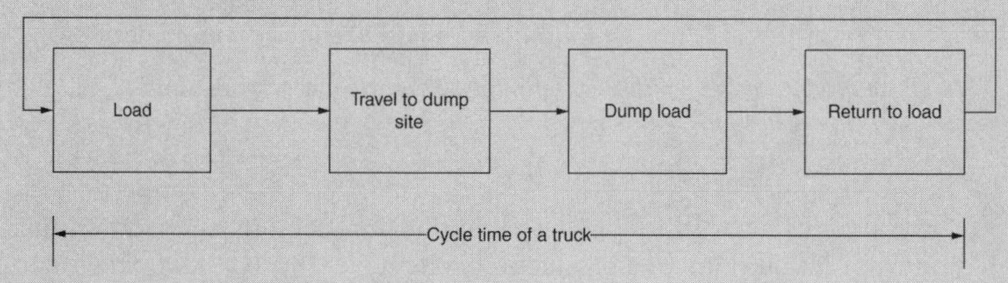

Figure 4.1 Truck Cycle Time

ASSERTIONS AND BACKGROUND

Mr. Jones of Smart Bidder Inc. analyzes overburden removal on a large highway interchange, whereby 63,360 m^3 of dirt is to be moved 12 km to a dump location according to these assertions:

1. Smart Bidder will use one medium backhoe (1.5 m^3 bucket capacity) operating at the workplace and six trucks (size 30 m^3) to haul material to the dumpsite.
2. From past experience Smart Bidder Inc. knows that, when removing soft material from this site, the excavator can produce 200 buckets per hour, or $200 \times 1.5 = 300$ m^3 per hour.
3. The truck cycle is 30 minutes, calculated as follows:
   a. 5 minutes for loading (entering the site, loading, and leaving),
   b. 10 minutes traveling loaded to the dump site.

c. 5 minutes to dump

d. 10 minutes to return, using same calculations as in b.

Calculations

To estimate the production from this operation, we calculate the production of the loader and the trucks:

Loader production: $1 \times 200 \times 1.5\ m^3/hr = 300 m^3/hr$

Truck production: 6 trucks $\times$ 60 min/hr $\div$ 30 min/hr truck cycle $\times$ 30 m^3 per truck = 360 m^3/hr.

It is obvious that with this arrangement the production will be limited to 300 m^3/hr since the backhoe can only produce that much in an hour (or at 8 hr/day = 2400 m^3/day). We can do the equipment balance to find out what the optimum number of trucks should be, given the number of excavators.

Smart Bidder, Inc. will be able to remove the overburden in 63,360/2400 = 26.4 days.

Cycle-Time Analysis

When Mr. Jones analyzes the bid requirements, he determines that the schedule is perhaps too long. He then sets out to determine what can be done to get the work done in 15 days. The number of trucks cannot be increased, as the ceiling on production is 300 m^3/day (determined by the excavator), which is reached when there are five trucks in the process. Table 4.1 shows the production he can get using simple calculations, as before. More than five trucks will not affect production, since the maximum the backhoe can produce is 300 m^3/hr.

Table 4.1 Production of the process

Trucks	Production per Hour	Production per Day	Days Required
1	60	480	132.0
2	120	960	66.0
3	180	1440	44.0
4	240	1920	33.0
5	**300**	**2400**	**26.4**
6	360	2880	22.0
7	420	3360	18.9

(continued)

Table 4.1 (*Continued*)

Trucks	Production per Hour	Production per Day	Days Required
8	480	3840	16.5
9	540	4320	14.7
10	600	4800	13.2
11	660	5280	12.0
12	720	5760	11.0

Discussion

The only way to increase production and reduce the overall time frame is to increase the production output of the backhoe. Smart Bidder can either use more backhoes or a larger-sized one.

For now, assume that 26.4 days were sufficient to estimate this job. When Smart Bidder Inc. was awarded the work and production commenced, a co-op student was assigned to complete a time study on the operation to see how the student's calculations would compare to the company's estimate. In particular, the student was asked to determine the following times:

1. Cycle time of the backhoe (time to load one bucket). The estimate was 200 cycles/hour, or 60/200 minutes per bucket.
2. Back-cycle time of the truck, calculated as the difference between the time it arrives at the loading location and the time it arrives at the next cycle point. Estimate was 30 minutes.
3. The student noticed that the cycle time of the backhoe did not vary much: around 0.4 minutes (24 seconds) per bucket. He stopped recording after 24 cycles. The truck cycles, however, varied from a low of 28.1 minutes to a high of 59.2 minutes. The student collected 32 cycles of those.

All the data is shown in Table 4.2.

Table 4.2 Sample Data for Truck and Backhoe Cycle Time

Cycle	Backhoe	Truck
1	**0.46**	**39.76**
2	**0.44**	**52.43**
3	**0.39**	**53.34**
4	**0.37**	**46.84**
5	**0.40**	**45.35**

6	**0.40**	**51.39**
7	**0.43**	**53.68**
8	**0.46**	**28.11**
9	**0.40**	**59.22**
10	**0.41**	**32.50**
11	**0.40**	**50.51**
12	**0.38**	**43.52**
13	**0.45**	**53.36**
14	**0.40**	**35.61**
15	**0.38**	**30.77**
16	**0.45**	**51.13**
17	**0.41**	**33.60**
18	**0.38**	**49.75**
19	**0.34**	**53.59**
20	**0.43**	**40.15**
21	**0.37**	**39.62**
22	**0.39**	**42.62**
23	**0.44**	**40.81**
24		**52.82**
25		**45.11**
26		**36.15**
27		**38.84**
28		**46.93**
29		**52.33**
30		**34.41**
31		**32.36**
32		**41.12**

(continued)

Table 4.2 (*Continued*)

Cycle	Backhoe	Truck
Minimum	**0.34**	**28.11**
Maximum	**0.46**	**59.22**
Average	**0.41**	**43.99**
Standard Deviation	**0.03**	**8.30**

The estimate for the truck cycle was 30 minutes, while the cycle time was on average 44 minutes. Why do you think the cycle time was that high? Did Mr. Jones underestimate the back-cycle time? Or is the work simply too hard and complex to complete?

In fact, the answer is quite simple. Mr. Jones did not account for the fact that with so many trucks there might be queuing at the loading location. This arises from the simple fact that when there are six trucks waiting for the same loader, it will take them more than five minutes to load. Thus, their back-cycle time will increase by the waiting time. It appears in this case that, on average, each truck waits for 15 minutes in a given cycle. This adds considerable time to the process, and the job in general.

Smart Bidder Inc. asks an expert how to improve his estimating techniques so that they are more accurate.

The hypothetical problem described here is best analyzed as a dynamic system. Not only will there be waiting time to account for—which is not always easy to estimate—but time to load, haul, dump, and so on, need to be accounted for as well. Dynamic systems are also random in nature; they change from cycle to cycle, due to many factors that are not always easy to model.

In such cases, when a system is dynamic and random, we have analytical tools useful for modeling and analysis. This is the science of *computer simulation*—more precisely, process interaction discrete event modeling. Discrete event simulation (DES) involves modeling a system (such as the earthmoving problem described here) as it evolves over time, using variables that change at discrete points in time.

In actual terms, DES can be more complex than we need in this chapter, as it requires programming code to model a system. There are many simulation tools that make it easier to create a DES model using simple modeling elements connected with each other. This is similar to building a CPM network, except that we have greater control and can include more details in the models, which allows us to better study production systems.

[1]Halpin, D. W. (1976). "CYCLONE: Method for Modeling of Job Site Processes" *Journal of the Construction Division*, ASCE, 103(3), 489–499.

4.1 BUILDING SIMULATION MODELS

In this chapter, we will learn how to build a simulation model to analyze production operations. The approach we will use is known as *process interaction simulation modeling*, where a model is composed of basic building blocks called *modeling elements*. Each element describes a specific modeling situation. When elements are connected together, they form a representation of a construction process. Modeling elements are generally connected with arrows that represent the directional flow of virtual entities in the model. In general terms, models include the following basic building blocks:

1. Modeling elements
2. Entities
3. Directional arrows
4. Containers to hold information

Modeling elements vary from one simulation system to the other, but most simulation systems include elements that represent work tasks, queuing of entities, and collections of statistics.

Entities are virtual objects that are essential to modeling dynamic systems, such as those we are interested in. The entity may represent a "customer" requiring service (e.g., a truck that requires loading); or a communication message between various elements to regulate flow in the model (e.g., "All precast materials required to start installation have been delivered"; "Send a signal to the installation submodel that installation can commence"). To build a model, we generally need to describe the life cycle of the entity as it navigates from one modeling element to the next in the model. In general terms, when a model is created, we should be able to describe the general workflow of the real construction process simply by following the journey of the entity within the model.

Directional arrows indicate the direction the entity follows in the model. The entity emanates from one modeling element, and generally flows to another element as per the direction of the arrow connecting the elements.

Containers, as the name implies, are used to hold information pertinent to the model, but where entities generally do not go. An example of this might be a container to hold statistics, which might be required to define which statistics need to be collected by other elements in the model. When observations are collected by other elements, they are simply stored in this container for analysis after the simulation is complete.

The expert hired by Smart Bidder, Inc. decided to model the excavation process using a simulation toolkit called Simphony. First, he will learn how it works, and then he will model the situation at hand and recommend the solution to Smart Bidder, Inc.

The expert starts to build a model of this dynamic system (the earthmoving operation) where trucks load dirt at an excavation location, travel to a dumpsite, dump their loads, and return for more loads. The trucks are loaded by one backhoe at the loading location. To create this model, he discovers that he needs some basic modeling elements to create entities, to represent tasks, and to record production. He quickly builds the system, which is shown in Figure 4.2.

The expert starts by defining the truck as the main entity in the model, since it can be thought of as a customer requiring service. To do this, he uses the NewEntity element shown in Figure 4.2. Then he sets up the elements required to model the life cycle of the truck. First, the entity (truck) should load dirt. The loading activity can be modeled with a Task element. The task has one server specified, since there is only one backhoe. This type of "constrained" task forces the trucks to wait in a queue if another truck is being served by the backhoe at the time they arrive, until the server becomes available. Once the truck finishes loading, it passes on to another Task element, which models the truck back cycle (travel, dump, and return). This task is not constrained by how many trucks are traveling, and, as such, it has an unlimited number of servers (there will be no queuing at the task). Upon finishing the back-cycle task, the truck goes to the production counter to register that one truck load has completed its cycle and returns for another load.

This simple model demonstrates the functionality of key elements, which can be summarized as follows:

- *NewEntity* creates new customers and allows them to flow through the model.
- *Task* models a specific activity by holding the entity for a period of time as specified in its duration property. The time may be constant, random, or a mathematical function, as required. If the task has a limited number of servers, it is associated with its own queue file, where it forces the entities to wait until all servers are occupied with other entities. If it is unconstrained, it releases the entity after the specified duration.
- The *production counter* records that an entity has passed through it, signifying an important milestone in the life cycle of that entity. It is used to record production in the model and so should be strategically positioned in the model to reflect the completion of a unit of production.

The best way to learn simulation is to visualize the dynamics of the model. Unlike a CPM network where we would normally follow the logic of the model by following the activities in the network, with a simulation model we have to follow the journey of the entity as it traverses various elements in the model. For the simple model we created, we should be able to look at the model and see that trucks are created at the NewEntity element; the truck then goes to load. Since the task has a limited number of servers (one backhoe, in this case), the truck checks to see whether the server is available for loading, or busy with other entities. When the backhoe becomes available, the truck starts the loading task. In 10 minutes, the loading is complete and the truck starts on its back cycle. The back cycle is unconstrained, therefore the truck commences the task it is scheduled to complete in

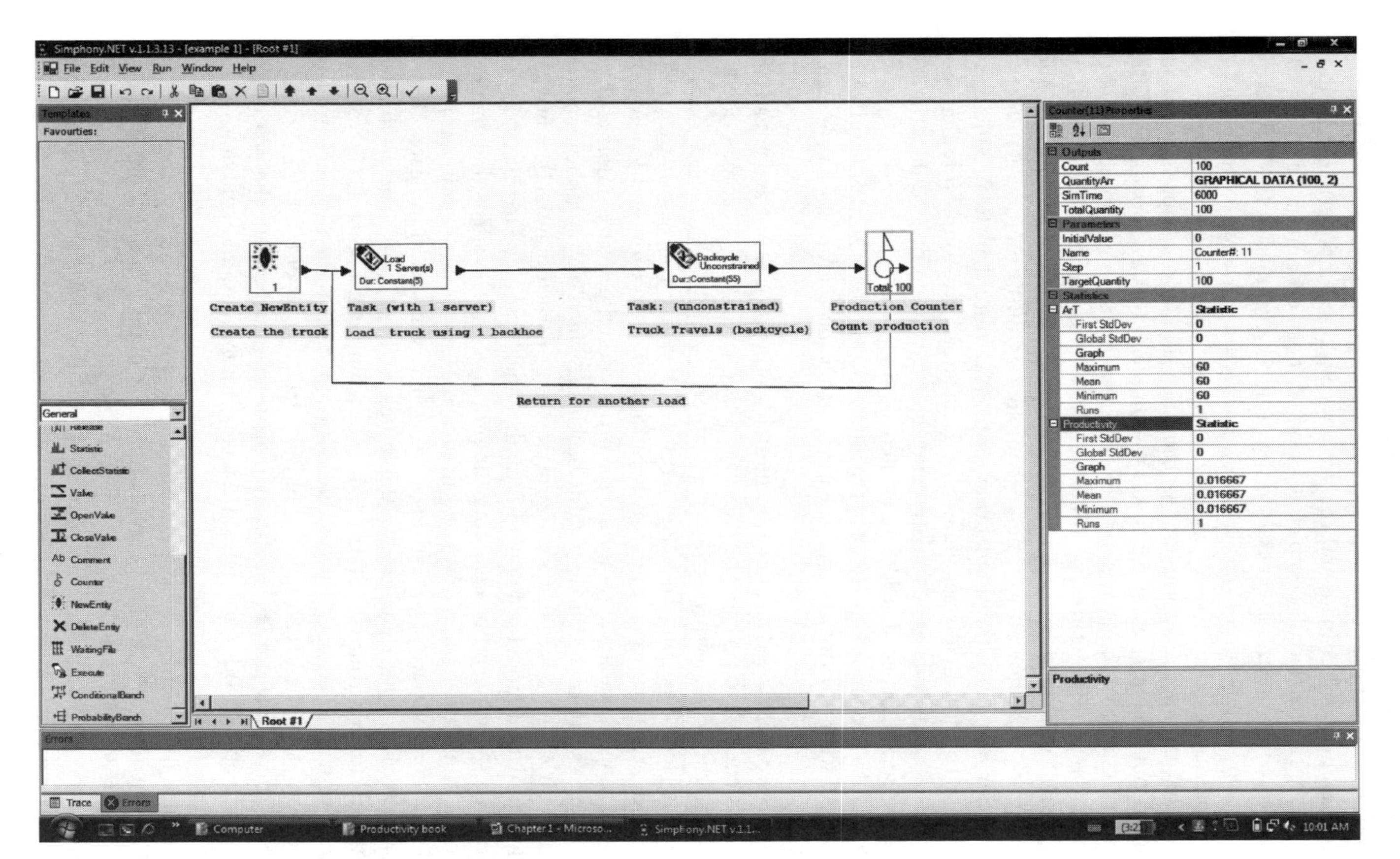

Figure 4.2 Process Interaction Model in Simphony

50 minutes. When it completes this task, it passes through the production counter, which records that one truckload of dirt has been produced. The truck then proceeds to pick up another load, and the process continues until certain conditions are met (e.g., 100 truckloads completed, or one day of work simulated).

The preceding process is generally done using a computer. In this chapter, we used Simphony to develop the model, as seen in Figure 4.2, and to conduct a simulation experiment.

Now that the model is complete, we can run a simulation. Suppose we would like to find out how many hours it takes to complete 100 truckloads. We set up this model to make it simple and intuitive. The loading time is 5 minutes and the back-cycle time is 55 minutes, so the truck completes one truckload per hour. Thus, for 100 truckloads, it should take 100 hours. Notice that the right-hand side of Figure 4.2 shows the time to complete as 6000 minutes (100 hrs × 60 min/hour) as part of the simulation output. Also note that the productivity is shown as 0.0167 truckloads per minute, or 1 truckload per hour. Note that, generally, simulation results are calculated and reported in minutes, unless the user converts those using special modeling elements.

The results shown in Figure 4.2 confirm that the model is accurate and logical to follow. All we have done is added the time it takes the truck to complete each of its tasks.

But do we really need a simulation model for this process, or are spreadsheet calculations enough to go on? The answer to this question will become evident as we return to this scenario throughout the chapter.

For now, let's make our model more realistic. We have 10 trucks in the process (all the same size, for now), plus two backhoes for excavation and loading. Although we can still manage to compute the production time using simple arithmetic, we have to account for queuing at the backhoes, which may complicate matters (in practice, it's a good idea to try this process, to verify the potential benefits).

The simulation model in Figure 4.2 can be quickly adjusted to reflect the new situation. This is achieved by changing the properties of the affected elements in the model; the right-hand side of the model in Figure 4.3 illustrates this. The model and its results are shown in Figure 4.3. The simulation indicates that we can complete the 100 truckloads in 620 minutes, or 10.3 hours. The productivity is 8.3 truckloads per hour (0.138 truckloads per minute in Figure 4.3). Note that although we have 10 trucks, the production is limited by the production of the two backhoes, thus the maximum output is 8.3 truckloads per hour.

Using this simple model, we can investigate many aspects of the process, including estimating production rates, balancing equipment, and studying the impact of various factors on production.

One final improvement before we leave the simple introductory model we created. Construction activities are generally uncertain in their timing. The back-cycle time of the truck will not always be 55 minutes, even along the same route. To account for this, we can model variability in the duration of work tasks using statistical distributions. In our example, we can use an exponential distribution with a

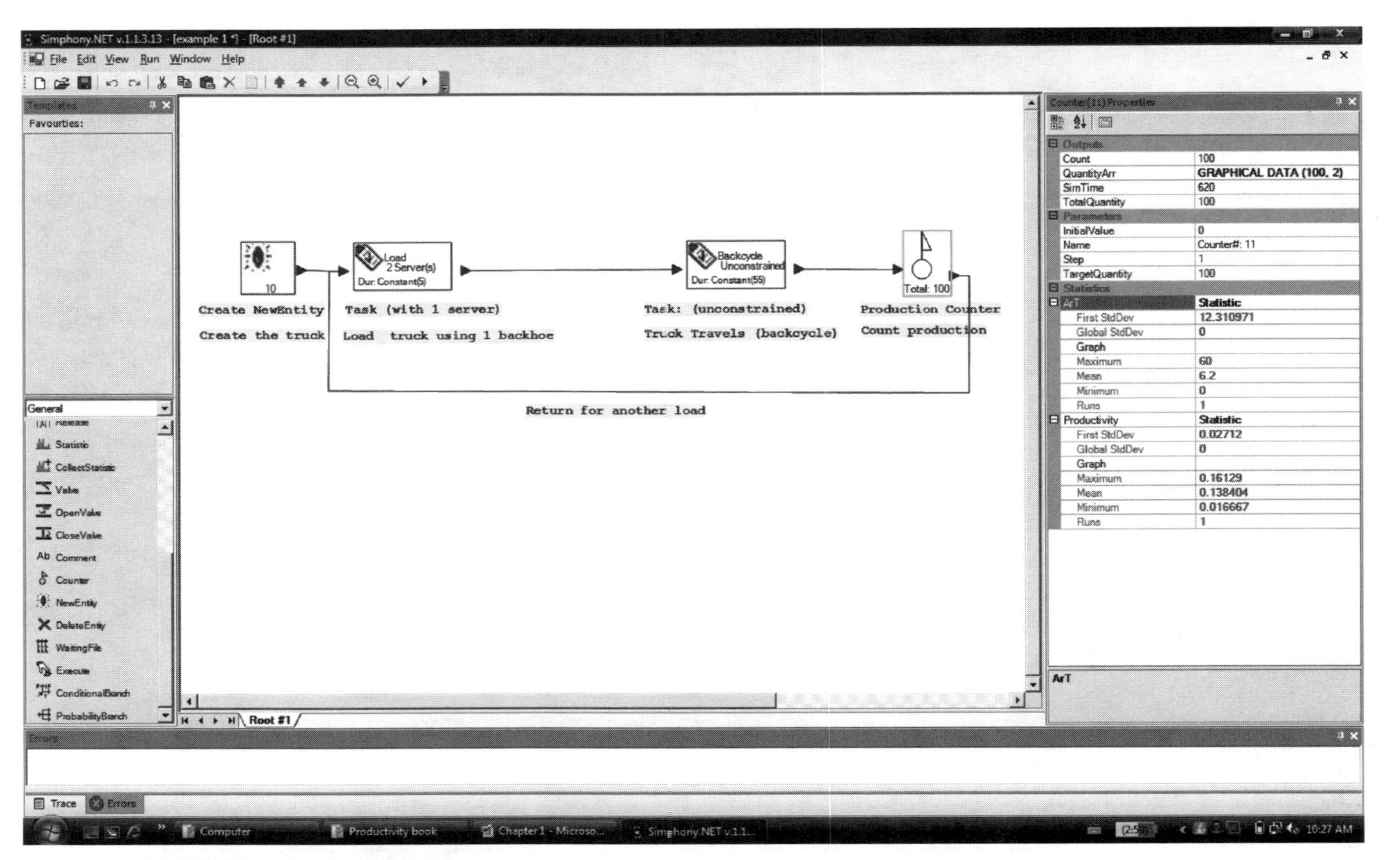

Figure 4.3 Modified Process Model in Simphony

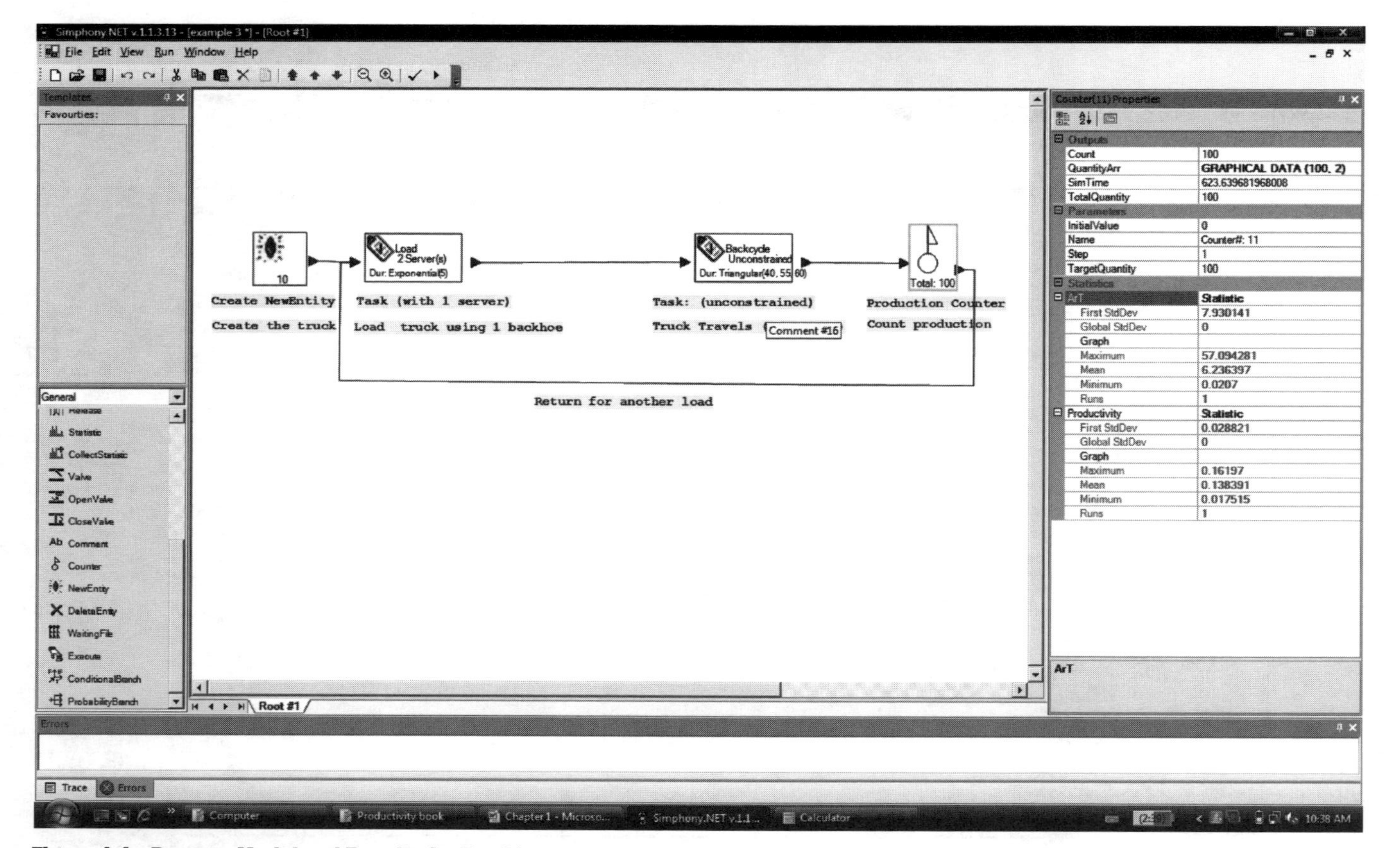

Figure 4.4 Process Model and Results for Trucking

mean of 5 minutes to model the loading task, whereas the back-cycle time may have a triangular distribution, with a minimum of 40 minutes, maximum of 60 minutes, and most likely a value of 55 minutes (similar to a PERT duration estimate). This gives us a more accurate representation of our actual construction process (although still simplified). The revised model and its results are shown in Figure 4.4.

Simulation models are fairly rich in information related to the process. For example, the production counter tracks production throughout the simulated time frame. The chart in Figure 4.5 shows the number of entities over the simulated time frame, reflecting an hourly production chart. The chart also illustrates that the process took roughly 400 minutes (over 6 hours) to reach a steady state of 0.16 trucks per hour. The mean production rate is reported as 0.13, reflecting lower production in the first 6 hours of the simulated process.

The model shown in Figure 4.2 can now be expanded to simulate more realistic situations. Previously we used a server within a task to model identical backhoes, and all the trucks we generated were identical. While this sufficed for the purpose of our introductory model, we may need more details in our representations for more complex situations. Therefore, we'll introduce a few more concepts before we detail all simulation modeling elements in Simphony.

Let us assume we have the following situation:

- The trucks in the model are of two types: three small and four large trucks.
- The backhoe can be engaged in more than one task and, therefore, is not a "slave" resource. For example, the backhoe can break down and be serviced.
- The back-cycle task includes separate tasks for travel, dump, and return.

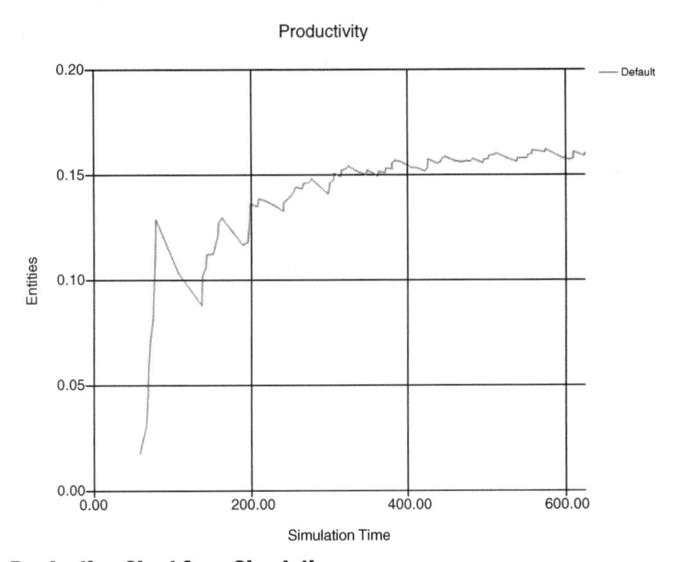

Figure 4.5 Production Chart from Simulation

To differentiate between the two types of entities, we need attributes added to the entity. We can accomplish this by having the entity pass through the SetAttribute element.

When a resource serves more than one task, we need to explicitly define the resource in the model. In Simphony, we can do this using the Resource element.

To divide the back-cycle time into three separate tasks is straightforward. We simply define three tasks and connect them with arrows.

Let's build the model.

First we define the trucks as entities, since they can be thought of as customers being served by the backhoe. Again, tracking the trucks' life cycle will allow us to completely model the earthmoving operation. Trucks complete four work tasks in this example, namely: loading, traveling to the dump site, dumping, and returning for more loads. The backhoe is the only resource required in our example. We need modeling elements to create the entities, specify their attributes, model work tasks, and collect statistics. The elements and the model are discussed here to illustrate.

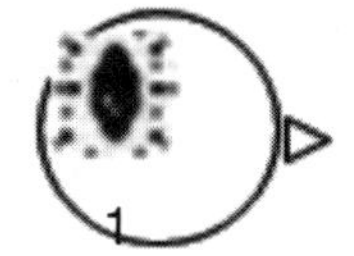

Figure 4.6 NewEntity Element in Simphony

Second, we create the entities. To do this, we use the NewEntity element, shown in Figure 4.6. When a simulation commences, this element generates entities as specified in its Properties box. The NewEntity element has three properties:

1. The number of arrivals that need to be generated (i.e. how many entities arrive into the model through the NewEntity element). We will specify three entities, since we need three trucks.
2. The time between arrivals. This specifies when the arrivals are scheduled to happen in the model. In our example, we want all seven entities to be created at the beginning of the simulation (to arrive at time 0); therefore, the time between arrivals is zero.
3. Time of the arrival of the first entity. Again, in our example, we want this to be time 0. Since we are trying to specify two types of trucks, we can use different NewEntity elements (to simplify the model).

SetAttribute(64) Properties

Parameters

Attr1Name	Capacity
Attr1Val	5
Attr2Name	StartTime
Attr2Val	0
Attr3Name	Hauling Time
Attr3Val	30
Attr4Name	Res
Attr4Val	Loader
NumofAttributes	4

Figure 4.7 SetAttribute Element and Properties

Now that we have created the three entities, the third task is to specify their attributes. The SetAttribute element (see Figure 4.7) allows us to specify various properties to distinguish between entities in the model. For example, when an entity arrives at this element, we can give it an identity attribute to help us track it in the simulation. We can set the attributes by specifying the property name and its value. The value can be changed in the model. In our example, we specify two attributes: ID and Size. For the small trucks' cycle, the size is set as "small," and for the large ones it is set as "large." Next we need to define the resources required to serve the trucks. To do this, we first need to specify a resource container using the Resource element shown in Figure 4.8. This element requires that we define the name of the resource and the number of resources available at the beginning of the simulation. The element is updated during the simulation to reflect the resource's assignment. For example, when the resource is busy loading a truck, the container will have 0 in it. When the resource is idle, it will have a value of 1.

Figure 4.8 Resource Element

For every resource in the model, it is generally good practice to specify its WaitingFile. This is the element where entities queue while waiting for the resource to become available (i.e., when the resource is busy, entities wait in this specific file). While it is not required that each resource have its own waiting file, it is considered good practice, since statistics are reported by file (e.g., queue length, average waiting time etc.). The WaitingFile modeling element is shown in Figure 4.9.

Figure 4.9 WaitingFile Element

Once the trucks are created and their properties are specified, they can load. For this to happen, however, the trucks need to check whether the backhoe is available for loading. To achieve this, we use the CaptureResource element, shown in Figure 4.10.

Figure 4.10 CaptureResource Element

Using the Capture element we specify which resource to capture (the resource's name), and how many of that resource. We also specify where to queue the entity if the resource cannot be captured. In our example, we instruct the entity to capture one backhoe, as shown in the Properties box in Figure 4.11.

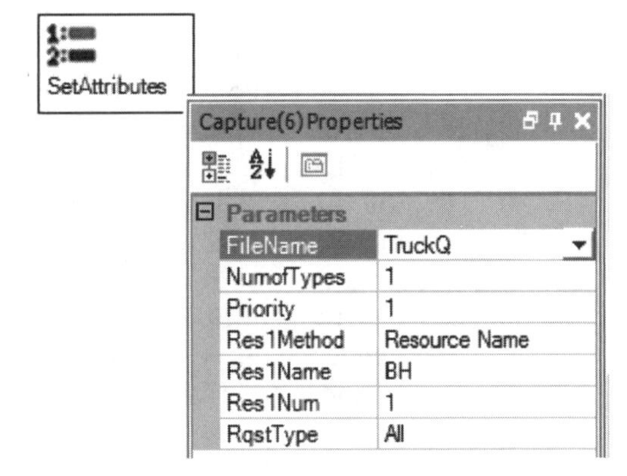

Figure 4.11 Properties Box of the Capture Element

When the capture of the backhoe is successful, the truck can be loaded. The loading is represented with a Task element, as shown in Figure 4.12. The task requires us to specify the amount of time it takes to complete the work in minutes. The duration can be a constant (e.g., every time we load, it is exactly 5 minutes); it can be a distribution (e.g., when a truck will load, a random sample from a distribution specified by the user is obtained); or be user-defined, using Visual Basic code embedded by the user in the element.

Figure 4.12 Task Element

Once the loading is complete, the truck can commence its travel to the dump location. Prior to that, it must release the backhoe

Figure 4.13 Release Element

resource, otherwise no other waiting trucks can be served. This can be achieved in the model by going through the Release resource element, shown in Figure 4.13. In our example, we specify that we will release one backhoe, since we captured only one prior to loading.

The truck can now go through the travel, dump and return tasks. In our model, these are unconstrained, and therefore no resources need to be captured in the process. The task elements for those three activities are straightforward—each specifies the amount of time required for the activities. The only additional element required to complete our model is a production counter, as shown in Figure 4.14 (indicated by the golf flag in the figure). The counter records the entities that pass through it, and therefore can measure production.

The model for the revised process is shown in Figure 4.15.

Notice that the NewEntity and SetAttribute elements for each of the small and large trucks merge at the Capture element. The cycles of the large and small trucks can now merge, because at this point we can refer to the attributes of the entity while processing it within any of the elements. Up until the merge point, we needed a way to specify the Size = S and Size = L properties for the entity we produced. That was achieved in the separate branches.

At the Capture element, we specify which resources are required, as shown in Figure 4.15. We need one resource, BH, and we will specify the request using the name of the resource (see the Properties box on the right-hand side of Figure 4.15). The file where the trucks will queue for the resource is specified as TruckQ in the same figure.

Also note that the backhoe resource container has already been specified in the Resource element called BH.

The release node shown in Figure 4.15 allows the entity (truck) to "release" the resource BH when the truck passes through the node. The remainder of the model is similar to the previous section, except for the loading task, which depends on the size of the truck. For small trucks, the duration has an average of five minutes to load. For the large truck, the average is 10 minutes. In Simphony, we can define the duration of the tasks as a user-defined formula. To achieve this, we write simple code in Visual Basic for Applications. In our example, the loading task has a duration specified as a formula in the Properties box of the loading task. The duration will then be calculated each time according to how we describe it in this model using the embedded code. Assume that we want the duration to be distributed exponentially with a mean duration of 5 minutes for small trucks and 10 minutes for large trucks. We achieve this by writing user code as shown here:

```
Select case Ob.CurrentEntity("Size")
        case "S"
               Dim X As Double
                     X = 0
                     X = SimEnvironment.Sampler.Expntl(5)
```

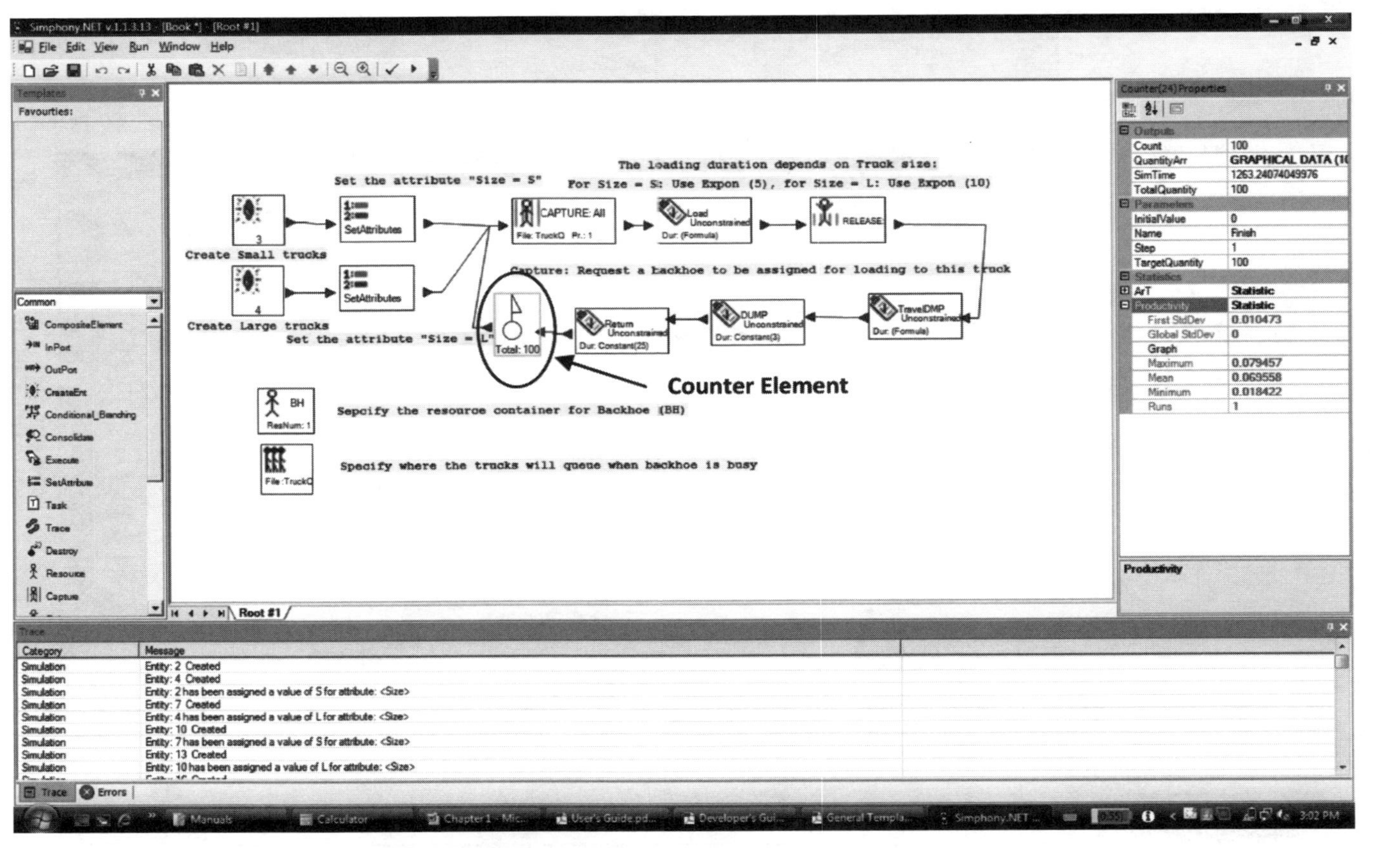

Figure 4.14 Sample Model for Trucking Scenario with Counter Element

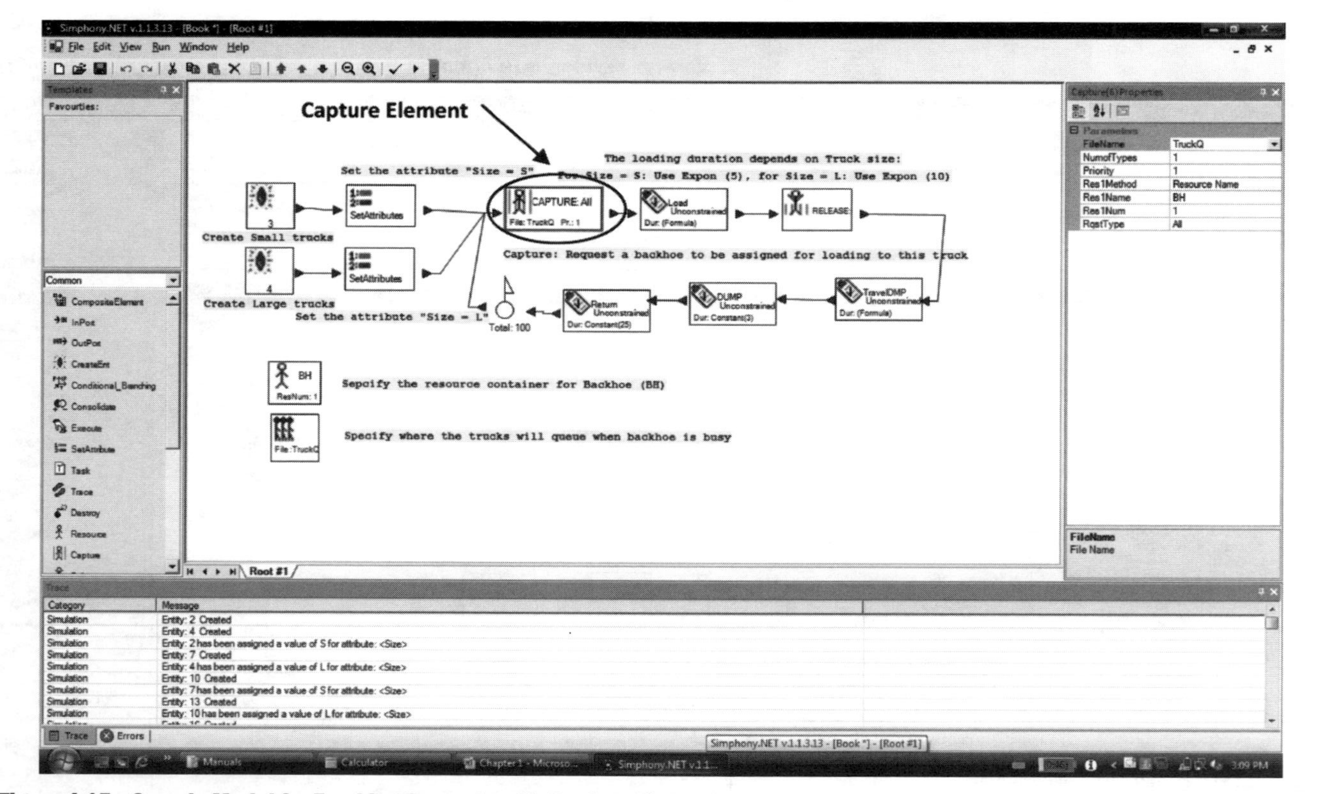

Figure 4.15 Sample Model for Trucking Scenario with Capture Element

```
        case "L"
                DimX As Double
                        X = 0
                        X = SimEnvironment.Sampler.Expntl(10)
        End select
ReturnX
```

First, we find the property of the entity that holds information about the size of the truck. The SetAttribute element previously specified that the size is kept in the Size property. In Simphony, we can retrieve this information by querying the incoming entity using the object: ob.CurrentEntity ("Size"). Second, we use a simple VB construct like: "If then Else or Select Case" to properly route the truck to its distribution sampling portion. Once there, we use the Sampler function to randomly sample an exponential distribution, as shown in the code.

CHAPTER REVIEW

Journaling Question

1. The fabrication of pipe spools involves several quality control (QC) tests. Spool material and pipes arrive to the fabrication shop at randomly distributed interarrival times. Fitting and welding (F/W) start after the arrival of the spool material and pipes. Assume one crew is used for performing both the fitting and welding activities. The duration for F/W is a function of the diameter-inches (DI) of the spool. After the spool welding is complete, one of the common tests is hydrotesting (HT). QC crews perform this test. If a spool fails the test, the spool is returned to F/W crews to fix it. Otherwise, the spool is shipped out of the fabrication shop.
   - Spool material and pipe interarrival time is uniformly distributed, at a minimum of 1 hour and a maximum of 8 hours.
   - Diameter-inches of the arriving spools is exponentially distributed, with a mean of 30 DI.
   - F/W duration in hours $= 0.5 + 0.2 \times \text{DI}$.
   - HT duration is distributed triangularly—1, 2, and 4 hours.
   - There is a 0.15 chance that a spool may fail the HT and return to F/W. In such a case, the time to fix it is 0.2 of its original F/W time. After fixing the spool, HT is performed on it again.
2. Develop a model using the General Purpose Modeling Template in Simphony to simulate this system. Assume 2 F/W crews and 1 HT crew; use the given values for the different input parameters, and report on the output of running the model for 2000 hours.

Traditional Homework

1. The City of Edmonton Public Works Department is planning to construct a tunnel that is 2.9 meters in diameter (finished diameter with precast concrete segment liners) and 2.1 kms long, for water storage in North Edmonton. A pump house will be constructed at 1.25 km from the main shaft. The City has requested your services to advise them to preplan the project. The preliminary project planning work was conducted by the program manager of the City of Edmonton, using Simphony. He determined it is not efficient to excavate the tunnel from the main shaft to the removal shaft, due to loss of productivity at the latter stage of the project.

   The program manager has suggested that you visit an ongoing tunnel project in Mill Creek in Edmonton. At the tunneling site, you speak to the tunnel supervisor. Some of your questions and the tunnel supervisor's comments are given below:

   What are your work hours?

   7:00 A.M. to 3:30 P.M.

   Do you stop work for lunch?

   Of course.

   Can you tell me the approximate time required to finish 1 meter of tunnel?

   That depends on various factors. We did about 3 to 5 meters per day last week. But when we started this project, we only managed to install 1 linear meter per day, due to the sandstone. The mole couldn't dig by itself. Now we are doing all right.

   How many meters do you think you could achieve if you don't have any problems?

   That also depends. On past projects we sometimes finished about 8 to 10 meters a day.

   Next, you decide to observe the tunnel work for a full day. The Mill Creek tunnel is 2.3 meters in diameter, finished with precast liners. The following are some of the observations you make during your site visit.

   - The construction project uses two trains. Each train has five muck cars. The capacity of one muck car is 0.70 m^3. Approximately 12 full train car loads were required for 1 meter of tunnel.
   - The undercut of this project is very small due to the site restrictions.
   - The tunnel achieved 4 meters of tunnel production (the total is 130 meters now) during the day you observed.
   - Average times for activities and train travel time were noted:

Train Travel Time to the Tunnel	
40–49 seconds	2 times
50–59 seconds	3 times
60–69 seconds	4 times
70–79 seconds	2 times
80–89 seconds	1 time

Train Travel Time from the Tunnel Face	
40–49 seconds	1 time
50–59 seconds	4 times
60–69 seconds	3 times
70–79 seconds	2 times
80–89 seconds	2 times

Liners

After the boring was complete, it took some time for the tunnel boring machine (TBM) to retract its position. The observed times were 1 minute 20 seconds, 2 minutes 3 seconds, 2 minutes 27 seconds, 3 minutes 02 seconds. Then the liners were installed, and the observed times were 16 minutes 22 seconds, 18 minutes 03 seconds, 16 minutes 50 seconds, 17 minutes 12 seconds. When the train arrived with the liners, it took about 3 minutes to unload them.

Dirt Removal

The average unloading time per train was between 18 and 20 minutes. The cleaning time per car was 30 seconds. The dirt cars were lifted about 20 meters above ground level and then moved to the disposal truck, which was placed about 20 meters from the shaft; then the car was lowered to the truck for dumping. The dumping time was approximately 45 seconds. The depth of the shaft was 35 meters. Hoisting from the undercut to ground level was very fast (typically, about 40 seconds with dirt and 30 seconds without dirt, with a variation of 5 seconds).

The proposed tunnel project will be bored in sandstone, except for the last 300 meters, which will be bored in clay. The soil investigation report indicates that bedrock is found at the following distances. The main shaft is 40 meters below ground level.

Bedrock Locations

Between 275 and 310 meters

Between 810 and 830 meters

Between 1150 and 1165 meters

Between 1600 and 1640 meters

a. What is the expected loss of productivity after the tunnel reaches 1400 m?

b. At what distance do you notice a drop in productivity, and why?

c. Propose different options for excavating the tunnel project. Test your proposed options using the TBM tunnel model. State your assumptions clearly for each option.

Open-Ended Question

1. In the city where you are located, there are many construction projects taking place. Visit one of those projects, interview the superintendent, and select a process to help him or her analyze and improve productivity. Ask the superintendent to define what he or she would like you to analyze. Examples include the earthmoving operation discussed previously, a tunnel operation, a pavement operation, a multistory building, an MSE retaining wall, and so on.

   a. Develop a simulation model of the process.

   b. Develop a systematic way to analyze the production process and establish a benchmark for where it is today.

   c. Develop recommendations for how to improve production from the benchmark you established. Simulate those recommendations and then share your recommendations with the superintendent.

Chapter 5

A Case Study: Applying Simulation to Tunnel Construction

In this chapter we will study how to create a simulation model that can be used in a real-life setting to assist in making decisions. Our focus will be on estimating the rate of production for a tunneling project in linear meters per shift, as illustrated in Figure 5.1. This will enable us to then estimate project schedule and cost.

We will demonstrate how we can derive an estimate of time for the project based on the production rate of various interacting cycles. We will also demonstrate how we can analyze the execution plan and address areas of productivity improvement.

5.1 PROJECT BACKGROUND

The project involves the construction of a water transmission main linking a water treatment plant (E.L. Smith Water Treatment Plant in Edmonton, Alberta Canada) on the north bank of a river (the North Saskatchewan River) to the southwest communities, as illustrated in Figure 5.2.

The scope of our project is limited to tunneling under the river and the installation of 1500-mm diameter steel pipe through the tunnel.

We are assuming the role of a project management service provider (CM Consultant), which is assisting the contractor with the preconstruction planning of the project, to ensure its successful completion. The scope of our activity includes production planning and constructability reviews. We will use simulation modeling to analyze and plan the production on this project.

As a consultant to the contractor, first we need to familiarize ourselves with the design of the facility. The contractor has negotiated this work with the owner on the following basis:

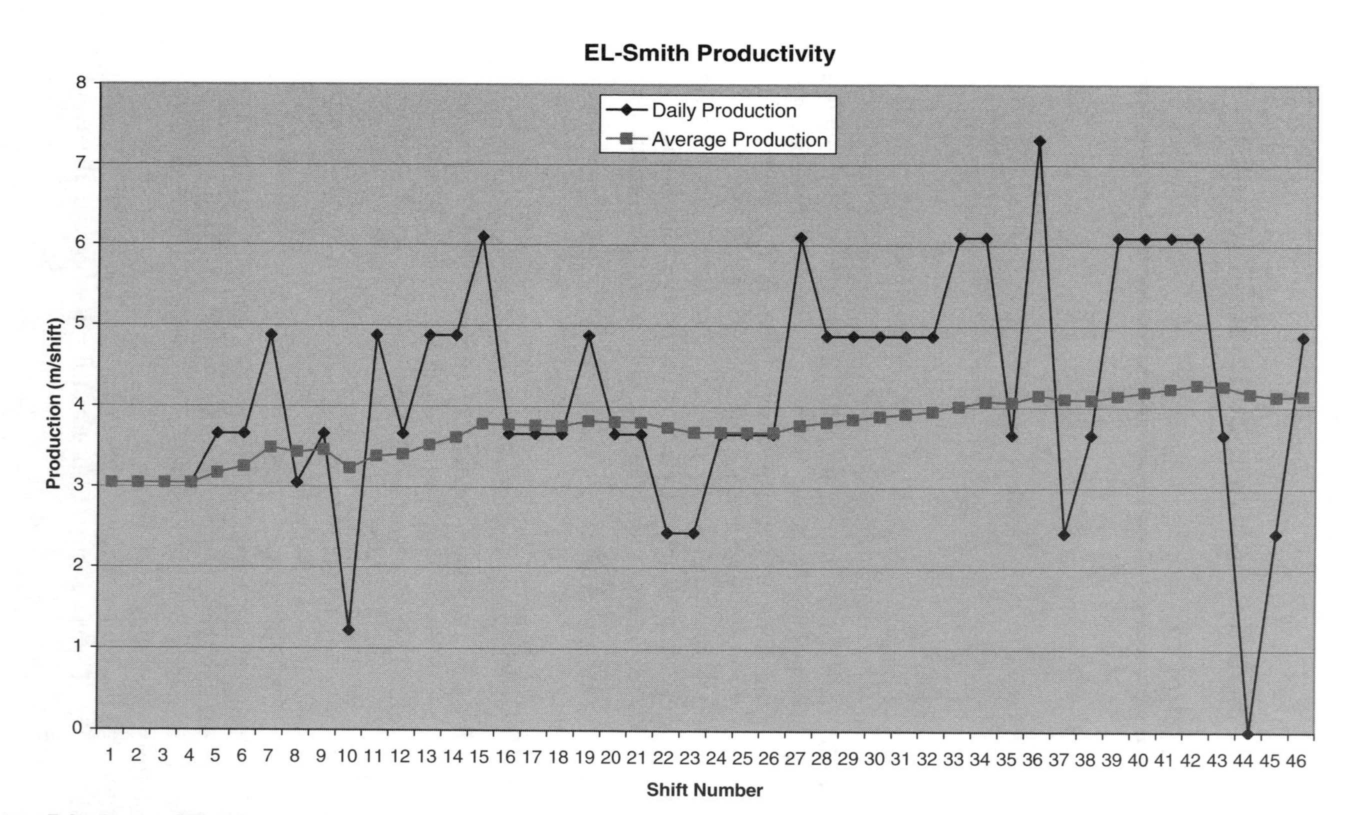

Figure 5.1 Production Rate for a Tunneling Process

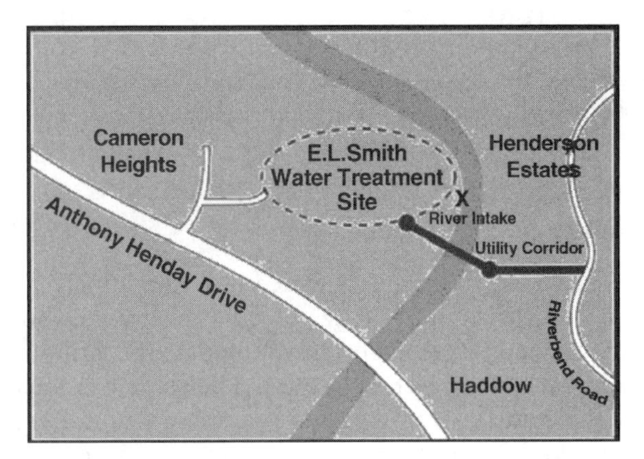

Figure 5.2 Project Location
***Source*: EPCOR Bulletin, February 3, 2006.**

1. The design is not 100 percent complete—the reason being that the design requires a contractor on board to finalize the methods of construction that will affect the final design. Therefore, the contractor is working with incomplete design information and will have input into the design.
2. The project will be completed by September 200X but no later than March 200Y. The contractor will confirm the date to the owner once the analysis is complete and all parties are satisfied that the plan is reasonable.
3. The project budget will be fixed to an upper limit once all design is complete, the construction analysis and plan are firmed up, and our analysis is complete. All parties will then work toward that fixed upper limit, with controls in place to ensure that the owners get value for the money they put into the project.
4. A geotechnical investigation was completed as part of the preliminary design study. The geotechnical report is not included in here for brevity.
5. A preliminary drawing showing the project is given in Appendix 5.1. The area maps show the available construction space, the potential site layout locations, and other pertinent information, such as potential access routes to the various components of the project.

5.2 PREPARATION WORK: UNDERSTANDING THE CONSTRUCTION PROCESS

For this case study we need to understand the methods used for constructing the shafts, tunneling, installing the steel pipe, grouting, and testing prior to modeling the production of this operation.

The following provides a brief overview of each method. Note that this is not comprehensive coverage of the various topics, as many different methods exist for accomplishing those operations. Rather, this is a concise description to facilitate your understanding of the case study and your learning of the various simulation functions we are introducing. We recommend that you independently research each to develop a complete understanding of the process, however.

5.2.1 Shaft Construction

The sinking of a shaft (excavating it to the depth of the tunnel) is usually the first operation in tunnel construction. The size of the shaft depends on its function. If the shaft will be permanent, its size must accommodate the facility it is intended to house (e.g. a pump station, a wet well, etc.). The shaft size must also be large enough to allow the contractor to set up the machine used for excavation (e.g. a TBM or a small backhoe), and to facilitate the entry of all material needed (i.e. liner segments, finishing pipe sections, etc.) and the movement of material and people during the excavation process.

In this case study we will sink our shaft using machine excavation. Circular shafts with large diameters or rectangular shafts can be excavated using backhoes and dozers, with cranes used for muck hoisting. Smaller shafts in soft ground are normally excavated by augers (drilling) and bucket excavators. For this case study, we opted to create a structural support for the shaft using soldier piles (by drilling and casting piles in situ) as demonstrated in Figure 5.3.

The piles can reach a depth of 18 m after which we simply use hand excavation and rib and lagging to complete the shaft to the desired depth. The general experience is that we can normally complete two piles per day. We need 30 piles of approximately 1 meter diameter each to complete the shaft circumference. The shaft itself is 30 meters deep, and therefore the remaining 12 meters will be excavated using mechanical equipment (small backhoe or digger), jackhammers, and manpower; the support is rib and lagging, as shown in Figure 5.4. The rate is approximately 1.2 meter per 8-hour shift.

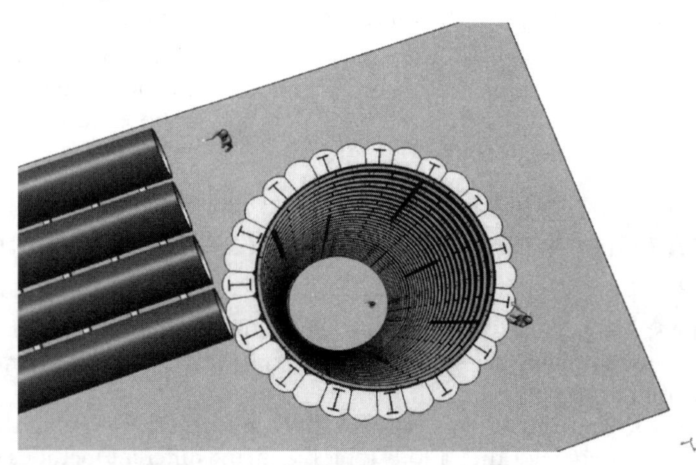

Figure 5.3 Structural Support for the Shaft Using Soldier Piles

Figure 5.4 Ribs and Lagging Shaft Construction
***Source*: Courtesy City of Edmonton.**

Once a shaft is excavated, an enlargement at the base is constructed by sequential excavation (hand tunneling and support with ribs and lagging). The enlargement is called an *undercut*.

5.2.2 Tunnel Construction

Utility Tunnel Construction

The tunneling process described here is based on methods used by the City of Edmonton (COE[1]), which are typical of most utility tunneling methods. These are not universal, however.

The layout for a typical tunneling operation is shown in Figure 5.5. The process is generally composed of:

- Excavation at the face of the tunnel by hand or tunnel-boring machine. (Our case study uses a tunnel boring machine similar to the one shown in Figure 5.6.)
- The tunnel face is supported by tunnel liner, as shown in Figure 5.7.
- Transportation of the excavated muck by cart or conveyor belt,

[1]Most of this information is provided by Mr. Siri Fernando and Mr. K.C. Er of the City of Edmonton. We are grateful for the contributions these two individuals have made throughout the years toward the authors' understanding of tunneling operations, and the information they have provided along the way.

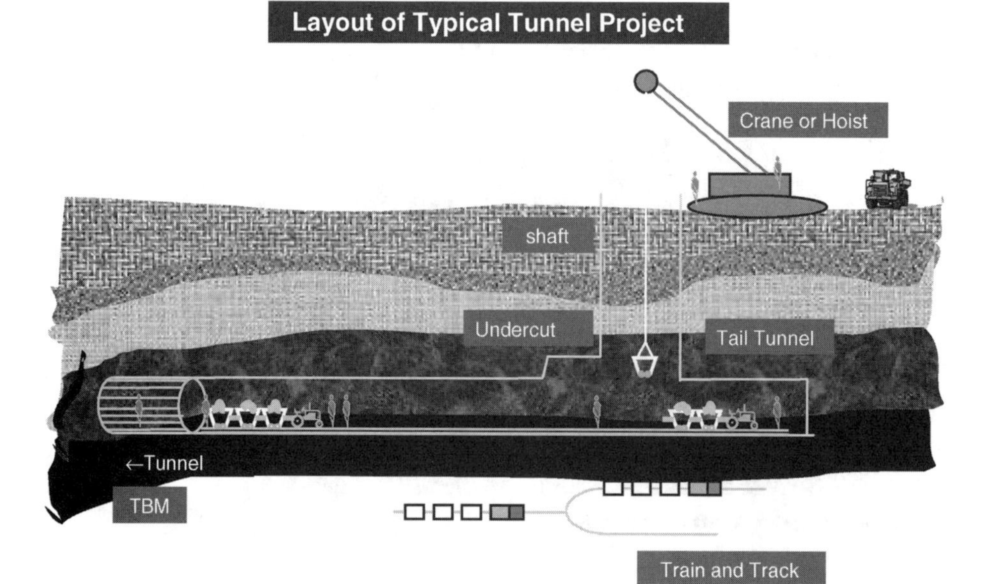

Figure 5.5 Tunnel Overview
(Source: J. Ruwanpura, S. AbouRizk, S. Fernando, and K. C. Er. "Experiences in Implementing Special Purpose Simulation Tool for Utility Tunnel Construction Operations." Proceedings of the 16th Annual Conference of the Tunneling Association of Canada, Montreal, Quebec, September 2000.)

- Vertical lifting of the muck cars through a working shaft, using a crane or a hoist, and using the same process to deliver tunnel liners to the face of the tunnel to support the excavation.

An effective way to study the production of a tunneling operation is to analyze the interacting cycles of various resources. In general, two cycles can be readily identified:

Figure 5.6 Close-faced tunnel boring machine (TBM)

Figure 5.7 Tunnel Liner Installation

1. The excavation (boring machine) and lining cycle at the face of the tunnel
2. The material-handling cycle (from the face to the shaft and up to the ground level)

The first cycle is composed of activities that take place at the tunnel face, including excavation, resetting the TBM, and lining (shown in Figures 5.8 and 5.9). They are repetitive and use the TBM as the main resource, with a small crew to operate the TBM, manage the material handling at the face, and install the liners.

The second cycle is composed of activities involved in the handling of the excavated material and the segments between the tunnel face and the ground level. Those include:

- Train traveling from the face to the shaft
- Hoisting of the carts to the ground level to empty the carts

Figure 5.8 Excavation by TBM

Figure 5.9 Precast Concrete Tunnel Liner Installation

- Unloading the dirt at the surface
- Loading the carts with segments or track extensions
- Lowering the carts to the shaft, as shown in Figure 5.10
- Traveling back to the face of the tunnel.

To have a productive operation, any superintendent will tell you that they would like to keep the leading resources (the TBM) as busy as possible. In other words, the TBM should not wait for other resources, and its cycle should be the leading one. In our situation, material handling should be optimized to synchronize its cycle with that of the TBM.

The typical tunneling activities are summarized in the bar chart given in Figure 5.10.

Pipe Installation

Once tunneling is complete and the tunnel is ready, a steel pipe will be installed to carry water under pressure. The annular space between the steel pipe and the tunnel will be grouted so that, in the long term, the space will not collapse on the pipe and cause problems.

Installing the required 1500-mm steel pipe through the finished tunnel is challenging work both in terms of welding and material handling. Although we considered two sizes of steel pipe sections, we will only discuss the shorter sections of 20 feet (the 40-foot section can be done as a separate assignment).

The section size primarily affects the material handling, the size of the shaft, and the amount of welding required.

A conceptual model of the installation of the pipe is shown in the series of Figures 5.11 to 5.13.

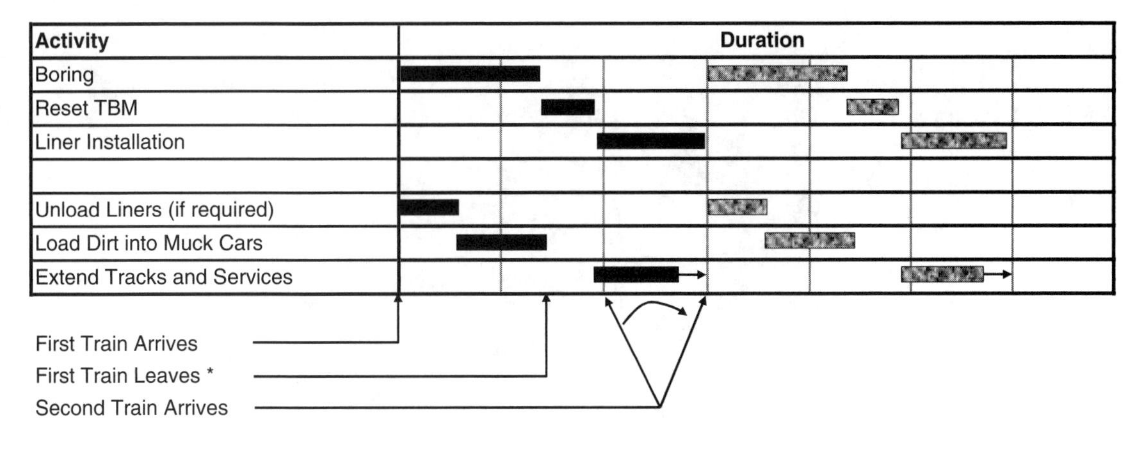

* The leaving of the train depends on the capacity of muck cars.

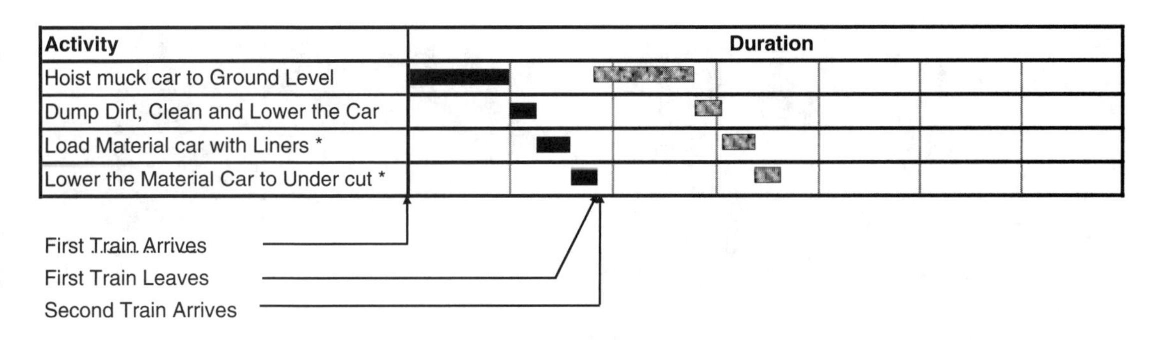

* These activities can be performed before muck cars are unloaded, if time permits.

Figure 5.10 Summary of Tunneling Activities

1. Steel pipe in 20-foot section arrives at the site. It will be lowered into the shaft.
2. The pipe is lowered using a mobile crane and workers. It must be angled and set in place prior to welding
3. Once inside the tunnel, it must be positioned so that it can be welded to the previous section and then pushed into the tunnel so that another pipe section can be installed.
4. After the final piece of pipe has been installed, the space between it and the tunnel liner must be grouted.

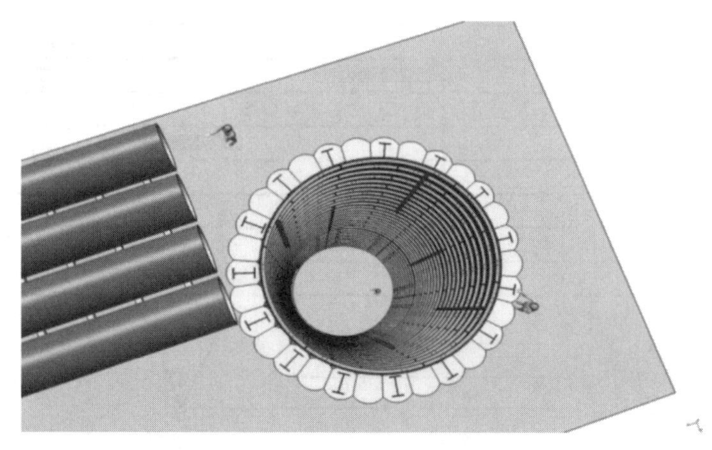

Figure 5.11 Lowering Pipe Section into the Tunnel

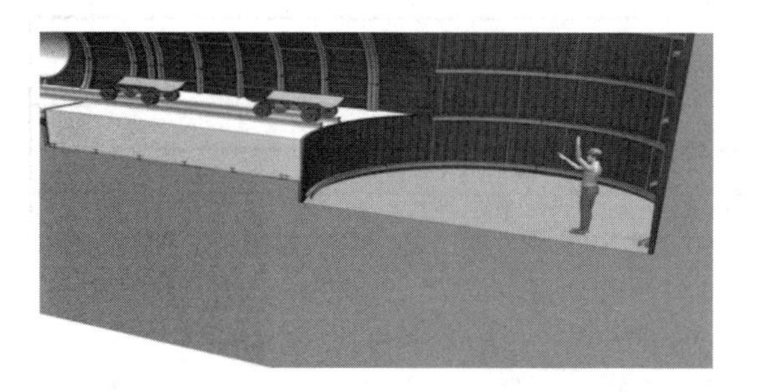

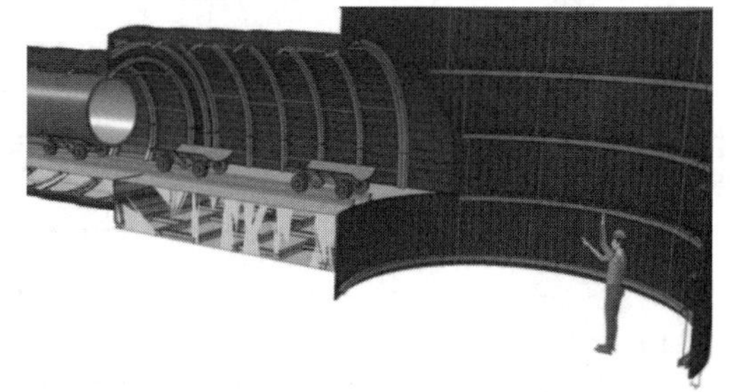

Figure 5.12 Positioning Pipe Sections into the Tunnel

Once the final piece of pipe is installed, the space between it and the tunnel liner must be grouted

Figure 5.13 Grouting Space between Liners and Pipes

5. The process of welding the pipe is critical for safety purposes and from a productivity standpoint, due to the constrained environment. After careful investigation of constructability it was determined that the most efficient and safe method would be to carry the pipes into the shaft area, set them on a platform, weld the sections in the shaft area, and push the pipes into the tunnel after the sections are welded, using a pipe carrier such as the one shown in the preceding figures. The advantage of this approach is that the pipe can be welded from the outside, which is the normal way. The logistics are also easier for setting up and prepping the pipe, welding, and inspecting the weld quality. The productivity will likely be in the neighborhood of a half pipe section per day. Productivity upto one pipe section per day is possible, if the welding process is automated.
6. After the welding and inspection/testing are complete, the annular space between pipe and tunnel must be grouted.

When we analyze constructability we can think of using 40-foot sections of pipe to reduce the amount of welding that takes place, as this is labor intensive, costly, and time-consuming. The problem is, in order to use a 40-foot section, we need to have a larger shaft built to facilitate movement of the pipe section to the tunnel.

Complete a cost-benefit analysis of this setup. Modify the models in this section to facilitate this analysis, compare and contrast the results.

5.3 DEVELOPING THE SIMULATION MODEL

Developing a simulation model for this process follows these steps:

1. Understand and document the process to be modeled.
2. Develop the objectives of the simulation and the scope of it.
3. Collect all relevant data and state all required assumptions. This includes all activities, resources, and the durations required for various tasks. State any assumptions made to simplify the model.
4. Create the model in Simphony and complete a trial run.
5. Verify that the model is producing the required results by tracing its steps and ensuring that the logic is maintained through a few cycles of the tunneling process.
6. Complete a simulation run once the model has been verified and then validate the simulation by comparing it to previous results and discussing it with the superintendent.
7. Develop runs to achieve the simulation objectives and then share them with the decision maker.

5.3.1 Assumptions and Input

We collected information and made various assumptions based on interviews with the superintendent of tunneling for the contractor. Table 5.1 lists the tasks required to build the project, along with their durations.

Table 5.1 Tunneling Tasks and Their Durations

Activity	Duration (in minutes, unless otherwise noted)
Excavate Working Shaft	
Drill piles (2 piles)	480
Excavate the sets	3360
Install walers	4800
Excavate sections	960
Install the ribs	1440
Install lagging	1920
Install walers	960
Break out of trench and pour mud slab	
Break out shaft	960
Pour mud slab	2400
Install mole—M126 rib and lagging	
Lower mole	480
Install and set up	2400
Cleaning interval	1200
Clean TBM (uniform)	120, 240
Excavate by TBM 455 m of rib and lagging soil, type 1	
Travel track 1 or 2	1
Travel track 1 or 2	Function of the speed
Case "train full"	See train load time formula
Case "length done"	See layer length formula
Case "section done"	See section time formula
Install tracks	15
Survey	120, 180
Unload muck car (triangular)	3, 5, 7
Install liners in train	1

Load train with liners (triangular)	3, 5, 7
Dismantle and remove mole	7 days
Clean tunnel and remove track	20 days
Level tunnel and install new track	10 days
Place 1500 mm pipe in tunnel (20-foot steel pipe)	
Move pipe	10
Attach rigging mechanism	15
Hook up crane	10
Hoist hook down and up	17
Weld	120
Conduct ultrasound test	45
Repair	30
Pull pipe	15
Coat exterior	90
Weld and install pipe 30 m (working shaft)	
Lower pipe	20
Weld	300
Conduct X-ray test (15)	15
Conduct X-ray test (60)	60
Re-weld	60
Weld and install pipe 60 m (exit shaft)	
Lower pipe	20
Weld	300
Perform X-ray test (15)	15
Perform X-ray test (60)	60
Re-weld	60
Conduct hydro test	4200

Other helpful calculations used in the model are the following:

- **Time to complete tunnel section**:

$$\text{Section Time (min)} = \frac{\text{Total Length(m)} - \text{Finished Length(m)}}{\text{Rate}\left(\frac{\text{m}}{\text{min}}\right)}$$

- **Train loading time:** this is the time it would take to load all the muck cars of the current train.

$$\text{Train Load Time (min)} = \frac{(\text{Muck Capacity} \times \text{NumCarsDirt}) - \text{CurrentLoad}(\text{m}^3)}{\text{SwellF} * \pi \times 0.25 \times \text{Tunnel Diameter}^2(\text{m}^2)\text{Rate}\left(\frac{\text{m}}{\text{min}}\right)}$$

- **Time to complete one push of the TBM:** This is the time that it would take to finish one section (1 m) of the tunnel, which is equal to the time to perform the working length attribute of the TBM (one meter).

$$\text{Push Time} = \frac{\text{Working Length(m)}}{\text{Rate}\frac{(\text{m})}{(\text{min})}}$$

The resources required to complete the project are given in Table 5.2.

5.3.2 Simulation Model

There are two simulation models developed for this process. Both are included in the CD/website. The first model uses the tunneling template of Simphony which

Table 5.2 Resources Used in Tunneling Project

Resources	Quantity
Piling Crew	1
Excavation Section	1
Crane	1
TBM with crew	1
Track	1
Welding crew	1
Movers with crew	1
Coating crew	1
Crane for pipe handling	1

was used for actual decision making during the course of this project (described as this case study). It is included for the interested reader who decides to delve deeper into Simphony and its varying templates. The tunneling template is a special purpose modeling template that is not described in this book but can be learnt from the Simphony documentation.

The second model was built using the same building blocks described in Chapter 4 (the general purpose modeling template in Simphony). This model is simplified (but comprehensive and realistic) for easier comprehension by a beginner in simulation. It was created by a graduate student at the University of Alberta Mr. Eduardo Sosa to mimic the original model. This model is summarized in this Section.

In order to be able to describe the model in this Chapter we broke it down into components (reader should note that the model supplied with the software will be one large model rather than the pieces shown in here).

Part 1: The Working Shaft and TBM Setup.

The first part of the model shows the working shaft and TBM setup model component.

Recall that the shaft is approximately 30 m deep. It will be excavated using two stages. The first stage is the shallow excavation where ground will be supported by soldier piles up to the depth of the drill we are using (18 m). The process is to drill a 1 m diameter pile, install reinforcement and cast it in place. We need 30 piles to complete the required circumference of the shaft. Two piles can be completed in a day using one piling crew of one rig and support labor.

The second stage of the shaft will be from depth 18 m to a depth of 30 m. This stage will be excavated without the drill (ground is hard and the reach is too far). The ground will be supported using rib and lagging and walers to cross brace as needed. The excavation is tedious and production will be slow.

First we model the project starting by creating an entity to identify the project and get the model started. We set the attributes of the project to the entity being created (see the label "1" on top of the element in the Figure above). The first thing we do is generate pile entities from the 'project entity' using the Generate/Consolidate node labeled as node 2. We generated 15 piles from this since our production is assumed to be two piles per day. We then capture a piling crew (including the drill rig) and proceed to complete the piling activity. When a piling task is completed, we release the piling crew and consolidate one set of two piles as shown. Once the piling is done we will commence with the excavation of the shaft. The shaft is broken down into four benches each of which is excavated using the drill rig and a small backhoe/small Brock machine. The figure shows the modeling strategy, which is very similar to the pile drilling process. Here we have only four entities to work with, since we have defined four sections. The durations should reflect the production rate for each of the sections.

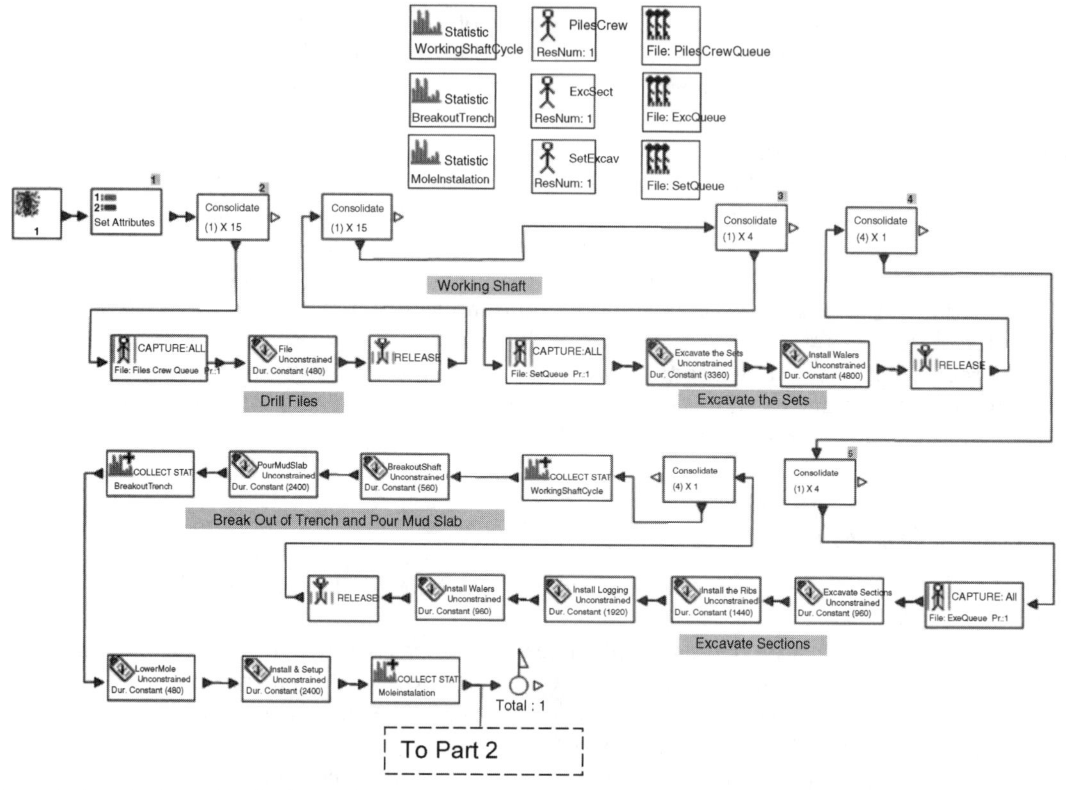

Figure 5.14 The Working Shaft and TBM Setup

The strategy for the second excavation stage is similar to the previous two. The distance remaining from the end of the first stage (18 meters) until the bottom of the shaft (12 meters) was also divided into four sections of 3 meters each (node 5).

The last set of activities represents the breakout from the shaft to the tunnel. This process was abstracted to one set of tasks, rather than in detailed cycles, for brevity. Once the breakout is complete, the model proceeds to the tunnel excavation (part 2).

Part 2: Tunnel Excavation

After the construction of the working shaft, we proceed to the excavation of the tunnel. Figure 5.15 shows all the resources we created for the simulation. It also shows two processes to model breakdown of the TBM, which normally requires a cleaning and maintenance process to follow. The model assumes this will take place every 2.5 days, but this can be a random process. A breakdown process can be modeled by virtue of having a breakdown event (random or scheduled) take place, and requests the resource that is to be taken out of service (TBM, in this

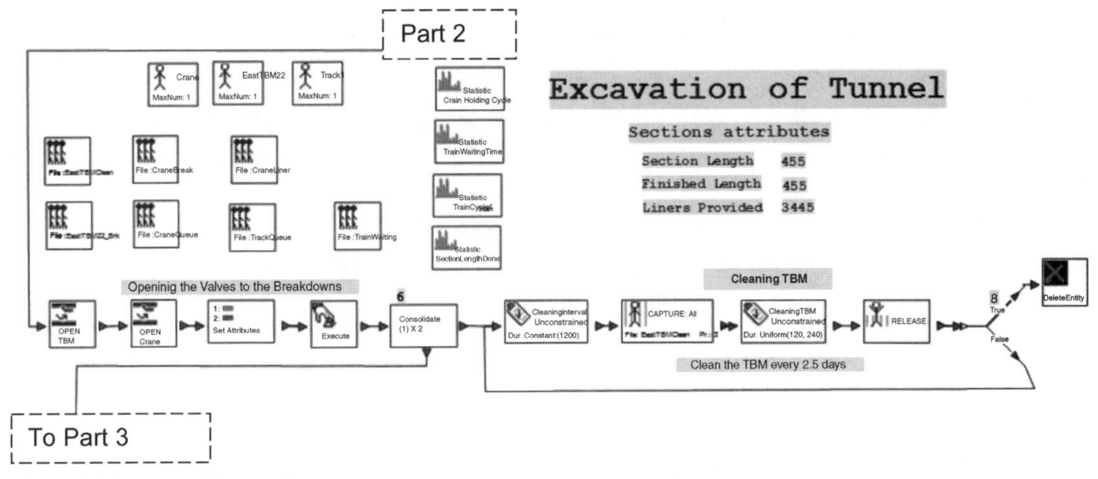

Figure 5.15 Tunnel Excavation

case). Once the resource is captured by this breakdown entity it will not be released until maintenance and repair is complete, at which point the resource becomes available again for other processes.

In our case study, we will use two trains to complete the material handling, the normal process for such tunnels. This approach is generally known to keep the TBM utilized and the production higher, since material handling can constrain the production for such a linear process. We start by creating two train entities, as detailed in part 3 (node 6 in part 2).

Part 3: Train Model

A train travels the first part of the rail, which includes a switch to facilitate movement of two trains on one track. At the switch, we use a conditional branch to check whether the tunnel is complete. If the tunnel length was accumulated through the previous iteration, we can stop the simulation; otherwise, we proceed to part 4, where we proceed with calculating a series of variables important for the simulation.

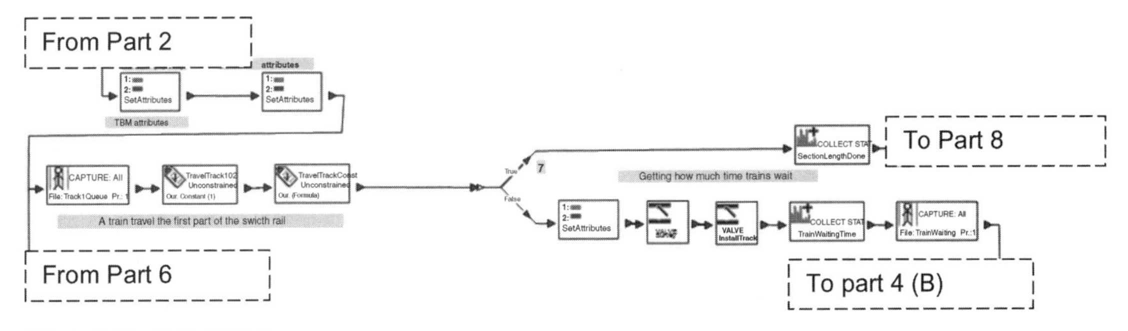

Figure 5.16 Train Model

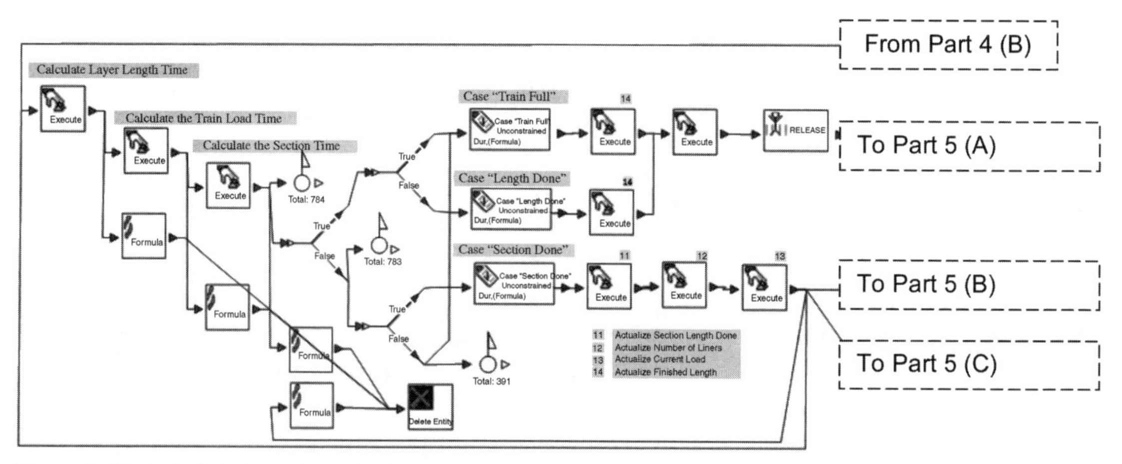

Figure 5.17 Calculate Length and Trainload

Part 4: Calculate the Length and Trainload

The variables are all calculated in Execute statements, as shown in Figure 5.17, and are self-explanatory. (Note: Check the contents of each of these modeling elements in the actual model, supplied on the CD/website.) Once variables have been calculated and updated, the train entity goes through decision loops to determine what to do next. For example, if the simulation is not complete, the entity takes the uppermost task (Train Full). When a train is loaded (or when the length of the tunnel is completed) a train is sent back to the shaft to unload, as shown in part 5(A) of Figure 5.18.

Part 5: Survey Process and Track Extension

The first part of this model deals with the completion of a section of tunnel and the collection of statistics relevant to it. The second and third sections determine whether surveying is required, and when the track extension is required. See Figure 5.18. In both cases, this is a simple matter of keeping count of the sections installed.

Part 6: Material Handling

After a train is full, we proceed to unload each muck car (each train has three muck cars). Then we load liners. See Figure 5.19.

Part 7: Liner Supply

If there are no liners remaining, we proceed to supply liners to the same train (Figure 5.20).

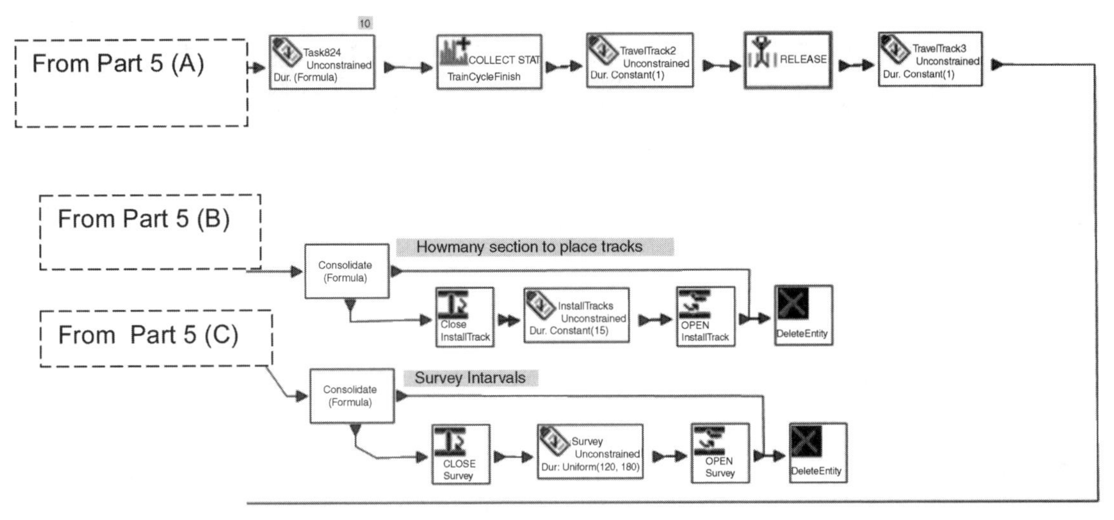

Figure 5.18 Survey Process and Track Extension

To model the breakdown, we use a separate model, as shown in Figure 5.21. In this model, we randomly sample a point in time when we create a breakdown event. When this event takes place, it captures the required resource, holds it for repair, and releases it when it is done. In this sense, the resource is not available to do productive work.

Part 8: Remove Shaft

After performing the excavation of the tunnel, we proceed with shaft removal, which takes three days to complete (Figure 5.22). The TBM removal and cleaning takes place afterward. Since those activities can be done in parallel to tunneling (for

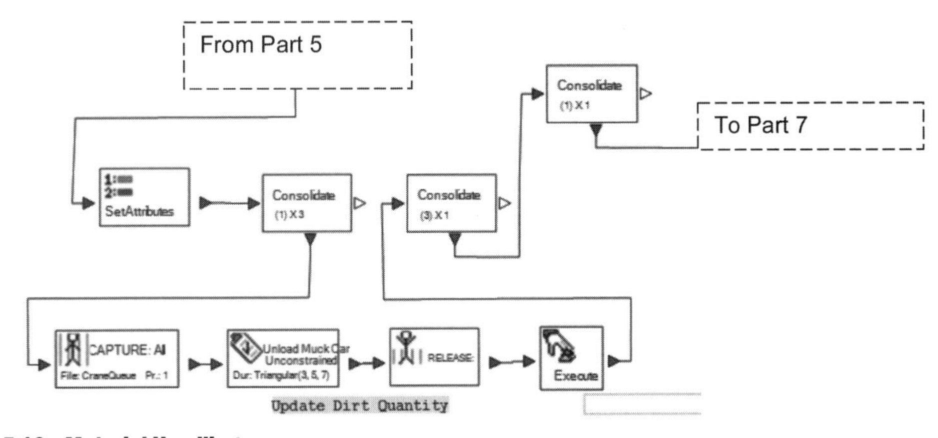

Figure 5.19 Material Handling

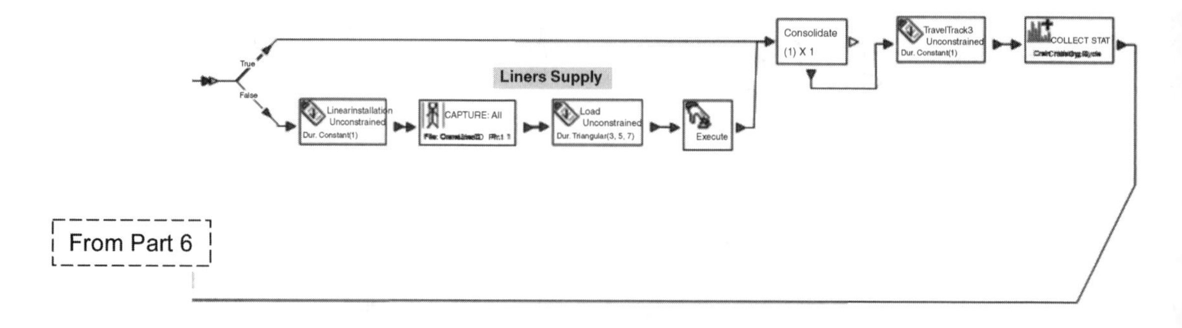

To Part 3

Figure 5.20 Liner Supply

the most part), they are not modeled in detail. We therefore assume that the TBM dismantling and removal takes 7 days, cleaning the tunnel and removing the track takes 20 days, and leveling the tunnel and installing new track for pipe installation takes 10 days.

Part 9: Pipe Placement After the TBM has being removed, we proceed with the installation of 20-foot pipe sections inside the tunnel. For this task we generate 76 entities (455 meters of the tunnel divided by 6 meters of each pipe). Using these entities, we capture each resource required for the respective activity and release them upon completion of the required tasks.

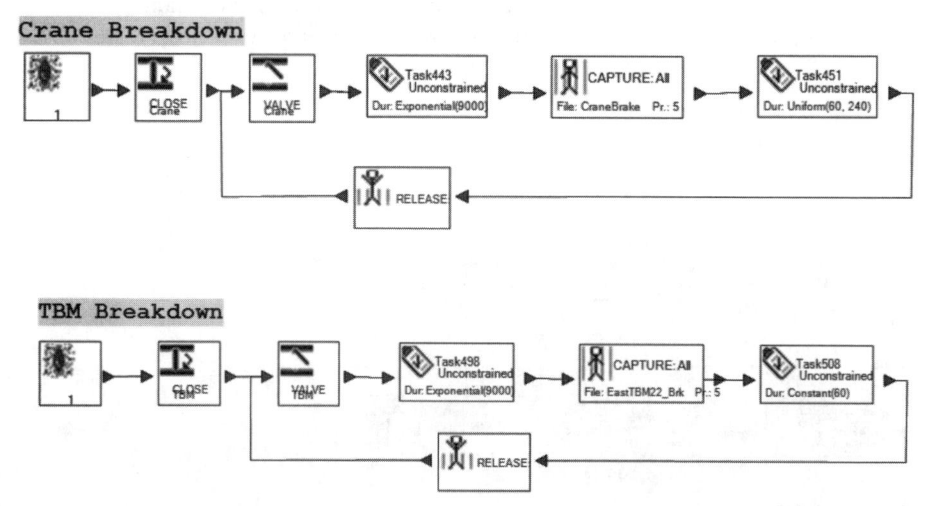

Figure 5.21 Liner Supply (continued)

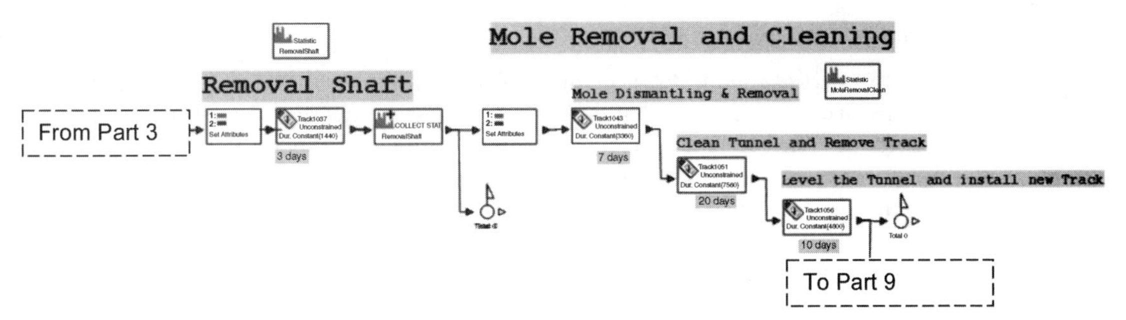

Figure 5.22 Shaft Removal

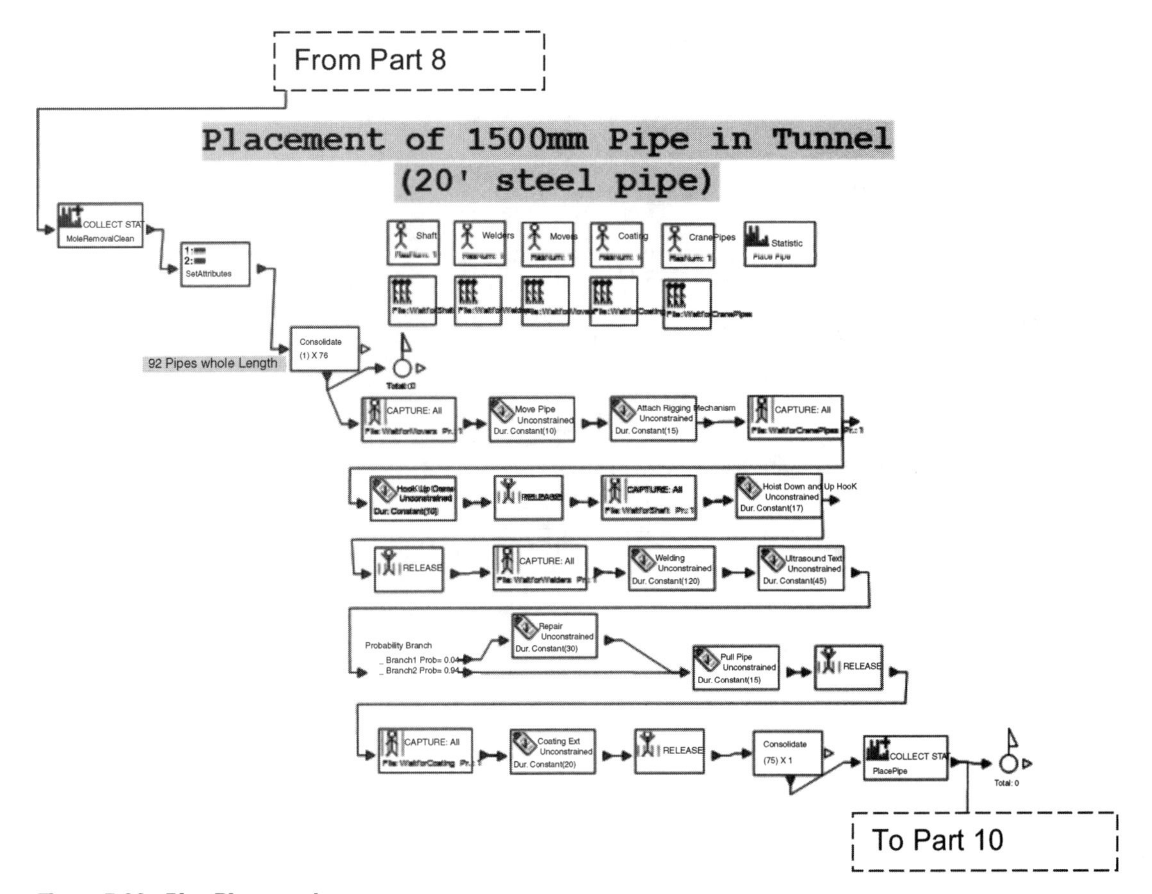

Figure 5.23 Pipe Placement

Part 10: Pipe Placement in the Shafts

In this part of the model we generate five entities corresponding to the number of times we need to lower and weld the pipes in the shaft (working shaft/pipe length). When we are done lowering and welding the pipes, we continue with the x-ray test, whose duration will vary depending on how many times the test is performed in a specific entity. The probability of an entity failing the test is equal to 1 percent.

The previous task was performed for pipes both in the working shaft and in the exit shaft (Figure 5.24).

The simulation ends when an entity reaches the final counter, whose target quantity is equal to 1.

5.4 RUNNING THE MODEL AND DERIVING RESULTS

The model just described was executed using Simphony 1.1.13. The results collected included those shown in Table 5.3.

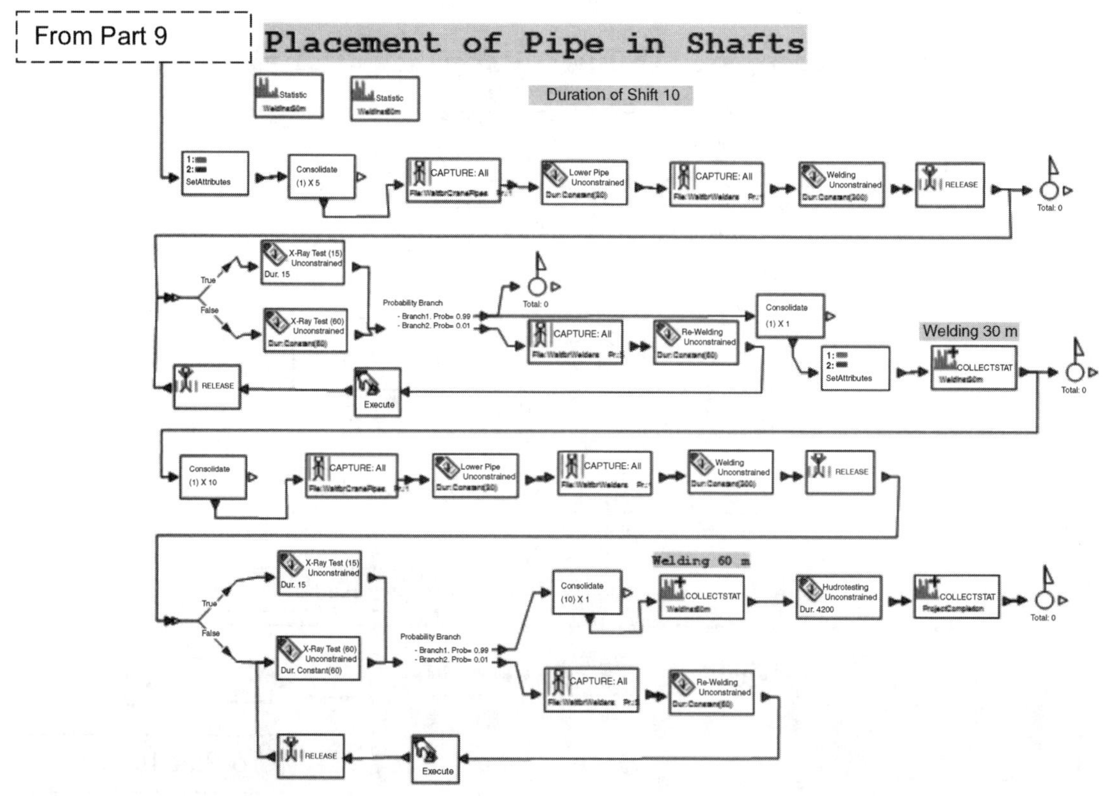

Figure 5.24 Pipe Placement in the Shafts

Table 5.3 Results from Running the Model

Activity	Durations (Days)
Excavate working shaft	127.00
Break out of trench and pour mud slab	7.00
Install mole—126-m rib and lagging	6.00
Excavate by TBM 455 m	51.97
Remove shaft	3.00
Remove and clean mole	37.00
Place 1500 mm pipe in tunnel (20-ft steel pipe)	31.64
Weld and install pipe 30 m (working shaft)	2.77
Weld and install pipe 60 m (exit shaft)	5.43
Total project time	283

The processes most relevant to the completion of the project, according to our superintendent, are the tunnel excavation and the steel pipe installation. He requested the information given in Table 5.4 and confirmed that it is within his expectations for this project.

Keep in mind that these results correspond to one specific run of the simulation. For decision-making purposes we need to complete multiple runs and derive our conclusions from the output distributions. Although the deterministic approach is not completely accurate for decision making, it is illustrative of the entire process. All that needs to be done to take it to the next step is add random process models to replace the deterministic ones.

5.5 ANALYZING THE OPERATION

Now that we have a model and have predicted the overall time frame and production rates of the operation, we can make use of our model to improve production and reduce delivery time. We can produce a wealth of information that is useful in assessing productivity. The first, basic information in this regard is the production

Table 5.4 Duration of Major Processes

	Meters	Time (days)	Rate (meters per day)
Tunnel excavation	455	51.97	8.76
Steel pipe installation	455	31.64	14.38

Most utility tunnels run over a few kilometers in length. In such a case, we would expect a significant drop in production as we pass a certain threshold.

Assuming this tunnel is 1.5 kilometers in length, design a process to keep production at around 8 meters/shift as the tunnel moves away from the working shaft.

Note: Normally a switch is introduced by enlarging the tunnel at the threshold point. It will take four weeks to complete a switch.

rate of the tunnel, measured at the end of each day. Figure 5.25 shows the production of the process predicted for each day from the simulation model.

Note the following:

- Production varies from day to day, because: (1) we included TBM cleaning every 2.5 days or so; and (2), we need to extend the track every so often.
- Production drops as we get closer to the end of the tunnel, because the time it will take to get to the TBM will increase, forcing the TBM to wait for the train, which waits at the switch located at the working shaft.

Now that we have the basic information, we can proceed with the productivity analysis on this project. In this section, we will consider a few productivity improvement issues and use our model to simulate them.

1. What opportunities are there to enhance the productivity of the tunnel excavation?

The key resource for excavation is the TBM. We seek to make its utilization as high as possible. When analyzing this, we note this is a function of the TBM's penetration rate and its interaction with the material-handling cycle. The penetration rate is usually a function of the soil conditions and the TBM itself, and for our case study

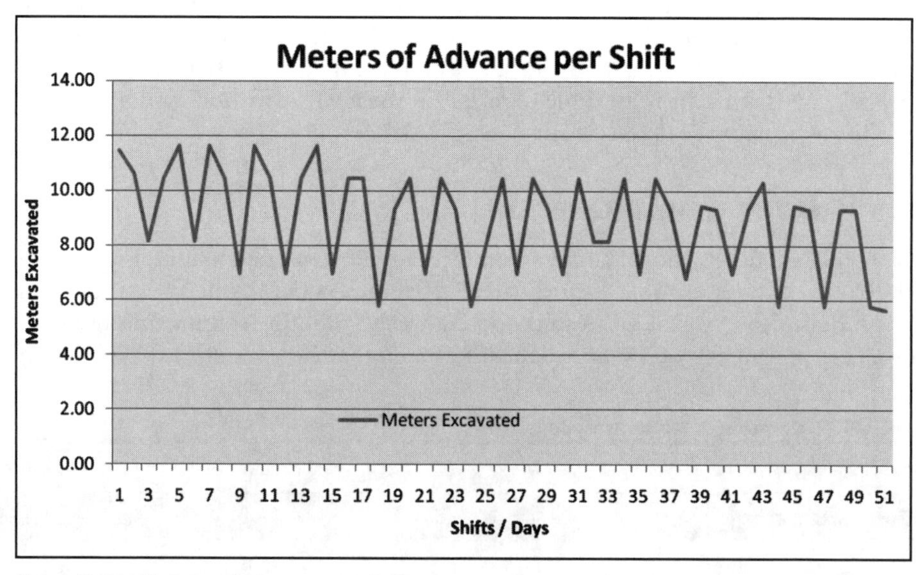

Figure 5.25 Meters of Advance per Shift

Table 5.5 Doubling the Train Speed

	Meters	Time (Days)	Rate (Meters per Day)
Tunnel excavation (10 m/min)	455	47.83	9.51
Tunnel excavation (5 m/min)	455	51.97	8.76

we will assume they are fixed. Let us then look at the material handling. Will it make a difference if we double the speed of the train? To do so, we change the variable in the model related to train speed (in the travel task) from 5 meters/minute to 10 meters/minute. The results, given in Table 5.5, show that we can shorten the duration of the tunneling activity by three days. This, however, may come at the expense of the safety of the crew and so must be valued in this regard first.

> Can you list ways by which soil conditions can be manipulated (e.g., by ground treatment)? If we use a different TBM with a higher penetration rate, we may improve production. Perform a cost/benefit analysis of such a situation by evaluating different TBMs, their costs, and the impact on overall project cost and schedule.
>
> Examine the simulation model and all statistics. Can you identify areas of improvement based on this information (other than the one noted here)?

2. The installation of the pipe takes 31 days to complete. Would it be possible to improve the installation process?

We can first investigate the major constraints to the pipe installation by examining the utilization of various resources. This can be easily inferred from the model, where we note the observations shown in Table 5.6.

These observations show that the pipe movers experience the longest waiting time. Now we can proceed to investigate what might be done to reduce this waiting time.

Two of the obvious courses of action in such a process are (1) to add resources, or (2) redistribute flow to ensure that delays are minimized. In reality, when we investigate this phenomenon, we note that—given the linear nature of the pipe installation—there is not much we can do to shorten the time the movers have to wait for the pipe to be ready. There are no opportunities for increasing the rate at which pipe is supplied to the tunnel with the crane; the welding and prepping activities are linear, and cannot be doubled due to space constraints. One obvious

Table 5.6 Resource Waiting Time

Resource	Original Number of Resources (crews)	Maximum Waiting Time (minutes)
Welders	1	260.00
Movers	1	14,563.00
Coating extension	1	0.00
Crane for pipes	1	2880.00

Can you design a different pipe-handling system to make the process flow smoother and to increase productivity? Model your new process and compare your results with the base-case scenario here.

course, therefore, is to try to use more experienced welders and different shifts for the crews involved. The crew used for moving the pipe can, for example, be working on other activities on the project.

CHAPTER REVIEW

Journaling Questions

Once a model is built it is possible to perform many experiments on the process without having to set foot on-site. Using the model described in this chapter, answer the following questions:

1. The superintendent tells you that he does not want any breakdowns in equipment. If we were to use equipment in top shape, and maintain it properly, we could eliminate as much as possible all breakdowns in equipment. Change the model parameters, rerun the simulation, and discuss the impact eliminating breakdowns has on production.
2. The superintendent asks you to investigate what would happen if he could decrease the duration of welding by 25 percent by using experienced welders or offering monetary incentives. Quantify the impact to production from such an improvement and discuss the potential of applying such a strategy, including all advantages and disadvantages.
3. The welding foreman notes that your assumption of 1 percent weld rejections is too low. What happens to your production if the rejection rate is 5 percent? Discuss how you can maintain a low rejection rate for welding.
4. The tunneling foreman indicates that aside from the material handling and the process design, the tunneling process depends on the TBM penetration rate, which greatly depends on the soil conditions encountered.
   a. What penetration rates were assumed in the model, and for what ground conditions (from the geotechnical report given).
   b. If we could manipulate ground conditions to increase the assumed rates by 20 percent, what increase in production could we expected?
   c. What can be done to the ground to enhance penetration rate?

Traditional Homework

1. Develop a modeling strategy that does not use any of the consolidate/generate modeling elements in the model described in this case study. Rerun the model and confirm that you are obtaining the same results.
2. What are the limitations of the breakdown process model described/used in this case study? How can it be made more realistic?

S:\Cadd\Geo\EG09200\EG09255\EG09255-003.dwg – Layout1 – Feb. 21, 2006 5:23pm – tim.guenther

Appendix 5.1 *(continued)*

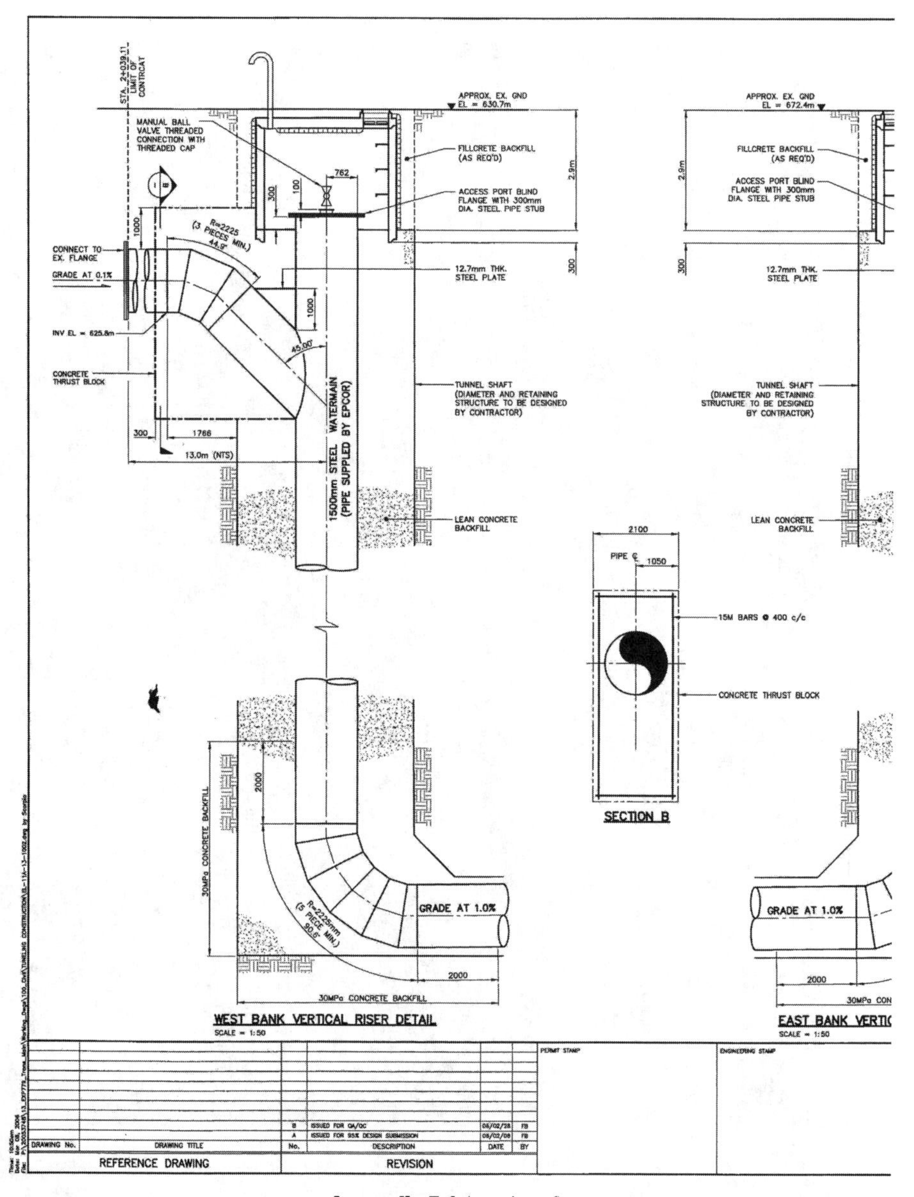

Appendix 5.1 *(continued)*

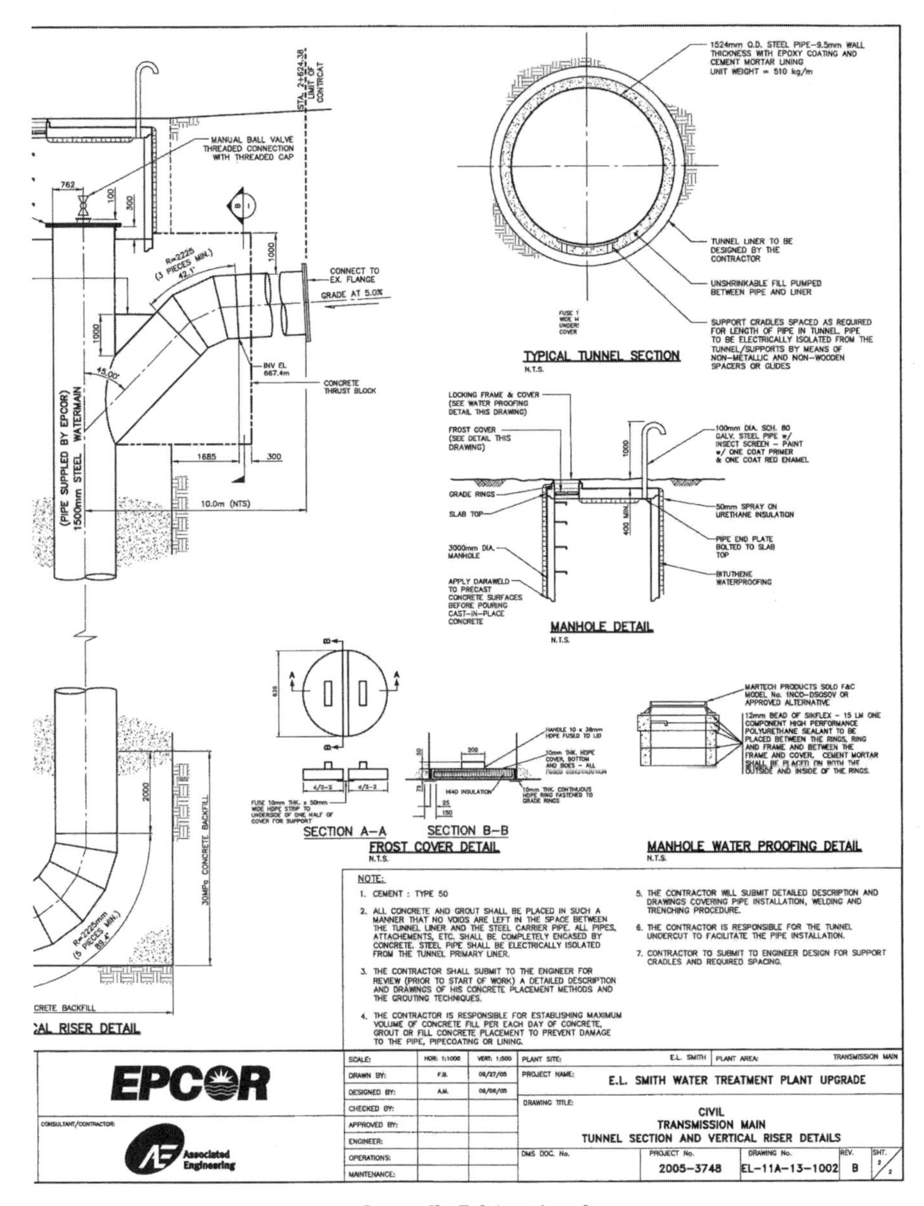

Appendix 5.1 (*continued*)

Chapter 6

Competencies That Drive the Company

The problem with incompetent people is the fact that they don't know that they are incompetent.

—Anonymous

Service enterprises, such as project management or design companies, sell the special expertise of their employees, rather than tangible items such as cars or computers. The management of construction projects requires a long list of specialists, and they all rely on their individual skills and knowledge to succeed in carrying out a particular assignment or solving a complex problem for a customer. The company itself, however, depends on its *internal* management expertise to stay competitive. We refer to these internal "service providers" as *competent persons*, whose competencies enable them to carry out specific tasks required in their respective positions within the company.

Craftsmen or equipment operators in construction, of course, require their own set of skills and knowledge, and competencies, which are gained through work experience—referred to as "expertise." Usually, after an initial period of education and training (e.g., an apprenticeship) an expert emerges after 10 years of practice in his or her field. What separates an expert from a technician is that the expert has developed a capability to apply his or her knowledge creatively to solve new and unique problems.

From Apprentice to Journeyman to Master Craftsman

During the medieval period, guilds controlled business in most urban centers. As chartered monopolies, they also controlled the apprenticeship programs, in which one of their own, a master craftsman, would provide room and board to an *apprentice* for seven years to test his ability to learn—and keep—trade secrets. After passing this basic schooling, the apprentice became a *journeyman* (from the French *journee*) and could ask for a day's wages and the right to travel from town to town. Only if he was accepted by a guild, and proved his skill (typically by submitting a high-quality example of his work), was he allowed open his own shop. Today's PhD thesis is a carryover from that time.

6.1 GENERIC WORK COMPETENCIES FOR THE TWENTY-FIRST CENTURY

The information economy of the late twentieth century is slowly being replaced by the knowledge economy. Even in the traditionally conservative construction industry, changes are taking place that require new skills, especially in information literacy. What are these new skills that will enable people to be productive within these changing paradigms? What does it mean in terms of education and training of new competencies?

By the 1980s many professional associations and agencies worldwide started to draw up lists of the essential skills they believed employees in their areas needed to develop. Not surprisingly, and in response to an increasingly mobile workforce, efforts were made to create lists of generic competencies that were readily transferable from one employer to another, even from one industry to another. What rose to the top were creativity, personal development/continuing education, and leadership skills. Table 6.1 summarizes seven cluster competencies that are considered basic and generic for future management professionals in the construction industry. Of course, any field-specific competencies would have to be added, such as project or equipment managers. It is apparent that many competencies, as never

Table 6.1 Generic Competencies for Management Professionals in Construction

Cluster	Competency
1. Self-directed learning	Effective learning skills (reading, writing, sketching, computation, time management, summarizing), metacognitive capabilities, problem-oriented learning, motivation to learn
2. Information literacy	Ability to find needed information efficiently and critically evaluate it and its sources; understand information storage and retrieval, appreciate the ethics and legality of information use
3. Problem solving	Problem breakdown and analysis, definition of goal criteria, creative design solution discovery, evaluation and refinement of best solution, "marketing" of solution to the customer
4. Communication	Professional emails, reports, and oral presentations; ability to listen, and understand nonverbal signals; comfort with foreign languages and cultures
5. Teamwork	Brainstorming, tolerance of others, conflict resolution, goal setting, consensus building, respect of diversity, domain expertise
6. Leadership	Understanding people's needs, original thinking, motivational ability, domain expertise, empowering others, measuring performance, decision making, trust building, planning and control
7. Personal development	Personal hygiene, ability to adapt to changing environments, lifelong learning, physical health, ability to balance work/personal life, stress management skills

before, depend on the personal initiative of the individual. In addition, the new field of information literacy feeds into others such as problem solving, self-directed learning, leadership, and communication.

As mentioned, Table 6.1 is limited to the basic and generic competencies, consisting of knowledge and skills that will make a manager productive. In the following sections, we will investigate the background, and the breadth, of the competencies needed today by managers and craftspeople working in the construction industry.

6.2 MANAGERIAL COMPETENCIES OF PRODUCTIVE ORGANIZATIONS

As noted in Chapter 1, the Scottish philosopher and economist Adam Smith argued for the need to move from work done by generalists who produce a single product, toward work by various specialists. As an example he used pinmaking. One pin-maker could fabricate 20 pins in one day. He demonstrated that 10 workers, working simultaneously but each specializing in one of the steps needed to make one pin, could make 48,000 pins in one day. This increase in productivity from 20 pin per man-day to 4,800 pins per man-day, would allow nations to create wealth for its citizens.

The emergence of the steam engine in the workplace, however, led to widespread unemployment in the lower classes, demonstrating the negative aspects of rapid increases in productivity. Nevertheless, economic growth did indeed rely heavily on improving productivity and the ability of large organizations to synergize the specialized capabilities of technology and people to produce ever-more complex and sophisticated outputs.

One "victim" of this unavoidable development was the construction architect. Until the Middle Ages, the architect (from Greek *arkhi*, for "chief," and *tekton*, for "builder, carpenter") or Master-Builder, had become a functional designer who coordinated engineers and contractors. During their heyday, architects were also engineers and contractors who hired the workforce on construction projects.

6.2.1 Hierarchical Structure of Construction Companies

As we've discussed before, among the earliest and well-known construction projects requiring a large organization were the Egyptian pyramids. It is estimated that the Great Pyramid at Giza, built around 2500 BC, required some 100,000 workers every winter for over 20 years, paying their taxes in form of service to the pharaoh. One of the most remarkable aspects was the management of food and water supplies, housing, and medical care for this army of laborers. The Master-Builder of the complex interacted directly with superintendents and technical foremen. There existed a functional hierarchy of the workforce. Gangs of slaves worked in

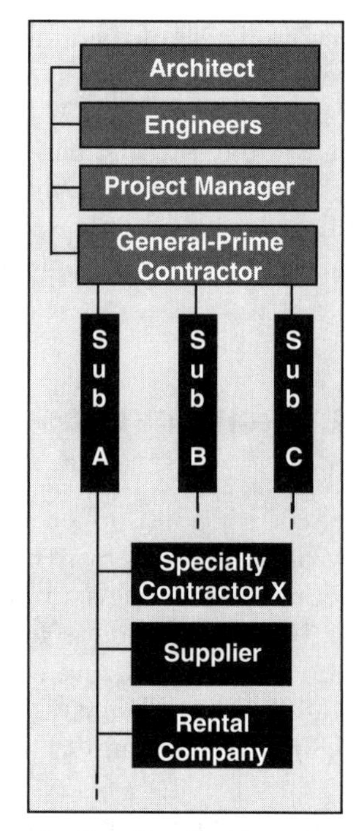

Figure 6.1 Structure of an Organizational Project Breakdown

the quarry, hauled stone, or labored at the construction site. The most skillful, however, were the stonemasons, who were trained to shape and polish the granite for sculptural figures, doorjambs, and lintels.

If we were to arrange such a workforce according to their jobs, starting with common laborers in the front row and progressively more skilled workers in successive rows behind them, we would end up with a horizontal representation that very much resembled what they were building vertically—a pyramid. Obviously, their lack of equipment compared to more recent times meant that they required a large number of people to move, for example, a mass of stone that today can be done with a backhoe or a crane.

Adam Smith would certainly be pleased to observe how his arguments have been embraced over time by the construction industry, which relies on specialty contractors, or subcontractors, to do the physical work, while general contractors have taken over the function of managing projects, deploying their own specialists to plan and oversee the work of the subcontractors. And depending on the needs of a given project, sometimes subcontractors will hire their own specialty subcontractors and contract with suppliers, equipment rental, or leasing companies, and so on. Even specialty contractors have to establish organizations that allow them to function efficiently. Figure 6.2 illustrates a generic structure that contractually interlinks all project participants. As shown, the general contractor normally has a

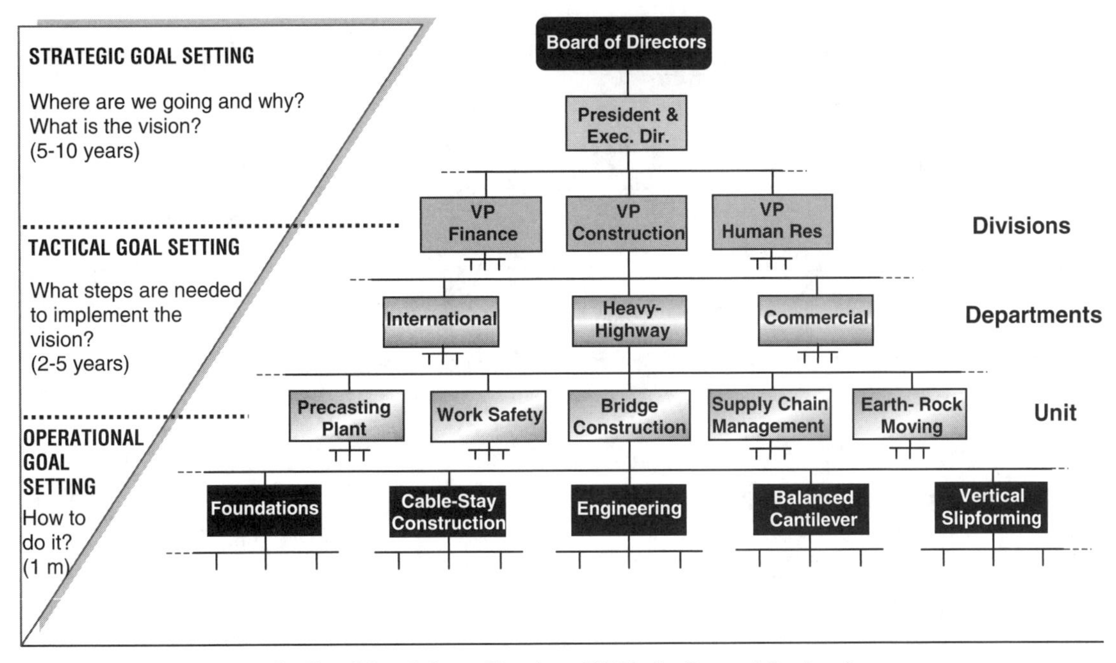

Figure 6.2 Generic Organizational Breakdown Structure (OBS) of a General Contractor

special relationship to the project, in that he has to the responsibility to build the physical product according to the designs of the architect(s) and engineers for a specified price and during a specific time period.

It is striking how today's structure of a large productive organization, such as that of a general contractor, mirrors that of a pyramid construction project: the board of directors can be thought of as the pharaoh, with his advisors on the top levels. It is easy to recognize how the functions between the levels differ; likewise, it is clear how the specializations break out, by comparing the boxes at each level.

One of the key features identifying a given level in an organization is the time horizon used for goal setting. Those at the top have to decide on the dimensions of the pyramid, the time of the year the 100,000 laborers should be called upon, and where to get the stone. The middle managers organize the work-force into gangs, supervise expert stonemasons carving the delicate contours, and organize the food and health care. Finally, the foremen in charge of the gangs who do the work decide on the operational details, such as how many laborers are needed to pull and how many to push a sled carrying a block of stone.

Let's use the Precasting Plant that is being run by the Heavy Highway division, shown in Figure 6.3, as a case study to review the relationships between a company owner and the people he hires with the necessary competencies to meet his project's needs, we'll use a precast plant that is being run by the Heavy Highway Division, shown in Figure 6.3. The plant is charged with providing

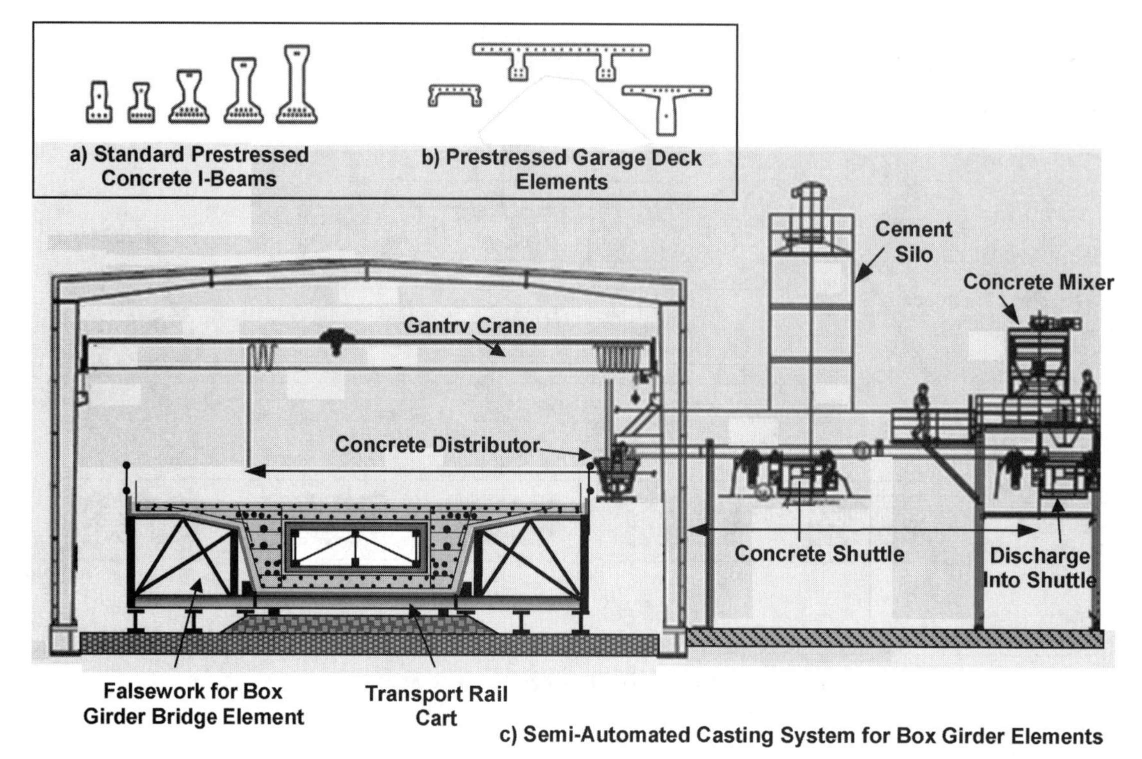

Figure 6.3 Overview of Technologies, Processes, and Products at a Hypothetical Precast Plant

high-quality elements for the bridge building department, as well as competing for contracts outside the firm. The plant specializes in prestressed I-beams for girder bridges and prestressed deck elements for garages. Both require strict adherence to standards, such as those of the ACI and ASTM, creativity in engineering, and specialized equipment for casting and tensioning the cables.

6.2.2 Diverse Expectations of a Precast Plant Manager

In order to understand the logic behind the expectations a precast plant manager might have, we first have to understand the inner workings of such a production system. The schematic in Figure 6.3 sketches the operation to fabricate a large box girder element in a steel building equipped with gantry crane and overhead concrete distributor. The concrete is supplied via a shuttle from a batch plant outside. The falsework is mounted on the ground, and a heavy-duty rail cart that transports the entire element after the formwork has been stripped. Figure 6.4 then shows the point at which the completed element is pulled out of the steel building to make space for forming the next element. The company's strategy—to

Figure 6.4 Transport of Elements to the Storage Area

provide high-quality elements, and compete for contracts—should ensure that the factory stays operational, without major ups and downs.

The manager of this hypothetical factory is expected to:

1. Optimize material supply logistics.
2. Sustain a zero accident plant.
3. Cultivate a highly motivated workforce.
4. Implement ISO 9000 Total Quality Management (TQM) standards.
5. Submit competitive bids and prices.
6. Develop new business opportunities.
7. Be result-oriented and self-confident
8. Adopt high-value return on investment (ROI) and profit targets.
9. Forecast accurately the economy and its impact on precast construction.
10. Maintain friendly relationships with customers and suppliers.
11. Be responsive to the needs of other departments and divisions in the corporation.
12. Ensure a smooth administrative process, and submit timely budgets and quarterly/yearly reports, and so on.
13. Interact effectively with division leaders, VPs, and the board of directors.
14. Engage in problem-solving sessions with engineering, and do operational planning, when necessary.
15. Maintain a 99 percent availability of equipment, tools, and facilities.
16. Foster a culture of change.

Frederick Winslow Taylor (1856–1915)

F.W. Taylor was an American mechanical engineer, educated entirely by what is today called "distance learning." He is widely considered the father of scientific management, a theory that focuses on improving work efficiency through the use of scientific procedures. For example, he found 21–1/2 pounds to be the most efficient load for hand shovels, and based on that discovery designed shovel sizes for different materials to fit their individual target loads.

Taylor defined four principles:

1. Optimize work methods using scientific procedures.
2. Select and train employees based on the optimized methods.
3. Measure the task performance of the workers.
4. Managers need to plan work by applying principles of scientific management.

17. Communicate well with external parties (banks, regulatory authorities, media, etc.).
18. Enjoy communal relationships with employees.

On examination, it is clear that the list includes items with vastly different characteristics. In contrast, a similar list for a division leader 120 years ago would probably have contained items such as:

- Pour an average of *X* yd^3 (m^3) of concrete per day
- Work six days per week.
- Provide quarterly reports.
- Keep operating costs at *X*.
- Maintain a productivity rate of *X* yd^3 (m^3) of concrete per laborer.

It is obvious that the second list of functions consists of specific, measurable target values that the plant manager was expected to achieve. The underlying belief was that the result of good management had to be clearly measurable. In turn, a plant manager was considered highly competent if he was able to meet his target values by working a specific amount of time. This was, however, a misguided application of Frederick Taylor's principles of scientific management, meant to improve the output of individual workers and work groups. Taylor called on managers to use calculations and objective measures, such as time, forces, distances, and so on, to structure an optimal work procedure. There was no talk about motivating the workforce as a way to improve productivity. The expected output was established using scientific management. As we will see a little later, Taylor's principles were later counterbalanced by Henri Fayol's five management functions.

At this point, we can consider five of the functions on the list of the modern--day plant manager as measurable job duties:

1. Sustain a zero-accident plant.
2. Implement ISO 9000 TQM standards.
3. Adopt high-value ROI and profit targets.
4. Ensure a smooth administrative process; submit budgets, quarterly or yearly reports, and other relevant documentation.
5. Maintain a 99 percent availability of key parts, tools, and facilities.

The first, sustaining a zero-accident plant, is a straightforward target, but its implementation will require an in-depth risk analysis of all the hazard spots in the plant, followed by creative engineering to protect laborers from harm, as well as orientation of new employees and providing an ongoing safety training program, accompanied by an incentive plan, to reward laborers for accident-free work time. Clearly, this duty is much harder to achieve than simply coming to work six days per week.

In the same vein, implementing ISO 9000 requires a long and sustained process of review and evaluation, after detailed plans and rules have been established. The manager will depend on input from the entire staff, demanding motivational and coordination skills.

Targeting high ROIs, on the other hand, requires a solid knowledge of business accounting principles. Achieving a 99 percent availability of resources will most certainly require the proactive use of a computerized maintenance system that tracks the life cycle of key components and parts, in order to replace them before they break down during operation.

The next set of very different characteristics has nothing to do with skills and duties; rather, they describe social and behavioral traits the successful plant manager should have.

- Cultivate a highly motivated workforce.
- Be result-oriented and self-confident.
- Maintain friendly relationships with customers and suppliers.
- Interact effectively with division leaders, VPs, and the board of directors.
- Foster a culture of change.
- Communicate well with external parties (banks, regulatory authorities, media, etc.).
- Enjoy communal relationships with employees.

What distinguishes these desired characteristics of a plant manager is the fact that they address behavioral traits—; specifically, the ability to interact with the large network of people surrounding the position. These competencies go beyond the ability to communicate effectively or make presentations to large groups of people; they require the ability to motivate others, to reach out and relate to employees on a personal level. In fact, the plant manager is expected to take on the role of a coach. Still, there is a difference between motivation and possessing the skills for turning that motivation into actions. What is also required are self-confidence, the ability to give and solicit feedback, and the psychomotor capabilities to speak clearly, hear accurately, and recognize and understand nonverbal messages.

Finally, the plant manager's communication skills have to be underpinned by an understanding of what is appropriate and effective in a given context. This involves knowing the correct channels of communication and chains of command, as well as knowing the organizational standards for communication. And if the manager is required to communicate with entities outside the company as well (such as news media), he or she also must understand how to communicate successfully with parties outside the organization.

Overall, the seven items in the second set of competencies indicate that the manager is expected not only to have the ability to communicate, but also a set of personal attributes and values such as self-confidence, conscientiousness, empathy toward employees, and a innovative spirit.

What is not on the list is any mention for the plant manager to have the knowledge and skills that are related to the core business of the company. As a matter of fact, interviews with high-level managers confirm again and again that the "soft skills" that are most needed for success. In fact, it is the functions and products that comprise the content knowledge of the manager, some of which are hidden behind the desired competencies of the plant manager:

- Optimize material supply logistics (e.g., supply-chain management, material-handling technologies).
- Submit competitive bids and prices (e.g., cost estimating, supply-and-demand economics).
- Develop new business opportunities. (e.g., strategic investing, risk management).
- Adopt high-value ROI and profit targets (e.g., accounting, tax strategies, financial management).
- Forecast accurately the economy and its impact on precast construction (e.g, macroeconomics).
- Engage in problem solving with engineers, and design operation plan (e.g., concrete, structural design, production).
- Maintain a 99 percent availability of equipment, tools, and facilities (e.g., information computer technology, statistics, inventory management).

It needs to be emphasized that knowledge alone of, for example, supply-chain management, will not guarantee that the manager can actually optimize the material supply. Content knowledge has to be paired with attention to the issue, as well as a vision and the self-confidence to make changes when necessary.

Now we are ready to turn the 18 capabilities and traits described in this section into the core competencies of a plant manager.

6.2.3 Core Competencies of a Precast Plant Manager

Before attempting to itemize core competencies, we want to ask ourselves what such a list would be used for. A good place to start to answer that question is the Competency-Based Approach (CBA), which is widely used today by many successful organizations. It cites three different purposes:

1. Organizational integration and value alignment
2. Recruitment and promotion
3. Training and HR development

Especially large organizations find it difficult to ensure that employees at all levels adopt the vision and value system established by top management. The CBA has been found effective at aligning managers at all levels to its strategies, an important undertaking since, typically, these executives are not tied to the specific goals of a particular unit within the organization. This is accomplished not by sending out yearly mission statements but by expecting position holders to embody the qualities and attitudes that are the essence of the company.

Implementing the CBA both for recruitment and promotion is straightforward—when it is accompanied by standards of relative importance and scales for performance evaluation. A key benefit of the CBA is the internal training program for candidates eligible for key positions:

First it provides the basis for identifying persons who exhibit the desired traits and have the knowledge and skills that are closest to what is required of the job.

Second, it helps identify any gaps that need to be filled. In turn, this enables training and development strategies to be designed on an as-needed basis.

While it is relatively easy to list the necessary knowledge base and skills a manager needs; defining the human traits is less so. That said, research and experience emphasize the importance of empathy and trust-building in the way managers communicate internally with the workforce. Based on these observations, competencies—and their level of importance—can be established for our hypothetical plant manager. They are listed in Table 6.2.

Competency

To date, the literature does not offer a standard definition of *competency*. In this book, we use the term to mean the following:

Managerial competence is characterized by a weighted set of *technical/managerial knowledge* and *skills* accompanied by *personal traits* necessary to excel at a high-level position within an organization.

Table 6.2 Competencies of a Hypothetical Precast Plant Manager

Desired Competencies: Precast Plant Manager	IMPORTANCE				
1 = Elementary; 2 = Essential; 3 = Vital; 4 = Extreme; 5 = Absolute;	1	2	3	4	5
1. Technical/Production Knowledge and Skills					
Supply-chain management			x		
Bulk storage technologies and automated inventory control				x	
Concrete materials and prestressed concrete design					x
Plant facilities design and control principles			x		
Estimating precast concrete production cost					x
Information computer technology (ICT)		x			

(continued)

Table 6.2 (*Continued*)

Desired Competencies: Precast Plant Manager 1 = Elementary; 2 = Essential; 3 = Vital; 4 = Extreme; 5 = Absolute;	IMPORTANCE 1	2	3	4	5
ISO 9000 and Total Quality Management (TQM)				x	
Safety regulations related to precast industry					x
Electronic data collection and statistical data analysis	x				
Risk management			x		
Macroeconomics and economic forecasting			x		
Ergonomics and safe rigging methods					x
2. Managerial Knowledge and Skills					
Organizational and administrative theory					x
Financial tax management				x	
Marketing			x		
Recruitment and salary structuring			x		
Brainstorming, creative problem solving					x
Business correspondence and public speaking				x	
Human factors					x
3. Communication					
Capability to listen and to mediate social conflicts and grievances; build trust					x
Use two-way communication channels with employees, customers, suppliers, etc.			x		
Willingness to be a strong steward for the company with media, banks, etc.				x	
Operate a fair performance appraisal and performance feedback system			x		
Coach the workforce and provide developmental opportunities for entire staff				x	
Adopt corporate-wide goals, translate them into objectives specific to the business					x
4. Personal Attitude and Self-Efficacy					
Display an innovative spirit that values critical thinking and creativity					x
Smart worker, with conscientious approach to work					x
Self-motivated: accepts challenging goals and takes calculated risks			x		
Presents a self-confident image to lead a confident team				x	

Header Problem 6.1: Results of a 360-Degree Performance Assessment

Mr. Lee, the plant manager of Presco, Inc., a concrete precasting company, is committed to helping the company's employees advance. To that end, Mr. Lee hired a human resources (HR) company to design and execute a 360-degree survey of Presco's fabrication manager, Ms. Lukas, who oversees forming, prestressing, casting, curing, and yard storage. Mr. Lee asked us for help in analyzing the results of the survey. In particular, he is interested in learning what the company can do to improve Ms. Lukas's effectiveness as a manager, by addressing both her weaknesses and strengths. Presco is a company that believes in continuous enhancements. Here is the partial result of the survey that was completed by 8 people. (We will return to this problem later in the book.)

The 360-Degree Survey

A 360-degree survey (introduced in Chapter 1) is a job performance survey sent to 8 to 10 persons at higher, equal, and lower levels in the hierarchy (not just to superiors.) A questionnaire asks the respondents to assess, by means of a Likert scale, how well the core competencies required for a job are being met.

Table 6.3 Results of a 360-Degree Job Evaluation

	Accumulated Responses from Eight Persons				
Ms. Lukas of Presco, Inc.	Highly Agree	Agree	Disagree	Highly Disagree	Don't Know
Informs the Quality Control Lab in a timely manner about upcoming casts.	6	2			
Alerts laborers when she sees an unsafe situation in the plant.		2	4		2
Assures that plant has a 95% availability of hardware and tools.			2	4	2
Organizes a monthly safety (rehearsal) meeting with everyone at the plant.	7	1			2
Ensures that all cranes, ropes, and slings are inspected daily.	7				1
Certifies proper technical training of an employee before assigning a new job.		1	4	2	1

Table 6.3 (*Continued*)

Ms. Lukas of Presco, Inc.	Accumulated Responses from Eight Persons				
	Highly Agree	Agree	Disagree	Highly Disagree	Don't Know
Has a sufficient number of forklifts, trucks, and elevating tables available.				6	2
Requires new hires to pass an intensive safety program.	8				
Is up to date on the newest precast technologies.		3	5		
Is extremely responsive to the requests of her subordinates.	8				
Writes fair performance evaluations for her subordinates.	8				
Actively participates in solving difficult problems.	2	4	2		
Offers training opportunities for those who plan to advance.		2	2	2	2
Makes sure each individual and all crews have an even workload.	5	2			1
Creates clear job descriptions and establishes work standards for everyone.	8				
Adheres to procedures to minimize waste (time, materials, etc.).		2	2	4	
Follows accepted procedures, keeps up-to-date and accessible data records	8				
Meets with each subordinate four times per year to discuss her vision, company progress, and goals, and invites ideas for improvements.	6	2			
Initiates personnel actions such as promotions, performance awards, or demotions, according to performance in the plant.	8				

Meets with other managers to discuss goals and needs.	4	2			2
Mentors promising individuals in her department.		4	2		2
Praises regularly the best innovative idea from her personnel in the monthly company newsletter.	2	2	2		2
Is self-motivated and conscientious in her work.	4	4			
Fosters a spirit of constant change and proactive thinking.		4	4		
Inspires everyone to excel beyond the expected.	2	4			2
Represents the department in a positive and professional manner.	4	4			

6.3 GAINING COMPETENCY THROUGH LEARNING AND TRAINING

Lifelong learning, or continuing education, emphasizes the need for individuals to upgrade their knowledge and skills regularly after they leave behind full-time education. The prime reason is that progress requires individuals to constantly expand their knowledge base to include new methods and technologies, enabling them to do their jobs better, more effectively, and, ideally, with fewer resources. The same principle applies to business organizations, which depend on constant improvement in productivity and managerial and technical abilities.

Although only individuals are able to gain new knowledge and change their behavior, it is necessary to recognize that an individual is a member of at least one team, which in turn is part of an even larger organization. Therefore, knowledge and skills development must have different components at each level, which means different methods have to be applied in each case. For example, whereas an individual can learn with the help of programmed packages, and be tested and rewarded based on the achievement of desired outcomes, teams consisting of many individuals have to collaborate as a unit to achieve the desired new competencies (see Figure 6.5). Here, the alignment of motivations and goals among the various individuals is critical to the transfer of knowledge and the achievement of targeted team performance. Each individual must acquire and contribute competences that will help the team and not necessarily the individual.

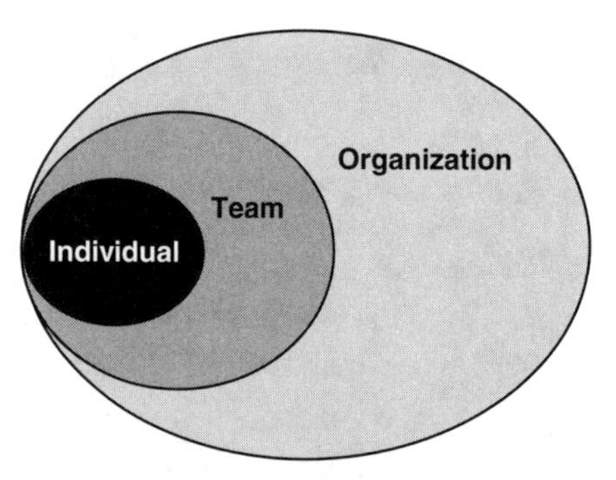

Figure 6.5 Levels of Learning

While learning at the organizational level is far removed from the skills training, it is the individuals who will actually transform the company. As in any living system, organizations must ensure that all its teams and the individuals that comprise them understand and agree to pursue the same vision, in a cooperative and collaborative spirit. In an environment that demands ongoing adjustments and realignment of goals and visions, organizations need to gain the competences necessary to enable them to adapt automatically and constantly to these changes.

6.3.1 The Learning Organization (LO)

Every productive organization has at least three managerial levels, each with different responsibilities: strategic, tactical, and operational. While different, they all share the need for continual improvement of the various competencies. A company could apply a passive strategy, assuming that each employee will engage in lifelong learning. For sure, a great deal of learning about business-specific issues happens in small groups working all across the company, making the transfer of knowledge from one group to another sometimes somewhat random. A more efficient approach is based on the recognition that the institution must purposefully manage the diffusion of knowledge from within.

A *learning organization* (LO) establishes an internal system for capturing, disseminating, and making accessible new knowledge, from either internal or external sources. Most crucial is *tacit knowledge*, generated by its employees, as it goes about the practices unique to the industry in which it operates. Construction businesses are traditionally reluctant to buy into the idea of developing their employees, pointing out the cost and the risk of enhancing the attractiveness of its highly trained personnel to competitors. Ironically, the long list of successful companies that consider themselves LOs, and which face similar risks, attribute their success precisely to the benefits of competence development. Companies that have implemented effective training programs report not only a reduction in costs, but improved performance and productivity as well, leading to savings that provide large ROIs of the expenses.

There are many motivations for an LO to invest in the development of its employees, but three stand out:

1. Adjustment to industry or governmental norms
2. Creation of a competitive advantage, and
3. Radical transformation of an entire sector.

An example for the first objective could be safety training for all field personnel; the second might include the training of the entire design team on new CAD software; an example of the third might be the switch to a paperless organization.

The success of all training efforts depends heavily on the motivation of the involved personnel. Equally important, however, is the recognition that learning takes place at many levels and, as such, requires both excellent teaching and learning skills.

Tacit versus Explicit Knowledge

Michael Polanyi (1891—1976) was a philosopher and physical chemist, born in Budapest but later taught at the University of Manchester. He discovered that people possess knowledge or skills that they cannot explain. *Tacit* (meaning silent or implied, not expressly stated) knowledge is highly personalized and generated inductively, by developing habits or beliefs that shape the person's environment. *Explicit* knowledge, on the other hand, is easily transferred, in the form of textbooks, manuals, films, and so on.

An expert who possesses tacit knowledge typically cannot easily communicate the tacit expertise, so special techniques have been developed to allow its codification into explicit knowledge, so that it can be shared more readily.

6.3.2 Taxonomies for Learning and Training

Organizational learning is different from that gained in educational institutions, in that it focuses on the distribution and adoption of new knowledge, skills, and behaviors specific to the operation of a business. But it has in common the same cognitive, psychomotor, and affective mechanisms that underlay the capabilities of each individual. Many psychologists have studied those mechanisms, but one man is of special interest to us here: Benjamin Bloom. In 1956, he published his hierarchical taxonomy of the human cognitive domain. Later, he modified his original taxonomy; it is shown in Figure 6.6.

- **Remembering:** Recalling of facts from memory.
  Examples: Names of capital cities, phone numbers.
  Key actions: Label, list, match, name, recall, state.
- **Understanding:** Constructing meaning from oral, written, and graphic messages.
  Example: Convert an equation into a spreadsheet.
  Key actions: Comprehend, convert, estimate, explain, generalize, infer, interpret, paraphrase, predict, rewrite, summarize, translate.
- **Applying:** Using what has been learned in a novel situation in the workplace.
  Example: Apply statistics to evaluate the reliability of a new test sample.
  Key actions: Construct, demonstrate, identify, manipulate, modify, predict, relate, show, solve.

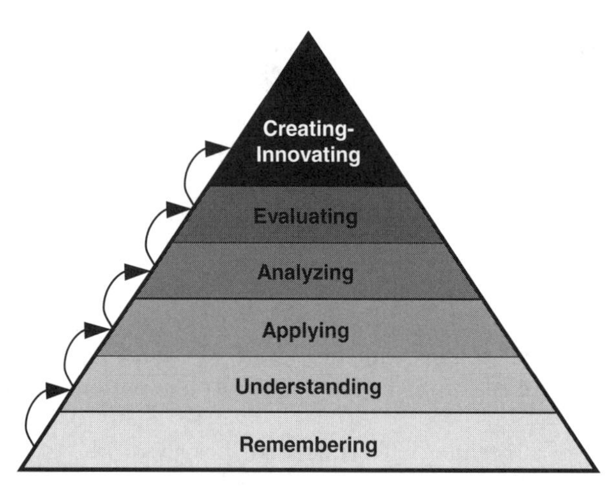

Figure 6.6 Bloom's Taxonomy of Thinking

- **Analyzing:** Organizing problems into constituent parts, determining how the parts relate to one another and to an overall structure or purpose.
  Example: Troubleshoot a piece of equipment.
  Key actions: Analyze, break down, compare, deconstruct, differentiate, discriminate, distinguish, identify, infer, relate, select, separate.
- **Evaluating:** Making judgments about relevance and importance of information based on criteria and standards through checking and critiquing.
  Example: Select the most effective solution.
  Key actions: appraise, compare, conclude, contrast, critique, defend, evaluate, explain, justify.
- **Creating:** Putting elements together to form a new coherent or functional entity.
  Example: Devise a car steering system that is smart enough to recognize if the driver is drunk.
  Key actions: Create, devise, discover, generate, innovate, plan, produce, remodel, reorganize, restructure, substitute.
  (Adapted from Anderson and Krathwohl [2001], pp. 67–68)

The arrows in Figure 6.6 indicate the cognitive development of a learner, starting with remembering facts such as locations, names, or phone numbers. In the next stage, the learner is able to associate meaning with the remembered facts. For example, the city name Paris is associated with France, in the same manner that Washington, DC, is associated with the United States. They are both capital cities where the federal government resides. Furthermore, France played a critical role in the founding of the United States. In October 1781 French troops under the command of General Comte de Rochambeau, along with the navy under the

Comte de Grasse, helped George Washington trap the British army at Yorktown under General Lord Cornwallis, capturing 8,000 troops.

This example also demonstrates how new facts can be linked in such a way to better anchor new knowledge. Using this understanding, a learner should be able to recall that the name of the U.S. capital relates to General George Washington. By analyzing the main commanders who participated in the relevant battle, the learner notices that no mention was made of the British navy. In fact, the large French fleet surprised the British, who were unable to reinforce their troops in Yorktown by sea, thus contributing to its fall.

A similar taxonomy, shown in Figure 6.7, has been established for affective or emotional behaviors during learning. They include issues such as motivation, interest, conscientiousness, and the ability to enjoy, listen, respect, and interact with others. This taxonomy includes the following five levels:

Benjamin S. Bloom (1913–1999)

Benjamin Bloom was an American educational psychologist interested in understanding how human thinking develops. In particular, he wanted to devise a method for assessing how far along in the thinking process a student was. He headed a team that designed a hierarchical taxonomy, now referred to as Bloom's Taxonomy, for teachers and instructional designers to use to classify instructional objectives and to measure the outcome of the instruction.

- **Receiving new material:** Awareness, willingness to hear with selected attention. Making eye contact.
  Example: Listening with respect. Taking notes.
  Key actions: Ask, choose, confirm, enjoy, follow, give, hold, laugh, note-taking, offer, point to, sit, respect, reply, use.
- **Responding to new material:** The learner participates actively in the learning process.

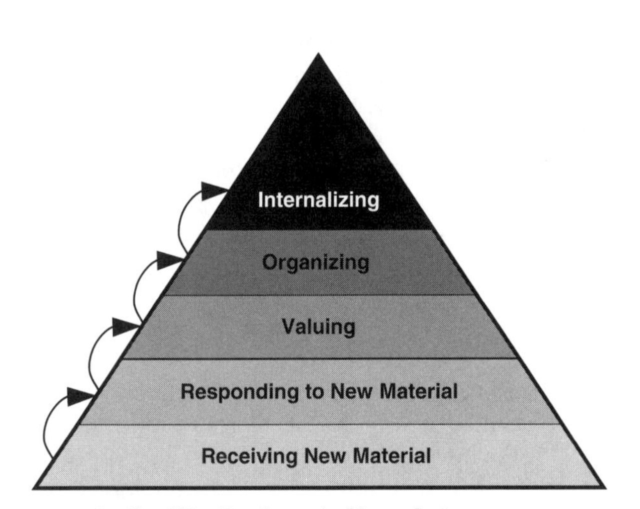

Figure 6.7 Taxonomy for the Affective Aspect of Learning

Example: Asking questions. Participating in class discussions. Giving a presentation.

Key actions: Answer, discuss, find, help, perform, practice, present, question, read, report, search, write.

- **Valuing:** Process of internalization or acceptance of the new phenomena. Learner expresses the valuing through overt behavior that are often identifiable.

  Example: Showing the ability and willingness to solve problems. Proposing a plan to improve and follows through.

  Key actions: Demonstrate, differentiate, explain, follow, form, initiate, invite, justify, propose, report, share, study, work.

- **Organizing:** Adopting the new phenomena and possibly resolving conflicts between the old and the new knowledge.

  Example: Taking up a systematic method to solve problems. Creating a prioritized time management plan that meets the needs of the organization, family, and self.

  Key actions: Adhere, alter, arrange, combine, explain, formulate, integrate, order, organize, synthesize.

- **Internalizing:** New behavior is pervasive, consistent, predictable, and characteristic of the learner.

  Example: Showing self-reliance when applying learned material. Participating in problem solving requiring teamwork. Revising judgments and changes behavior in light of new knowledge.

  Key actions: Argue, discriminate, display, influence, modify, perform, practice, propose, revise, verify.

Learning Psychomotor Skills

This comprises the development of organized patterns of muscular activities guided by signals from the environment. Behavioral examples include driving a car, and using eye-hand coordination for tasks such as sewing, throwing a ball, typing, operating a lathe, and playing a trombone. In research on psychomotor skills, particular attention is given to learning coordinated activities that involve the arms, hands, fingers, and feet.

The third element considered important by Bloom was the psychomotor aspect of learning new capabilities. Psychomotor abilities are essential for managers who have to engage with customers or make public appearances or presentations, activities for which nonverbal body communication is critical. On the production level, good examples are the operators of large cranes or barges, who need both adequate knowledge and sufficient eye-hand coordination to guide the movement of large equipment without damaging the surroundings.

Again, adopting Bloom's cognitive growth model, several hierarchies have been established for the psychomotor aspect of learning new capabilities. One example is presented in Figure 6.8. The pyramid implies that learning that involves psychomotor skills progresses in phases or steps, starting with sensory perception, which enables one

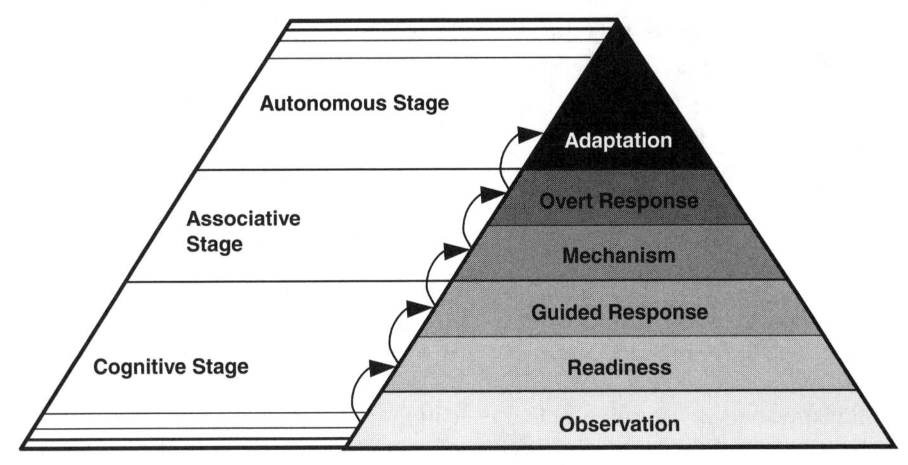

Figure 6.8 Taxonomy of Psychomotor Aspect of Learning

person to observe another perform, then be ready to copy the observed motions or follow a coach's instructions. The learner begins to execute the rudiments of the skill by imitating or following directions.

- **Observation:** Watching others demonstrate expertise.
  Example: Watching use of a forklift to remove a pallet from a high stack.
  Key actions: Detect, identify, isolate, relate, see, watch.
- **Readiness:** Willingness to act.
  Example: Showing desire to try a new process or method.
  Key actions: Begin, display, explain, experiment, move, proceed, ready, show, state, test, volunteer.
- **Guided response:** Imitation and trial and error.
  Example: Following the signals of an expert crane operator.
  Key actions: Copy, trace, follow, react, reproduce, respond
- **Mechanism:** Learned responses have become habitual and the movements can be performed with some confidence and proficiency.
  Example: Using of a personal computer. Driving a stick-shift car.
  Key actions: Assemble, calibrate, construct, dismantle, display, fasten, fixe, grind, heat, manipulate, measure, mend, mix, organize, sketch.
- **Complex overt response:** Making judgments about relevance and importance of information based on criteria and standards through checking and critiquing.
  Example: Selecting of the most effective solution.

Key actions: Appraise, compare, conclude, contrast, critique, defend, evaluate, explain, justify.

- **Adaptation:** The learner modifies and fine tunes movement patterns to fit special requirements.
  Example: Performing a task with a machine that it was not originally designed for.
  Key actions: Alter, change, modify, rearrange, reorganize, revise, try.

Fitts (1964) described skills learning as involving three stages: (a) cognitive, (b) associative, and (c) autonomous. These three separate steps are equivalent to the single *cognitive stage* in Bloom's model; Fitts's model, however, emphasizes the fact that the three stages show recognizable traits. For example, a beginner or novice thinks about each motion before executing it. As the beginner becomes more adept at a task, he or she needs less cognitive or conscious thought. For example, a novice backhoe operator will pay attention to many cues such as a supervisor's instructions or feedback from the environment (e.g., scratching, noise from the bucket). In the associative stage, the cues and the feedback are now more directly linked to appropriate actions, and no instructions are needed. Through the repetition of motion sequences some parts of the task are becoming habitual. Rasmussen (1986) says that in this phase, the trainee demonstrates a rule-based behavior, which is "guided by conscious control, but in the form of stored rules, rather than in the form of knowledge of a system." Skill-based, or autonomous, behavior is characterized by "smooth, automated, and highly integrated patterns of behavior."

6.3.3 Teaching and Learning around the Circle

The essence of the following section was famously expressed by the Chinese philosopher Confucius around 450 BC: "Tell me, and I will forget. Show me, and I may remember. Involve me, and I will understand." What Confucius highlights is the effectiveness of what today is called *experiential learning*, wherein the student is expected to actively participate in the learning process and use analytical and reflective skills to abstract and internalize the new knowledge.

One of many proponents of this *constructivistic* approach is David Kolb (1984) who offered a simple but effective circular model made up of four axes representing four principal stages of experiential learning:

- Concrete experience (sensing/feeling)
- Reflective observation (watching)
- Abstract conceptualization (thinking)
- Active experimentation (doing)

Figure 6.9 displays the model that has emerged, in which four types of learners are organized into four quadrants, bracketed by the four experiential learning stages.

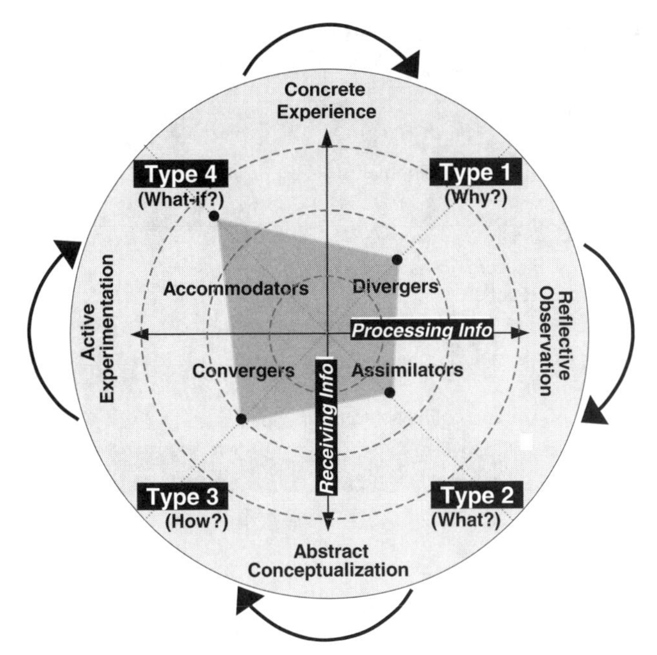

Figure 6.9 Kolb's Elements of Learning and Preferences

According to Kolb's model, a Type 1 (*diverger*) learner perceives the new experience concretely and processes it through reflective observations (theoretically). In contrast, a Type 2 (*assimilator*) grasps new information abstractly (e.g., by reading a textbook). A Type 3, learner (*converger*) perceives abstractly but transforms new experiences actively (e.g., by doing homework). Type 4 learners (*accommodators*) prefer to receive new information concretely (e.g., by observing an experiment).

Each of the four types has been characterized using a variety of different traits, such as their favorite among the questions *Why? What? How?* and *What-if?* Kolb and others developed and validated questionnaires that can be used to identify the learning preferences of a person. It is important to understand that every person has learning capabilities in each of the four types, but also has a "strong suit" in one or more quadrants. As an example, Figure 6.9 profiles one person, indicated by the corner points of the gray-tinted polygon. The data plot indicates that the person is a type 4/type 3 learner, meaning that his or her preference is to learn by actively experimenting and using his or her senses to experience the new knowledge.

David A. Kolb (b. 1939)

David Kolb is an educational theorist whose interests and publications focus on experiential learning, individual and social change, career development, and executive and professional education. He developed the Experiential Learning Model, which is composed of four elements:

a) Concrete experience

b) Observation and reflection,

c) Formation of abstract concepts, and

d) Testing of concepts

Table 6.4 Learning Activities Preferred by the Four Learning Types

Type 1	Type 2	Type 3	Type 4
Discussions	Formal lecture	Worked-out problems	Group discussion
Journal writing	Reading assignments	Homework	Oral presentations
Group projects	Individual projects	Individual projects	Group projects
Role-playing	Lecture notes from teacher	Specific guidelines	Role-playing
Group problem solving	Seminar	In-class demonstrations	Simulations
Experiments	Library/online search	Trips	Report writing
Subjective tests	Objective tests	Objective tests	Subjective tests

Kolb's model shows us something else, too. Because a class will surely consist of people with different learning preferences, so the teaching should be planned in such a way as to allow all students to use their individual strong suits. This can be accomplished through the use of activities specifically designed for each of the four quadrants—around the circle, so to speak.

A class plan designed in this way has another, and possibly critical, effect: Every student will receive and process the new material by addressing the four critical questions: *Why?*, *What?*, *How?*, *What if?*

Table 6.4 offers some learning activities that we have found effective for each type.

6.3.4 The Learning Curve

Frederic Taylor was the first to study how the setup of the workplace could impact the base productivity of a factory process. He believed it was the responsibility of management to design the work layout for the laborers (i.e., scientific management). According to the researcher Elton Mayo and his team, other equally important factors affect the output or individuals and teams, among them job satisfaction, motivation, and an opportunity for professional growth. At the Hawthorne Works between 1924 and 1932, they found that learning new tasks, to add variety to otherwise repetitive work, is crucial to operational growth.

In order to predict the expected output of both new hires or current personnel after instituting a new process, the concept of learning curve was introduced.

Psychologists have shown that the human brain improves its response to stimuli that are repeated. For example, the time it takes to put together the same puzzle generally drops with every iteration. Of course, the original time as well as the reduction in time with each trial, varies with the size and complexity of the puzzle. Nevertheless, by plotting the total times versus the number of trials, a well-known curve develops, which follows the power law function, shown here:

$$T_i = T_1\, x_i^{-a} \text{ (Power law function)}$$

where:

$-a$ = a constant that mirrors the complexity of the work
$1T_1$ = time of the first cycle
T_i = time of the *i*th cycle
x_i = cycle number

Figure 6.10 presents the graphical and tabular results of learning curves for a task, with an initial duration of 500 and with different levels of difficulty for improving through the learning process. They are based on indices that show improvements when the base cycle number is doubled. If:

$$x_2 = 2 \times x_1\ T_2 \div T_1 = T_1\, 2\, x_1^{-a} \div T_1\, x_1^{-a} = 2^{-a} = \text{Progress Index (PI)}$$

Now it becomes clear that the key number for representing learning is the factor *a*, as it describes the amount of reduction in time between two cycle numbers, which differ by a factor of 2, as in 2 – 4, 4 – 8, or 8 – 16. The Progress Index (PI) indicates the amount of time reduction in terms of percentage.

Using a puzzle as an example, one can easily forecast that simple puzzles will allow only small improvements, whereas more complex ones will allow for significant improvements.

Figure 6.10 illustrates that the power function produces rapid improvements at the beginning of the learning process, when the brain is eliminating the most wasteful actions or the time it takes to figure out intermediate problems the first time. Those large time savings are possible because the human memory is storing the solution from the first trial, then recalling it quickly the second time around. Naturally, the time savings grow smaller over time, once the memory has stored all the main problem solutions. Still, further improvements are possible, through practice or by making small changes determined after several what-if trials.

The mathematical representation of this slowdown is shown by the data in Figure 6.10. For a PI of 85 percent, it takes only two trials to reduce the cycle time from 425 to 361 (= 63.7), but seven trials to lower it from 316.9 to 269.4 (= 47.5). Not only is the number of trials between the 85 percent improvement getting larger, but the time reductions themselves are getting smaller.

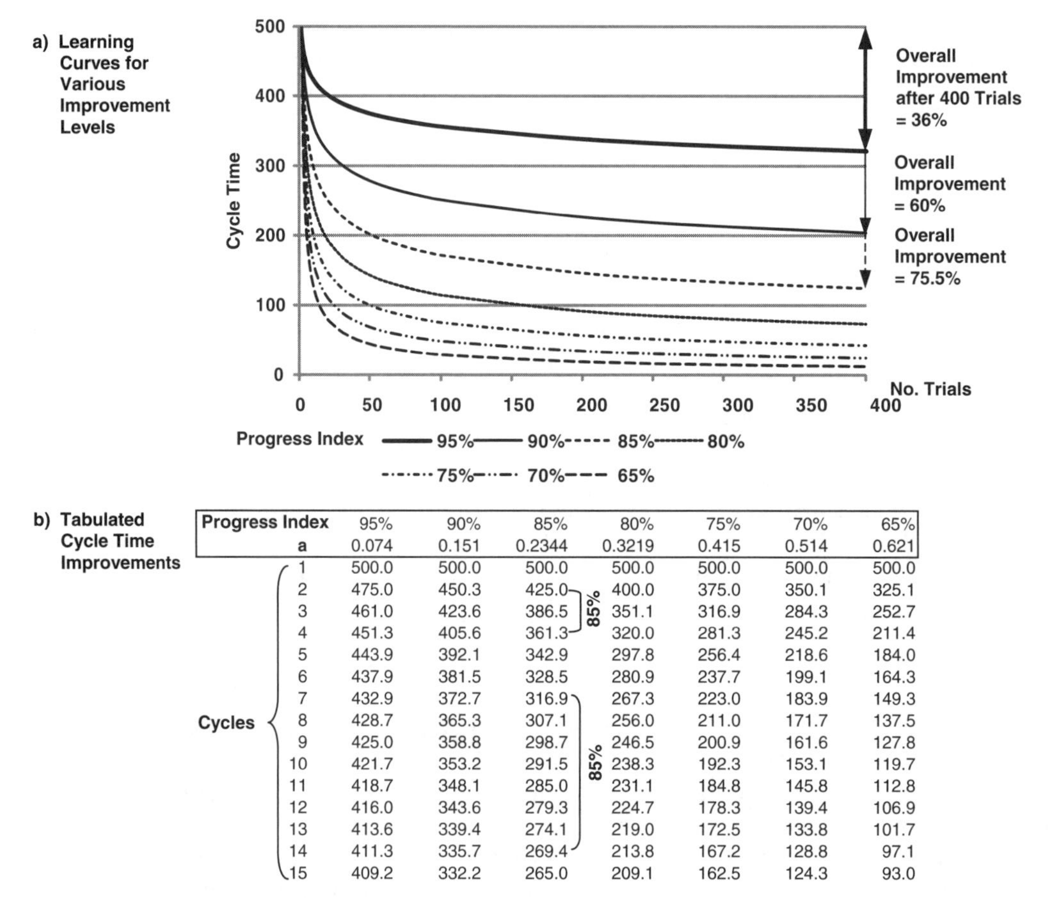

Progress Index	95%	90%	85%	80%	75%	70%	65%
a	0.074	0.151	0.2344	0.3219	0.415	0.514	0.621
Cycles 1	500.0	500.0	500.0	500.0	500.0	500.0	500.0
2	475.0	450.3	425.0	400.0	375.0	350.1	325.1
3	461.0	423.6	386.5	351.1	316.9	284.3	252.7
4	451.3	405.6	361.3	320.0	281.3	245.2	211.4
5	443.9	392.1	342.9	297.8	256.4	218.6	184.0
6	437.9	381.5	328.5	280.9	237.7	199.1	164.3
7	432.9	372.7	316.9	267.3	223.0	183.9	149.3
8	428.7	365.3	307.1	256.0	211.0	171.7	137.5
9	425.0	358.8	298.7	246.5	200.9	161.6	127.8
10	421.7	353.2	291.5	238.3	192.3	153.1	119.7
11	418.7	348.1	285.0	231.1	184.8	145.8	112.8
12	416.0	343.6	279.3	224.7	178.3	139.4	106.9
13	413.6	339.4	274.1	219.0	172.5	133.8	101.7
14	411.3	335.7	269.4	213.8	167.2	128.8	97.1
15	409.2	332.2	265.0	209.1	162.5	124.3	93.0

Figure 6.10 Effect of Learning Patterns on Cycle Time

Worked-Out Problem 6.1: Creating Learning Curves for Individuals

Imagine that you are working for a construction company that values training, not only to raise productivity but also to create a safer and quality-conscious work environment. Your boss, Mr. Jack Merk, has been charged with developing a set of training units to educate supervisors and managers about how to improve productivity, safety, and quality along the learning curve. He asks you to come up with a simple exercise to use in class that would allow the trainees to create their own curve and use it to predict a cycle time after 16 repeats of a certain operation.

Concept

The simplest way to show the effect of learning on the brain is to pick a task that requires that the brain, hands and eyes interact in order to detect patterns during an assembly operation, such as building formwork, using a construction crane, or putting together a puzzle. The latter could be a 2D image that can fit on a classroom table, showing pictures with various complexities and size. If the puzzle is not too complex, it can be repeated several times in class by the same person, while the second student keeps track of time, and the coach tabulates all the times on the board.

Design of Exercise

It is proposed that two sets of puzzles be prepared, one simple and one more complex, approximately 15 x 15 cm (6 x 6 in.). Puzzle images can be created by printing photos on a black-and-white printer and cutting them into pieces with scissors or a sharp knife, in patterns similar to those shown in the two example pictures in Figure 6.11. Each puzzle should end up with around 28 pieces. Using more pieces will, obviously, require more time in class to complete the experiment, and the learning effect will surely show more dramatic effects.

Image A Image B

Figure 6.11 Using Puzzle Pieces to Measure Learning

During the class, organize students in pairs: one as the timer and the other as the "puzzler." (A variation would be to give one member of the pair a copy of the final picture, while the other would not know what the puzzle represented.) The timer notes the time and watches the progress of the puzzler, and lists any special problems that sections of the puzzle cause. After the first cycle is finished (it is not a race), the timer writes the cycle times on the board.

Discuss differences. What created problems? The same puzzler should repeat the experiment four or six times, with the timer writing down the time after each cycle: T_1, T_2, T_3, T_4, T_5, T_6.

(*continued*)

(continued)

CALCULATIONS

Naturally, in this situation, the number of experiments is very small. But using the averages for one puzzler and for the entire class, you should end up with a good idea about the effect of the brain on learning.

Step 1 : Progress Indices (PI) $= 2^{-a}$ $PI_1 = T_2/T_1$, $PI_2 = T_4/T_2$, $PI_3 = T_6/T_3$
Step 2 : Average PI for first sets : $PI_1 + PI_2 + PI_3/3 = PI_{Ave}$
Step 3 : $-a = \log_2 (PI_{Ave})$
Step 4 : $T_{16} = T_1^* \, 16^{-a}$
Alternative Step 3 : $\underbrace{\overbrace{\underbrace{T1 * (1 - PI_{Ave})}_{T2} * (1 - PI_{Ave})}^{T_4} * \overbrace{(1 - PI_{Ave})}^{T_8} * (1 - PI_{Ave})}_{} = T_{16}$

DISCUSSION

Significant differences should emerge between the two image groups, resulting in different PI averages. The puzzlers will also end up with different times, proving that everyone's brain has its own pace. By averaging the times, it is possible to simulate a crew that has to work together.

6.3.5 The Relearning Curve

It is in the nature of construction that the number of repetitions where the situation and the task does not vary is small, when compared to the manufacturing industry. On the other hand, as in building high-rises or digging tunnels, the same task reoccurs after others have been completed. For example, the formwork for a deck and elevator shaft has to be moved up after the rebar has been placed and the concrete poured. The times for setting up 19 new formwork tables or wall segments will follow the power curve function (let's assume a PI = 85%). However, in the break between two stories, a crew will have to "unlearn" then relearn what they did before. Figure 6.12 illustrates the effect of unlearning and relearning.

Here two more factors are introduced. First, after 80 cycles, a steady-state phase replaces the power function learning. This acknowledges that the large number of ever-changing situations and conditions on a construction site make it impossible to continue along the power function forever. The second element takes into account the randomness of life. While the power function results a single curve, it must be recognized that the actual performances will be distributed around the mean, or average, represented by the curve. Naturally, the deviations will tend to get smaller with practice.

For our assumed case, the duration for moving and assembling the first formwork table, T_1, is 500 units (i.e., person minutes). This amount of time is reduced to 425 units for the second table. After 19 tables, the cycle time is at 250 units. Now the crew is finished and moves on to other work, causing a break in the

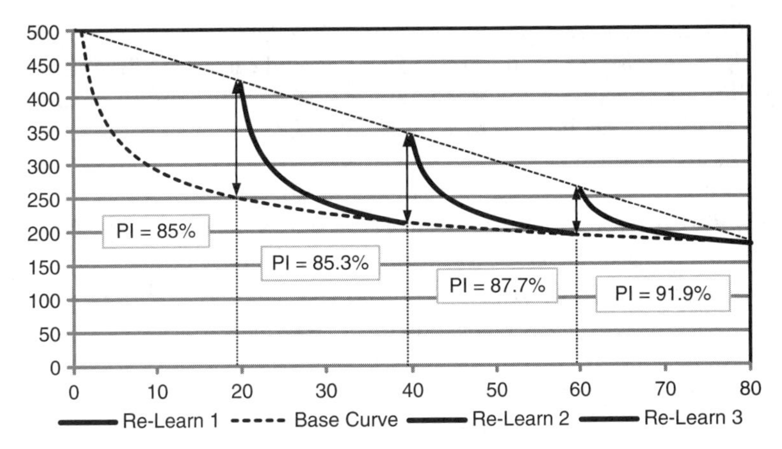

Figure 6.12 The Effects of Breaks on the Learning Progress

learning curve. Since the break is only one week, we can assume that the crew still remembers a lot about moving the tables, although they will have forgotten some aspects of the task. The line connecting T_1 with T_S (steady-state) represents the improved memorization, or the reduction in forgetting, after repeated breaks.

At the end of the first break, the crew will not start at 500 units but, as indicated by the dotted line, at $T_{20} = 420$ units. In other words, the crew forgot how to do some key steps since before the break, resulting in a loss of 420 units –250 units = 170 units. Their new PI is now 85.6 percent, slightly larger than 85 percent, meaning they learned more quickly than the first time. As it turns out, by the time they finished the second 19 formwork table, they caught up with the original learning curve, ending with a cycle time of 214 units. Still, the relearning after the break resulted in a time loss, indicated by the darkened area, which can be calculated by subtracting the integrals for the curves for the base learning and the relearning. In fact, the average T, using the base curve, is 228 units, while the average time for relearn 1 is 265. Thus, the average cycle loss due to the first relearning equals

$$265 \text{ units} - 228 \text{ units} = 37 \text{ units, or a total of } 703 \text{ units.}$$

As we would expect, repeated relearning phases lead to less forgetting and quicker relearning, as indicated by higher PIs. The results are ever lower losses, ending at the steady-state phase. Still, the amount one forgets depends not only on the number of repetitions of a given activity, but also on the length of the break from the task and its complexity. More complex work will result in lower initial PIs, but also more forgetting during breaks.

To apply the power curve function in a meaningful way to calculate the learning effect, at least two values have to be known or assumed, namely ***PI*** and the expected start- or steady-state cycle duration (T_1 or T_S). Of course, establishing those values takes a lot of practice, and will not be totally accurate, as they are impacted by many unknown factors. However, as soon as the absolute first or the first four cycles have been completed, it is possible to immediately adjust the assumed PI and T_1.

Finally, the concept of the learning curve is valid outside the realm of crew production. Graphs can be used to show skill progression against the time required for such mastery.

- **Equipment-tool operation**—An operator develops skills through practice and on-the-job training.
- **Technological innovation**—As a company acquires new equipment and tools, or more efficient management software, it learns to constantly improve its efficiency
- **Supply chain effects**—Experience curve effects are not limited to one company. Suppliers and distributors are also able to improve the learning curve, making the whole value chain more efficient.

All in all, the concept of the experience curve is an interesting tool on which to model the learning and relearning effect for individuals, crews, operators, projects, companies, and even supply chains.

Addressing Header Problem 6.1: In Search of Improvements

Assumptions

First, it is assumed that the respondents who used the Don't Know column have not made a sufficient number of observations to address a particular question. Second, the Likert scale does not allow the respondents to be neutral. Thus, answers in the middle (Agree and Disagree) can mean that a person leans toward one side or the other, without strong convictions.

Analysis of Survey Data

Ms. Lukas is clearly a strong communicator, with strong behavioral qualities. Her weak points seem to be in her technical management of the operation. One area stands out: namely, the sufficiency of equipment and hardware, combined with a possible inadequacy in maintenance and technical training. In sum, three issues might be linked together, in that lack of training might lead to misuse, resulting in frequent breakdowns, thus overwhelming the maintenance department (see Figure 6.13).

The hardware needed most in prefabrication are:

- Mobile and rail gantry cranes
- Heavy forklift, pallet lift mover, and hand carts
- Mobile hydraulic pump and prestressing jack

- Concrete placing equipment, panel lifter, and hydraulic panel tilting table
- Concrete saw and sandblasting machinery

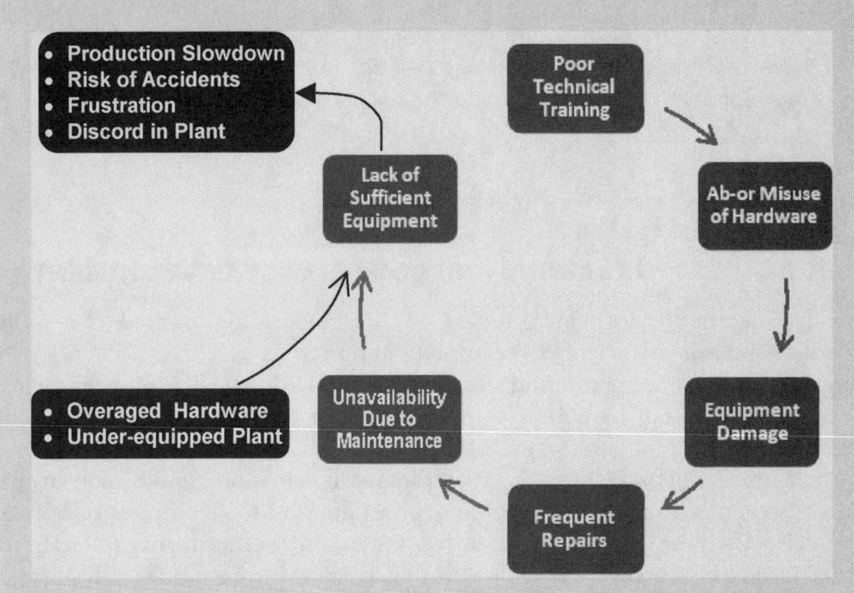

Figure 6.13 Cause and Effect Related to Poor Technical Training and Underequipped Plant

It is apparent that the majority of the hardware is dedicated to dealing with the heavy concrete elements, requiring proper maintenance, as well as skill and understanding of force vectors, moments, and torques. Specialty equipment, such as the hydraulic pump, requires extensive understanding of pressure-force relationships, as well as the design of hoses and connections. All in all, the technology in a precast plant has made great strides in the last 20 years, which requires more and more highly technical knowledge for the efficient and safe handling of equipment. This fact is easily forgotten in the heat of work pressures when, for example, someone unqualified impulsively tries to "help out" and takes over the controls.

Recommendations:

Priority 1: Development of an in-house standard technology training program for each major equipment type that contains several levels, from novice to expert.

(*continued*)

(*continued*)

Priority 2: Establishment of an incentive system to motivate the operational personnel to participate in the in-house training. Each manager should become competent on about 50 percent of the hardware.

Priority 3: Identify operational bottlenecks caused by an insufficient equipment park. Study of cause-and-effect relationships as a basis for recommendations to improve operational hardware in the future.

6.4 JOB-ORIENTED TRAINING AND COMPETENCY DEVELOPMENT

The 360-degree performance survey of Ms. Lukas highlighted the multidimensional effect that a lack of technical training can have. This is not a new phenomenon in the construction industry, as it is assumed that, like a journeyman in the Middle Ages, staff have gone through sufficient training. On the other hand, technology is changing more rapidly than ever before.

In a precast plant, where the workforce is much more stable than in general construction, an ongoing training program could offer opportunities for personal growth and professional advancement in the company, while eliminating large amounts of waste—waste that costs a lot of money. How might we devise a training program that builds on what we've been discussing about learning and teaching?

Figure 6.14 offers a basic model for coached competency building. Learning is shown as a cyclical process in which a student advances from novice to expert. At the center of the model is the learner who is motivated to acquire new capabilities by adding to his or her present level of understanding and skills. As shown, the learner's predisposition toward learning is affected by many factors, such as the person's motivation, physical dexterity, facility for learning, and so on. Some people will learn quickly, while others will need more time.

Experience with this concept shows that motivation is by far the most direct predictor of success. Personal motivation is linked to personal ambition and, very importantly, to the belief that the organization will reward the effort.

6.4.1 Personal Learning Strengths Dictate Progress

Many types of construction skills can be acquired through training. For example, operating a backhoe excavator safely and effectively requires not only psychomotor motion skills but also a knowledge of soil behavior, hydraulics and mechanics. Welding requires, besides a steady hand, knowledge of how different steels behave when heated, and the physics of various welding technologies. A manager not only needs knowledge about economics; he or she also

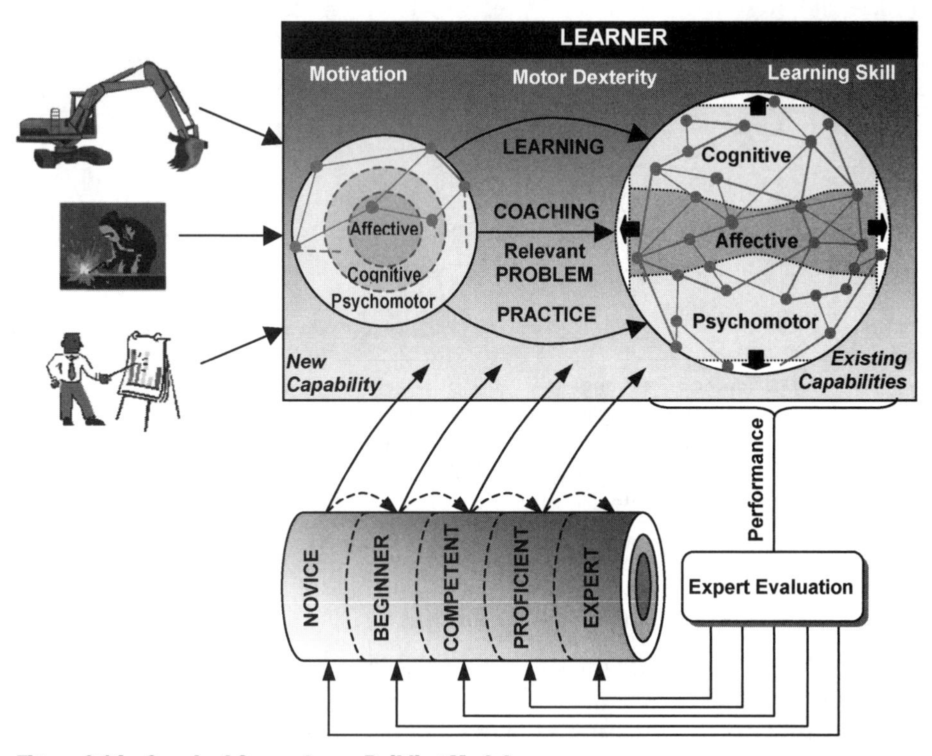

Figure 6.14 Coached Competency Building Model

needs to be an effective public speaker, a skill that requires substantial practice. Figure 6.15 presents a conceptual graphic showing the result of practice and confidence-building on efficient completion of a problem-solving exercise to reach a desired goal (e.g., complete a puzzle). Motivated to try, a novice copies and executes small "safe" steps during a process, with many interruptions to check progress and plan—or replan—the next step.

The learning curve effect results in reduced interruptions along the way and, at the same time, increased confidence in the rapidly improving ability. Climbing Bloom's pyramids of cognitive, affective, and psychomotor knowledge, the novice slowly develops into an expert, who can efficiently develop and execute his or her own plans. True experts are able to solve unfamiliar problems by recognizing familiar patterns, quickly devising and executing efficient solutions based on knowledge stored away after intensive practice.

Metacognitive Knowledge

Personal metacognitive awareness was a concept recognized by the Greek philosopher Aristotle (384–322 BC). It relates to the learner's ability to monitor and control learning. Students who possess metacognitive skills—such as selecting the best reading method for a subject, or self-testing—perform better on tests and writing assignments. Metacognitive knowledge helps the learner to become self-regulated, to know for example, how to pick the right tool for the job.

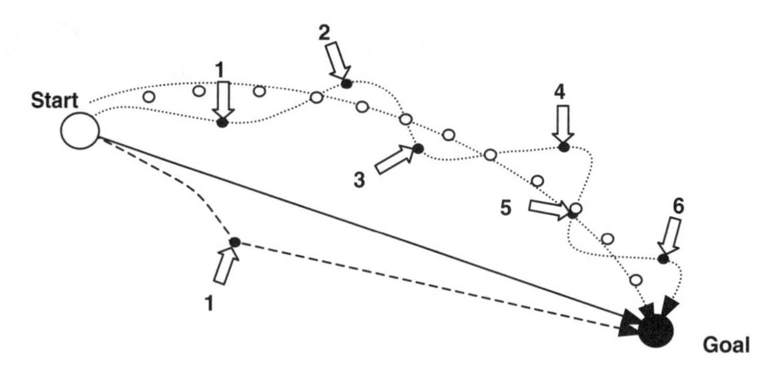

Figure 6.15 Individual Learning Paths, from Novice to Expert

6.4.2 Designing a Confidence-Building Process

Gaining knowledge or expertise is a process, not a product, one that has to be performed by the learner. That said, it is the coach or teacher who designs not only the problem(s) but also determines the teaching strategy that takes into account every learning preference, so that learners can "lock on" and practice as much and as long as they desire, in the way best suited to their individual styles of learning.

The expert fulfills another critical function in the learning process: He or she provides invaluable feedback about the learner's progress. This evaluation is extremely valuable to both the coach and the trainee in engineering the most effective training program. The matrix shown in Table 6.5 builds on the learning model shown in Figure 6.9. It aligns the competency levels and the characteristics of the four major learning preferences with specific activities that a trainer or teacher could design.

What better motivator to move from one competency level to the next, and continue the learning process, than the feeling that one has really mastered the necessary information and achieved in prescribed skill level during the previous phase of the process? Along with a proven learning method is, as just mentioned, an evaluation by an expert or master. The goal of such an evaluation is not to find out what the students do *not* know, but to help them prove to themselves that they did learn according to expectations, thus gaining the confidence to build on and strengthen competency.

As noted earlier, the benefit of confidence-building evaluations will increase dramatically if the different learning strengths exhibited by the trainees are taken into consideration: A critical thinker might be asked a more

Table 6.5 Growth-Directed Competency Levels of Trainees

Level	Descriptor	Learning Preference	Characteristic Activities of Trainee/Learner at Each Level
1	Novice	Why? What? How? What-if?	Discussing problems; watching coach; developing own plan; taking first stepsStudying written materials/ manuals; watching relevant videos; copying example; taking first stepsSearching for example solutions; making exact copy of example; taking first stepsQuestioning coach; planning a possible path using previous knowledge; taking first steps
2	Advanced beginner	Why? What? How? What-if?	Recognizing cause-effect relationships between certain functions and successful actionsGaining ability to select and execute successful functions to solve general problem sets
3	Competent	Why? What? How? What-if?	Acquiring multiple sets of scripted success functions to apply toa multitude of situationsAnalyzing a problem and selecting scripted functions to execute actions
4	Proficient	Why? What? How? What-if?	Learning to analyze a situation holistically, and separate problems according to prioritySelecting best plan of action instinctively and quickly
5	Expert	Why? What? How? What-if?	Grasping new situation based on tacit knowledgeHas assimilated rules and regulations, leading to fluidity, flexibility, and high proficiency

theoretical *Why?* question, whereas a *How?* oriented student might be queried on more practical issues.

Eventually, there comes a time for the final assessment, when the journeyman has to present his "masterpiece" in order to be declared a master. At this time, the his everyday skills and knowledge have to be second nature; at the same time, future improvements to become a true expert are viewed as desirable and achievable.

Table 6.6 details how to link the core competencies of a backhoe operator with the everyday activities or knowledge that a new "master operator" should be able to demonstrate to an encouraging expert.

At this juncture, it is essential to restate the importance of a learner's metacognitive abilities, to maintain confidence and a high level of motivation. According to the definition given earlier, metacognition is based on the "availability" of a range of learning skills to choose from. Worked-out-Problem 6. What Does It Take to Learn? lists necessary study skills; they can be expanded to include: (a) concentration/focusing, (b) note taking and summarizing, (c) Internet literacy, (d) SRQ3 (reading-to-understand), (e) self-evaluation, (f) test-taking strategies, (g) mind

Table 6.6 Job-Specific Competency Matrix for Training and Evaluation of a Backhoe Operator

	Issues	Operator Activities Linked to Safe and Productive Operation
	1. Equipment selection	Verifies that a backhoe has the capacity to perform the necessary digging and lifting tasks.
	2. OSHA rules in work zone	Ensures that the site layout provides sufficient access and protections according to OSHA 29 CFR 1926.800 (Excavation).
Job planning	3. Safety equipment	Checks the availability of appropriate safety equipment (e.g., ladders, barricades, trench boxes).
	4. Line of authority	Identifies the person who is competent and authorized to start the work.
	5. Notification of one-call	Follows OSHA 29 CFR 1926.651.
	1. Walk-through	Reviews utility markings with available as-built drawings and layout of new job. Establishes action plan to protect against overhead power lines (OSHA 29 CFR 1926.550(a)(15)). Finalizes plan of work with authorized/competent person.
	2. Laborer safety training	Interviews workers about their knowledge of safety procedures and attitudes. Initiates additional training if necessary.
Preparation	3. Inspection of equipment	Inspects the delivered equipment, including maintenance record.
	4. Work zone safety installations	Inspects barricades around the work site and the backhoe (back-swing).
	5. Inspection of add-on devices	Checks that all the devices are working properly.
	1. Enforcement of safety standards	Insists on adherence to safety procedures, from the beginning of the project.
	2. Equipment setup	Travels at appropriate speed; levels the machine, activates outriggers, and ensures that attachments are secured.
Operation	3. Equipment use	Avoids abusing the equipment; stays within load capacity.
	4. Energy efficiency	Moves material the shortest distance, without rehandling and wasteful machine idleness; ensures overall smooth operation.
	5. Ergonomic and drug-free workplace	Avoids risky behavior caused by work-induced fatigue and use of drugs.

mapping, (h) anxiety/stress management, and (i) study aids. Metacognitive capabilities are especially helpful to self-directed students operating within an adult learning framework.

Concept/Mind Mapping Boosts Learning and Thinking

This multipurpose yet simple tool helps learners transform new information into personal knowledge. Instead of taking written notes or writing summaries, learners represent new information as webs of connected nodes, similar to the way the brain works. Instead of listing facts, they construct and add new facts to what is already known. Illustrations give the learner the opportunity to use colors, draw meaningful pictures, or, using computerized mapping, include sounds, thus invoking the affective power of constructive learning. And example is shown in the figure below.

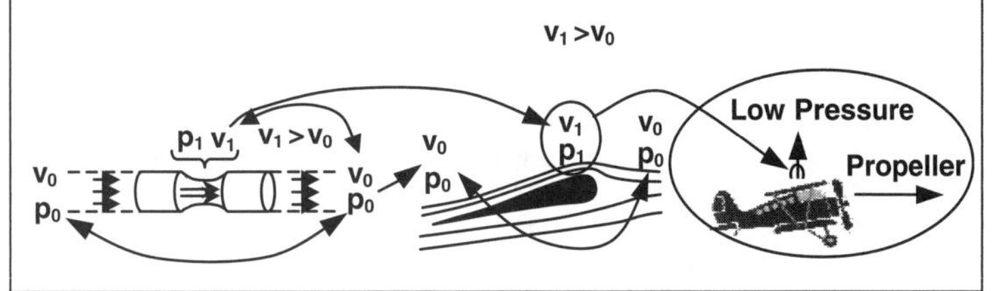

6.5 BECOMING A LEARNING ORGANIZATION (LO)

As previously mentioned, training of field and headquarters managers is not as widely implemented in construction as in other industries. Traditionally, it relies on the individual to acquire "whatever it takes" through trial-and-error and learning through job rotation. This approach may be sufficient for a company that does not believe that a culture of continuous improvement is necessary. But experience revealed a number of benefits to an organization that considers itself an LO—one that encourages its managers to participate in training programs—that are summarized in Table 6.7.

Training needs can be identified within each of the three main knowledge areas: technical, managerial, and affective. Each company will, of course, have its own specific requirements in the technical arena, whereas managerial competencies typically are more similar among companies.

For example, many surveys have been conducted to establish a prioritized list of the knowledge and skills that young engineers often lack when they first enter the work force, and later when they move on to managerial positions. Such data could provide the impetus to provide entry-level training, as well as for continuous competency development for an entire professional group—in this case, engineers. But beyond individual-level training, there may also be a need for training teams or the entire organization.

Still, even if we establish the best training program in the world, if nobody signs up for it, the effort will fail. This is the primary incentive for surveying the needs of the present workforce, from bottom to top. The inevitable question, "What's in it for me?" can be easily answered when employees realize they will learn something that will streamline their work or provide professional development opportunities.

For a company that already uses a competency-based evaluation system, identifying weaknesses may also reveal other sets of needs. Figure 6.16 illustrates that a multilayered needs assessment should be the first step in establishing a successful

Table 6.7 Learning Organization Benefits

Source of New Ideas	Managers in a line-function tend to become isolated from other parts of the company. Managers in a LO company strive to learn from others who have developed new and effective practices that are easily adapted from one area of the organization to another.
Reenergize Performance	Well-organized programs expose trainees to a variety of different experiences that promote lateral thinking. Discussions, role-playing, brainstorming, and so on allow managers to take on different functions and explore different viewpoints.
Guided Evaluation of Own Management	With the ability to "step outside" daily responsibilities, the trainee will be able to use his or her own operation as a test case to be reviewed in class. Guided by trainers, the manager evaluates his or her management tools, such as communication or performance assessment. Presented with options he or she might be motivated to adopt proven alternative methods.
New Support Network	Meeting other managers with whom to share both good and bad experiences opens a wide network, one that can be accessed even after the training session is over.
Pool for Internal Promotion	Participating managers show they are committed to the company, and demonstrate their willingness to learn. Those showing a marked improvement in performance can serve as model for others, and as candidates for in-house promotion.

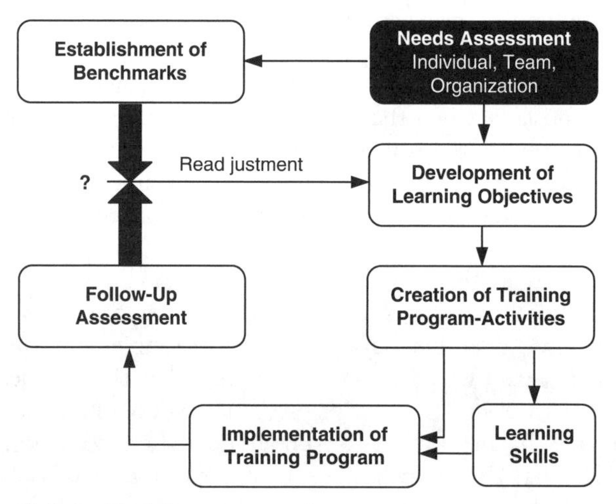

Figure 6.16 Needs-based Training

Table 6.8 Recognized Weaknesses of Engineers

Personal Work Skills	Communication, time management, organization, delegation, emotional intelligence, social skills
Management 101	Decision making, accounting, management styles, scientific management, organizational theory, cost control, marketing
Human Resource Management	Human factors, motivation, learning, conflict management, performance evaluation, training, recruitment
Finance	Financial management, budgeting, financial accounting, balance sheet analysis, ROI analysis
Project Management	Project selection, contracting, estimating, scheduling, financing, supply chain, JIT

training program; Table 6.8 lists the perceived weaknesses, to be used as a foundation of needs training for that group.

As indicated in Figure 6.16, the list of needs is the basis for two equally important steps: (1) definition of learning objectives and (2) establishment of benchmarks to measure the success of the training program.

Success is, of course, determined by comparing the outcome of the training program with the desired impact, as represented by the benchmarks. Benchmarks may be a 10 percent year-on-year reduction in the incidence of all reportable accidents, or an increase in equipment availability by 10 percent. Without measurable benchmarks, training programs not only will lose credibility but cannot be improved strategically. This will frustrate those who were motivated to participate in the first place, and lend credence to those who questioned the value of the expense to implement the program.

As discussed earlier, learning objectives include skills, knowledge, behavior, or a combination of those three. They have to be aligned with the desired outcomes or benchmarks, and be anchored in the list of needs.

Ensuring that the training modules are indeed relevant to the trainees, learning objectives should be shaped into example problems that an attending manager typically faces. Problem-oriented learning of this sort is more than just an appropriate pedagogical strategy for adult learners; it also demonstrates the immediate applicability of the lesson. A wide range of classroom and on-the-job training methods are available, which can be woven into a rich learning experience that addresses all four learning types, as well as variations in knowledge backgrounds. New material should be processed around the circle in such as way to address the four key questions: Why?, What?, How?, and What-if?

Before launching a training module, it is essential to ensure that everyone involved has the necessary learning capabilities. Worked-out Problem 6.2: What Does it Take to Learn?, can be used as a guide to test and amend necessary skills (a variety of assessment tools are available online). The learning model in Figure 6.16

stresses not only the need for the trainees to practice what they learn, but also the importance of expert assessments as feedback to the students. In this way, the coach can gauge how far along each student is in the learning process, and more important, the students then have a performance report on their progress toward greater competency. Such awareness is very useful as a motivational tool, especially after initial enthusiasm has waned.

Is location important—classroom versus on the job site? Learning can take place anywhere, at least potentially, as long as the motivational effort and learning skills of the trainee match the program contents. Consider the coach charged with safety training site personnel: He or she may create prepackaged "toolbox" sessions that can be held by a foreman or on-site safety officers. New personnel could get their initial instruction in headquarters workshops, equipped with the necessary hands-on learning tools, followed up with on-site assessments performed by safety officers. Again, a motivational system that encourages each individual to improve, from novice to expert, will lead to a reduction in accidents and an improvement in productivity. Here, "Safety First" is not just a slogan. Planning each process with safety as a key goal will lead to new ideas for actually *improving* safety, not just following already-existing rules; and with this comes both increased productivity and higher quality.

Worked-Out Problem 6.2: What Does It Take to Learn?

OnTime Inc., a company offering project management services, plans to launch an online training program to develop the knowledge of its newest hires. Later, it plans to offer advanced programs in leadership to employees who are slated for managerial positions in the company. Brent Padgett, the director of the HR department, is concerned about the effectiveness of online education; specifically, he questions the ability and willingness of young college graduates to learn on their own. Mr. Padgett asks for a list of competencies that could provide the basis for assessing and training the learning skills that each participant is lacking.

Background

A review of the current literature in education highlights a series of issues to consider in regard to the online, or e-learning, approach that OnTime Inc. intends to launch. There are many ways these types of programs can be implemented. All state-of-the-art applications are, however, founded on the *constructivist* view of learning, which is considered a social and cognitive process whereby students need, in addition to the mental skills to learn, a social environment for human interaction, as a means to create and evaluate the acquired knowledge. Furthermore, and equally important, is the efficacy to take on and sustain the demands that a self-directed learning experience will demand. Required are self-confidence, desire to learn, self-motivation, discipline, perseverance under stress, goal-setting, self-control, curiosity, positive attitude, ambition to succeed, and action orientation.

Cognitive abilities that are especially important in self-directed learning, such as e-learning, comprise a series of learning or study skills. They include the capability to apply effective study aids to identify and process important concepts from a large amount of materials in a short amount of time. To ensure the effectiveness of this process, the students need to be able to set aside study time and establish a method of testing themselves on the information they are learning.

COMPETENCIES FOR E-LEARNING

Although the literature on this topic does not provide a validated and proven tool for use in evaluating e-learning, Table 6.9 attempts to cover the most important aspects; it is not intended to be comprehensive.

Table 6.9 Traits of a Competent Learner

Category	Specific Traits
Learning Skills	Time management skill
	Note taking and summarizing
	Overcoming procrastination
	Internet literacy
	Capability to concentrate
	Study strategies
	Self-evaluation/self-control
	Test-taking strategies
	Metacognitive awareness
	Problem-solving capability
Knowledge	Prerequisite knowledge
Self-Efficacy/Personal Traits	Self-motivation; expectation of future benefits
	Willingness to do the work and prioritize learning
	Realistic overall plan
	Self-confidence about self-learning capabilities
	Goal orientation
	Perseverance under pressure
	High level of curiosity

(*continued*)

(continued)

Discussion

As important as student preparedness to the success of an e-learning program is the method of teaching. It is assumed in this case that the online course will be built on constructivistic principles.

The level of expected proficiencies by a prospective e-learning student can be evaluated using validated assessment tools that are available on the market today. It is proposed that the HR department buy several of the preformulated questionnaires, to be filled out by the course candidates. The results of the evaluation should be used to remedy deficiencies.

CHAPTER REVIEW

Journaling Questions

1. This chapter presented a list of generic competencies that every employee should possess in order to be successful in the twenty-first-century business world. It also provided a list of skills that today's engineers show weakness in. Can you identify any important connections between the two lists? Discuss two avenues that could be taken to improve the present situation. Which one would you recommend?
2. Refer to the learning-curve experiment described in Worked-out Problem 6.1. Can you identify what is changing between trials? In other words, what capacity of the "puzzler's" brain seems to allow him or her to speed up certain parts of the puzzle while struggling with others? You may want to focus on how the puzzler works on the same three or four pieces time over time.
3. The university is the quintessential learning organization (LO). Do you think that you have been properly trained to be a successful learner? From the material presented about learning, what do you think would help you in your studies?

Traditional Homework

1. A company that manufactures pneumatic nail guns field-tested a new gun with a built-in sensor, used to detect the location of a stud behind a plywood piece. It was designed to make a difference in the speed at which a novice carpenter can properly nail down a 12 × 12-foot section of roof sheathing. To save time for testing, the new device was tested on small number of houses.

Assume that the first sheet will take an average of 14 minutes, regardless whether a novice uses a traditional or the new sensor-equipped gun. Additional data points for the new gun are:

Time for second sheet = 9.4 minutes

Time for third sheet = 7.4 minutes

Time for fourth sheet = 6.0 minutes

Assumptions for traditional nailing:

Progress Index = 80 percent

Time for first sheet = 14 minutes

Based on these data:

a. What will the duration be for the novice's fiftieth sheet using the traditional nail gun?
b. How many trials will be needed using the traditional nail gun to reach a duration that matches that using the new gun?
c. Plot the results of your comparison on an appropriate graph, putting the number of trials along the x-axis.

2. You are asked to design a one-hour class module on competency-based precast plant management for your classmates, who represent all four learning styles.
   a. What will be the sequence of material/subjects you will cover?
   b. What support materials will you prepare for everyone to read before coming to class (the topic is too big to cover in an hour).
   c. What will you do to teach the main subjects from "around the circle?"
   d. Create a matrix that shows how your planned activities (on the y-axis) will meet the strengths of each learning style (on the x-axis).

Open-Ended Questions

1. The list of competencies expected of a precast plant manager extends far beyond technical knowledge about concrete and cranes.
   a. What background would better prepare someone to become a plant manager: a business major, a construction management major, or a construction engineer? Why?
   b. What are the competencies that your recommendation is missing?
   c. What sequenced training levels would you propose are necessary to provide the missing competencies of your favorite choice?
   d. How will you make sure that a potential candidate will succeed after starting your proposed program? Assume that it will take probably

several years of the person's free time and a large amount of the company's money.

BIBLIOGRAPHY

Aghazadeh, S. M. Re-Examining the Training Side of Productivity Improvement: Evidence from Service Sector. *Inter. J. Prod. Perform. Mgmt.*, vol. 56, no. 8, 2007.

Alavi, M., J. L. Cook, and D.E. Leidner. Knowledge Management and Knowledge Management Systems: Conceptual Foundation and Research Issues. *MIS Quarterly*, vol. 25, no. 1, 2001.

Aleven, V., E. Stahl, S. Schworm, F. Fischer, and R. Wallace. Help Seeking and Help Design in Interactive Learning Environments. *Rev. of Educational Res.*, vol. 73, no. 3, 2003.

Anderson, L. W., and D. R. Krathwohl (eds.). A Taxonomy for Learning, Teaching, and Assessing: A Revision of Bloom's Taxonomy of Educational Objectives. New York: Addison Wesley Longman, 2001.

Anton, J. What is the Role of Academic Institutions for the Future Development of Six Sigma? *Int. J. Prod. Perform. Mgmt.*, vol. 57, no. 1, 2008.

Azmi, F. T. Mapping the Learn-Unlearn-Relearn Model Imperatives for Strategic Management. *European Bus. Rev.*, vol. 20, no. 3, 2008.

Berik, G., and C. Bilginsoy. Still a Wedge in the Door: Women Training for the Construction Trades in the USA. *Inter. J. of Manpower*, vol. 27, no. 4, 2006.

Bernold, L. E. Applying Total-Quality-Management Principles to Improving Engineering Education. *J. Prof. Issues in Engrg. Educ. and Pract.*, vol. 134, no. 1, 2008.

______. Early Warning System to Identify Poor Time Management Habits, *Int. J. of Engrg Edu.*, vol. 23, no. 6, 2007. (www.ijee.dit.ie/)

______. Quantitative Assessment of Backhoe Operator Skill. *J. Constr. Engrg. Mgmt.*, vol. 133, no. 11, 2007.

______. Preparedness of Engineering Freshman to Inquiry-Based Learning. *J. Prof. Issues Engrg. Edu. Pract.*, vol. 133, no. 2, 2007.

______. A Paradigm Shift in Education is Vital for the Future of Our Profession. *J. Constr. Eng. and Mgmt.*, ASCE, vol. 131, no. 5, 2005.

Bernold, L. E., J. Spurlin, and C. M. Anson. Understanding Our Students: A Longitudinal Study of Success and Failure in Engineering. *J. of Engrg. Edu.*, ASEE, vol. 96, no. 3, 2007.

Bibby, L., S. Austin, and D. Bouchlaghem. The Impact of a Design Management Training Initiative on Project Performance, *Engrg. Constr. Arch. Mgmt.*, vol. 13, no. 1, 2006.

Binde, J. What Future is There for Work? *Foresight*, vol. 7, no. 4, 2005.

Brandenburg, S. G., C. T. Haas, and K. Byrom. Strategic Management of Human Resources in Construction. *J. Mgmt. Engrg*, vol. 22, no. 2, 2006.

Caligiuri, P. Developing Global Leaders. *Human Res. Mgmt. Rev.*, vol. 16, 2006.

Capaldo, G., L. Iandoli, and G. Zollo. A Situatationalist Perspective to Competency Management. *H. Res. Mgmt.*, vol. 45, no. 3, Fall 2006.

Carayanisa, E. G., D. Popescub, C. Sippc, and M. Stewart. Technological Learning for Entrepreneurial Development (TL4ED) in the Knowledge Economy (KE): Case Studies and Lessons Learned. *Technovation*, 26, 2006.

Cardinali, R. Women in the Workplace: Revisiting the Production Soldiers, 1939–1945. *Work Study*, vol. 51(3), 2002.

Carnahan, B. J. Identifying Training Needs of Logging Truck Drivers Using a Skill Inventory. *J. Agri. Safety Health*, vol. 10, no. 4, 2004.

Charness, N., and M. Tuffiash. The Role of Expertise Research and Human Factors in Capturing, Explaining, and Producing Superior Performance. *Human Factors*, vol. 50, no. 3, June 2008.

Dattée, B., and H. B. Weil. Dynamics of Social Factors in Technological Substitutions. *Techn. Forecast Soc. Change*, 74, 2007.

Debra, Y. A. and G. Ofori. The State, Skill Formation and Productivity Enhancement in the Construction Industry: The Case of Singapore. *Int. J. Human Res. Mgmt.*, vol. 12, no. 2, 2001.

Derrick, M. G. Creating Environments Conducive for Lifelong Learning. *New Direction Adult Cont. Edu.*, no. 100, Winter 2003.

De Toni, A. F., A. Fornasier, M. Montagner, and F. Nonino. A Performance Measurement System for Facility Management. *Int. J. Prod. Perform. Mgmt.*, vol. 56, no. 5/6, 2007.

Doolen, T., E. Van Aken, J. Farris, J. Worley, and J. Huwe. Kaizen Events and Organizational Performance: A Field Study. *Int. J. Prod. Perform. Mgmt.*, vol. 57, no. 8, 2008.

Drexler, J., T. A. Beehr, and T. A. Stetz. Peer Appraisals: Differentiation of Individual Performance on Group Tasks., *Hum. Res. Mgmt.*, vol. 40, no. 4, Winter 2001.

Ellis, R. E., G. D. Wood, and T. Thorpe. Technology-Based Learning and the Project Manager., *Engrg. Constr. Arch. Mgmt.*, vol. 11, no. 5, 2004.

Enshassi, A., S. Mohamed, P. Mayer, and K. Abed. Benchmarking Masonry Labor Productivity., *Int. J. Prod. Perform. Mgmt.*, vol. 56, no. 4, 2007.

Epaarachchi, D. C., and M. H. Stewart. Human Error and Reliability of Multistory Reinforced-Concrete Building Construction. *J. Perform. Constr. Fac.*, vol. 18, no. 1, 2004.

Fitts, P. M. Perceptual-Motor Skill Learning. In *Categories of Human Learning*, New York: Academic Press, 1964.

Ginzburg, S., and E. M. Dar-El. Skill Retention and Relearning—A Proposed Cyclical Model., *J. Workplace Learning*, vol. 12(8), 2000.

Guri-Rosenblit, S. "Distance Education" and "E-Learning": Not the Same Thing. *Higher Education*, vol. 49, no. 4, 2005.

Haas, C. T., A. M. Rodriguez, R. Glover, and P. M. Goodrum. Implementing a Multi-Skilled Workforce. *Const. Mgmt. Eco.*, 19, 2001.

Halachmi, A. Performance Measurement is Only One Way of Managing Performance. *Int. J. Prod. Perform. Mgmt.*, vol. 54, no. 7, 2005.

Hanna, A. S. and K. T. Sullivan. Impact of Overtime on Construction Labor Productivity. *Cost Engrg.*, vol. 46, no. 4, 2004.

Hanna, A. S., P. Peterson, and M. Lee. Benchmarking Productivity Indicators or Electrical/Mechanical Projects. *J. Constr. Engrg. Mgmt.*, vol. 128, no. 4, 2002.

Hansen, M. T. Knowledge Networks: Explaining Effective Knowledge Sharing in Multiunit Companies. *Org. Sci.*, vol. 13, no. 3, 2002.

Haponava, T., and S. Al-Jibouri. Identifying Key Performance Indicators for Use in Control of Pre-Project Stage Process in Construction. *Int. J. Prod. Perform. Mgmt.*, vol. 58, no. 2, 2009.

Harney, B., and C. Jordan, Unlocking the Black Box: Line Managers and HRM-Performance in a Call Centre Context. *Int. J. Prod. Perform. Mgmt.*, vol. 57, no. 4, 2008.

Harris, E. G., and J. M. Lee. Illustrating a Hierarchical Approach for Selecting Personality Traits in Personnel Decision: An Application of the 3M Model. *J. Bus. Psych.*, vol. 19, no. 1, Fall 2004.

Harzallah, M., G. Berio, and F. Vernadat. Analysis and Modeling of Individual Competencies: Toward Better Management of Human Resources, IEEE Trans. Syst. *Man Cybernetics*, vol. 36, no. 1, January 2006.

Heron, R. *Job and Work Analysis: Guidelines on Identifying Jobs for Persons with Disabilities.* Geneva, Switzerland: Int. Lab. Org (ILO), 2005.

Hunt, J. W., and Y. Baruch. Developing Top Managers: The Impact of Interpersonal Skills Training. *J. Mgmt. Develop.*, vol. 22, no. 8, 2003.

Huselid, M. A. The Impact of Human Resource Management Practices on Turnover, Productivity, and Corporate Financial Performance., *Acad. Mgmt. J.*, vol. 38, no. 3, 1995. <www.jstor.org/stable/256741>

Ipe, M. Knowledge Sharing in Organizations: A Conceptual Framework. *Hum. Res. Developmt. Rev.* 2, 2003.

Judge, T. A., and C. A. Higgins. The Employment Interview: A Review of Recent Research and Recommendations for Future Research. *Hum. Res. Mgmt. Review.*, vol. 10, no. 4, 2000.

Karuppan, C. M. Strategies to Foster Labor Flexibility, *Int. J. Prod. Perf. Mgmt*, vol. 53 no. 6, 2004.

Kearns, P. *Generic Skills for the New Economy, Review of Research.* Australian National Training Authority, Kensington Park, Australia, 2001. <www.ncver.edu.au>

Knuf, J. Benchmarking the Lean Enterprise: Organizational Learning at Work. *J. Mgmt. Engrg.*, vol. 16, no. 4, 2000.

Kolb, D. A. *Experiential Learning*, Englewood Cliffs, NJ: Prentice-Hall, 1984.

Konrad, A. M., and J. Deckop. Human Resource Management Trends in the USA—Challenges in the Midst of Prosperity. *Int. J. Manpower*, vol. 22, no. 3, 2001.

Kosbab, D. Dispositional and Maturational Development Through Competency-Based Training, *Edu. + Train.*, vol. 45(8/9), 2003.

Krathwohl, D. R., B. S. Bloom, and B. B. Masia. *Taxonomy of Educational Objectives, the Classification of Educational Goals. Handbook II: Affective Domain.* New York: David McKay Co., Inc., 1973.

Kruger, J., and D. Dunning. Unskilled and Unaware of It: How Difficulties in Recognizing One's Own Incompetence Lead to Inflated Self-Assessments. *J. Personality Soc. Psych.*, Vol. 77(6), 1999.

Kulatunga, U., D. Amaratunga, and R. Haigh. Performance Measurement in the Construction Research and Development. *Int. J. Prod. Perform. Mgmt.*, vol. 56, no. 8, 2007.

Kumar, U. D., H. Saranga, J. Ramirez-Marquez, and D. Nowicki. Six Sigma Project Selection Using Data Envelopment Analysis. *TQM Magazine*, vol. 19, no. 5, 2007.

Lam, A., and B. A. Lundvall. The Learning Organization and National Systems of Competence Building and Innovation. In N. Lorenz and B-A Lundvall (eds.), *How Europe's Economies Learn: Coordinating Competing Models*, New York: Oxford University Press, 2006. <http://mpra.ub.uni-muenchen.de/12320/>

Lapré, M. A., A. S. Mukherjee, and L. Van Wassenhove. Behind the Learning Curve: Linking Learning Activities to Waste Reduction. *Mgmt. Science*, vol. 46, no. 5, 2000.

Leidner, D., and S. L. Jarvenpaa. The Use of Information Technology to Enhance Management School Education: A Theoretical View. *MIS Quarterly*, vol. 19, no. 3, 1995.

Lengnick-Hall, M., C. A. Lengnick-Hall, L. S. Andrade, and B. Drake. Strategic Human Resource Management: The Evolution of the Field. *Res. Mgmt. Review*, 2009.

London, M., and V. Sessa. How Groups Learn Continuously. *Human Res. Mgmt.*, vol. 46, no. 4, 2007.

Love, P. E., and P. Josephson. Role of Error-Recovery Process in Projects. *J. Mgmt. Engrg.*, vol. 20, no. 2, 2004.

Major, D. A., D. D. Davis, L. Germano, T. D. Fletcher, J. Sanchez-Hucles, and J. Mann. Managing Human Resources in Information Technology: Best Practices of Higher-Performing Supervisors. *Hum. Res. Mgmt.*, vol. 46, no. 3, 2007.

Man, J. Six Sigma and Lifelong Learning. *Work Study*, vol. 51(4), 2002.

McInerney, C. Knowledge Management and the Dynamic Nature of Knowledge. *J. Amer. Soc. Info. Sc. Tech.*, vol. 53, no. 12, 2002.

Mellander, K. Engaging the Human Spirit: A Knowledge Evolution Demands the Right Conditions for Learning. *J. Intellectual Capital*, vol. 2, no. 2, 2001.

Min, W., and L. S. Pheng. Re-Modeling EOQ and JIT Purchasing for Performance Enhancement in the Ready Mixed Concrete Industries of Chongqing, China and Singapore. *Int. J. Prod. Perform. Mgmt.*, vol. 54, no. 4, 2005.

Mowen, J. C., S. Park, and A. Zablah. Toward a Theory of Motivation and Personality with Application to Word-of-Mouth Communications. *J. Bus. Res.*, 60, 2007.

Murray, B., and B. Gerhart. Skill-Based Pay and Skill Seeking. *Human Res. Mgmt. Rev.*, vol. 10, no. 3, 2000.

Murray, L. W., and A. D. Efendioglu. Valuing the Investment in Organizational Training. *Indust. Commercial Training*, vol. 39, no. 7, 2007.

Nygaard, C., and P. Bramming. Learning-Centered Public Management Education. *Int. J. Public Sect. Mgmt.*, vol. 21 no. 4, 2008.

Othman, R. Enhancing the Effectiveness of the Balanced Scorecard with Scenario Planning. *Int. J. Prod. Perform. Mgmt.*, vol. 57, no. 3, 2008.

Paajanen, G. E., T. L. Hansen, and R. A. McLellan. Employment Inventory Research. *Pers. Dec. Int.*, 1999.

Pathirage, C. P., D. G. Amaratunga, and R. P. Haigh. Tacit Knowledge and Organizational Performance: Construction Industry Perspective. *J. Knowledge Mgmt.*, vol. 11, no. 1, 2007.

Pew, R. More Than 50 Years of History and Accomplishments in Human Performance Model Development. *Human Factors*, vol. 50, no. 3, June 2008.

Pollitt, D. Training Means Zero Accidents for FHM: Program Exceeds All Expectations. *Inter. Digest Human Res. Mgmt.*, vol. 14, no. 4, 2006.

Priddy, L. The View Across: Patterns of Success in Assessing and Improving Student Learning. *On the Horizon*, vol. 15, no. 2, 2007.

Quek, A. Learning for the Workplace: A Case Study in Graduate Employees' Generic Competencies. *J. Workplace Learning*, vol. 17, no. 4, 2005.

Raiden, A. B., and A. R. Dainty. Human Resource Development in Construction Organizations An Example of a "Chaordic" Learning Organization? *The Learning Org.*, vol. 13, no. 1, 2006.

Radnor, Z., and D. Barnes. Historical Analysis of Performance Measurement and Management in Operations Management. *Int. J. Prod. Perform. Mgmt.*, vol. 56, no. 5/6, 2007.

Ramirez, Y. W., and D. A. Nembhard. Measuring Knowledge Worker Productivity: A Taxonomy. *J. Intel. Capital*, vol. 5, no. 4, 2004.

Rasmussen, J.*Information Processing and Human-Machine Interaction: An Approach to Cognitive Engineering.*, Amsterdam, Netherlands, 1986.

Ravensteijn, W., E. De-Graff, and O. Kroesen. Engineering the Future: The Social Necessity of Communicative Engineers. *Eur. J. Engrg. Edu.*, vol. 31, no. 1, 2006.

Roberts, R., and P. Hirsch. Evolution and Revolution in the Twenty First Century: Rules for Organizations and Managing Human Resources. *Hum. Res. Mgmt.*, vol. 44, no. 2, 2005.

Robotham, D. Developing the Competent Learner. *Indu. Com. Training*, vol. 36 (2), 2004.

______. Learning and Training-Developing the Competent Learner. *J. Euro. Ind. Training*, vol. 27(9), 2003.

Rosti, R. T., and F. Shipper. A Study of the Impact of Training in a Management Development Program Based on 360 Feedback. *J. Managerial Psych.*, vol. 13, no. 1/2, 1998.

Sanchez, J.I., and E. L. Levine. What Is (or Should Be) the Difference Between Competency Modeling and Traditional Job Analysis? *Hum. Res. Mgmt. Rev.*, 2008.

Schlosser, F., A. Templer, and D. Ghanam. How Human Resource Outsourcing Affects Organizational Learning in the Knowledge Economy. *J. Labor Res.*, vol. 27, no. 3, 2006.

Scott, E. D. Just (?) a True-False Test Applying Signal Detection Theory to Judgments of Organizational Dishonesty. *Bus. in Soc.*, vol. 45, no. 2, 2006.

Serpell, A., and X. Ferrada. A Competency-Based Model for Construction Supervisors in Developing Countries. *Personnel Review*, vol. 36, no. 4, 2007.

Simpson, E. J. *The Classification of Educational Objectives in the Psychomotor Domain.* Washington, DC: Gryphon House, 1972.

Smith, P. J. Workplace Learning and Flexible Delivery. *Rev. Edu. Res.*, vol. 73, no. 1, 2003.

Stevens, J. A.*The Motivations-Attributes-Skills-Knowledge Competency Cluster Validation Model and Empirical Study.* Ph.D. dissertation, Texas A&M University, College Park, TX, August 2003.

Stoman, S. H. Effective Management Style. *J. Mgmt. Engrg.*, vol. 15, no. 1, 1999.

Stone, D. L., and K. M. Lukaszewski. An Expanded Model of the Factors Affecting the Acceptance and Effectiveness of Electronic Human Resource Management Systems. *Hum. Res. Mgmt. Rev.*, 2009.

Styhre, A. Peer Learning in Construction Work: Virtuality and Time in Workplace Learning. *J. Workplace Learning*, vol. 18, no. 2, 2006.

Tangen, S. An Overview of Frequently Used Performance Measures. *Work Study*, vol. 52, no. 7, 2003.

Theeranuphattana, A., and J. S. Tang. A Conceptual Model of Performance Measurement for Supply Chains. *J. Manu. Tech. Mgmt.*, vol. 19(1), 2008.

Thompson, P. How Much Did the Liberty Shipbuilders Learn? New Evidence for an Old Case Study. *J. Pol. Eco.*, vol. 109, no. 1, 2001.

Tangen, S. Demystifying Productivity and Performance. *Int. J. Prod. Perform. Mgmt.*, vol. 54, no. 1, 2005.

Ulrich, D., and N. Smallwood. HR'S New ROI: Return on the Intangibles. *Hum. Res. Mgmt.*, vol. 44, no. 2, 2005.

Valle, R., F. Martin, P. M. Romero, and S. L. Dolan. Business Strategy, Work Processes and Human Resource Training: Are They Congruent? *J. Org. Behavior*, vol. 21, no. 3, 2000.

Vermunt, J. D. Metacognitive, Cognitive and Affective Aspects of Learning Styles and Strategies: A Phenomenographic Analysis. *Higher Education*, vol. 31, no. 1, 1996.

Walker, G., and J. R. MacDonald. Designing and Implementing an HR Scorecard. *Human Res. Mgmt.*, vol. 40, no. 4, 2001.

Whiting, H., T. Kline, and L. Sulsky. The Performance Appraisal Congruency Scale: An Assessment of Person-Environment Fit. *Int. J. Prod. Perform. Mgmt.*, vol. 57, no. 3, 2008.

Winch, G., and B. Carr. Benchmarking On-Site Productivity in France and the UK: A CALIBRE Approach. *Constr. Mgmt. Eco.*, 19, 2001.

Yankov, L., and B. H. Kleiner. Human Resources Issues in the Construction Industry. *Mgmt. Res. News*, vol. 24, no. 3/4, 2001.

Yeo, R. K. Liberating Murphy's Law: Learning from Change. *Ind. Commercial Train.*, vol. 41, no. 2, 2009.

______. Change In(ter)ventions to Organizational Learning: Bravo to Leaders as Unifying Agents. *Learning Org.*, vol. 14, no. 6, 2007.

Zolin, R., P. J. Hinds, R. Fruchter, and R. E. Levitt. Interpersonal Trust in Cross-Functional, Geographically Distributed Work: A Longitudinal Study. *Info. Organ.*, 14, 2004.

Chapter 7

Productivity in a Healthy and Safe Work Environment

The construction industry has one of the most unsafe work environments, challenging construction managers every day to protect its workforce from harm. One such tool, the science that studies the human factor in the work environment, or *ergonomics*, concerns itself with the physical as well as cognitive capabilities while defining the limitations of humans operating within a productive system. It provides valuable guidelines for construction in setting up healthy and safe work stations.

Physical health can be destroyed by more than just physical factors, however. Work stresses that result in ulcers and insomnia are as destructive to productivity as accidents and overexertion on the construction site. Today's workplace is full of mental stressors, imagined or real, that are tolerated well one day but may breach the individual's psychological stress limits the next, eventually resulting in a physical reaction of the body. Finally, people themselves can create an unhealthy and unsafe workplace through harassments and bullying of colleagues.

As we will learn in this chapter, all of these stressors can have a dramatic negative effect on productivity.

7.1 TWO HEALTH STRESSES THAT AFFECT PRODUCTIVITY

It hardly need be said that human effort is at the heart of construction, and a worker's ability to perform at the level required of the job at hand depends on his or her overall condition. Since the experiments conducted at the Hawthorne Works in the 1920s we have known that productivity is linked to factors that cannot be improved by time studies. For example, even a highly competent and motivated scraper operator will be rendered immobile if one of his spinal discs tears or herniates, as the result of intense shock caused by "speeding" his vehicle repeatedly over a rutted road. A very different, but equally debilitating, example is the accidental death of a

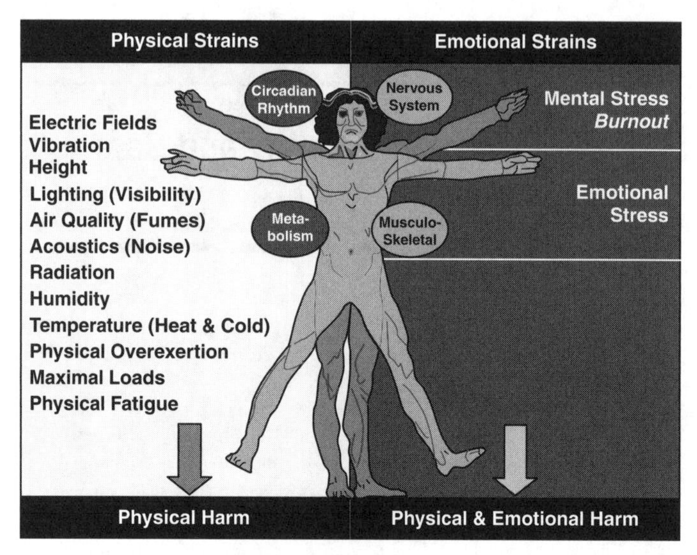

Figure 7.1 Strain and Stresses Affecting Humans

construction manager's son on a project he was managing, an event that understandably, would have a major impact on the manager's capability to perform.

These are but two examples of potentially devastating strains and stresses that can have a negative impact on job performance. Figure 7.1 graphs some of the most important of these to keep in mind.

The human body is an extremely complex and multifaceted system, but four elements are most critical to the understanding of physiological factors on productivity:

- The nervous system, which supplies the brain with information about its surroundings, and, when necessary, interacts with other nerves to instruct the muscles or organs to react and respond accordingly.
- The circadian rhythms, connected to the 24-hour day-night cycle of light and darkness, determine many human bodily cycles as well,, such as sleep, body temperature, and brain wave activity.
- The musculoskeletal system fulfills several functions that are critical to the performance of construction workers. The skeleton provides structural support, while the muscles make it possible to move about and to exert force on the environment.
- Finally, the chemical reactions of metabolism provide the body the with the energy to perform physical and mental work. As we will see later, it also fulfills other functions key to survival and the performance of heavy work.

All organizations have a moral and legal obligation to protect their workers from harm by providing a safe workplace. Construction, due to its many physical risks, focuses its efforts mainly on physical dangers and the avoidance of potentially

The Hawthorne Effect

In the 1920s, Hawthorne Works, a company that produced electronic parts, commissioned a major scientific study to determine whether its workers would work more productively under higher or lower levels of light. The employees were eager to participate in the study, and wanted to make it a success, resulting in a high increase in productivity. After the researchers left at the end of the project, however, productivity quickly returned to the level it had been before the study. This led to the realization that worker performance during the study had nothing to do with lighting, but rather with the desire of the workers—conscious or otherwise—to meet the perceived expectations of the researchers.

The *Hawthorne effect* refers to the changes in behavior by people taking part in a study that are an effect of that participation in itself, rather than of the actual subject under study.

deadly accidents. As Figure 7.1 shows, however, most dangers are non-lethal (at least in the short run) and are not always immediately visible. Factors that have the potential to cause death, injury, illness, disability, or simply a reduction in job performance include acoustic energy, chemical substances, oxygen deficiency, radiation energy, shock, temperature extremes, and humidity.

Unexplained disappearances from work or a sudden verbal outburst against a coworker are non-physical expressions of harmful stresses associated with work. Figure 7.1 lists these factors under emotional strains. The word "emotion" is from the French *emouvoir*, meaning to stir up or displace. Many human behaviors are outward manifestations of a person's subjective experiences and can be extremely self-destructive, with suicide being the action of last resort. If left unmanaged, work stresses can trigger emotional harm that can cross over into the physical realm and cause, for example, a heart attack.

7.2 THE ENGINE THAT DRIVES—AND LIMITS—HUMAN WORK

Anyone who studies the way the human body generates the energy it needs to operate and sustain itself is amazed at its complexity, and the array of chemical processes needed for it to function smoothly.

The central process of the body that maintains its operation is called the *metabolism*, consisting of two parts: *catabolism*, which is involved in breaking down organic matter to process it into energy, and *anabolism*, which transforms energy into life-sustaining cells such as proteins and nucleic acids.

7.2.1 A Look Into the "Boiler Room"

Our body needs a minimum amount of energy to sustain itself. All the vital organs in the human body, such as the heart and lungs, require some amount of energy to power their many muscles, 24/7.

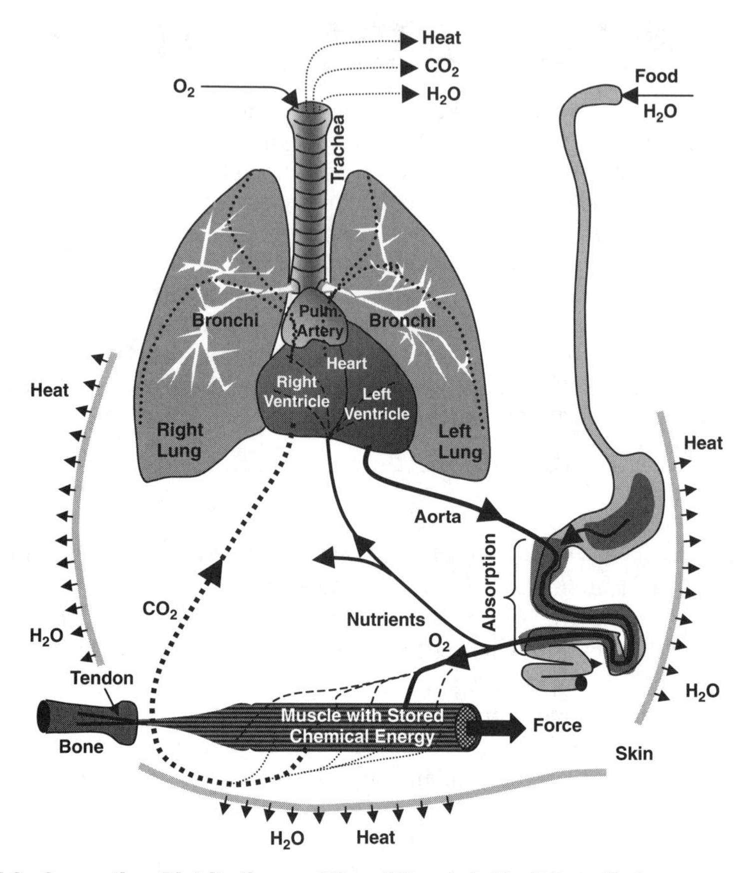

Figure 7.2 Generation, Distribution, and Use of Energy in the Human Body

The amount of basic energy conversion needed is called the *basal metabolic rate (BMR)*. The average BMR for 30-year-old males is approximately 70 kcal/hour; and for females, 38 kcal/hour. The human body creates energy according to the basic steps shown in Figure 7.2.

The metabolism depends on two basic ingredients: oxygen and nutrients. The body takes two separate paths to get these ingredients to the point of need: the muscles. One path starts at the human nose and mouth, which take in oxygen (O_2) through the breathing action of the lungs. The second path starts with the intake of food and water through the mouth, which travels to the stomach, where it is "treated" by chemicals before entering the digestive system. Through absorption, the nutrients inside the digestive system (e.g., intestines) are transferred into the bloodstream through osmosis, active transport, and diffusion. The resulting nutrients, together with the O_2 added to the bloodstream by the lungs, are pumped by the left ventricle of the heart to the organs and muscles, as needed. In turn, the muscle cells produce the adenosine triphosphate (ATP) molecules critical to the energy-producing process. Muscles can store a limited amount of energy in chemical form, which can be rapidly converted to glucose when energy is called for. This oxygen-free, "energy on demand" anaerobic (i.e., absence of oxygen) metabolic process gives way to a second method of producing energy, one that does depend on oxygen: the highly inefficient aerobic metabolic process. The body reacts to its varying needs by increasing the heart rate, blood pressure, and breathing rate, as more O_2 is required while pushing out H_2O. The heat generated is what causes hot and sweaty skin.

7.2.2 Calculating Sustainable Energy Use

In aerobic energy production, along the metabolic pathway, a chemical conversion takes place, of carbohydrates, fats, and proteins into carbon dioxide and water, to generate a form of usable energy. This process also produces lactic acid, a waste product, which is carried away by the blood. But if more of this acid is produced than can be carried away by the blood, it accumulates in the muscles and can eventually shut it down. The following equation provides the basic metabolic energy function:

$$\underbrace{\text{Food Energy}}_{\text{Input(kcal)}} = \underbrace{\text{Metabolic Energy Produced}}_{\text{Metabolic Rate(kcal/hr)}} + \underbrace{\text{Heat}}_{5-10\%} = \text{BMR} + \underbrace{\text{Productive Work}}_{\text{Output(kcal/hr)}} + \text{Heat} + \text{Storage}$$

Research has found that a person's BMR and the sustained metabolic energy production rate are related, though they vary slightly among groups of people. The general population can generate 2.4 times more energy than the BMR needs; in contrast, for long-distance runners, this factor rises to 4.0 to 5.0.

Table 7.1 Energy Required for Human Activities

Human Activity	Metabolic Energy (including BMR)	
	Female (kcal/hr)	Male (kcal/hr)
Sitting	65	90
Standing	85	100
Resting after work	66	100
Office work	95	110
Walking briskly	205	240
Shoveling sand	—	400–600
Bricklaying	—	150
Hammering, sawing	410	490
Digging in heavy soil	355–370	505–535
Long-distance running	530	600

Worked-Out Problem 7.1: Time it Takes to Run Out of Energy

Your boss has learned that you have studied metabolism and the length of time a worker can sustain a steady state of energy. She knows that the time until exhaustion occurs depends on the type of work performed, the fitness of the body, and the surroundings. She would like to know the length of time that work can be sustained before a break is needed, how long the break should be, and why. She asks you to compare a male office worker with a shoveling laborer and a runner.

Additional Information and Assumptions

During sustained physical effort, the anaerobic energy stored in the blood and muscle is depleted quickly, while the aerobic energy production takes up more and more of the load. We have learned, however, that the production of aerobic energy also produces lactic acid, a waste product that interferes with muscle use. The deposit of significant lactic acid is one critical factor in fatigue after sustaining effort for a long period. Thus, the rest period should be long enough to give the muscles time to get rid of the waste and to replenish the spent aerobic glucose.

We assume that the initial amount of stored glucose (ready-to-use energy) is the same for the office worker, laborer, and runner—approximately 25 kcal. We also know that a laborer's body typically produces metabolic energy at a rate that is 3.8 times higher than the BMR. Therefore, we conclude the rest periods should be taken after 60 minutes for the office worker, 15 minutes for the shoveling laborer, and 15 minutes for the runner. This is shown as a graphic in Figure 7.3.

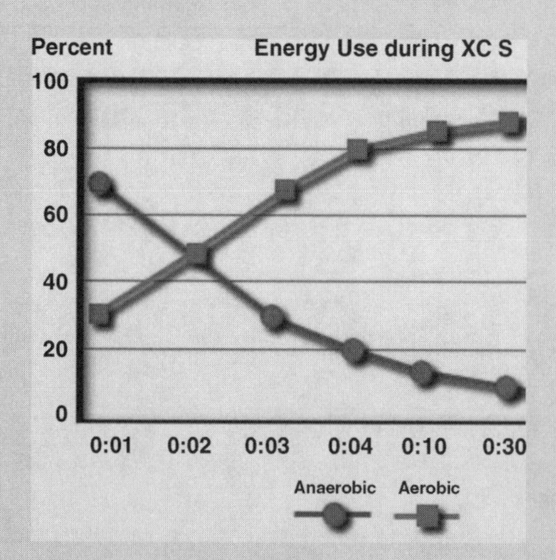

Figure 7.3 Long-Term Energy

Table 7.2 Comparing Energy Needs for Different Kinds of Work

Basal metabolic rate (kcal/hr)	70	70	70
Energy need rate for work (kcal/hr)	110	500	600
Metabolic energy production factor	2.4	3.8	4.5
Initial energy storage (kcal)	25	25	25
Time until next rest period (hr)	1	0.25	0.25
Maximum energy production (kcal/hr)	$2.4 \times 70 = 168$	$3.5 \times 70 = 245$	$4.5 \times 70 = 315$
Maximum energy production at end of work (kcal)	168	$0.25 \times 245 = 61$	$0.25 \times 315 = 79$
Energy spent during time of work (kcal)	110	$0.25 \times 500 = 125$	$0.25 \times 600 = 150$
Energy imbalance at rest time (kcal)	$168 - 110 = +58$	$61 - 125 = -64$	$79 - 150 = -71$
Quick energy available at beginning (kcal)	25	25	25
Energy deficiency at rest time (kcal)	0	$25 - 64 = -39$	$25 - 71 = -46$
Energy need at rest (kcal/hr)	0	100	100
Rest time to replenish energy deficiency (hr)	0	$39 \div (245 - 100) = 0.27$	$46 \div (315 - 100) = 0.21$

Calculations

The calculations used are presented in a matrix format in Table 7.2.

Discussion of Results

As expected, the office worker does not need a rest period to replenish energy, since the expenditure of 110 kcal/hr is lower than the metabolic energy production of 168 kcal/hr by an average person. Also as expected, the other two use significantly more energy than they produce, thus leaving

(*continued*)

(continued)

an energy deficit after 15 minutes, which the stored energy of 25 kcal cannot cover. The rest-time energy need is slightly higher than the BMR, since some of it is used to remove the lactic acid, breathe heavily, and replenish the anaerobic energy of 25 kcal. As a result, the rest time for the laborer is calculated as 0.27 hour, or 16.2 minutes; and for the runner, 12.6 min.

It is interesting to note that although the runner spends more energy than the laborer, his higher rate of energy generation while running and resting enables him to recuperate much more quickly.

Note: Resting does not necessarily mean that the worker sits down. It can mean engaging in activities that require very little energy, such as walking, cleaning, and so on.

7.3 ERGONOMICS IN CONSTRUCTION

Ergonomics

From the Greek words *ergon* + *nomos* = work (by) natural laws. The objective of ergonomics is to match, or "fit," workplace conditions and job 'demands with the capabilities of a human body. An ergonomically designed workplace helps to avoid illness and lowers the risk of injuries and accidents, resulting in more efficient work, fewer errors, and higher-quality performance and workplace morale.

In 1997, the U.S. General Accounting Office (GAO) declared, "Ergonomic programs can substantially reduce workers' compensation costs, with savings of up to 60 percent–80 percent over a 4- to 5-year period." One of the prime beneficiaries of ergonomics is the construction industry, with its physically demanding work that, as statistics show, has a high rate of injuries, from strains and sprains to work-related musculoskeletal disorders (MSD).

Injuries among construction workers became so prevalent in the 1990s that they were responsible for almost half of the safety-related costs in the industry during that decade. The costs of every accident or injury are high: in the form of pain and recovery for the injured worker; emergency and subsequent medical treatments; and workplace disruptions, for investigations, paperwork, repairs, and the like. Not surprisingly, one of the negative impacts of MSD was lower productivity.

Header Problem 7.1: What Happened to Benny?

Benny grimaces every time someone asks him about the day he could not get out of bed and could not move his arm in a circle, due to pain in his shoulder. The day before, he had started a new job as a ironworker (rod buster), after leaving the construction industry, where he had installed sheetrock. He felt that sheetrock installation had become too hard on his body.

Then, yesterday, as he was laying rebar on a bridge job, he had to interrupt his work many times, due to pain. That evening, Casey, his wife, saw immediately that he was hurt and rubbed some alcohol on his back before he went to sleep. The next morning, when Benny tried to get out of bed at 5:30 A.M., he discovered his back had locked up and the pain in his shoulder, thighs, and back was excruciating. He was not even able to visit a doctor that day.

7.3.1 Biomechanics

No matter what we do—walking, shoveling, or even sitting and typing—requires the work of a series of muscles in our body. For example, when the quadriceps shortens when we lift our leg, it meets with resistance caused by the gravitational force created by the leg mass. As the retracting muscles expend energy to build up forces, reaction forces in other parts of the body come into play. In case of walking, the weight of the lifted leg has to be carried by the other leg, thus increasing the stress in its hip joint and knee. While these forces don't amount to much, other muscles and skeletal locations will experience large increases.

Let's examine the forces created by lifting a concrete block from a stack with one hand. An average block weighs 12 kilogram (26 pounds). Figure 7.4 presents both parts of the arm and the free-body diagram involved in this activity.

To understand the necessary muscle force to produce movement and the forces at work on the elbow joint we can refer to good old Isaac Newton's equilibrium condition for bodies at rest. But in order to make his classic laws work in our case we need to introduce four assumptions:

1. The elbow joint between the ulna and the humerus is frictionless
2. No dynamic motions are considered

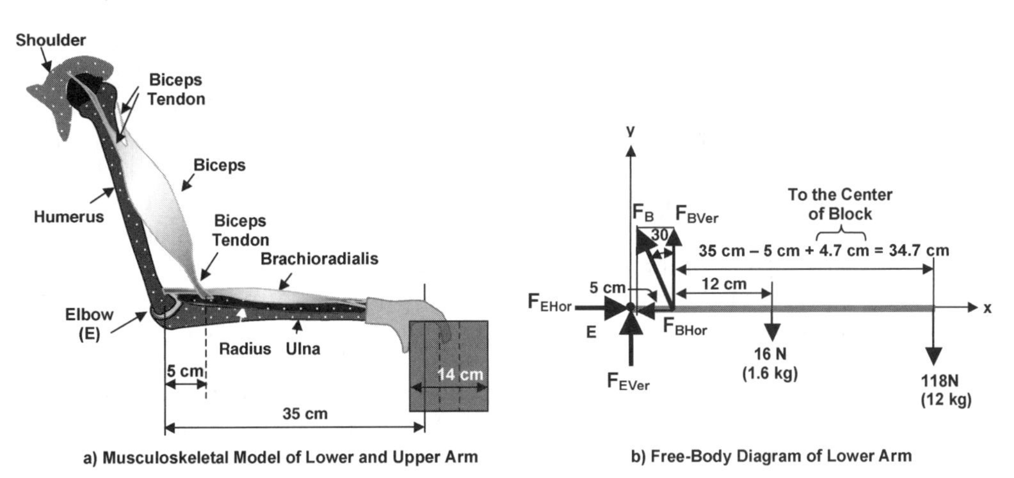

Figure 7.4 Biomechanical Models of a Hand Lifting a Concrete Block

3. All the forces are on the same vertical plane
4. The dimensions and weights for the human body are averages

Now we can establish the first function by setting the "moment in elbow" as:

$$M_E = 0 (= \text{frictionless})$$

M_E has to balance the moments created by the gravity of the two and the vertical portion of the biceps. We know all the distances, which leaves only one unknown remaining, which we can find with the moment function for M_E.

$$\begin{aligned} \Sigma M_E &= M_{Biceps} - M_{Arm} - M_{Block} \\ &= (F_{BVer} \times 5\text{cm}) - (16\text{N} \times 17\text{cm}) - (118\text{N} \times 39.7\text{cm}) = 0 \end{aligned}$$

$$\begin{aligned} F_{BVer} &= (16\text{N} \times 17\text{cm}) + (118\text{N} \times 39.7\text{cm}) \div 5\text{cm} \\ &= (272\ \text{Ncm} + 4685\ \text{Ncm}) \div 12\text{cm} = 991\ \text{N} \end{aligned}$$

$$F_B = F_{BVer} \div \cos 30^\circ = 991\text{N} \div 0.866 = 1,145\ \text{N}$$

$$F_{BHor} = FB \times \sin 30^\circ = 1145\ \text{N} \times 0.5 = 572\ \text{N}$$

Finally, we can use the function that demands that the vector sum of the forces in one direction be zero.

$$\Sigma\ \text{Forces in y-direction} = 0 \quad F_{Ever} + F_{BVer} - 16\text{N} - 118\text{N} = 0$$

$$F_{Ever} = 1,145\ \text{N} - 134\ \text{N} = 1,011\ \text{N}$$

$$\Sigma\ \text{Forces in x-direction} = 0 \quad F_{EHor} - 572\ \text{N} = 0$$

$$F_{EHor} = 572\text{N}$$

The approach we just used to calculate the forces in the musculoskeletal system of the arm is called *biomechanics*. Biomechanics uses the laws of physics and engineering mechanics to describe the motions of various body segments (*kinematics*) and explain the effects of forces and moments acting on the body (*statics*). Using biomechanics we can prove quantitatively how much effort can be saved by applying one of the rules set by Frank Gilbreth's motion studies of the 1910s: the elimination of strenuous motions. Naturally, this approach can be extremely useful in reducing worker fatigue in the construction industry, which is characterized by repetitive lifting, carrying, and pushing and pulling very heavy materials and loads. It is known that lifting excessive weights may become hazardous after a single incidence. Workers may pull muscles or strain their backs by attempting to lift something whose weight is above their body's capacity to do so. For example, an average mortar-mix bag weighs 94 pounds (42.5 kilograms), and heavy rebar carried by two or three can quickly double the load onto the back.

7.3.2 The Endangered Human Spine

Lower back pain (LBP) affects 75 percent of people in the developed countries. According to the U.S. Department of Labor's Statistics of 2007, 30 percent of all nonfatal occupational injuries and illnesses in the construction industry were related to the back and shoulder. It is believed that over half of these injuries are the result of lifting excessive weights, or lifting incorrectly. A long list of publications from the National Institute for Occupational Safety and Health (NIOSH) and the Occupational Health and Safety Council of Ontario (OHSCO) emphasize the dramatic personal impact such injuries can have on the human trunk; they also point out how both LBP and MSD can be controlled, through correct body posture, correct lifting techniques, use of lifting equipment, and the limitation of weight.

Still, the incidence rate of this extremely painful and costly type of injury continues to be high, an indicator that a primary focal point of every ergonomic effort should be a review of lifting needs and procedures.

The human spine is a "heavy-duty" system, in that it is strong enough to enable us to walk and carry loads yet is flexible enough to allow the upper body to bend forward and rotate. What makes this possible is the architecture of the spine, which is composed of bony vertebrae, with and elastic fibrous discs sandwiched in between. Nestled in close is the cord of the spinal nerves, which run from the brain to the legs, and a set of strong muscles holding and "powering" the vertebrae.

The images in Figure 7.5 depict the deformation that a disc experiences during loading, twisting, and bending; (c) shows a herniated disc resulting from the

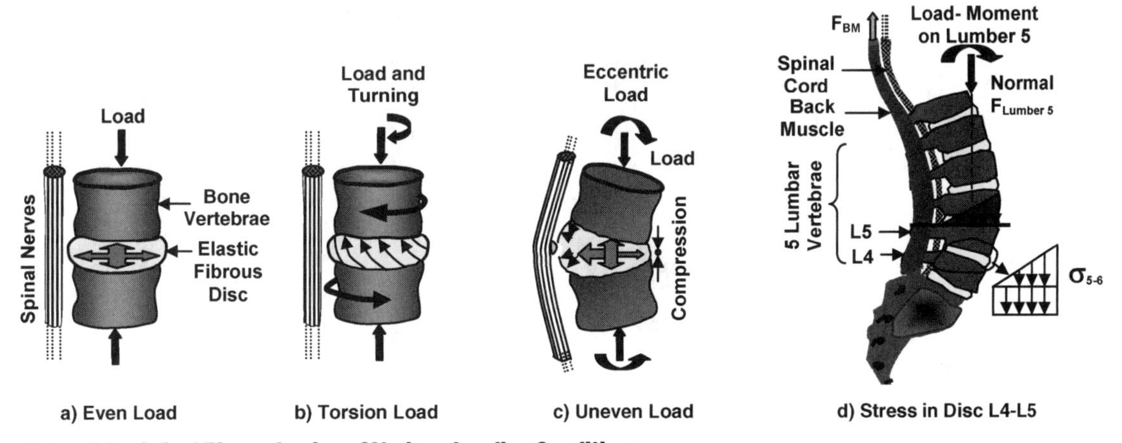

Figure 7.5 Spinal Biomechanics of Various Loading Conditions

disc being overloaded, or possibly a shock load, causing a small rupture at the tension side where a small part of the fibrous material separates and pushes against the spinal cord. The cord has nowhere to go—it is held in place by bones and muscles—so the disc rubs against it, causing excruciating pain. If the body is still young enough, the pain may subside after months of treatment, but will commonly come back as the disc fragment and the spine grow older and less flexible.

As indicated in Figure 7.5 (d), the disc most likely to become damaged is between lumbar region 4 and 5; lumbar regions 1 to 3 have grown together into a solid bone. When lumbar 5 experiences stress from the upper body, it acts on disc L4–L5, causing compression, with maximum stress toward the front. Figure 7.5 (c) indicates that the pressure inside the disc results in a dramatic deformation of the disc, into the shape of a wedge. There are also large horizontal stress components toward the backside, toward the spinal cord. The only reason the disc does not tear in tension is because the strong back muscle keeps the delicate joint in balance.

To study the forces that act on the lumbar section of a construction laborer lets observe the work done by Fred, a mason, who is presently laying the first course of a brick wall. Figure 7.6 lays out the location of the brick cube, the mortar pan on the ground, and the brick course close to ground.

Fred is assumed to be 70 inches (1.92 m) tall, with a skeleton measuring as shown in Figure 7.6.

Since all the forces experienced by Fred's lumbar area are gravity based, we need to find the horizontal distance of each to the fulcrum at the hip. In addition, we must assess each major body part's center of gravity (CoG). With the exception of the upper body, with CoG 15 inches (41 centimeters) from the hip, the CoG of the upper and lower arms is assumed to be in the middle.

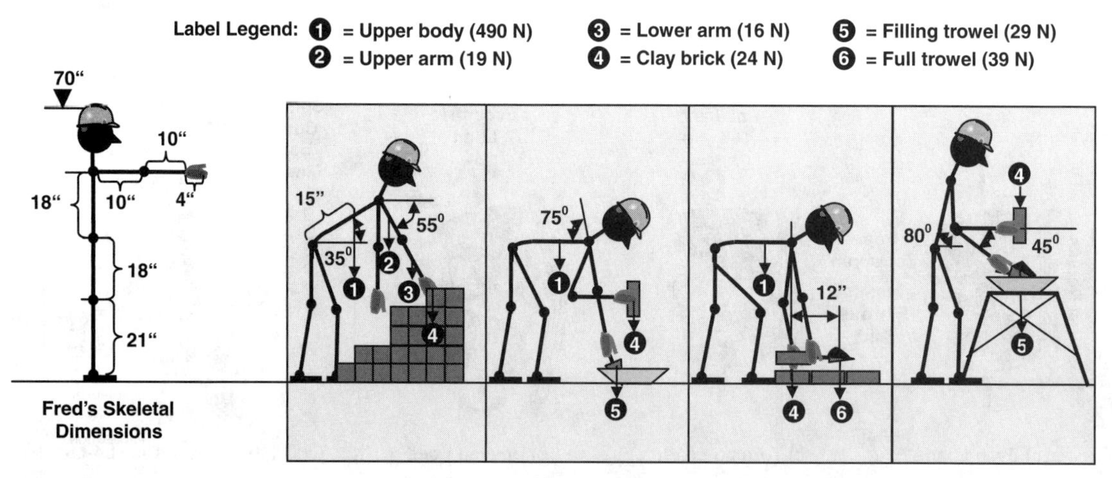

Figure 7.6 Biomechanical Load Comparison of Repetitive Bending Motion During Bricklaying

Reviewing the result of the calculations, total moments, and total loads, it's clear that the loads do not change significantly. On the other hand, the moment due to bending shows marked increases from picking up a new brick (8,020 N-in) to getting mortar from the pan on the ground (10,071 N-in) to applying the bed mortar (10,348 N-in). As we learned, a slowly rising wall will require less bending over time, whereas the brick cube will require deeper bending, as more and more layers of bricks are removed. In this case, we are not too worried about a single motion, but rather the repetition that tires Fred out, that fatigues his body. And we can appreciate the effect of dispensing with the bending motion while fetching mortar with the trowel. The apparent reduction to 3,374 N-in represents a 66 percent drop, an effect that could be repeated by elevating the lower-level bricks. All in all, the biomechanical evaluation supports quantitatively the presumed ergonomic improvements by eliminating the bending motion.

NIOSH advocates the use of a lifting equation it has developed for the design and evaluation of manual lifting tasks. The result of the equation is a recommended weight limit (RWL) for a specific task, to be lifted by a healthy worker for a substantial period of time without increasing the risk of developing lifting-related LBP. The equation consists of several factors that have to be multiplied, with an assumed base load of 23 kilograms (51 pounds):

$$RWL = LC \times HM \times VM \times DM \times AM \times FM \times CM$$

Whereas:

		METRIC	U.S. CUSTOMARY
Load Constant	LC	23 kg	51 lb
Horizontal Multiplier	HM	(25/H)	(10/H)
Vertical Multiplier	VM	1-(.003 |V-75|)	1-(.0075 |V-30|)
Distance Multiplier	DM	.82 + (4.5/D)	.82 + (1.8/D)
Asymmetric Multiplier	AM	1-(.0032A)	1-(.0032A)
Frequency Multiplier	FM	From Table 5	From Table 5
Coupling Multiplier	CM	From Table 7	From Table 7

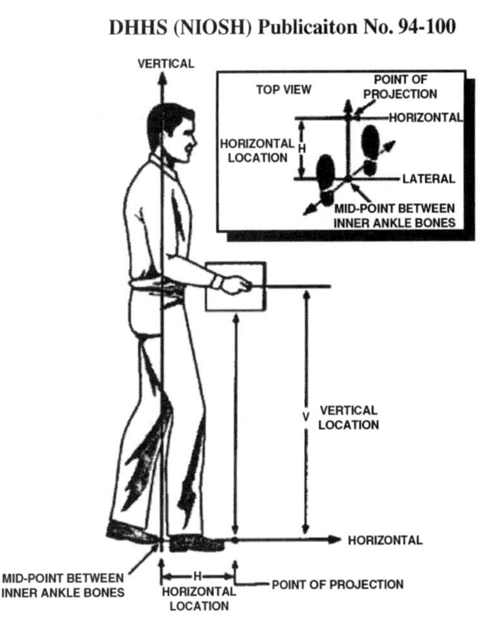

Figure 7.7 Definitions of Key Dimensions

In order to use the lifting equation, we have to establish several measures:

- H: The horizontal distance of hands on the load from the midpoint, between the ankles at the start and end of the lift
- V: Vertical locations of the hands at both the start and end of the lift/lower action
- D: Vertical travel distance of the hands from the start to the end of the lift
- A: Angular location of the load relative to a line extending from the worker's navel, if the worker were standing in a neutral posture, in degrees, at both the start and end of lift
- F: Frequency of lifts (average number of lifts per minute and total duration of lifting)
- C: How well the load can be grasped (based on presence and type of handles)

NIOSH's publication No. 94-110, available online at www.cdc.gov/niosh/docs/94-110/, provides the necessary tables for some of the latter two multipliers.

Finally, by dividing the actual load weight by the RWL we can establish the level of physical stress of a lifting task, defined as:

$$\text{Lifting Index(LI)} = \text{Load Weight} \div \text{RWL}$$

This can be used to determine whether the lifting task is safe or needs to be modified. We will discuss how to do that after we address another health threat to construction workers: vibration.

7.3.3 Hand-Arm and Whole-Body Vibrations

Operators of heavy construction equipment such as bulldozers, backhoes, scrapers, haulers, loaders, skid-steer vehicles, and excavators experience high-frequency vibrations, combined with large shocks transmitted to the whole human body, an effect called *whole-body vibration* (WBV). Similarly, laborers who work with large power tools, such as demolition hammers, drills, hammer drills, angle grinders, chainsaws, and handheld circular saws, that require extreme physical control are exposed to *hand-arm vibration* (HAV).

The effects of vibration are exacerbated when a tight grip is required and the air temperature is low. Employees who are regularly exposed to this kind of vibration may suffer from hand/-arm vibration syndrome, which causes neurological and motor disorders in the hands and fingers, as well as circulatory disorders in the fingers and of the musculoskeletal system. Vascular disorders also may occur, in the form of "white finger," which is caused by inadequate circulation. This phenomenon is generally more prevalent in colder weather. Depending upon the duration and intensity of the exposure to vibration, it may affect only the fingertips or the entire fingers.

Neurological disorders are experienced as tingling and numbness in the fingers, which become more severe with increasing exposure. Examples of musculoskeletal disorders related to hand-arm vibration include wrist tendinitis and tenosynovitis.

The three main sources of WBV from heavy equipment are:

- Low-frequency vibration, caused by tires and terrain
- High-frequency vibration, from the engine and transmission
- Shock, caused by running into potholes or obstacles

Hand-Arm Vibration

A building worker who has worked on construction sites since the age of 14, using drilling and pneumatic hammers three or four days a week for several hours each day, now suffers from an occupational disease. At the age of 42, the pain in his wrist is so severe that he can no longer work.

Source: European Agency for Safety and Health at Work (http://osha.europa.eu).

Measurements taken using vibration sensors show, as might be expected, that different machines vibrate at various levels and intensities. Figure 7.8 shows that mobile cranes, set up on outriggers, vibrate minimally, while skid-steer vehicles and scrapers develop mean accelerations that reach above 1 m/s^2. Even after exposure to a vibration magnitude of only 1 m/s^2, workers reported such symptoms as abdominal and chest pain, headaches, lower back pain, and loss of balance.

An effect that exacerbates the problem is the resonance between the equipment and certain body parts. Every object (or mass) has a resonant frequency, and if an object is vibrated at its resonance frequency, the maximum amplitude of the

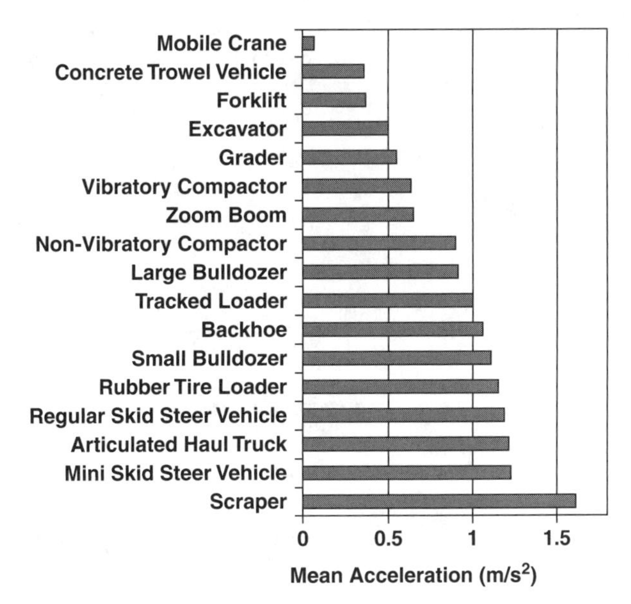

Figure 7.8 Common Equipment Vibration

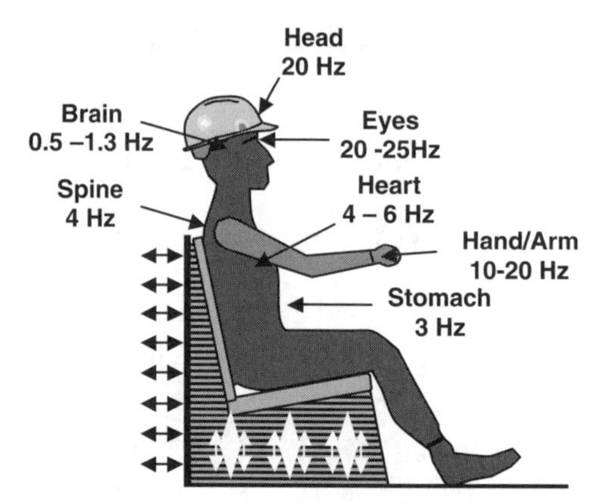

Figure 7.9 Whole-Body Vibration and Dangerous Resonance

vibration will be greatly amplified. Further complicating the matter is the fact that individual body parts and organs have their own resonant frequencies, and do not vibrate as a single mass. For example, the acceleration forces that impact the spine at 4 Hz are amplified by as much as 200 percent. Even worse, the head, eyes, and arms vibrated at 20 to 30 HZ experience an amplification of 350 percent. According to ISO 2631, "Mechanical Vibration and Shock—Evaluation of Human Exposure to Whole-Body Vibration," permissible exposure limits are a function of acceleration, frequency, and length of exposure (1 minute to 24 hours). As

Table 7.3 Vibration Rates of Average Tools

Power Tool Type	Dimension	Lowest	Typical	Highest
Jackhammer	m/s^2	5	12	20
Demolition hammers	m/s^2	8	15	25
Hammer drills/combination hammers	m/s^2	6	9	25
Needle scalers	m/s^2	5	—	18
Scabblers (hammer type)	m/s^2	—	—	40
Angle grinders	m/s^2	4	—	8
Chipping hammers (metal)	m/s^2	—	18	—
Stone-working chisels	m/s^2	10	—	30
Chainsaws	m/s^2	—	6	—
Sanders (random orbital)	m/s^2	7	—	10

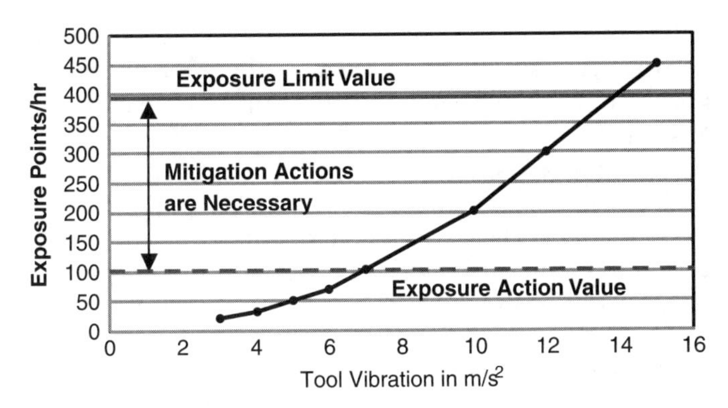

Figure 7.10 Time-Dependent Danger

expected, frequencies between 4 and 8 Hz have to be counterbalanced with lower accelerations, since the human body is most sensitive to WBV at these frequencies. For example, a piece of equipment that vibrates at a frequency between 4 and 8 Hz should not accelerate above 0.315 m/s^2 over an eight-hour day. If the operator works only four hours per day, the acceleration may reach 0.54 m/s^2.

British Health and Safety Executive (HSE) leaflet INDG242 (revision 1) addresses the dangers of vibrating hand-tools and provides guidance on maximal exposure to avoid health problems, as shown in Table 7.3.

According to the table, operating a chainsaw causes 6 m/s^2 vibration of the body. One hour of chainsawing "earns" 60 points of exposure points (see Figure 7.10). With the maximum exposure action value at 100 points, an operator should stop after approximately 1.5 hours. Any longer use needs to be accompanied by mitigation measures. It is apparent that the highest vibrations of many power tools are far beyond the hourly exposure limit (even if mitigated) of 15 m/s^2 and so require extensive controls, such as remote control, if used for any length of time.

7.3.4 Factors Leading to Hearing Loss

The average human ear is capable of detecting air that vibrates at frequencies in the range of 20 to 20,000 Hz. Prolonged exposure to intense noise measured in decibels (dB) on a logarithmic scale can limit or even destroy the sensitivity of the ear—or, more accurately, the inner ear. For example, an explosive noise registering 150 dB may cause permanent damage to several sensitive parts of the middle and inner ear. Longer durations exposed to 120 dB will cause discomfort and, over time, destroy the fine hair cells inside the cochlea which will not grow back.

According to the Federal Occupational Safety and Health Administration (OSHA) the permissible exposure limit (PEL) allows for an eight-hour, full-shift average exposure of 85 dBA, defined as an eight-hour time-weighted average (TWA) of 85 dB (action level). Note that this number contradicts many state OSHAs, and even the recommendation by NIOSH, which has set the threshold level at 85 dBA. Regardless, in all cases hearing conservation programs must be implemented when employees are exposed to 85 dB or more in an eight-hour day. These programs include annual audiometric testing, and require hearing protection devices, such as earplugs.

Engineering or administrative noise controls are required when exposure exceeds 90 dB. Engineering controls include redesigning the space to reduce machinery noise, replacing machinery with quieter equipment, and enclosing the noise source or enclosing the noise receiver. Administrative controls include mandating the length of time an employee may be exposed to a particular noise source. Since TWA is calculated by summing exposures multiplied by the durations, even changing the work assignments of a laborer or operator may be considered. Another alternative is to shift an exposed person during the day from high- to low-exposure areas. For every 5 dBA increase in exposure level, the allowable time is cut in half.

To determine whether a set of multiple exposures at different dBA levels is permissible, we calculate the noise "dose" (D), representing the total exposure to any sound above 80 dBA during an eight-hour day as a percentage. The value of D needs to be below or equal to 100 percent to be considered permissible. The equation for evaluating the noise dose is as follows:

$$D = 100 \times (C_1 \div T_1 + C_2 \div T_2 + \ldots C_n \div T_n)$$

where:

D = noise dose during an eight-hour day as a percentage

C_i = Hours spent at a given noise level

T_i = Hours permitted at noise level

Note: Workers must *always* use hearing protection when noise levels exceed 115 dBA.

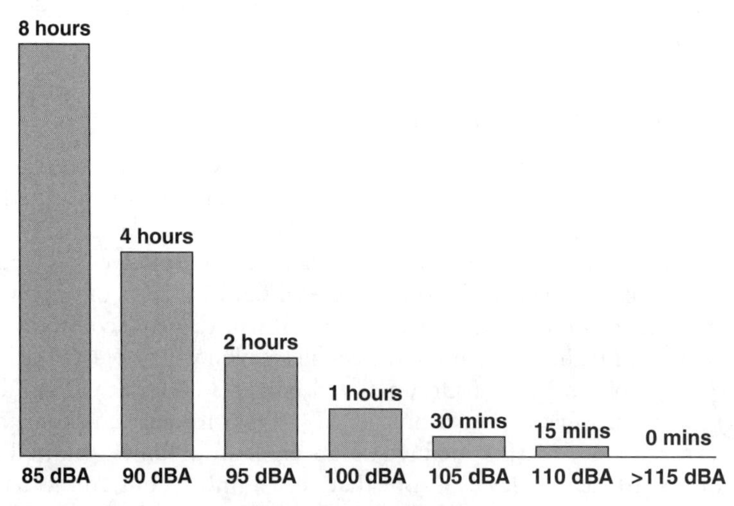

Figure 7.11 Permissible Exposure Durations for 85 PEL

Construction employees are prone to developing permanent hearing loss, because many of the industry's heavy equipment depend on large diesel engines for power. Furthermore, power tools use pneumatic power to drive hammering, cutting, or chiseling actions, to penetrate rock, hard soil, concrete or even steel. A more recent source of intense noise on construction sites comes from the load warning signals emitted by equipment that is backing up; Perversely, this noise—intended as a safety feature—is cancelled out by wearing noise protection devices to protect hearing.

A recent study conducted by Eillmorth Spencer and Peter Kovalchik of NIOSH, "Heavy Construction Equipment Noise Study Using Dosimeter and Time-Motion Studies," showed that bulldozer operators consistently were exposed to the highest levels of noise, ranging from 92 to 109 dbA, and truck drivers without air conditioning, 90 to 92dBA. The study also found that newer equipment, which comes with mufflers and air-conditioned operator cabins, further reduces the noise level by approximately 3 dBA. Still, one road grader participating in the study had such a cab, and even with an idling engine, was still exposed to 91 dB(A). The highest sustained dose was recorded at 104 dBA, when the grading was done uphill.

In 2004, the School of Public Health and Community Medicine at the University of Washington published "Noise and Hearing Loss in Construction" (http://depts.washington.edu/occnoise), based on six years of research in the construction industry, starting in 1997. Previously, it had been found that in

Worked-Out Problem 7.1: Noise Exposure

Frank, a laborer/operator on a commercial project, worked at several jobs in the same day. In the morning, he cut masonry blocks for a half hour; then he operated the forklift for one hour, lifting bricks onto the scaffold. In the afternoon, he helped the formwork crew for an hour, cutting wooden studs and plywood using the circular saw. He would like to know whether he should have worn ear protection.

Assumptions

The noise level surrounding Frank all day was high, due to other operations going on around him. We assume that those were below 85 dBA and, so, permissible. For the calculations, we will use average dBA values. Exposures: (1–3)

.5 hours at 100 dBA (chip)

1 hour at 97 dBA (forklift)

1 hour at 94 dBA (circular saw)

(continued)

(continued)

Calculations

$$D = 100 \times (C_1 \div T_1 + \ldots)$$

For PEL 90 dBA:

T_1 for 100 dBA = 2 hr T_2 for 97 dBA = 2 hr T_3 for 94 dBA = 4 hr

Noise Dose $D_{90} = 100 \times (.5 \div 2 + 1/2 + 1/4) = D90 = 100\%$ (OK)

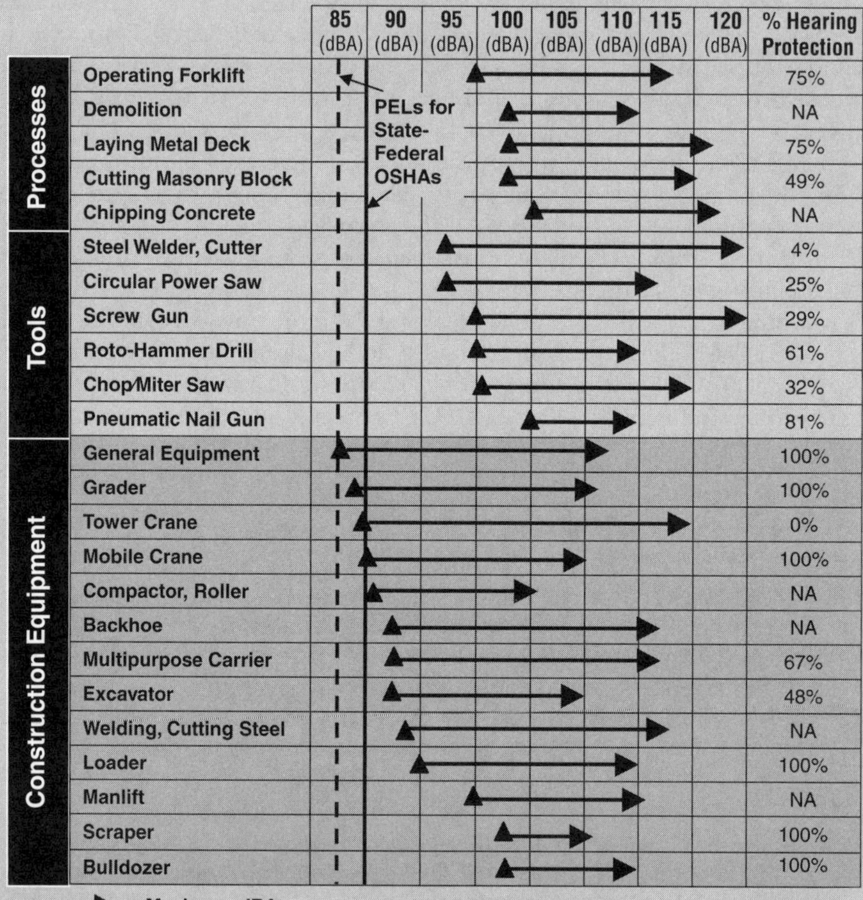

Figure 7.12 Noise Levels of Various Construction Tools

For PEL 85 dBA:

T_1 for 100 dBA = 1 hr, T_2 for 97 dBA = 1 hr, T_3 for 94 dBA = 2 hr

Noise Dose $D_{85} = 100 \times (.5 \div 1 + 1 \div 1 + 1/2) = \mathbf{200\%}$

DISCUSSION OF RESULTS

The PEL calculations reveal that if Frank was working in an area where the rules of OSHA apply, he was fine. Anywhere else in the United States or Europe, where PEL = 97 dBA, he was required to protect, and should have protected, his hearing. In any case, when he was working with the masonry saw, Frank was exposed to maximum dBA values above 115 percent. This means he needed to wear ear protection during that operation.

It is recommended that Frank immediately join a hearing protection training program and take an audiometric test. This will enable him to assess the hazards he is facing.

Washington State, construction workers were five times more likely to file workers' compensation claims for hearing loss than workers in all other occupations combined. Figure 7.12 charts some of the group's findings in terms of average and maximum dBAs, as well as the percentage of time construction workers were wearing their hearing protection as required.

7.3.5 Impact of Heat and Cold Stress on Productivity

One unique condition facing those working in the construction industry is exposure to the elements. While rain and snow will normally cause an operation to shut down, extreme high or low temperatures alone are not typically considered cause for stoppage. Regardless, the human brain will continue to try to maintain a constant body temperature of about 98.6°F (37°C). As anybody who has worked in hot weather will attest, heat will cause the body to sweat, in order to cool the skin, leading to water and salt depletion. Conversely, when the temperature drops, the body will raise its internal temperature through the metabolic process, even as the extremities start to get numb. The latter results in a dramatic loss of muscle strength, without providing much warning to the person, who will, unaware, overestimate his or her capabilities.

The next sections review more systematically the effects of heat, cold, and humidity on the productivity of construction workers.

Heat Stress: The key factor related to heat stress it the *net heat load* (NHL) on the body, resulting from the heat generated by the metabolic process and the heat exchange with the environment.

The NHL can be calculated as:

$$\mathrm{NHL} = \mathrm{M} + / - \mathrm{C} + / - \mathrm{R} - \mathrm{E}$$

where:

M = heat gain from metabolism

C = heat gained or lost from convection

R = heat gained or lost due to radiant energy

E = heat lost through evaporation of sweat

Converting Centigrade to Fahrenheit

Degrees F = 9/5 Centigrade + 32 degrees F

Evaporation has a cooling effect on the body, because the transformation from liquid to gas requires some small amount of energy, which comes from the heat. The amount of evaporation is closely related to air temperature and humidity. Humidity is a ratio that compares the present water vapor in air with the maximum possible at a specific temperature. When humidity is low, more perspiration evaporates from the skin than when the humidity is high. This phenomenon has been captured in a heat index, presented in Figure 7.13. The horizontal dashed line is related to an air temperature of 90°F. At a humidity index of 25 percent, the body is perspiring at an expected rate. When the humidity rises, the amount of evaporation falls, thereby reducing its cooling effect. The result is a higher net heat load for the body at the same metabolic rate. This is translated into an apparent temperature that is higher than the air temperature, and needs to be considered when making decisions about the stress load caused by work.

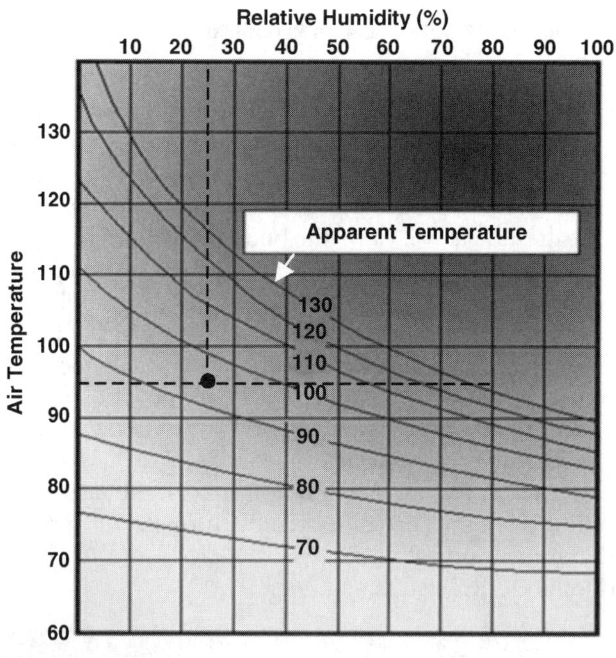

Figure 7.13 Heat Index Curves

Table 7.4 Permissible Work Temperatures

Work	Rest	Work Load		
		Light	Moderate	Heavy
%	%	Degrees F	Degrees F	Degrees F
100	0	86	80	77
75	25	87	82	78
50	50	89	85	82
25	75	90	88	86

A healthy body temperature, fueled by the metabolism, is considered to be 98.6°F. If body temperature passes above 102.0°F for more than a short period, measures should be undertaken to lower it; a prolonged temperature above 105.0° F will cause the body to stop functioning. Note that these numbers are averages; other factors come into play, as well, such as age, general health, and the acclimatization level of the body.

Heat stress on the body is a gradual process that results in increasingly severe reactions of the body. First, the muscles may start cramping up, followed by general exhaustion and fatigue. Extensive sweating will induce dehydration of the body, which induces thirst, weakness, and headache. If the cooldown and rehydration processes are insufficient, continued heat stress will lead to fainting, possibly heat stroke, and in the worst case, death.

Table 7.4 offers a general guide on how workload and apparent temperatures are related to the length of exposure time, in percent of work during an hour, for a healthy body. Light work can be done continuously at a maximum temperature of 86°F, but heavy work can be at this pace only if the temperature is below 77°F. Again, we are talking about the apparent temperature, which considers the humidity index. As the table shows, the allowable temperature increases when rest periods are introduced.

Wind Chill: Human skin loses most heat through evaporation and convection. Both are dramatically accelerated by cold wind blowing over skin. Enhanced by the contrast between a cold wind and a warmer body, the wind is able to "harvest" more energy from the warmer body, and thus cool its surface and the layers below. The rate of heat loss depends on the wind speed above that surface: the faster the wind speed, the more readily the surface cools. For inanimate objects, the wind chill reduces any warmer area to the ambient temperature more quickly.

The common method to account for windchill is to adjust the air temperature by a factor that is related to the wind speed. Figure 7.14 presents a graphical means to convert air temperature to a windchill-equivalent temperature value. Two wind speeds of 10 and 20 mph are indicated by the vertical dashed lines intersecting two

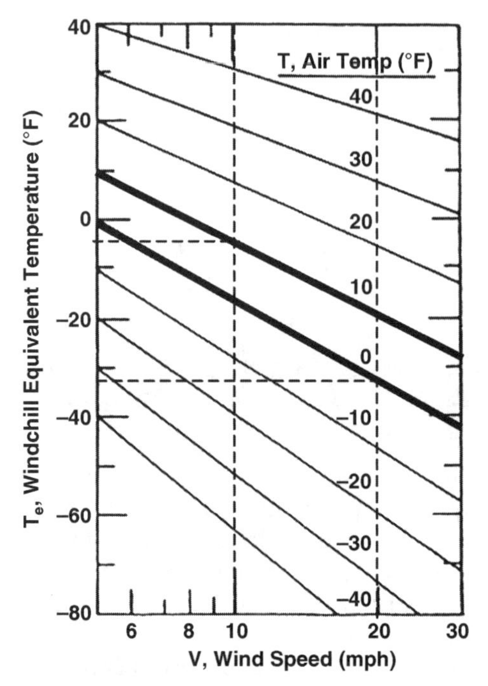

Figure 7.14 Windchill-Corrected Ambient Temperature

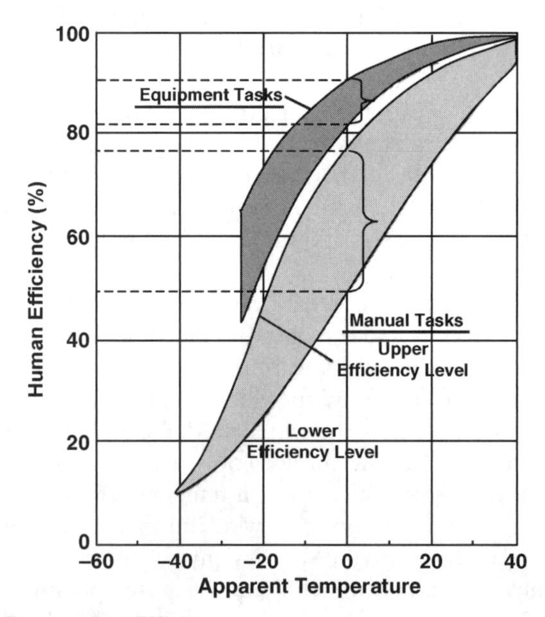

Figure 7.15 Low-Temperature Effect

sloped solid lines, for 0°F and 10°F. The circled points indicate that a 10-mph wind at a temperature of 10°F converts to an apparent temperature of about −4°F, a 14°F drop. At 0°F and 20 mph, the drop reaches −33°F.

Similar to hot temperatures and humid conditions, cold temperatures also will have an impact on the efficiency of the humans working on construction sites. In the cold, construction equipment can help warm the body via the heat the engine is generating, and so lower the impact of the cold air. In the figure, the area for manual tasks is bracketed by two S-curves for lowest and highest impacts. At 0°F, the manual task of a human is between 50 to 75 percent efficiency level. In other words, work that typically takes two hours will require up to four hours instead. For an equipment operator, the effect of the cold is decreased to "only" 82 to 90 percent efficiency.

It is apparent that equipment cannot been operated at temperatures below −25°F, whereas humans can work until the temperature reaches −40°F—albeit only at a 10 percent efficiency level.

7.4 A MODERN DEBILITATING DISEASE: JOB STRESS

NIOSH defines stress as: "Harmful physical and emotional responses that occur when the requirements of the job do not match the capabilities, resources, or needs of the worker." It is the cumulative emotional and/or physical strain on a person, caused by pressure from the "outside world," resulting in muscle tension, irritability, and an inability to focus. More advanced physical reactions to continued mental stress include headache, elevated heart rate, ulcers, or even total col lapse. In the twentieth century, job stress became a major workplace problem and concern, especially in construction.

In a survey conducted by the British Institute of Building in 2006, 847 construction industry professionals were questioned about job stress. According to the survey, 68 percent reported suffering from stress, anxiety, or depression as a direct result of working in the construction industry. This was broken down further to show that 61.9 percent of respondents had experienced stress, 48.4 percent had experienced anxiety, and 18.5 percent had experienced depression. And a NIOSH survey (DHHS Publication No. 99–101, "Stress at Work") reported that 40 percent of workers think their job is very or extremely stressful, while 26 percent assert they are often or very often burned out or stressed by their work.

The effects of stress include the following:

- Cause chronic fatigue, digestive upsets, headaches, and back pain.
- Affect the blood cells that help individuals fight off infection, making them more susceptible to colds and other diseases.
- Increase blood pressure and increase the risk for stroke.
- Increase the danger of heart attacks, particularly for those who are often angry and mistrustful.

- Make an asthma attack worse.
- Trigger behaviors that contribute to death and disability, such as smoking, alcoholism, drug abuse, and overeating.
- Makes it harder to take steps to improve health, such as giving up smoking or making changes in diet.

7.4.1 Modeling Stressors

The saying that "pressure will create results" implies that stress is good for productivity. In fact, that is confusing *stress* with *challenge*. When we successfully meet a challenge, we feel relaxed and satisfied. Thus, challenge is an important ingredient for healthy and productive work. In contrast, ongoing or chronic stress extends its influence from the mind to the body, where it wreaks havoc and, as noted before, may even cause death.

Figure 7.16 provides a schematic overview of the many factors that come to play when a person is on a stressful work path. Each person will have unique responses along the way, but NIOSH and other health groups have identified three phases that all people pass through: 1) resistance, 2) burnout, and 3) exhaustion-collapse.

The figure makes apparent that job stress cannot be compared to a stress imposed by a unidirectional force. The model emphasizes that job stress is an

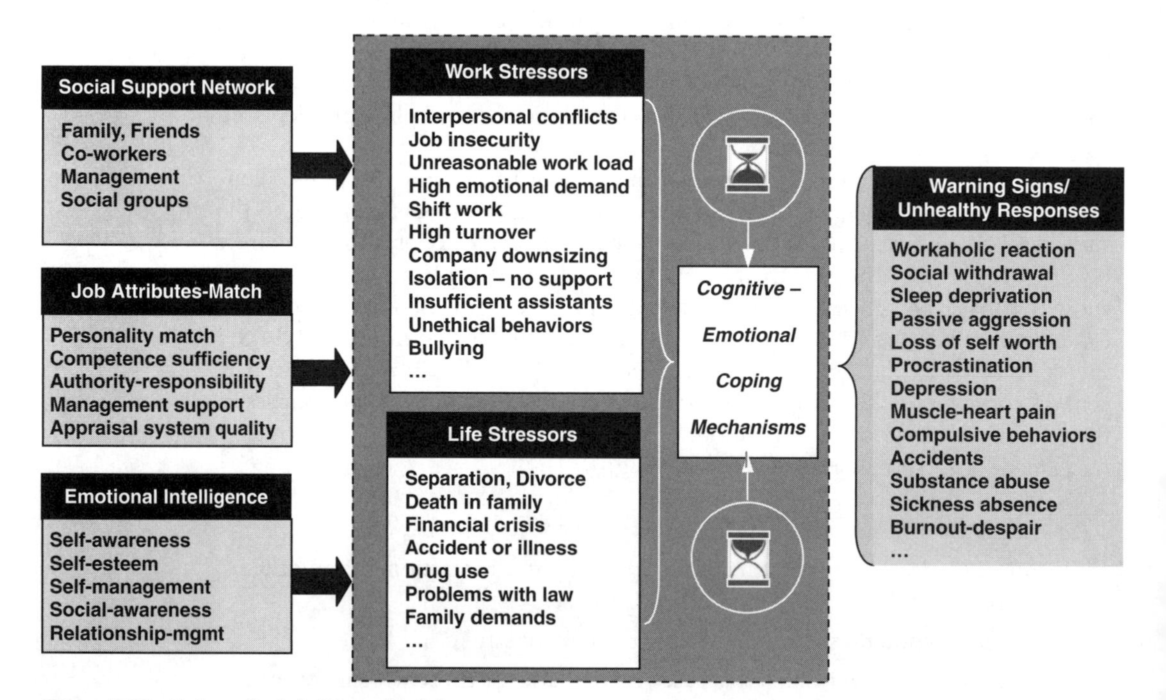

Figure 7.16 Schematic Job-Stress Model

aggregate of stressors, which have their roots outside the realm of the workplace. For example, humans bring their personal problems to work, meaning that job stressors have important partners in life stressors, and these need to be added into the equation.

7.4.2 The Coping Mechanism

The sum of the stressors is "processed" by the cognitive-emotional coping mechanism, supported by emotional intelligence and an individual's social support network. The two hourglasses in Figure 7.16 point out that the build-up of unhealthy responses to an accumulation of stresses takes time, giving a conscientious supervisor at the workplace not only recognizable signs that an employee is in trouble, but also the opportunity to intervene.

Phase 1: Passive Resistance

Stressors typically mushroom slowly in an unsuspecting person's psyche—as if entering through a secret entrance without making any noise. Therefore, in the beginning of phase 1, a stressor is often impossible to discern by an outsider or even consciously by the sufferer. What typically triggers the start of resistance is the person's sudden realization that he or she can no longer meet the cumulative demands.. The origin of such demands can be anything from an unrealistic boss, family pressures, a personal crisis, or some combination.

This build-up of a expectations by others and oneself arouse an instinctive response, production of additional adrenalin and the instinct to "fight back." The image of a Japanese samurai employing jujutsu to defeat an armed and armored opponent without weapons comes to mind. As a result, the stressed person concentrates all his or her energy and attention on defeating the growing mountain of demands and expectations, but neglects everything else, most detrimentally the very things that provide a person's social supports, such as family, friends, social groups, and even the help of coworkers. The sole focus becomes on finding a way to scale that mountain.

Phase 2: Fatigue and Burnout

A supervisor attuned to others should be readily able to recognize the signs of exhaustion and stress in an employee in this state. Unfortunately, too often the common response of a manager is to add still more demands, making it more overwhelming. It is at this critical juncture that the employee's level of emotional intelligence plays a crucial role in terminating or continuing the toxic behavior. Other needs, listed on Maslow's hierarchy come into play, as well. For example, perceived risks of losing employment will serve as an inhibitor to intelligent, healthy responses, forcing the person to stay and continue to deplete his or her energy. Emotional intelligence allows the person to objectively assess his or her strengths and weaknesses, followed by determining an optimal adaptation to outside demands. Essential, however, is the proper valuing of the emotional needs of human beings, without which a healthy life is impossible.

If an individual's emotional intelligence is not highly developed enough, the endless "fight" to meet the demands will slowly erode energy, leading to irritation and fatigue. Friends, family, and job will become less important, day by the day. The person will begin to experience headaches, muscle aches, literal heartache, sleep deprivation; alcohol or drug abuse and other compulsive behaviors may start to manifest. Erratic outbursts of anger, loss of memory, inability to concentrate, and insensitivity toward loved ones are also common. So is "calling in sick," or even questioning the purpose of life. Over time, the person becomes caught in a downward spiral.

Phase 3: Exhaustion/Collapse

By this time, no one—not a supervisor or colleagues—can ignore that the person is in a desperate situation. If no intervention occurs now, the downward spiral will speed up, due to a malfunction of the cognitive/emotional coping mechanism. Feelings of despair and abandonment will bring about a total loss of self-esteem and confidence. Similar to an insufficiently trained runner who enters a marathon, the overstressed employee's physical organs will, one after the other, stop working properly. In other words, the emotional distress begins to change its sphere, to impact the physical organs. Muscle pain will, perhaps, be followed by stomach ulcers, and possibly a rapid weight loss or weight gain.

In any event, the unsustainable drain on energy will eventually result in a physical collapse, and potentially require hospitalization. Collapses attributable to job stress result, on average, to twice the number of days taken off work than physical injuries.

7.5 THE SILENT EPIDEMIC: WORKPLACE HARASSMENT

In Europe it's called mobbing; in the United States, bullying. These two terms have been added to the more familiar one, sexual harassment; but by any name, harassment in the workplace is associated with millions of attacks on employees and 1,000 deaths in the United States alone. Sexual harassment is well known, and still rampant, while mobbing and bullying in the workplace is a more recent phenomenon, one now being studied.

7.5.1 What Are Mobbing and Bullying?

Various definitions of mobbing and bullying have been put forth in the literature, from which we can discern a characterization that applies to the construction industry:

> Persistent, offensive, abusive attacks that are deliberate, repeated, and sufficiently severe to harm the targeted person's health or economic status, through social

isolation, untrue rumors, and intimidation by one individual with the help of willing, or unwilling, participants abusing power as well as official regulations. Perpetuated in a climate of silence and deception, the mobbers follow a strategy that circumvents accountability and ignores any kind of professional ethics.

Most vulnerable to bullying/mobbing are high-level performers working in secure positions and who exhibit some characteristics that are seen as different (e.g., being a "foreigner"). After the target becomes aware of the "hidden operation" against him or her, he or she seeks the help of the authorized and educated officers, who may or may not take action, only to be circumvented by the schemers. Feeling unfairly treated and let down by the organization, the victim might retaliate, also out of view of authorities. Normally, bullying goes on for months or even years, making the cost to employers high indeed, in the form of untold hours in lost employee work time, lower productivity, a higher incidence of accidents, and escalating health care costs.

7.5.2 The Common Pattern of Mobbing

According to the Workplace Bullying Institute and many other organizations, bullying and mobbing follow a predictable process, though with some variations imposed by existing barriers. Figure 7.17 presents the five steps in the process, which are explained in greater detail in the following sections.

Step 1: Conflict

The initial incident may be prompted by an argument, a criticism, or even an award given to the future victim and that the bully thinks should have gone to him

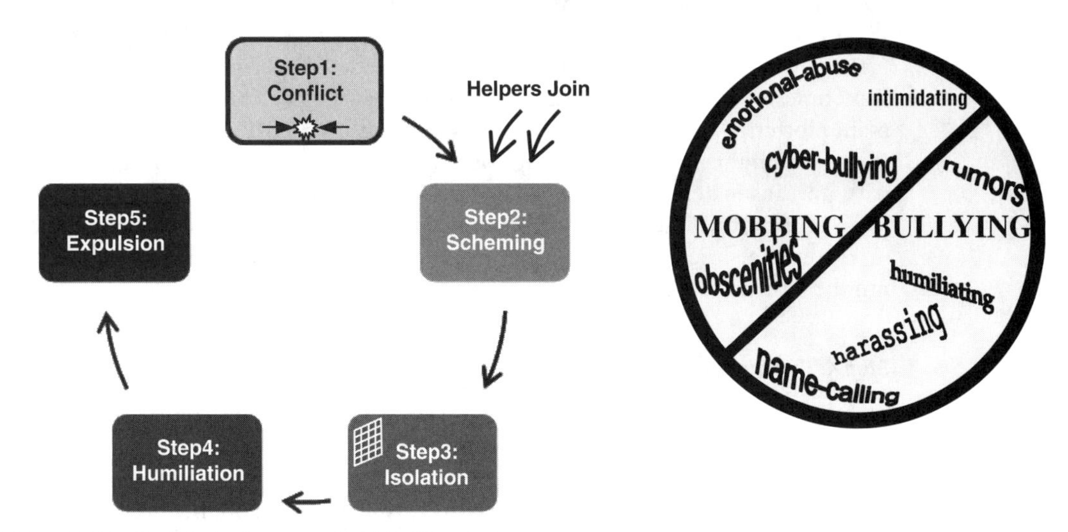

Figure 7.17 Model of a Common Mobbing Process

or her. The mobber/bully does not act out of an inferior complex; rather, just the opposite. Research has shown that the bullies who start trouble generally have a high level of self-confidence, typically based on a vastly exaggerated sense of accomplishment. Bullies begin to challenge the exemplary quality of their victim's work ethic and achievements, and initially, the victim usually is totally unaware of any conflict.

Step 2: Scheming

With the help of willing participants, who often exhibit the same level of undeserved self-esteem, a plan is made to "put the victim in the proper place." A supervisor might be convinced to retroactively enforce or make up new rules that the target could not possibly have followed. The bullies may start rumors or spread gossip about the victim, leading to doubts among colleagues about the integrity, honesty, and performance of the victim. The plotters may also start to quietly assemble a list of false, unsubstantiated accusations in a combined effort to "bring down" their victim. Eventually, the list is shared with other coworkers and to "receptive" high-level managers, in order to encourage them to join in.

Step 3: Isolation

In this phase, the victim might be made aware of the scheming operations by "quiet" friends. Feeling helpless, the victim might request help from a department whose nominal purpose is to deal with grievances and harassments. Often, the victim will be told nothing can be done because no guidelines exist to address workplace harassment. At some point the schemers will certainly become aware of the complaint, and, instead of stopping, simply change tactics.

Any effort made by the target to disprove the charges with the support of colleagues will usually be met by them with unease, empty promises to help, and sometimes outright refusal, but rarely any real assistance. This then encourages the main plotter to escalate, since no one is backing up the victim. Bullying tactics at this stage might include arranging for the target to be assigned undesirable work with unrealistic demands (workload, deadlines, special duties). Due to the by-now large group of willing or unwilling participants, upper management may tend to interpret the situation in their favor, since the company is more dependent on numbers than on the single person.

Step 4: Humiliation

Now the floor has been cleared for aggressive acts, with hidden traps and revolving doors. While the schemers continue to plot their attacks, everyone else is focusing on the accused, who by now not only feels isolated but also robbed of any healthy self-esteem and emotional intelligence. Relatively benign statements or actions are deemed "inappropriate," and the target is labeled as uncollegial or unprofessional or mentally unstable. The perpetrators become more open, and add verbal attacks, ridiculing or making insulting remarks in public. Strong responses by the victim are added to the list of accusations as the newest evidence of unfriendly behavior and instability. Conversely, a lack of response is used as evidence of guilt.

Another favorite bullying tactic is to create as situation where the victim is asked to engage in an activity that is too dangerous, then calling him or her "insubordinate" should he or she fail to carry out the unreasonable command. Management is reluctant to intervene, as it knows it cannot operate without the large group involved in the scheme, even though by now the overall performance of the entire group has deteriorated.

Step 5: Expulsion

The final campaign by the mob is to oust the victim, who is close to exhaustion. It is at this point that about a third of bullying victims leave their companies, and 30 percent stop working, reportedly due to illness or disability. The rest are either laid off, leave with a negotiated agreement, or put in for early retirement. In most cases, victims experience posttraumatic stress disorder (PTSD), rendering them helpless and incapable of rejoining the workforce for an extended period of time.

Conclusion

There are many "losers" of bullying and mobbing, beyond the victim: the family, the company, and society as a whole also suffer. The victim not only will have to deal with the PTSD but also with the financial loss and the cost of medical care. The family will have to provide the necessary care for the emotional recovery of the victim.

The company incurs both direct and indirect costs, in the form of absenteeism, sick leave, recruitment and training of replacements, reduced productivity of both the attackers and their target, possible compensatory damages if the victim takes legal action, and loss of public goodwill. In addition to losing the skills of top performers who are now removed from the workforce, society as a whole may incur a financial burden for premature retirement on the grounds of ill health if the victim is eligible for public assistance such as long-term unemployment or welfare. All this so an employee can be "taught a lesson."

7.5.3 What Should Employers Do—and Not Do?

To control bullying/mobbing in the workplace, employers should:

- Treat all employees in a fair and respectful manner.
- Be aware of the signs and symptoms that a person may be having trouble coping with stress.
- Encourage managers to have an understanding attitude and be proactive by looking for signs of stress among their staff.

- Design jobs to allow for a balanced workload. Allow employees to have control over the tasks they do, as much as possible.
- Refuse to tolerate bullying or mobbing in any form.
- Do not ignore signs that employees are under pressure or feeling stressed.
- Do not forget that elements of the workplace itself can be a cause of stress. Stress management training and counseling services can be helpful to individuals, but do not neglect to look for the root causes of the stress and address them as quickly as possible.

7.5.4 Sexual Harassment

> Nothing strengthens authority as much as silence.
>
> —*Leonardo da Vinci*
>
> All that is necessary for evil to succeed is that good men [or good women] do nothing.
>
> —*Winston Churchill*

Sexual harassment is a special case of bullying, one that has a gender target. In the United States, the Civil Rights Act of 1964, Title VII, prohibits employment discrimination based on race, sex, color, national origin, or religion. The prohibition of sex discrimination covers both females and males. The United Nations (UN) General Recommendation 19 to the "Convention on the Elimination of all Forms of Discrimination Against Women" defines sexual harassment of women to include:

> . . . such unwelcome sexually determined behavior as physical contact and advances, sexually colored remarks, showing pornography and sexual demands, whether by words or actions. Such conduct can be humiliating and may constitute a health and safety problem; it is discriminatory when the woman has reasonable ground to believe that her objection would disadvantage her in connection with her employment, including recruitment or promotion, or when it creates a hostile working environment.

A 2008 telephone poll taken by Louis Harris and Associates of 782 randomly chosen U.S. workers revealed that:

- 31 percent of female workers reported they had been harassed at work.
- 7 percent of male workers reported they had been harassed at work.
- 62 percent of targets took no action.
- 100 percent of women reported the harasser was a man.
- 59 percent of men reported the harasser was a woman.
- 41 percent of men reported the harasser was another man.

Of the women who had been harassed:

- 43 percent were harassed by a supervisor.
- 27 percent were harassed by an employee senior to them.
- 19 percent were harassed by a coworker at their level.
- 8 percent were harassed by a junior employee.

According to statistics, no industry or profession is immune from sexual harassment; however, women working in fields that are traditionally male-oriented, such as mining, firefighting, and construction, are at greater risk of being harassed. Also, in work environments that condone the use of obscenities and sexual joking, women are three to seven times more likely to be sexually harassed than in an environment where such talk is not tolerated.

There are two main types of sexual harassment. The first is *quid pro quo* sexual harassment, "something for something." This is the "you do something for me and I'll do something for you" type of exchange. In this situation, typically an employer offers an employee certain benefits, such as a job advancement or raise, in exchange for verbal, nonverbal, or physical conduct of a sexual nature. Quid pro quo harassment is equally unlawful whether the victim resists and suffers the threatened harm or submits and to avoid the threatened harm.

The second type is *hostile workplace* sexual harassment: This form of sexual harassment occurs when an employee is forced to work in a hostile environment. Repeated lewd comments, sexist jokes, unwelcome and unsolicited propositions, and offensive sexual stereotyping are examples of hostile workplace sexual harassment.

It is estimated that only 5 to 15 percent of harassed women file a formal complaint to their employers or employment agencies such as the Equal Employment Opportunities Commission (EEOC), out of fear for losing their jobs, not being believed, shame, or guilt. On the positive side, the Society for Human Resource Management reports that 62 percent of companies now offer sexual harassment prevention training programs, and 97 percent have a written sexual harassment policy.

Morse v Future Reality Ltd. (United Kingdom, 1996)

Ms. Morse shared an office with a number of men who downloaded sexually explicit/obscene images from the Internet, which they shared and discussed openly. According to Ms. Morse, her coworkers also regularly used "bad language," and created what the plaintiff deemed a "general atmosphere of obscenity." Although these activities were not usually directed at her, they did cause her to feel uncomfortable. She complained to her supervisors, but after no action was taken to resolve the problem, she resigned.

The tribunal held that all the preceding factors had a detrimental impact on Ms. Morse, such as to constitute sexual harassment, and determined that the company was liable because no one had taken action after she complained.

In terms of the construction industry, it is clear that bullying and harassment can result in substantial losses to a company, not only in financial terms but also by denigrating ethics, morale, and human dignity. It is imperative that company managers treat sexual harassment as a potent type of social cancer, one that will slowly spread through the entire organization and eventually become known throughout the industry and community. Like any cancer, only drastic measures in the early stages to halt the spread of its toxic poison.

CHAPTER REVIEW

Journaling Questions

1. This chapter makes the argument that productive workers commonly face a variety of stresses and strains in the workplace that can dramatically affect their health and, with it, the capacity to be productive.
   a. What are the differences between the two major classes of strains/stresses to consider?
   b. Pick an example from each main class of strain/stress. What factors come to play in each?
   c. Develop a cause-and-effect diagram for each of the two examples, showing the relationships that could lead to a serious incident.
2. Back injuries in construction are considered epidemic. What contributes to this phenomenon? Sketch the biomechanics of an injury. There are very few older rodbusters or rebar ironworkers found in construction today. Can you predict why most of them have left the industry by the age of 40?
3. Harassments in the workplace can take many forms. Discuss one you are familiar with from personal experience or that of friends. Did it fit the standard theme of how harassment cases develop? What was management's role in it? Did you learn any lessons from the experience?

Traditional Homework

1. Consider that a mason, David Johnson, is facing the situation shown in the figure:
   a. Develop a biomechanical model for position B.

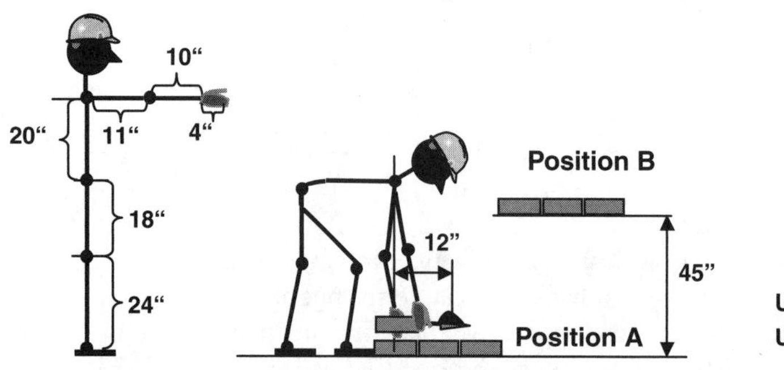

   b. What are the total bending moments for both positions in N-in?

c. What is the total load for each in N?

d. Explain the differences between the two positions and make three recommendations for the design and operation of mason scaffolds that would make David's work the easiest on his back.

2. You are asked to compare the hearing risks for Downey Prill, one of the carpenters who is operating several tools during an eight-hour day. On average, he uses a circular power saw for half an hour, producing 100 dBA. In addition, he uses a miter saw for one hour at 95 dBA, and a nail gun for one hour, generating 100 dBA.

   a. Calculate the noise dose for two PELs: 85 dBA and 90 dBA.

   b. What recommendations would you make if the PEL in your area were 85 dBA?

   c. What recommendations would you make if the PEL in your area were 90 dBA?

3. John Stonepick, the foreman of the pipe-laying crew, is checking the weather prediction for the day and learns that the air temperature will be 85°F, with relative humidity of 60 percent. The crew will start work at 5:00 A.M. and finish at 1:00 P.M. Starting at 10:00 A.M., John thinks he should factor in the weather forecast, since his crew has to do heavy work. What do you propose he should do?

   a. What is the apparent temperature?

   b. What should John plan for the time after 10:00 A.M. in order to keep his crew operating in safety?

4. An airport project was behind schedule, so it was decided to keep the runway construction going even when the temperature in January dropped to 15°F, because it stayed dry and the work involved no clay material. The average cycle time of an 8 yd^3 scraper at normal temperature is 18 minutes, of which 1.5 minutes is spent on push-loading.

   a. What will the upper and lower productivities of the scraper and pusher be when the wind speed is 6 mph?

   b. What will the average productivity be at 10 mph and 20 mph?

   c. What is a condition that would cause you to stop work? Explain your logic.

Open-Ended Questions

1. American companies and society in general seem to admire workaholics, equating the extra time they spend working with dedication to their firms.

   a. As a manager of a workaholic, would you be unconditionally happy with his or her behavior?

b. Why should a manager be concerned by a workaholic on staff? What other problems might workaholism be a symptom for?

c. What would you do to protect the worker, the company, and other employees from a potentially dangerous situation?

BIBLIOGRAPHY

Abele, G. *Construction-Effect of Cold Weather on Productivity*. Hanover, NH: U.S. Army Cold Regions Res. Enr. Lab, February 1986.

Albers, J. T. and C. F. Estill, *Simple Solutions, Ergonomics for Construction Workers*. Cincinnati, OH: National Institute for Occupational Safety and Health (NIOSH)—Publ. Diss, 2007. www.cdc.gov/niosh.

Ambrose, D. H. Developing Random Virtual Human Motions and Risky Work Behaviors for Studying Anthropotechnical Systems, Circular 9468. Cincinnati, OH: National Institute for Occupational Safety and Health (NIOSH)-Publ. Diss., March 2004. www.cdc.gov/niosh.

Australian Safety and Compensation Council. Overweight and Obesity: Implications for Workplace Health and Safety and Workers' Compensation, Commonwealth of Australia, August 2008. www.ag.gov.au/cca.

Bakker, A. B., M. Westman, and J. H. van Emmerik. *Advancements in Crossover Theory*. *J. Managerial Psych.*, vol. 24, no. 3, 2009.

Barger, L. K., B. E. Cade, N. T. Ayas, J. W. Cronin, B. Rosner, F. E. Speizer,. and C. A. Czeisler. *Extended Work Shifts and the Risk of Motor Vehicle Crashes among Interns*. *New England J. Med.*, vol. 352, no. 2, January 2005.

Bernold, L. E. Accident Prevention through Equipment-Mounted Buried Utility Sensors. *Int. e-Journal in Construction*. Gainsville, FL: M. E. Rinker School of Building Const., University of Florida, August 2005. www.bcn.ufl.edu/iejc/pindex/0831/b.pdf.

Bernold, L. E., S. J. Lorenc, and M. L. Davis. *Innovative Technology to Eliminate Back Injuries Caused by Nailing*. *J. Constr. Eng. and Mgmt., ASCE*, vol. 127, no. 3, 2001.

Bernold, L. E., S. J. Lorenc, and E. Luces. Intelligent Technology for Truck Crane Accident Prevention. *J. Constr. Eng. and Mgmt., ASCE*, vol. 123, no. 3, 1997.

Bernold, L. E., and N. Guler. *Analysis of Back Injuries in Construction*. *J. Constr. Eng. and Mgmt.*, vol. 119, no. 3, 1993.

Brown, M., N. Tsagarakis. and D. G. Caldwell. Exoskeletons for Human Force Augmentation. *Int. J. Industrial Robot*. vol. 30, no. 6, 2003.

California OSHA Consultation Service. *Ergonomic Guidelines for Manual Material Handling*. NIOSH Publ. No. 2007-131. Cincinnati, OH: National Institute for Occupational Safety and Health, Publ. Diss., 2007.

California OSHA, and NIOSH. *A Guide to Selecting Non-Powered Hand Tools.* Cincinnati, OH: National Institute for Occupational Safety and Health, Publ. No. 2004-164, NIOSH—Publ. Diss., 2004. www.dir.ca.gov/dosh/puborder.asp.

Campbell, F. *Occupational Stress in the Construction Industry Survey.* Ascot, UK: Chartered Institute of Builders (CIOB), 2006.

Caruso, C. C., E. M. Hitchcock, R. B. Dick, J. M. Russo, and J. M. Schmit. *Overtime and Extended Work Shifts: Recent Findings on Illnesses, Injuries, and Health Behaviors.* Cincinnati, OH: National Institute for Occupational Safety and Health, Publ. No. 2004-143, NIOSH Publ. Diss., April 2004.

Chang, F., Y. Sun, K. Chuang, and D. Hsu. Work Fatigue and Physiological Symptoms in Different Occupations of High-Elevation Construction Workers. *Appl. Ergo.*, 39, 2008.

Cohen, A. L., C. C. Gjessing, L. J. Fined, B. P. Bernard, and J. D. McGlothlin. Elements of Ergonomic Programs: A Primer Based on Workplace Evaluations of Musculoskeletal Disorders, NIOSH Publ. No. 97–117. Cincinnati, OH: National Institute for Occupational Safety and Health, Publ. Diss., 1997.

Construction Safety Association of Ontario (CSAO). Heat Stress: Guidelines for Recognition, Assessment, and Control in Construction, Construction Safety Association of Ontario, Etobicoke, Construction, Canada, 2000. www.csao.org.

Dowell, J., and Y. Shmueli. Blending Speech Output and Visual Text in the Multimodal Interface. *Human Fact.*, vol. 50, no. 5, October 2008.

Drury, C. G. Human Factors in Industrial Systems: 40 Years On, *Hum. Factors*, vol. 50, no. 3, June 2008.

Duffy, V. G., and A. Chan. Effects of Virtual Lighting on Visual Performance and Eye Fatigue. *Hum. Factors Ergo. Manuf.*, vol. 12, no. 2, 2002.

Dunston, P. D., and L. E. Bernold. Adaptive Control for Safe and Quality Rebar Fabrication. *J. Constr. Eng. and Mgmt., ASCE*, vol. 126, no. 2, 2000.

Gao, C., I. Holmer, and J. Jabeysekera. Slips and Falls in a Cold Climate: Underfoot Surface, Footwear Design and Worker Preferences for Preventive Measures, *Appl. Ergo.*, 39, 2008.

Govindaraju, M., A. Pennathur, and A. Mital. Quality Improvement in Manufacturing through Human Performance Enhancement. *Integrated Manuf. Syst.*, December 5, 2001.

Greco, M., N. Stucchi, D. Zavagno, and B. Marino. On the Portability of Computer-Generated Presentations: The Effect of Text-Background Color Combinations on Text Legibility. *Hum. Factors*, vol. 50, no. 5, October 2008.

Hanna, A. S., C. S. Taylor, and K. T. Sullivan. Impact of Extended Overtime on Construction Labor Productivity. *J. Constr. Eng. Mgmt.*, vol. 131, no. 6, June 2005.

Haslam, R. A., S. Hide, A. Gibb, D. Gyia, T. Pavitt, S. Atkinson, and A. Duf. Contributing Factors in Construction Accidents. *Appl. Ergo.*, 36, 2005.

Heames, J., and M. Harvey. Workplace Bullying: A Cross-Level Assessment, *Mgmt. Decision*, vol. 44, no. 9, 2006.

Helin, K., J. Viitaniemi, S. Aromaa, J. Montonen, T. Evilä, S. Leino, T. and Määttä.OSKU-Participatory Design Approach: A New Tool to Improve Work Tasks and Workplaces. VTT.Work.83, VTT Tech. Res. Centre of Fin., VTT, Finland, 2007.

International Labor Office (ILO). *Safety and Health in Construction.* Publ. Bureau, ILO, Geneva, Switzerland, 1992.

______. *Code of Practice on Workplace Violence in Services Sectors and Measures to Combat This Phenomenon.* Meet. Rep. Exp. Develop Code of Practice, Geneva, Switzerland, October 8-15, 2003.

______. *Ambient Factors in the Workplace.* Publications Bureau, 2001, ILO, Geneva, Switzerland.

Johnson, S., C. Cooper, S. Cartwright, I. Donald, P. Taylor, and C. Millet. *The Experience of Work-Related Stress Across Occupations. J. Mngr. Psy.*, vol. 20, no. 2, 2005.

Kalveram, K. T. How Acoustical Noise Can Cause Physiological and Psychological Reactions, Proc. 5th Int. Symp. Transp. Noise Vibr., St. Petersburg, Russia, June 6-8, 2000.

Kittusamy, N. K. *Ergonomic Risk Factors: A Study of Heavy Earthmoving Machinery Operators.* Cincinnati, OH: NIOSH Publications Diss., 2002.

Kleiner, B. M. Macroegonomics: Work System Analysis and Design. *Hum. Factors*, vol. 50, no. 3, June 2008.

Larson, J. RoboKent—A Case Study in Man-Machine Interfaces. *Ind. Robot*, vol. 25, no. 2. 1998.

Lee, J., S. J. Lorenc, and L. E. Bernold. Saving Lives and Money with Robotic Trenching and Pipe Installation. *J. Aerospace Eng.*, vol. 12, no. 2, 1999.

Lee, J. D. Review of the Pivotal Human Factors article: "Humans and Automation: Use, Misuse, Disuse, Abuse." *Hum. Factors*, vol. 50, no. 3, June 2008.

Lipscomb, H. J., J. E. Glazner, J. Bondy, K. Guarini, and D. Lezotte. Injuries from Slips and Trips in Construction. *Appl. Ergo.*, 37. 2006.

Lutgen-Sandvik, P. Take This Job and . . . : Quitting and Other Forms of Resistance to Workplace Bullying. *Communication Monographs*, vol. 73, no. 4, December 2006.

Maiti, R. Workload Assessment in Building Construction Related Activities in India. *Appl. Ergo.*, vol. 39 no. 6, 2008.

McLeod, R. Human Factors Assessment Model Validation Study, Health Safety Exec. Norwich, UK: Her Majesty's Stationery Office, 2004.

Mohamed, S., and K. Srinavin. Forecasting Labor Productivity Changes in Construction Using the PMV Index. *Int. J. Ind. Ergo.*, no. 35, 2005.

Moray, N. The Good, the Bad, and the Future: On the Archaeology of Ergonomics. *Hum. Factors*, vol. 50, no. 3, June 2008.

Muir, J. Alcohol and Employment Problems. *Work Study*, vol. 43, no. 7, 1994.

National Household Survey on Drug Abuse (NHSDA). *Survey of Substance Use, Dependence or Abuse among Full-time Workers.* Report, Office Appl. Stud. Subst. Abuse Mental Health Serv. Admin., September 2002. www.DrugAbuseStatistics.samhsa.gov.

Occupational Safety and Health Branch. *Air Impurities in the Workplace*, 3/2004-1-OHL29b. Hong Kong Occ. Safety Health Council., Hong Kong, 2004. www.labour.gov.hk/eng/public/content2_9.htm.

Occupational Safety and Health Institute of Ireland (OSHII). *Safety Behaviour in the Construction Sector.* Rep. Health Safety Auth., Dublin, and the Health Safety Exec., Northern Ireland, HSA, Dublin, Ireland, March 2002.

Osterman, P. The Wage Effects of High Performance Work Organization in Manufacturing. *Ind. Labor Rel. Rev.*, vol. 59, no. 2, January 2006. www.jstor.org/stable/25067516.

Parasuraman, R. Putting the Brain to Work: Neuroergonomics Past, Present, and Future. Hum. Factors, vol. 50, no. 3, June 2008.

Qiana, X., and J. Fan. Interactions of the Surface Heat and Moisture Transfer from the Human Body under Varying Climatic Conditions and Walking Speeds. *Appl. Ergo.*, 37, 2006.

Root, L. S., and L. P. Wooten. Time Out for Family: Shiftwork, Fathers, and Sports. *Hum. Res. Mgmt.*, vol. 47, no. 3, Fall 2008.

Rosa, R. R., and M. J. Colligan. Plain Language about Shiftwork. NIOSH Publication No. 97-145. Cincinnati, OH: U.S. National Institute for Occupational Safety and Health, July 1997.

Ryerson, M., and C. Whitlock.Use of Human Factors Analysis for Wildland Fire Accident Investigations, Proc. 8th Int. Wildland Fire Safety Sum., April 26-28, Missoula, MT, 2005.

Seo, N. J., and T. J. Armstrong. Investigation of Grip Force, Normal Force, Contact Area, Hand Size, and Handle Size for Cylindrical Handles. *Hum. Factors*, vol. 50, no. 5, October 2008.

Seppänen, O., W. J. Fisk, and Q. Lei. Effect of Temperature on Task Performance in Office Environment. LBNL-60946, Lawrence Berkeley Nat. Lab., July 2006.

Sheridan, T. B. Risk, Human Error, and System Resilience: Fundamental Ideas. *Human Factors*, vol. 50, no. 3, June 2008.

Toole, T. M. Construction Site Safety Roles, *J. Const. Eng. Mgmt.*, vol. 128, no. 3, June 2002.

_______. Increasing Engineers' Role in Construction Safety: Opportunities and Barriers. *J. Prof. Iss. Eng. Edu. Pract.*, vol. 131, no. 3, July 1, 2005.

U.S. Army Corps of Engineers. Safety and Health Requirements. EM 385-1-1. Washington, DC: U.S. Government Printing Office, November 2003. http://140.194.76.129/publications/eng-manuals/em385-1-1/2003_English/toc.htm.

U.S. Congress, Office of Technology Assessment. Biological Rhythms: Implications for the Worker,OTA-BA-463. Washington, DC: U.S. Government Printing Office, September 1991.

U.S. Department of Transportation-Federal Aviation Administration, Safety Risk Management Guidance for System Acquisitions. Air Traffic Org., Office of Safety, Version 1.5, December 2008.

Van Vuuren, B., E. Zinzen, H. J. van Heerden, P. I. Becker, and R. Meeusen. Work and Family Support Systems and the Prevalence of Lower Back Problems in a South African Steel Industry. *J. Occ. Rehabilitation*, 17, 2007.

Vi, P. Effects of Rebar Tying Machine on Trunk Flexion and Productivity. Ontario, Canada: Workplace Safety Ins. Board, 2006.

Chapter 8

The Complexity of Human Motivation

Construction is performed by laborers, foremen, technicians, managers, engineers, and other specialists. Each of these individuals has their own reasons for going to work, for performing in a particular way, to personal standards, with varying degrees of energy and enthusiasm. The many driving forces behind all of these is summed up in one word: *motivation* (from the Latin word *motivus* = a moving cause). What drives human behavior, what motivates it, has been the object of study, accumulating a correspondingly large body of knowledge, for the last 3,000 years. It is important that leaders in all walks of life—business, government, the military, education—have an understanding of this body of knowledge, as it will have a major influence on their success.

This chapter will provide a short overview of some of the theories of human motivation that have survived the test of time and must be understood by construction managers who are charged with forming teams and leading them.

8.1 BACKGROUND

Hedonist philosophers, starting with the Greeks, argued that the primary purpose of humans was to maximize happiness and minimize pain. They concluded that personal desires were the fundamental motivation behind all human actions. Buddhists, in contrast, point to the dehumanizing and negative effects of insatiable cravings, which can lead to addiction of all sorts. Such desires can be fulfilled only by hurting others, and so should be avoided.

At the dawn of civilization, basic survival needs motivated early humans to hunt and gather berries and other edible foodstuffs. The best hunters brought home the biggest pieces of meat, which provided an incentive for them to continue to excel the next time. But was the biggest piece of meat their only gratification? Or was there more? Did it also provide a feeling of being special, and earn the admiration of their family and community? We will never know, of course.

The great Greek philosopher Aristotle (384–322 BC) distinguished between two basic types of motives for people engage in a certain behavior: *ends* and *means.* If someone is motivated purely by the direct result of an activity, such as the pleasure derived from reading a book, he or she will be motivated to undertake the activity for its own sake. If, on the other hand, a person pursues a goal in order to gain something that has instrumental value—for example, studying medicine to become rich rather than to help others—the immediate pursuit is a means to achieve something that is not directly related.

Not much hunting and gathering goes on today, though. So what is it that motivates people to get up in the morning, possibly drive for hours on packed highways, to sit in totally enclosed building for hours every day? During Frederick Taylor's lifetime, the monetary reward was considered the only gratification the workers desired. Concepts such as a healthy work environment, protection against harassment, and professional pride were not on even on their personal radar screens. All that changed dramatically in the 1920s when the surprising results of the Hawthorne experiments demonstrated that people did, indeed, have very strong motivating desires that had nothing to do with money. This revelation was the starting point of a new thinking, called the Human Relations Movement, which called into question Taylor's Scientific Management principles.

Proponents of the Human Relations Movement considered the whole human being as a living organism, one that could not be moved around like a pawn on a chessboard by their employers. Rather, the question became: How can we create work environments where people feel they fit in and are satisfied and productive? In this chapter, we will cover some of the main motivational theories that have been proposed since the early 1950s. Table 8.1 offers the results of recent studies of the underlying principles of human motivation, which in many ways are still not fully understood.

Note that, though they are numbered in the table, the list of desires is actually in alphabetical order rather than any kind of priority order. This structure acknowledges the fact that each person will develop his or her own priorities and determine their relative importance. In fact, their relative importance will change over time, especially during the middle of life (e.g., midlife crisis). In other words, understanding the items on the "desire list" is not enough to determine what motivates an individual working for a construction company or anywhere else. We need more insight into human psychology.

8.2 BEHAVIORAL ASPECTS OF THE HUMAN MIND

Who has never been unable to "make up his/her mind" when trying to order food in a restaurant, or "changed his/her mind" after the waiter/waitress has already taken the order? What exactly is happening in our minds when we do these things, and where does it happen? The short answer is: We aren't really sure, but we have some ideas about how this mental process works.

Table 8.1 Sixteen Motivating Desires*

	Motivation	Definition Desire		Motivation	Definition Desire
1	Acceptance	For approval	9.	Physical activity	To exercise muscles; become strong, healthy
2.	Curiosity	For knowledge	10.	Power	To influence, lead
3.	Eating	For food	11.	Romance	For sex (including courting)
4.	Family	To raise own children	12.	Saving	To collect; value frugality
5.	Honor	To obey a traditional moral code	13.	Social contact	For peer companionship, play
6.	Idealism	To improve society; social justice	14.	Status	For social standing, attention
7.	Independence	For individuality, autonomy	15.	Tranquility	To prevent anxiety, fear
8.	Order	To organize; for ritual	16.	Vengeance	To "get even," to compete, to win

*S. Reiss, "Multifaceted Nature of Intrinsic Motivation: The Theory of 16 Basic Desires." *Review of General Psychology*, Educational Publishing Foundation, Vol. 8, No. 3, pp. 179–193, 2004.

It is generally agreed that the mind is one of the three fundamental aspects—along with body and spirit—that make us human. It constitutes the brain itself but also emotions and imagination. Figure 8.1 illustrates the three elements comprising the human mind. As shown, the spirit and body are both thought to be connected to the mind—although, as mentioned, this is still something of a mystery. In other words, the mind is influenced by both spiritual beliefs and the person's genetic history. Further, the mind is thought to encompass three components: (1) *cognition*, which perceives and stores data and processes information for thinking; (2) *affect*, which brings attitudes, predispositions, and emotions into the thinking

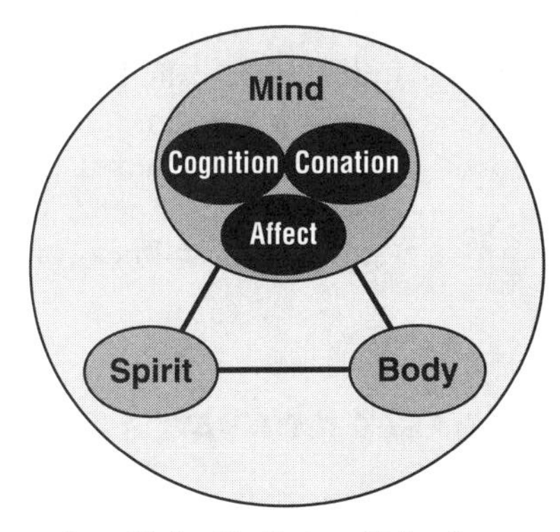

Figure 8.1 The Three Dimensions Linked to Human Behavior

process, and (3) *conation* makes decisions, is responsible for reasoning, and commits to a course of action. Finally, the body is called upon to execute the activity of the mind, through diverse outlets, such as speech, aggression,, and the fight-or-flight response. A full and complete understanding of how the human mind works, and what motivates a person to engage in a particular behavior still eludes us. But in the next section, we will attempt to dissect the process that motivates a high school student to choose a college major, or a construction manager to change jobs.

8.3 INTRINSIC VERSUS EXTRINSIC MOTIVATION

What motivates a student to spend hours playing computer games, while neglecting homework that would improve his grades? Why do people stay at stressful jobs that cause ulcers and insomnia? It is believed that the motivators are related to *natural human needs* or instincts that stimulate desires, which in turn build up internal energy or drive to satisfy those desires. Those needs may be of an *activational* or *avoidance* nature. This is easily understood in terms of food: Hunger will stimulate the desire to eat, which generates the motivation to actively seek food. Similarly, the fear of catching a deadly virus will motivate people to avoid contact with others to minimize their risk of becoming sick.

But let's return to the example of the student, as it illustrates several points. First, playing computer games seems to offer a lot of satisfaction to the student, but no apparent lasting value. Whereas failing to study and do the assigned homework will probably have a lasting effect on the student's grades and, possibly, job prospects down the road. However, the student chooses playing over studying because he perceives less value in the latter. When people do things that have no external inducement or consequence—such as play computer games—they are referred to as *intrinsically motivated*. In contrast, actions that are externally induced are called *extrinsically motivated*. As we will see, intrinsic motivation plays a very important role in the workplace today.

In summary, motivation is a product of the human mind, which originates in a physiological or psychological (personal) need or deficiency that then generates an energy or force in the body that triggers the fulfillment of the need or alleviation of the deficiency.

Affect, Emotion, and Feeling

Affect complements the rational cognition of the brain's capability to deal with the emotional aspects of life. Considered an intrinsic capacity, the brain reacts unconsciously to external stimuli much quicker than the cognitive part of the brain. It produces feelings and reactions that are not necessarily explainable (e.g., first impressions). *Feelings* can be displayed to others through body language, such as facial expressions, hand gestures, posture, tone of voice, and so on. Different cultures develop unique behaviors that are understood by members of those societies, such as the V-sign, to mean "victory."

8.4 MASLOW'S NEEDS-BASED MOTIVATORS

Until the onset of the twentieth century, instinct-based and hedonistic theories about human motivation were at the forefront of philosophical thought. Then, in

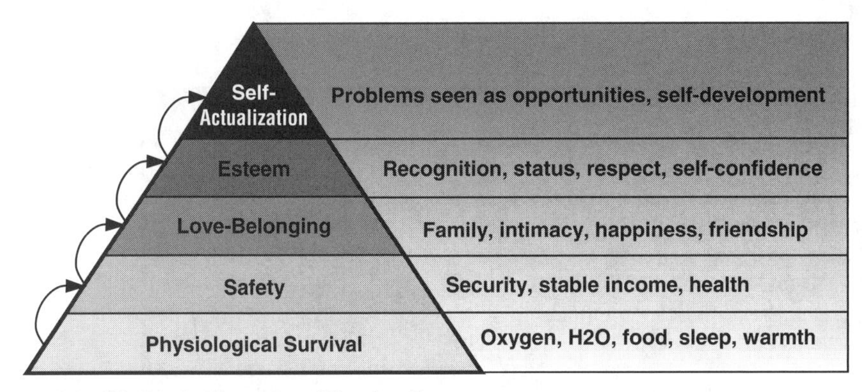

Figure 8.2 Maslow's Hierarchy of Needs

1943, Abraham Maslow (1908–1970) proposed his model of motivational needs. It became widely known as "Maslow's pyramid," due to the graphical presentation of his theory, as shown in Figure 8.2. The foundation of the hierarchy—the base of the pyramid—acknowledges basic physiological needs, without which a human being cannot survive. Thus, the need for food, water, sleep, and so on, outweighs all other desires. Only after these needs have been satisfied, at least temporarily, will other needs come to the fore, emerging as the drivers for action.

The need for safety or a feeling of security is a natural next level. Once assured of physical survival and gaining a sense of stability, human beings, according to Maslow, will strive for companionship and form relationships within a group. One might call this desire the "search for happiness." Of course, it goes without saying that should safety or a food source disappear, the person's focus will revert back to satisfying those basic lower-level needs.

The top two levels represent "mental" needs, which can be fulfilled after the person feels safe and protected. The fourth level, the need for esteem, respect, and appreciation, reflects the human desire to become self-reliant and to seek recognition by others for their efforts and contributions. Maslow calls this the need for *self-actualization*, since, as he put it, "What a man *can* be, he *must* be." This desire is the perfect example of an intrinsic need, for people will spend effort and time on issues that are freely chosen, even against strong extrinsic pressures to the contrary. For example, a person with a secure job and loving family might decide to give up that lifestyle and move to a dangerous place to help an endangered culture or environment.

8.5 VROOM'S EXPECTANCY THEORY

"It takes two to tango" is a popular adage, which is applicable here, in the sense that a leader is helpless without his or her followers. If the employees do not feel motivated to perform above the minimal acceptable level, the leader is helpless to improve productivity. Victor Vroom (b. 1932) was an early researcher who

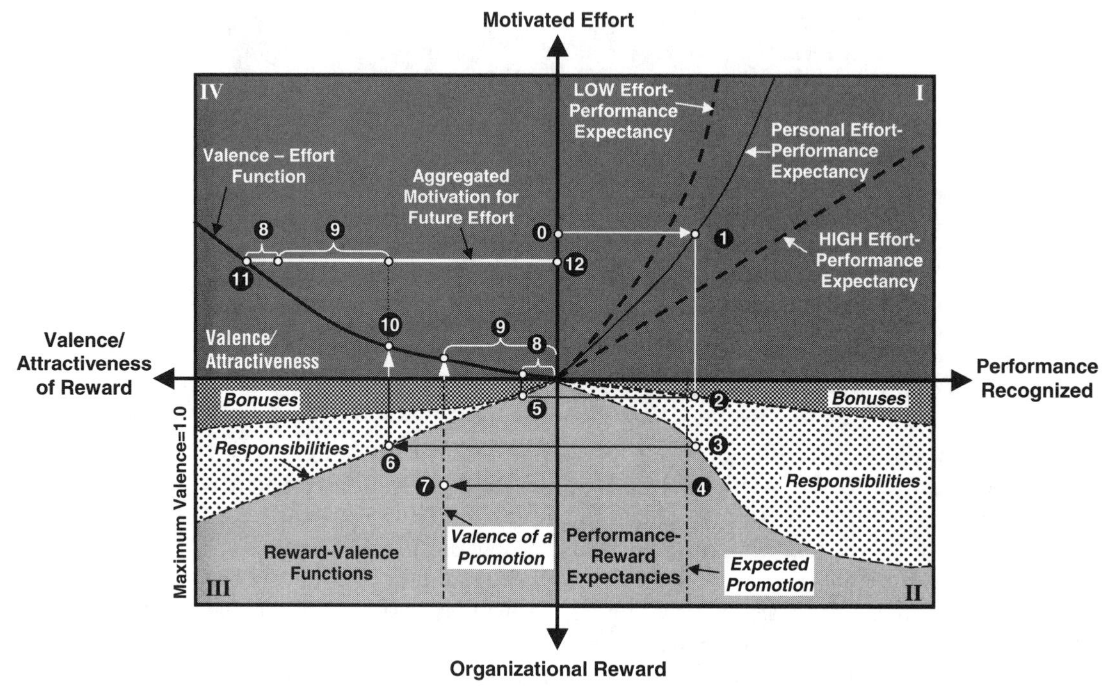

Figure 8.3 The Four Quadrants of Vroom's Expectancy Theory Model

recognized key relationships between motivation and performance. His theory is known as the *expectancy theory*. In this theory, motivation to exert effort is defined as a product of the likelihood that personal effort will elevate performance sufficiently to be rewarded by the company at a value equal to the effort expended. Key to the theory is the understanding of the employee's goals, and the linkages among and between effort, performance, and rewards, as presented in the four integrated quadrants in Figure 8.3.

Quadrant I correlates the employee's expectations that his or her personal effort will be noticed and recognized by the company. The two dashed lines indicate the highest and lowest levels of a person's expectations. Let's assume that Jane is working in the scheduling department at a construction company. She works hard to write detailed reports to accompany schedule updates. On her own time she improves her software skills, to make her process more efficient. The question is: Does Jane's employer recognize her personal efforts? If it does, then there is a direct and possibly linear relation between effort and recognized performance. If not, Jane's highest efforts will go for naught, and the curve will become vertical, meaning that no matter what Jane does above and beyond her job description, she will receive no additional recognition.

The point is, low expectations can have various causes, among them lack of support and doubt that the organization will assess and reward performance fairly.

Figure 8.3 shows an arbitrary personal effort/performance recognition expectancy curve, which is intersected by the horizontal personal effort line in ❶.

Quadrant II graphs the expected organizational rewards as a function of recognized performance. Depicted in Figure 8.3 are three rewards: a bonus, which increases linearly; an increase in responsibilities; and a promotion, which is a one-time event, upon attaining a certain performance threshold, whereas the others increase according to linear, nonlinear, or complex functions.

The graphical representation begins with a vertical line starting at ❶, which intersects with the bonus line at ❷, and with the responsibility curve in ❸. In the example, the promotion level is to the left, indicating that the person's effort will be rewarded with a promotion, the expected recognition.

Quadrant III illustrates unique personal priorities, which vary from person to person, as well as over time for each individual. As described earlier, Maslow offered a basic model useful in understanding the relationship between the motivation to work and rewards. To repeat, his pyramid of human needs is based on the observation that these needs are hierarchical, with survival at the base. Humans seek to reach the next higher need level while maintaining the conditions that guarantee the foundation-level satisfactions.

The three reward/valence functions shown in the figure signify that the person for whom they have been drawn considers the bonuses very attractive. Vroom refers to the attractiveness of a reward as its *valence*, an anticipated satisfaction, as opposed to its real value, the actual satisfaction it provides. The apparent attractiveness of bonuses, indicated by the steep slope of its function, is not surprising, given that monetary rewards would provide added financial security now and in the future—the second most important needs category according to Maslow's pyramid. Promotion that could result from high-level performance, adding to self-esteem and, possibly, a higher income, grants a certain degree of valence.

Finally, additional responsibilities, an example of self-actualization, results in a linear slope of valence increases, but ends at a far lower level, given the company may hand out a reward. Since additional responsibilities usually go hand in hand with greater stress, time away from family, and other disadvantages, a limited attractiveness indicates the presence of a person's strong lower-level needs, such as the love provided by family and friends, which would probably suffer. In this sense, the possible satisfaction from more responsibilities is reduced by the dissatisfaction caused by a diminishment in personal needs satisfaction. As we will see later, Fredrick Herzberg studied this dual-force concept, called the *two-factor theory*.

Figure 8.3 links the three expected rewards with the valence curves for a person via three horizontal lines starting at the intersections ❷, ❸, and ❹. Those lines create three more intersection points, with the valence curves for bonuses, responsibilities, and promotion, at ❺, ❻, and ❼.

Quadrant IV In quadrant IV, the anticipated satisfaction that a person might achieve from his or her original effort is connected to the personal energy that will be put forth in the future. Should the original effort be at a much higher level than the effort that correlates with the total valence, the imbalance would logically cause the employee to either reduce or increase effort until an equilibrium is achieved. Alternatively, the employee might request a change in the performance/reward functions, to more closely match the reward/valence functions. In the situation here, the employee might ask for a bonus increase, along with a reduction in responsibilities, because the increased bonus provides an extremely high valence multiplier (e.g., small increase bonus increase adds a lot of valence.) Reducing or increasing responsibilities will not change the relative attractiveness of the effort as long as the horizontal line, through point ⓭, intersects with the responsibility-valence function in Quadrant III. In other words, the misalignment between original and valence-effort is closed more quickly by increasing the bonus reward than by adding more responsibilities. This might be representative of someone started a new family needing money to buy a house, etc.

We will discuss later what a manager might do to further improve the attractiveness of an effort. For now, Figure 8.3 closes the loop in the example by adding all the valences—❽, ❾, and ❿—horizontally, and intersecting the line with the valence effort function in ⓫, which indicates how much effort would equate to that amount of total valence. As it turns out, ⓬ is lower than the starting point ⓿, indicating that the person puts in a greater amount of effort than he or she should have to, to achieve perfect balance.

Worked-Out Problem 8.1: Low Expectancy Causes Effort/Valence Imbalance

John, married and with a two-year-old son, has been working for some time as a rental agent/manager at a construction equipment rental company. He is not satisfied with his present job situation, and wants to assess his options for personal growth with his current company against those at a new firm. He is asking us to help him sort out the situation, so he can balance the motivational force with his desires and needs.

Background and Assumptions

Vroom's expectancy theory provides an excellent model to lay out John's options. Using the four-quadrant graph given in Figure 8.3, we will be able to plot his effort/expectancy/valence relationships.

Establishing Expectancy Functions

Quadrant I

This quadrant allows us to establish John's expectancy that his efforts will result in contributions to fulfilling the company's mission. John might feel that his hours of work correlate predictably and linearly with a performance level that is recognized and rewarded by the company—a line can be drawn through the origin with a slope represented by the ratio of Δ effort/Δ performance.

Most often, such a linear correlation is elusive, especially in situations where other people or resources are involved.

In the case of a rental agent for construction equipment, it is possible that spending more hours on the phone, writing advertisements, and so on, to promote new equipment or special deals will not equate to more business for John alone, but also other rental agents. Further, more rentals will also create additional work for the maintenance and repair shop, whose capacity is currently limited, unless more people are hired. As a result, the effort/performance expectancy might be best represented by a U-shaped quadratic or exponential curve. Here, a Δ effort will result in an ever-decreasing amount of Δ performance, characterized by a low effort/performance expectancy.

Quadrant II

The second quadrant is reserved to graph John's expectation that performances recognized by the company will equate to rewards. A common type of monetary awards, in addition to a base wage, are given as bonuses, which correlate to measureable outcomes such as, in John's case, a cumulative

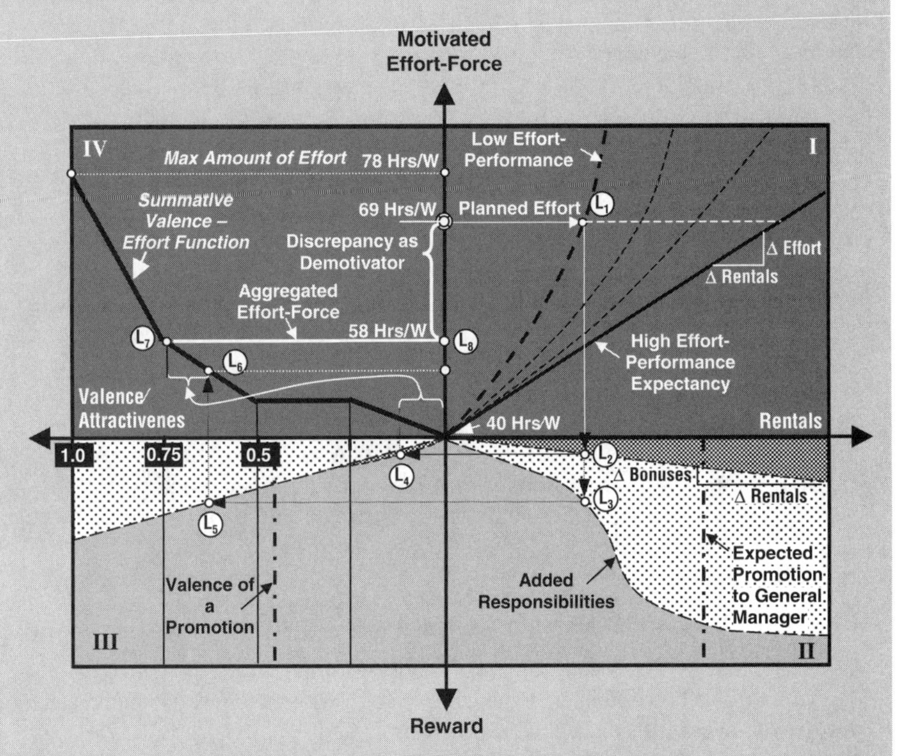

Figure 8.4 Demotivating Valences Resulting from a Low Expectancy

number of rentals. Again, this might be a linear function that provides Δ bonuses for Δ rental revenue. A second type of reward that John might expect is a promotion to, say, general manager of the rental store, should he be able to arrive at a fixed number of rentals for a year or more. Here, the expected reward would be immediate, if John reached a certain threshold.

Taking on more responsibility will lead to greater respect from others, the second-highest need on Maslow's pyramid. Thus, we should include John's expectancy of greater responsibility linked to performance. For example, an assistant or a mechanic might be directly assigned to him after he achieves a higher number of rentals. This would also give him greater flexibility.

Quadrant III

Not every reward is equally desirable. We need to query John about his feelings about the relative importance of the rewards options, on a scale from 0 to 1. What bonus amount that he could *reasonably expect* would be most attractive to him? How much would make it half as attractive, and so on? What would be the attractiveness on a scale of 0 to 1 of being promoted to general manager, considering all the additional risks? How about the desirability of added responsibility? How much responsibility would be ideal for him, to give him an optimal amount of support and self-esteem? Based on the answers, three functions can be plotted to relate rewards to John's assessment of their attractiveness.

Quadrant IV

The function that relates reward attractiveness to John's motivation to put effort into his work is probably the most difficult to elicit. First, we need to aggregate all the different valences into one value that expresses the motivational force it creates. Depending on the chosen scale of effort, such as hours of work per week, we need to ask John about the relationship between valence and his willingness to exert effort. John tells us he would never be willing to work more than 78 hours/week, which equates to the maximum summative valence. The lowest effort is 40 hours/week. How many hours would he work if the total valence were 0.5 of the maximum, and so on?

The sum of the individual valences can be scaled down if they exceed 1.0, or are plotted without scaling. Now we are ready to develop the initial expectancy plot.

Iteration to Compare Planned and Valence-Motivated Effort

We begin the with first querying John about how many hours he would be willing to work without pushing himself too much while receiving satisfying rewards. Let's assume the value is 69 hours/week (see Figure 8.4). Next, he needs to tell us which effort/performance function we should select (anything between high and low). If John is very pessimistic about how his effort would be rewarded we would pick L_1 as it lies on the lowest expectancy curve. Intersecting the vertical line with the two performance/reward expectancies in the second quadrant, we create L_2 and L_3. From these two points we draw horizontal lines into quadrant III to intersect with the related reward/valence functions, which produce L_4 and L_5, a valence of 0.12 for the bonuses and 0.625 for the added responsibilities. From these two points, vertical arrows lead to an intersection point with the summative valence/effort function in quadrant IV. L_6 identifies the effort/force of taking on more responsibility, which would motivate John to add approximately 10 to 11 hours to the 40-hour minimum. Summing it with the valence from the expected bonuses L_7 is calculated with the aggregated effort/force identified by L_8. According to the scale shown in Figure 8.5, the attractiveness of the bonuses combined with the added responsibilities does not, at 58 hours/week, produce the force that is close to the 69 hours/week that would be necessary to produce it.

Discussion of Results

Based on Vroom's expectancy theory, John is facing an imbalance between his planned effort, 69 hours/week, and the equivalent value he associates with the rewards he expects the equipment rental company will give him (58 hours/week). Obviously, working fewer hours will lower his performance and, subsequently, the expected rewards and summative valence. Working more hours would not result in much improvement, mainly because of the shape of the low-effort performance expectancy curve, with a steep slope at around 69 hours/week. For example, under the present assumptions, it would be impossible for John to ever be promoted, which would heighten the overall valence of his work.

Recommendations for John

We have three recommendations for John:

1. The single largest obstacle to improving John's situation is his expectancy regarding what the company would use to assess his performance. He should verify that his low expectancy is justified. Furthermore, he might suggest additional measures the company could use in addition to his rental numbers—for example, customer satisfaction or amount of repair cost from his rentals.

   The sole purpose of this attempt to change John's present situation is to move the point to the right, along the horizontal line in quadrant I, which identifies John's planned effort of 69 hours/week.
2. By comparing the valence/reward functions for bonuses and responsibilities it's easy to see that bonuses are much more attractive than the added responsibilities. Thus, a second option would be to

exchange responsibilities with the bonuses the company would provide.

3. Finally, John might reevaluate his valence/reward functions, especially that of responsibilities. The valence of increased responsibilities might be improved if John could link the additional work to a possible promotion, which seems to be solely linked to rentals. Maybe the company would recognize his greater responsibilities in the form of training to become a general manager in the future, thus making the additional work more attractive to him.

8.6 HERZBERG'S TWO-FACTOR THEORY

Frederick Herzberg (1923–2000), who earned his PhD in the area of human motivation, was one of the early members of the aforementioned Human Relations Movement, a behaviorist group that challenged the common assumption that money was the single most powerful motivator for performing work. Herzberg conducted extensive surveys and interviews to prove Maslow's motivational theory. The data did, in fact, confirm Maslow's hypothesis, but also brought to light a new insight: that besides intrinsic motivators, a set of demotivators played an important role in establishing an overall level of motivation.

> When our respondents reported feeling *happy* with their jobs, they most frequently described factors related to their tasks, to events that indicated to them that they were *successful* in the performance of their work, and to the possibility of professional growth . . . . When feelings of unhappiness were reported, they were not associated with the job itself but with *conditions* that surround the doing of the job. These events suggest to the individual that the context in which he performs his work is unfair or disorganized and as such represents to him an *unhealthy psychological work environment.* Factors involved in these situations we call *factors of hygiene*, for they act in a manner analogous to the principles of medical hygiene. Hygiene operates to remove health hazards from the environment of man. It is not a curative; it is, rather, a preventive. Modern garbage disposal, water purification, and air-pollution control do not cure diseases, but without them we should have many more diseases.
>
> —*F. Herzberg, B. Mausner, and B. B. Snyderman,* **The Motivation to Work** *(New York: John Wiley & Sons, Inc., 1959)*

Herzberg argued that both factors, motivators and psychological hygiene, are equally important in the workplace, but for different reasons. A good work environment will not serve as a motivator; but a poor work environment will serve as a demotivator.

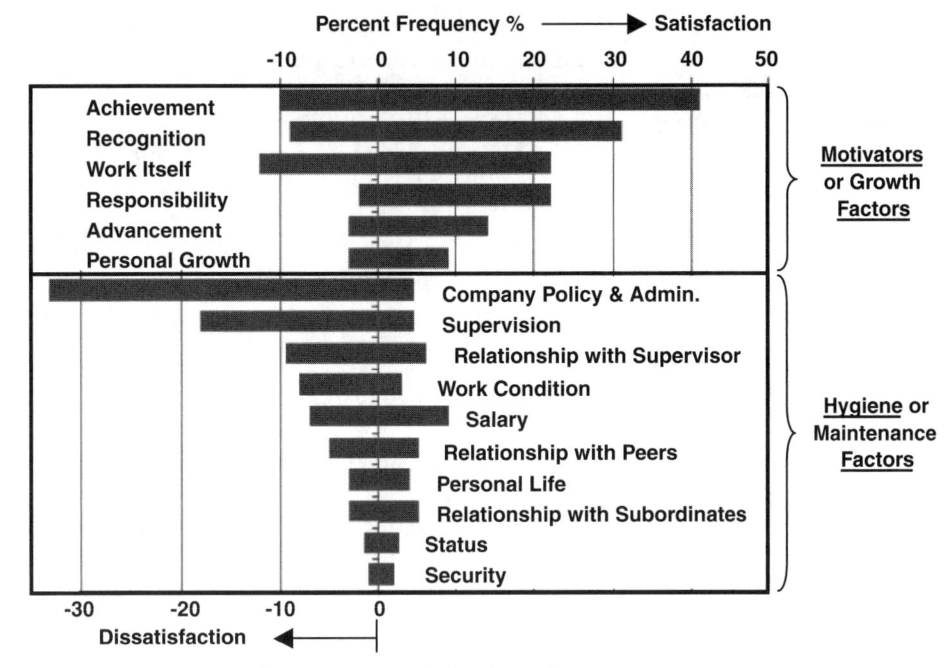

Figure 8.5 Herzberg's Most Important Motivational Factors

Figure 8.5 summarizes graphically Herzberg's findings (comprising a large data set), which have since been confirmed by other researchers. The analysis resulted in a list of 16 high-priority factors: 6 most responsible for providing satisfaction and 10 hygiene factors responsible for causing the most dissatisfaction.

Reviewing the graph we can see that the upper six factors align with Maslow's intrinsic values of self-esteem, which result from recognition, advancement, and responsibility, while achievement, work itself, and personal growth align with self-actualization. The bottom 10 are of a totally different nature. In addition, each factor can, at sharply different magnitudes, serve as a satisfier as well as a dissatisfier. For example, receiving recognition leads to, at a high frequency, a high level of satisfaction. The lack of recognition, however, was found not to produce the concomitant level of dissatisfaction. The opposite was true for hygiene factors. "Toxic" supervisors create a lot of dissatisfaction for a large number of people, whereas exceptional supervisors produce only a small amount of satisfaction. Most interestingly, salary shows up as a hygiene factor, *not* a motivator. Although higher salaries might align with motivators such as recognition and achievement, the failure to be paid a salary that is deemed necessary to sustain a minimal level of economic self-sufficiency or to match the work effort will produce an equally strong feeling of dissatisfaction among a high percentage of workers.

In summary, the large study that laid the foundation for Herzberg's two-factor theory showed that motivating factors were responsible for 81 percent of job satisfaction mentioned by the workers covered in the study. Hygiene factors, such as company policy and administration, supervision, interpersonal relationships, working conditions, salary, and so on, were mentioned by 69 percent of the workers polled as causing dissatisfaction at work.

Herzberg's observations can be summarized in two central motivational rules:

1. Managers who provide motivational satisfaction will boost the level of an employee's job satisfaction; however, they will not be able to prevent job dissatisfaction.
2. A management that focuses on minimizing the negative effects of hygiene factors will minimize job dissatisfaction; however, doing so will not generate job satisfaction.

Herzberg's two-factor theory has survived over time, but the validity of the listed factors and their assumed interdependence have met with criticism. But what's important for a construction manager to recognize is the importance of considering both satisfiers and dissatisfiers when motivating subordinates. One might cancel out the other, overriding any managerial efforts to improve job satisfaction.

8.7 MEASURING JOB SATISFACTION

It has been reported that in the United States up to 80 percent of employees are unhappy at work. It's also clear that the more job satisfaction people have, the better they feel about their jobs, the companies they work for, and the people they work with. As Herzberg pointed out, job satisfaction a combination of personal feelings about several intrinsic and extrinsic factors. Although job satisfaction does not result automatically in high productivity it is clearly linked to the level of motivation a person feels towards work. Extending on that, productivity is higher in organizations whose workforce is more satisfied; satisfied employees also have fewer absences, are less likely to quit, and are more willing to work overtime to meet an important deadline.

As workers' attitudes and perceptions about work are subject to change, companies that pay attention to their employees level of job satisfaction over time will generally see higher performance levels. A major effort to create a tool to measure job satisfaction reliably was launched in 1967, by Weiss, Dawis, England and Lofquist. The result the Minnesota Satisfaction Questionnaire, or MSQ, that queried the respondents on 20 five-item scales, including the job itself, pay, promotion opportunities, supervision, and coworkers. It revealed that high satisfaction with a supervisor does not necessarily imply a high degree of job satisfaction. In the banking sector, for example, the level of job satisfaction may be low due to the weak impact of most intrinsic motivation factors on overall job satisfaction.

In the 1980s a shorter 20-question version of the MSQ (1 item rather than 5 on each of the 20 scales), was developed each question addressing either an intrinsic or an extrinsic facet of the job situation (see Table 8.2).

Table 8.2 Minnesota Satisfaction Questionnaire (MSQ): Short Version*

Ask yourself: How satisfied am I with this aspect of my job?	
5 = Extremely Satisfied, 4 = Very Satisfied, 3 = Satisfied, 2 = Somewhat Satisfied, 1 = Not Satisfied	
1.	Being able to keep busy all the time
2.	The chance to work alone on the job
3.	The chance to do different things from time to time
4.	The chance to be "somebody" in the community
5.	The way my boss handles his/her workers
6.	The competence of my supervisor in making decisions
7.	Being able to do things that don't go against my conscience
8.	The way my job provides for steady employment
9.	The chance to do things for other people
10.	The chance to tell people what to do
11.	The chance to do something that makes use of my abilities
12.	The way company policies are put into practice
13.	My pay and the amount of work I do
14.	The chances for advancement on this job
15.	The freedom to use my own judgment.
16.	The chance to try my own methods of doing the job.
17.	The working conditions.
18.	The way my coworkers get along with each other.
19.	The praise I get for doing a good job.
20.	The feeling of accomplishment I get from the job.

*The MSQ is a copyrighted scale. Any use of the scale is prohibited unless purchased.

Based on the items on Herzberg's two-factor list in Figure 8.5, it is easy to see how the questions on the MSQ fit into the two main categories of satisfiers and dissatisfiers. For example, question 20, "The feeling of accomplishment I get from the job," relates directly to number 1 on Herzberg's list: *achievement*. Question 4, "The chance to be 'somebody' in the community," finds its basis in number 2 on Herzberg's list: *recognition*. This tool can be used to improve job satisfaction (i.e., low responses on any of the items may prompt remedial actions by the company).

Here are some possible interventions (parenthetical numbers refer to the MSQ in Table 8.2):

Dissatisfaction with:

- The praise I get for doing a good job (19)

  In situations where work is limited and automatic, offering praise might be harder for a supervisor than in situations where the type of work offers more opportunity to excel in productivity, quality, or safety. Nevertheless, the company might start a development program for its supervisors, to include brainstorming sessions to find ways to recognize and acknowledge the contributions of individuals. Such a program also could open avenues for employees to gain recognition outside their work environment (e.g., training courses).

Dissatisfaction with:

- The chance to do different things from time to time (3)
- The chance to do something that makes use of my abilities (11)
- The freedom to use my own judgment (15)

  The three questions from the MSQ referenced above address several of the growth factors or motivators listed in Figure 8.5: work itself, responsibility, recognition or a sense of achievement, and personal growth.

Classical Temperaments

Temperament is defined by a set of innate characteristics a person possesses that uniquely influence his or her behavior. The Greeks established the four temperaments model, comprising the following characteristics:

Sanguine: Light-hearted, entertaining, friendly, spontaneous, people-oriented

Choleric: Energetic doer, passionate, ambitious, can be pushy or domineering

Melancholic: Reflective thinker, kind and considerate, creative; likes it quiet

Phlegmatic: Consistently cautious and self-contained, rather shy but observant and curious

Monotonous, boring work can be perceived as demeaning to people who have reached high levels on Maslow's hierarchy of needs. Early on, Herzberg promoted the use of job enrichment methods—which does not mean adding new jobs—for workers to gain intrinsic rewards. The intent of job enrichment is to redesign the current situation, such as job sharing or job rotation—for example, assigning an estimator to work on the construction site to update cost figures with actual observations or measurements. Other methods might include training courses for different jobs, and increased autonomy regarding when or how to perform the job. Finally, it might be an excellent opportunity to challenge dissatisfied workers to reengineer their own work processes, by, for example, introducing new technologies.

8.8 JOB ENRICHMENT

Research has found that being satisfied with one's job does not automatically lead to high productivity, or that satisfaction and job performance are directly related. Rather, the link that has emerged is between job performance and personality and temperament. Similar to Kolb's four learning types presented in Chapter 4, there exists a classical taxonomy of four temperaments going back to the ancient Greeks: *sanguine choleric melancholic* and *phlegmatic.*

David Keirsey (1921) developed a system also consisting of four temperaments, which he called *artisan, guardian, idealist,* and *rationalist,* along with 8 roles and 16 subvariants, defined in Table 8.3. He focused his attention on describing how people operate and get things done. For example, he saw two

Table 8.3 Keirsey's Four Temperaments*

Temperament	Characteristics	Roles	Subvariant
Artisans	• Fun-loving, optimistic, realistic; focused on the here and now • Tend to be unconventional, bold, and spontaneous • Make playful mates, creative parents, and troubleshooting leaders • Excitable; trust their impulses, seek stimulation, prize freedom, and dream of mastering action skills	Operators	Crafters
		Entertainers	Promoters
			Composers
			Performers
Guardians	• Considered dependable, helpful, hard-working • Make loyal mates, responsible parents, stabilizing leaders • Tend to be dutiful, cautious, humble, and focused on credentials and traditions • Concerned citizens who trust authority, join groups, seek security, prize gratitude, and dream of meting out justice	Administrators	Inspectors
			Supervisors
		Conservators	Protectors
			Providers
Idealists	• Enthusiastic; trust their intuition, yearn for romance, seek their "true self," dream of attaining wisdom • Pride themselves on being loving, kindhearted, authentic • Giving, trusting, spiritual; focused on personal journeys and human potential • Make intense mates, nurturing parents, inspiring readers	Mentors	Counselors
			Teachers
		Advocate	Healers
			Champions
Rationals	• Tend to be pragmatic, skeptical, self-contained; focused on problem-solving and systems analysis; • ingenious, independent, strong-willed; • Make reasonable mates, parents, strategic leaders • Even-tempered; trust logic, yearn for achievement, seek knowledge, prize technology, dream of understanding how the world works	Coordinators	Masterminds
			Field marshals
		Engineers	Architects
			Inventors

[shaded] = Directors [unshaded] = Informers

*Adapted from www.Keirsey.com.

primary motivations for people to communicate: to inform or to direct. We can easily recognize something of the classical temperaments in those characterized by Keirsey. For example, artisan and the sanguine type have very similar characteristics.

The descriptions of these roles and temperaments can be very helpful in tailoring job enrichment plans that will fit the individual and thus have the best chance to improve job satisfaction and, at the same time, job performance. A company might ask its employees to fill out one of the questionnaires available from Keirsey.com and, subsequently, request a professional evaluation, which could then be discussed with the employees.

8.8.1 Enrichment Schemes

It should be emphasized again that job enrichment plans need to be tailored to the temperament of the individual. Giving an employee who is an "artisan" temperament (e.g., an entertainer) responsibility for supervising the equipment yard and authority to purchase equipment would add a lot of job stress and cause dissatisfaction. But if this same employee were given the responsibility to develop a plan for the company to publicize it efforts to adopt green technologies, he or she would probably be extremely happy.

From a large selection of possible actions, five basic areas of job enrichment have been defined:

Action Area 1—Increase skill variety

Possible actions are job rotation, addition of new responsibilities appropriate to the employee's temperament, or sponsorship of interesting training courses.

Action Area 2: Foster work identity

Create work groups or teams responsible for self-organizing their workflow. Rather than doing one specific task, the members are allowed to be more flexible and expand their roles.

Action Area 3: Improve work relevance

Encourage the employee to join an important task group being put together to solve an important problem—assuming he or she has relevant competence. Or invite the employee to join a long-term task force, to be the "new pair of ears and eyes."

Action Area 4: Allow more work autonomy

Increase the employee's degree of decision-making power, and give the freedom to choose how and when work is done. Expand the role of work groups in choosing their own team members and controlling their work. With this method, supervisory oversight can be reduced, and group members will gain leadership and management skills.

Action Area 5: Provide more personal feedback

Rather than having a supervisor point out mistakes, give teams the responsibility for their own quality control. In this way, workers will receive immediate feedback, and learn to solve problems, take initiative, and make decisions. People should learn as early as possible how well, or poorly, they're performing their jobs.

8.8.2 Designing a Job Enrichment Program

Before embarking on a job enrichment program, we should step back and review all the reasons why people go to work. Recall Vroom's expectancy theory, which stresses the differences in an individual's valence of money, promotion, time-off, and benefits. While some people work primarily to earn money to buy possessions or to gain financial security, others do so because they enjoy the interaction with others or to stay active physically, mentally and emotionally.

As in any other problem-solving or design project, a job enrichment program can be organized in phases.

Phase 1: Establishment of Motivational Profiles

This is the time to investigate why some people are dissatisfied and are not performing as expected. This phase should be accomplished with the full participation of those involved. Online tools are available, as are consultants, to help measure job satisfaction/dissatisfaction, as well as temperaments. It is a major fallacy to *assume* what people want; rather, it is necessary to get down to basics, and use that information to build your enrichment options.

Phase 2: Design of Interventions in Work Process

Based on the motivational profiles described in this chapter, it should be fairly easy to determine how each individual could be enriched at work. But this is also the time to ask employees directly what would give them greater satisfaction in the workplace. By considering the responses of many individuals, synergistic changes across different departments are allowed to evolve (e.g., cross-disciplinary work teams). Keep in mind that job enrichment options do not necessarily require drastic changes in the work process. This phase might also involve the creation of a "job enrichment task force," charged with making recommendations to management.

In summary, this is the phase that offers the opportunity to leverage the employees' aspirations with the goals of the company, to create a work environment that will enhance both.

Phase 3: Implementation and Assessment

This phase might require some training of those involved, in addition to an informational campaign if many people are going to be impacted by the change. At the same time, productivity measurements of the present system should be taken. Follow-up assessments should be done to document the effectiveness of the initiated changes.

CHAPTER REVIEW

Journaling Questions

1. There seem to be some common threads in the way the Greek philosophers categorized personalities, Maslow's needs-based motivations, Herzberg's two-factor theory, Keirsey's temperaments, and Reiss's motivational desires. Trace some of the commonalities and discuss how they show up in each model.
2. The most critical expectancy relationship is the one between personal effort and recognized performance. Assume that you adopted a very low effort-performance expectancy at a job you are working on. What are the actions you could imagine taking to change the exponential curve to a sloped line?
3. The two-factor motivations chart shown in Figure 8.5 resulted from surveying many professional people in the Pittsburgh area. What would the changes in the frequencies be (not the listed factors) if you were to redo the survey with construction workers (including foremen)?

Traditional Homework

1. It is desirable that the members of a team have similar motivational goals and expectancies. Draw the situations a-d into a four-quadrant expectancy map, and explain what the effect would be on the motivational efforts of two members of the same team, assuming they have:
   a. The same effort-recognized performance expectancy
   b. The same performance-reward expectancy
   c. Different reward-valence expectancies
   d. A different valence-effort function.

Open-Ended Question

1. A job enrichment program at any company could create an exciting and satisfying work environment. Unfortunately, not everyone will share this enthusiasm. Can you predict what personalities might most likely try to undermine its success after you reviewed Keirsey's temperaments or Reiss's desires? Propose a plan to circumvent internal "torpedoes" designed to sink the effort. Explain how your plan would alleviate possible negative feelings by employees or even upper management.

BIBLIOGRAPHY

Colombetti, G. Appraising Valence. *J. Consciousness Stud.*, vol. 12, nos. 8–10, 2005.

Fairbrother, K., and J. Warn. Workplace Dimensions, Stress, and Job Satisfaction, *J. Managerial Psych.*, vol. 18, no. 1, 2003.

Feldman, D. C. Toxic Mentors or Toxic Proteges? A Critical Re-Examination of Dysfunctional Mentoring. *Hum. Res. Mgmt. Rev.*, vol. 9, no. 3, 1999.

Fielden, S. L., M. J. Davidson, A. Gale, and C. L. Davey. Women, Equality and Construction. *J. Mgmt. Develop.*, vol. 20, no. 4, 2001.

Halepota, H. A., Motivational Theories and Their Application in Construction. *Cost Eng.*, vol. 47, no. 3, March 2005.

Kinjerski, V., and B. J. Skrypnek. A Human Ecological Model of Spirit at Work. *J. Mgmt. Spirit. and Rel.*, vol. 3, no. 3, 2006.

Lundberg, C., A. Gudmundson, and T. D. Andersson. Herzberg's Two-Factor Theory of Work Motivation Tested Empirically on Seasonal Workers in Hospitality and Tourism. *J. Tourism Mgmt.*, vol. 30, no. 6, 2009.

Lutgen-Sandvik, P. The Communicative Cycle of Employee Emotional Abuse. *Mgmt. Com. Quart.*, vol. 16, no. 4, May 2003.

Pollack, B. L. The Nature of the Service Quality and Satisfaction Relationship: Empirical Evidence for the Existence of Satisfiers and Dis-satisfiers. *Managing Serv. Qual.*, vol. 18, no. 6, 2008.

Reiss, S. Multifaceted Nature of Intrinsic Motivation: The Theory of 16 Basic Desires, *Rev. Gen. Psych.*, vol. 8, no. 3, 2004.

Savery, L. K. The Congruence between the Importance of Job Satisfaction and the Perceived Level of Achievement. *J. Mgmt. Develop.*, vol. 15, no. 6, 1996.

Varvel, T., S. G. Adams, S. J. Pridie, and B. C. Ulloa. Team Effectiveness and Individual Myers-Briggs Personality Dimensions. *J. Mgmt. Eng.*, vol. 20, no. 4, October 1, 2004.

Walsh, K., and J. R. Gordon. Creating an Individual Work Identity, Hum. Res. Mgmt. Rev., 18, 2008.

Zhang, P. and G. M. von Dran. Satisfiers and Dissatisfiers: A Two-Factor Model for Website Design and Evaluation. *J. Am. Soc. Info. Science*, vol. 51, no. 14, 2000.

Chapter 9

Performance Factors of Leaders and Teams

A leader is best when people barely know he exists. When his work is done, his aim fulfilled, they will say: we did it ourselves.

—Lao Tzu

The manager asks how and when; the leader asks what and why. The manager accepts the status quo; the leader challenges it.

—Warren Bennis

In Chapter 8, we learned that motivating people is a multifaceted undertaking, one that has to consider not only personal temperament but also individual expectations and desires. Now we are ready to study the competencies of the manager and leader who are able to instill a winning spirit in all the members of a team, department, or division.

Many of the great leaders in history are remembered mostly for their conquests of lands and people, for destroying cities and towns—citizenry included. Also remembered are the great explorers of the world, who led their followers into the great unknown. More recently, some corporate leaders became famous more for making fortunes than for their leadership capabilities. Since the early nineteenth century, people have been asking: What makes a good leader? What is good leadership? Early on, it was believed that leaders are "born, not made," meaning leader had certain intrinsic traits, which could not be taught nor learned. Later, it was recognized that even the born leader needed followers, to demonstrate his or her leadership abilities. Followers were expected to adjust to their own leadership styles accordingly.

Today, this view has been turned on its head as it is now understood that an effective leader is able to use many different leadership styles, adapting them to the circumstances and the people to be led. As this chapter will show, the quotations of Lao Tzu and Warren Bennis exemplify the foundation of modern leadership and teamwork.

Before we start, take a moment to mark your answer(s) to the three questions in Table 9.1. We will come back to them later in the chapter.

Table 9.1 Multiple-Choice Quiz: What Should a Good Leader Do?

A construction manager demands respect for racial, ethnic, and gender minorities. He hears someone in the office tell a racist joke. What should the manager do?

A. Ignore it, since the employees were just having fun.

B. Call the person into his or her office and address the issue privately.

C. Speak up on the spot, making it clear that such jokes will not be tolerated.

D. Enroll the person in an appropriate training program.

A change oriented construction manager is put in charge of a new team that is known to reject any attempts to modernize their procedures. What should the manager do?

A. Develop an agenda for initiating change and call a meeting.

B. Invite the team to a meeting at a nice location outside the workplace, where they can all get to know each other.

C. Ask each individual how he or she would improve the present situation.

D. Call for a brainstorming session, to develop new ideas for change.

A long-serving manager has been assigned a new hire who is not capable of making any decisions about his work without advice from the manager. What should the manager do?

A. Assign others on the team to do the new hire's work.

B. Ask the HR department to talk to the employee.

C. Assign the employee complex problems to solve, in an attempt to raise his self-confidence.

D. Develop a set of training exercises for the employee and offer help if needed

9.1 IS A MANAGER ALSO A LEADER?

We learned that the changes brought on by the Industrial Revolution and the complexities of today's large corporations require a large number of organizational entities that require an integrated managerial structure to align all departments and divisions behind the corporate vision and goals. To carry out their necessary functions, frequently companies will hire highly paid managers and charge them with "getting things done," even when faced with tight constraints of both time and money.

The general manager is responsible to the firm's shareholder/stakeholder; the middle-level managers are both subordinate to upper management as well as superior to their groups of subordinates. A manager is assigned the responsibility and authority according to his or her organizational position, to help the manager in planning, organizing, staffing, directing, and controlling the work of the given division/department in an optimal manner. In other words, managers are in positions of authority within the company; their subordinates work for them and, largely, fulfill their job requirements in exchange for a salary and other benefits (e.g., medical insurance, paid vacation, etc.). There are also specialty managerial functions, such as project managers who fill their roles only until the project is complete. Their responsibilities include bringing the project to a successful conclusion, typically measured in terms of cost, time, and quality.

Studies have shown that employees only use 10 to 15 percent of their abilities at work. More specifically, a survey by the Public Institute showed a deep chasm between subordinates and their managers in regard to what they motivates them on the job.

Figure 9.1 supports the findings by Herzberg, discussed in the last chapter, which concluded that hygiene factors are considered very important by employees. Also note that managers cite wages as the most important motivator, whereas employees indicated that feeling appreciated, acknowledged, or even helped with personal problems as much more important.

What do these facts tell us? Perhaps the key to answering this question is that the top seven motivators for managers are organizational factors that they control or expect, including loyalty to the company. The bottom three factors require managers to show affective behavior toward their subordinates, and recognize their personal contributions and efforts to stay "on top of things." As we will learn, the ability to recognize and respond to employees' needs on an emotional level has been found to be a key area where a true leader excels and a traditional manager

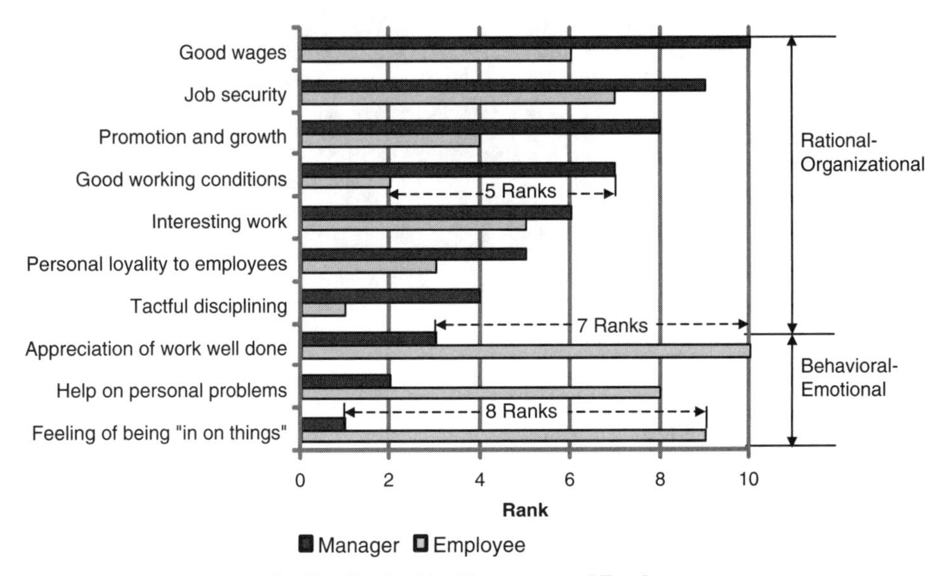

Figure 9.1 Effective Motivation Ranked by Managers and Employees

may not. *Emotional intelligence* is the term used to describe the ability to identify, evaluate, and handle emotions—one's own, those of others, and of groups.

There are many cases of engineers who were unable to succeed as managers within an organization because they lacked exactly these kinds of skills. Not surprisingly, engineering curricula rarely provide or encourage taking courses in behavioral science, which would provide students with the basis to avoid being a victim of the "Peter Principle."[1] Consider, for example, an excellent mason who is

Table 9.2 Differences between Managers and Leaders

Managers	Leaders
Administer	Involve, create, and innovate
Ask how and when	Ask what and why
Focus on business process	Focus on people
Do things right	Do the right things
Maintain old ways	Develop new ideas
Use organization to control	Inspire trust; show empathy
Accept status quo	Challenge status quo
Have eye on bottom line	Have eye on horizon
Imitate	Originate
Are good soldiers	Are their own person
Copy	Show originality
Keep the boat on even keel	Rock the boat
Dislike uncertainty	Love ambiguity
Send signals	Send messages
Focus on correct procedures	Focus on resolving issues
Rely on standard methods	Challenge creative minds

1. "The Peter Principle" was formulated by Dr. Laurence J. Peter and Raymond Hull in their book of the same name (*The Peter Principle*, New York: William Morrow and Co., 1969). It states that people tend to be continuously promoted as long as they can do their job. This pattern continues until they reach their "level of incompetence"—i.e., until they reach the level at which they are no longer able to perform the duties required of them. At this point the promotions cease and they remain in a position whose duties they are not competent to perform.

promoted to foreman, a position in which he fails miserably due to a lack of managerial skills. Similarly, an engineer is promoted to lead a group of engineers, a managerial position in which he fails. The underlying problem comes from assuming that those who perform well in one job will also perform well in higher-level positions requiring different or additional skills.

Table 9.2 lists a number of factors that show the differences between a manager and a leader: how they operate, how they decide what is important. Some of these have been mentioned by Warren Bennis (1989).

Header Problem 9.1 How Can John Become a Leader?

John has been working for J. A. Brothers, a bridge design firm, for more than eight years in the CADD department and has become a valuable asset. Excellent at his work and committed to the company, he has attracted the attention of upper management, since they plan to promote the present manager of the CADD department to director of the Information Communication Technology (ICT) department.

So it was decided to promote John, with a special charge to implement a new GIS software program. John was sent to a training program, which helped him to become very proficient with the new software. He loved the experience. Excited about his new responsibilities, he decided to train everyone in the department. This made sense, since the GIS program would be integrated into other software packages the staff used. John also believed his former co-workers (now his subordinates) would also welcome the new software.

Instead of hiring outside trainers, John took it upon himself to train the 12 departmental employees in 20 after-work training sessions (twice a week from 5:00 to 7:00 P.M.) His original enthusiasm was quickly eroded when people started to call in sick on training days, forcing him to postpone the sessions. He also noticed that no one was doing the practice assignments he gave them, claiming that they were too busy to find the time. At the same time, they started to make mistakes in their work, something that had never happened before. This forced John to spend more and more time checking their work and correcting errors. When he confronted those who had made serious errors, they complained that they did not have sufficient information from John to avoid their mistakes.

Needless to say, the performance of the CADD department started to go downhill, while John's stress level increased dramatically. At home, his wife started to notice that he began to drink and take sleeping pills, something he had never done before the promotion.

What went wrong? How can John's situation be "rescued"?

9.2 THEORIES ABOUT EFFECTIVE LEADERSHIP

As noted at the beginning of the chapter, it once was believed that effective leaders were born—such as Genghis Kahn, Mohammed, Alexander the Great, and Gandhi. This has theory has long been abandoned, since it is impossible to predict whether someone will become a leader or not. Over the years, research has focused on observing what successful leaders do, then mapping it onto a matrix with the axes labeled, respectively, *production-oriented* and *people-oriented*. For example, a manager focusing on production is concerned primarily with planning and controlling operations so that targets are met, regardless of personnel problems this approach might cause. This was emphasized by Taylor and the Scientific Management group, who saw the manager's role to plan the work and instruct the laborers on how to do things right. This was later challenged by the Human Resources Movement, a group whose members emphasized the emotional needs of people. A manager who follows this approach cares more about the needs of subordinates than about supervising the details of work. He or she spends much time listening to people and creating a satisfying work atmosphere for all. The manager who attends to both production and people recognizes that only motivated individuals, working as a team in a well-oiled production system, will perform above expectations.

9.2.1 The Managerial Grid Model

In the early 1960s, Robert Blake and Jane Mouton presented an interesting grid model as a means to differentiate between leadership styles that mixed both concern for people and production. Its referred to as the managerial grid model and is shown in Figure 9.2.

The axes of the grid consist of the two orientations of a manager, *production* and *people*, each with a scale from 1 to 9, from low to high.

Managers who belong to style group A (1,1), are indifferent; they try to stay out of trouble by doing nothing special on any project so as to avoid making mistakes. They might be close to retirement or have no challengers for their position.

Managers using style B (1,9), take the "country club" approach, focusing primarily on accommodating employees in order to make them feel good at work. As we learned in Chapter 7, satisfaction at work does not equate to high production.

Style C (9,1) managers are called dictatorial, since they don't worry about people; rather, they direct their complete attention to maximizing production at all times. Through pressure, incentives, and even punishment, employees who work for them will be driven to exhaust themselves, for the good of the company. As we will see in the next section, Douglas McGregor called these people "Theory X managers."

Managers using style D (5,5) try to balance between people and production, while not fully engaging in either.

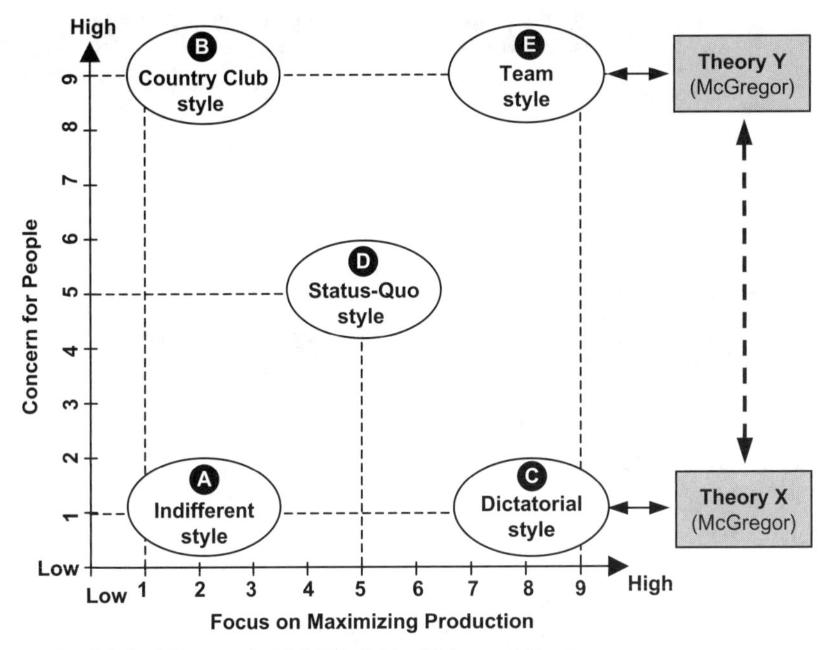

Figure 9.2 Original Managerial Grid Model by Blake and Mouton

Style E (9,9) managers are fully engaged and committed to both people and production. Employees are challenged to make decisions, encouraged to participate in problem-solving processes, are given the authority to address problems immediately on their level, and are called on to propose better ways to doing things (including criticize the manager).

9.2.2 Theory X and Theory Y

At the same time Blake and Mouton developed their approach to classify different managerial styles, Douglas McGregor, an American social psychologist, offered a much simpler bipolar model, which we now call Theory X and Theory Y. According to McGregor's model, leaders use either the X or the Y style to operate, each applying a totally different assumption of how employees view their work, especially their motivations.

The **X type leader** operates under the assumption that subordinates do not like to work, only do something when they are told to do it, and avoid taking any initiative and, thus, any kind of responsibility. Management is responsible for organizing and supplying all the resources—information, materials, equipment—and staff, who have to be directed, trained, coerced, controlled, and disciplined, according to the needs of the organization. On the other hand, a

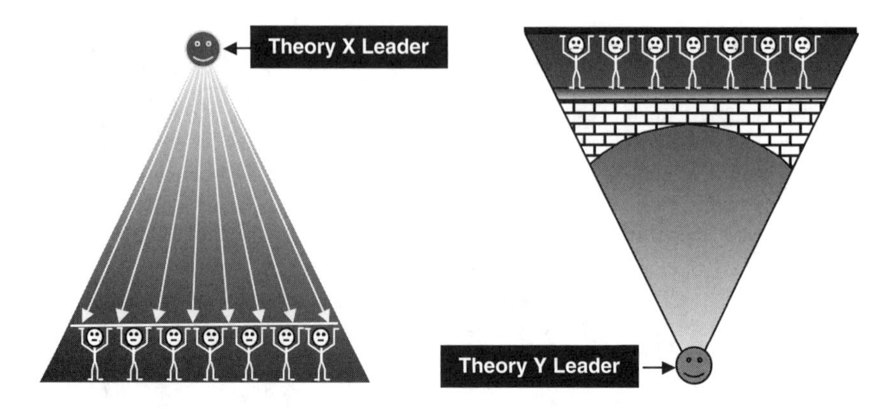

Figure 9.3 Essence of X and Y Leadership Styles

Y type leader believes that employees see work as natural as eating and drinking; they prefer self-direction over orders, and are willing to accept responsibility and enjoy using their imagination and creativity in pursuit of the company or department mission. The leader's key role is to supplement the employees' knowledge, skill, and confidence, to prepare them to adopt and work successfully toward the organizational goals.

Figure 9.3 illustrates key differences between the two leadership styles. The authoritarian Theory X leader enforces tight control over the workforce, whereas the Theory Y leader sees his or her role as the enabler, the person who provides the necessary support and removes obstacles that could hinder the effectiveness of his or her subordinates.

Working for a Theory X boss isn't easy for people who want to develop their own work method or who expect to be informed about issues that involve them, listened too when recommending improvement, or shown appreciation after achieving a major goal. Theory X leaders depend heavily on the authority provided by their position, as often they are insecure and intolerant.

The short questionnaire in Table 9.3 can be used to identify a Theory X or Theory Y supervisor.

It is apparent that McGregor's Theory X and Theory Y model did not take into account the fact that leaders face many different types of situations and personalities in their subordinates. This shortcoming was addressed by the contingency model proposed by Fred Fiedler, a psychologist born in Austria in 1922, who later taught at the University of Washington.

9.2.3 Fiedler's Contingency Model

According to Fiedler, there is no ideal manager or leader. He claimed that totally different situations require totally different leadership styles. Take, for example, the

Table 9.3 Survey to Determine Your Supervisor's Management Style

	Is your boss Theory X or Theory Y style?: Use following scale to score each of the 15 questions: (5 = always, 4 = mostly, 3 = often, 2 = occasionally, 1 = rarely, 0 = never)	Score
1.	My boss asks me politely to do things, gives me reasons why, and invites my suggestions.	
2.	I am encouraged to learn skills outside of my immediate area of responsibility.	
3.	I am left to work without interference from my boss, but help is available if I want it.	
4.	I am given credit and praise when I do good work or put in extra effort.	
5.	People leaving the company are given an exit interview to hear their views on the organization.	
6.	I am encouraged to work hard and well.	
7.	If I want extra responsibility, my boss will find a way to give it to me.	
8.	If I want extra training, my boss will help me find how to get it or will arrange it.	
9.	I call my boss and my boss's boss by their first names.	
10.	My boss is available for me to discuss my concerns or worries or suggestions.	
11.	I know what the company's aims and targets are.	
12.	I am told how the company is performing on a regular basis.	
13.	I am given an opportunity to solve problems connected with my work.	
14	My boss tells me what is happening in the organization.	
15.	I have regular meetings with my boss to discuss how I can improve and develop.	
	Evaluation of total score: 60–75 = representative of a strong Theory Y boss; 45–59 equates generally to Theory Y; 16–44 equates generally to X Theory; 0–15 points to a strong Theory X boss	

leader of an emergency-response team. The team leader can't wait for a group decision on what to do, or organize a brainstorming session by the team. In contrast, a construction site superintendent will want to explain in detail what he wants a crew to do because they need to know in detail. Thus, according to Fiedler, a leader's effectiveness depends on the interface of leadership style and situation.

Figure 9.4 highlights Fiedler's idea that a leader's level of performance is influenced by how well his or her leadership style matches the situation, both from the perspective of the employee and the task. In other words, both task-oriented and

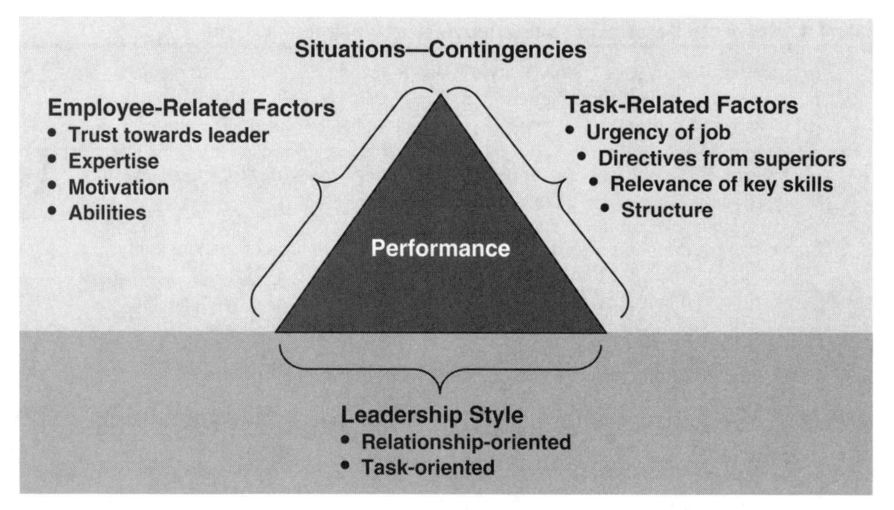

Figure 9.4 Basic Relationships in Fiedler's Contingency Model

relationship-oriented leaders can be effective, if their leadership style fits the situation. When there is a good leader/member relationship and a highly structured task, the situation is considered "favorable." Fielder did not necessarily propose that leaders could adapt their leadership styles to different situations, but that leaders with different leadership styles would be more effective when placed in situations that matched their style.

Based on his leadership model, Fiedler also created a simple self-assessment instrument, which he called the Least-Preferred Co-worker (LPC) scale, to assess what type of person someone would like to have as a leader: task-oriented or relationship-oriented.

Fiedler's LPC Instrument

Think of the *one person*—whether a peer, boss, or subordinate—with whom you work least well; that is, the person with whom you have had the most difficulty getting a job done. This is your least preferred coworker. Describe this person by circling numbers at the appropriate points on each of the pairs of bipolar adjectives. Do not spend too much time thinking about each question; trust your first instincts.

Table 9.4 Least-Preferred Co-worker Survey to Assess Personal Leadership Preference

Pleasant	8	7	6	5	4	3	2	1	Unpleasant
Friendly	8	7	6	5	4	3	2	1	Unfriendly

Rejecting	1	2	3	4	5	6	7	8	Accepting
Tense	1	2	3	4	5	6	7	8	Relaxed
Distant	1	2	3	4	5	6	7	8	Close
Cold	1	2	3	4	5	6	7	8	Warm
Supportive	8	7	6	5	4	3	2	1	Hostile
Boring	1	2	3	4	5	6	7	8	Interesting
Quarrelsome	1	2	3	4	5	6	7	8	Harmonious
Gloomy	1	2	3	4	5	6	7	8	Cheerful
Open	8	7	6	5	4	3	2	1	Guarded
Backbiting	1	2	3	4	5	6	7	8	Loyal
Untrustworthy	1	2	3	4	5	6	7	8	Trustworthy
Considerate	8	7	6	5	4	3	2	1	Inconsiderate
Nasty	1	2	3	4	5	6	7	8	Nice
Agreeable	8	7	6	5	4	3	2	1	Disagreeable
Insincere	1	2	3	4	5	6	7	8	Sincere
Kind	8	7	6	5	4	3	2	1	Unkind

Scoring: Compute your LPC score by totaling all the numbers you circled. Enter that score here: LPC = ________________

If your score is 73 or higher, you are more compatible with a "relationship-oriented" leader. If your score is 64 or lower, you prefer a "task-oriented" leader. If your score is 65 to 72, you are a mixture of both, and it is up to you to determine which leadership style is most compatible with your working style.

9.2.4 Burns's Transactional and Transformational Leadership Models

The next effort to improve the existing leadership models came in the late 1970s with James MacGregor Burns's 1978 book, *Leadership*. It recognized a key component that was missing from other models: namely, the need for leaders to inspire and transform people, to prepare them to reach a higher level of personal aspirations (e.g., Maslow). Burns contrasted the traditional manager, which he called the transitional leader, as he or she exchanged work for satisfiers, with the transformational leader, who broadens the capabilities and commitments of the followers/employees.

In his book, Burns presented four type of leaders that he regarded as predestined to change, to transform, the status quo:

1. **The Intellectual.** An intellectual leader is devoted to ideas and values that transcend the status quo. It is a person with a vision that can transform society by raising social consciousness.
2. **The Reformer.** Reformers depend on perfect timing and a large number of allies with various reform and non-reform ideas. Reform leadership, by definition, implies moral leadership, as a means of keeping the more self-serving participants from jumping ship they don't get what they want.
3. **The Revolutionary.** Revolutionary leadership demands commitment, persistence, courage, and sacrifice.
4. **The Heroic (Charismatic).** The heroic, charismatic leader is what is today most often cited as an example of transformational leadership.

In 1985, Bernard Bass extended Burns's work by including the effectiveness with which a leader is able to command trust, admiration, loyalty, and respect from employees. Bass also added other leadership styles between the transitional and transformational leader types.

The *charismatic coach* seems to be the most appropriate way to describe a transformational leader who possesses a clear vision of where the company, the division, or the department needs to go. He or she is able to communicate that vision and accompanying beliefs in a clear and understandable manner, presenting it as desirable for everyone, and achievable through hard work. Finally, the charismatic leader is able to be a coach, mentor, and teacher, the person who lays out a course for future change; encourages employees to think outside the box, critically review their old methods of doing things, learn new skills and competencies; and applauds individuals and/or teams who attempt to accomplish the goals he or she sets.

Extending on the work by Fiedler, a leader may have to be transformational at some times and transitional at others. For example, there will always be groups or individuals who are not able or willing to take initiative or risk, or lack what it takes to learn necessary skills. In such situations, the leader needs to direct people what to do, and train, evaluate, and reward good work while correcting and/or penalizing poor performance.

A survey instrument, developed by Bernard Bass and Bruce Avolio in the early 1990's, is called the Multifactor Leadership Questionnaire (MLQ). It has since been revised several times to improve its transparency as a tool to assess ones leadership style on a full-range scale between transformational and transactional. some Sample MLQ survey questions are given in Table 9.5.

9.2.5 Hersey-Blanchard Situational Theory

Paul Hersey, a behavioral scientist, and Kenneth Blanchard, a leadership specialist, created in the 1960s a contingency leadership model that extended Fiedler's by defining leadership styles that would fit certain situations. Hersey-Blanchard's

Table 9.5 Sample Questions from the Multifactor Leadership Questionnaire

As a Supervisor I . . .
Don't interfere until a problem becomes serious.
Focus my attention on irregularities, mistakes, and deviations from standards.
Avoid getting involved when difficult issues arise.
Reexamine the appropriateness of important assumptions.
Provide my employees with assistance in exchange for their efforts.
Talk about my most important values and beliefs.
Seek differing perspectives when solving problems.
Talk optimistically about the future.
Instill pride in my employees for being associated with me.
Discuss in specific terms who is responsible for achieving performance targets.
Wait for things to go wrong before taking action.
Am often absent when needed.

theory reiterates the fact that humans have needs, and that these will vary across geography, culture, context, company size, and so on. Most importantly, it emphasizes the need for the manager and leader to adjust and recognize that the employees or followers may not be prepared to do the same. That means an effective leader needs competencies that encompass those of a task-oriented manager as well as those of a transformational, people-oriented leader.

The Hersey-Blanchard theory also introduces specific analysis of employees' or followers' readiness to embrace the leader's vision. Hersey and Blanchard argue that effective leadership is possible only by selecting the right leadership style, which is contingent upon the acceptance of the leader by the employees, and the level of their ability and willingness to perform as expected.

Figure 9.5 presents a graphical representation of the Hersey-Blanchard situational leadership model. A bell-shaped curve passes through four leadership quadrants of a grid, with employee readiness to perform (high/moderate/low) on the x-axis, leader's concern for people (high/low) on the y-axis, and leader's concern for the task (high/low) along the upper x-axis.

Hersey and Blanchard characterized leadership style in terms of the amount of direction and support leaders provide to their employees. Each of the four quadrants represents one style, labeled S_1 to S_4. Here is a short description of each:

S_1—Telling/Directing: Because the employees are unable to follow the leader's vision on their own, due to lack of skill, confidence, or willingness (**R_1**),

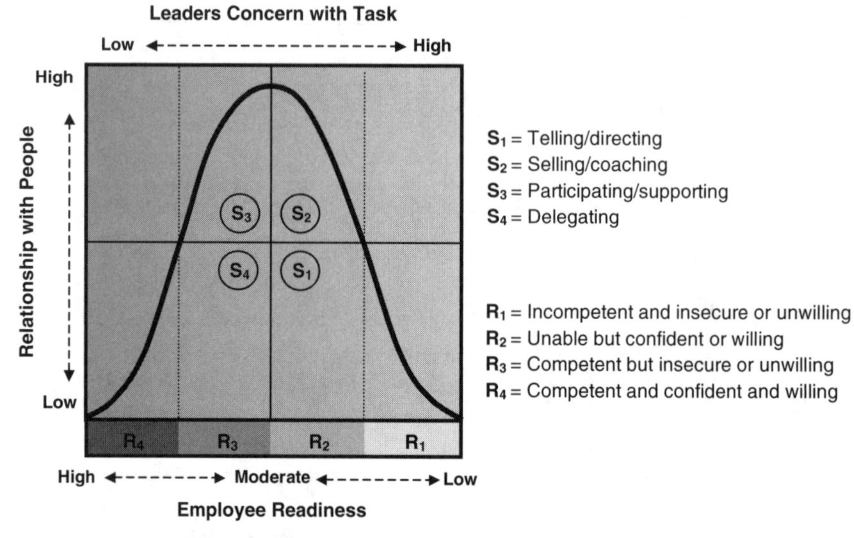

Figure 9.5 Hersey-Blanchard's Situational Leadership Model

the leader has to be very specific in giving direction, defining tasks for followers, and supervising them closely. Decisions are made by the leader and announced, meaning that communication is largely one-way.

S_2—Selling/Coaching: This quadrant covers the situation where the leader deals with employees who lack the competence and skill but are willing to follow (**R_2**). The role of the leader becomes that of a coach, who raises the skill level while providing praise and instilling self-confidence.

S_3—Participating/Supporting: Leaders supervising employees who possess the competence and skills to succeed but lack either confidence or willingness (**R_3**) can let go of detailed decision making and control. Instead, leaders can focus on personal relationships, to support employees gaining confidence, through appreciation and praise for work well done.

S_4—Delegating: Here, employees are competent and willing to follow their leaders' vision (**R_4**), leaving the leaders free to focus on strategic planning while still being involved in critical decision making. Little supervision or encouragement is needed, since the employees are enabled to make decisions with confidence.

9.2.6 Managers with Emotional Intelligence

Herzberg showed that, for employees, many motivators, such as receiving recognition, and hygiene factors, such as the relationship with a supervisor, are of an emotional nature. In fact, psychology puts the affective domain right next to the cognitive capabilities, as being key to a healthy life. Expert managers believe that the ratio of importance between emotional and technical competence is approximately two to one. That is not to say that technical expertise is unimportant. However, junior managers and professionals serve primarily as technical experts, who plan projects and keep them on track. As managers climbing the organizational ladder, they take on more responsibility for leading, motivating, internal and external collaboration, and team building. Consequently, emphasis is placed on strategic thinking, rather than technical expertise.

The array of noncognitive capabilities, competencies, and skills that help us mastering environmental demands and pressures has been defined as *emotional competence* or *emotional intelligence* (EI). Emotional competence or intelligence means that one understands that, for example, anger is a normal reaction to aggression and gives a person the strength to put up a defense. Fear, as another example, is a response to danger that results in sharpening of the senses and heightened reactions.

When Salovey and Mayer (1990) coined the term *emotional intelligence* (EI), they defined it as consisting of a set of skills:

> . . . relevant to the accurate appraisal and expression of emotion in oneself and in others, the effective regulation of emotion in self and others, and the use of feeling to motivate, plan, and achieve in one's life (Salovey and Mayer, 1990, p. 185)

Daniel Goleman's 1995 book, *Emotional Intelligence*, contends that emotionally intelligent people will experience greater success and satisfaction in life. EI provides a concept to study one's ability, capacity, or skill to identify, assess, and manage the emotions of oneself, of others, and of groups.

Table 9.6 summarizes the concepts currently used to describe the components and relationships of EI.

Can emotional intelligence be learned, or is it intrinsic? The apparent answer is both. Furthermore, some of the underlying emotions, such as fear or anxiety, have an indirect impact on emotional intelligence. That said, there are a number of items listed in Table 9.6 that can be "trained" more or less successfully. But we have to recognize that emotional intelligence training is much different from gaining cognitive intelligence through a college course or by applying effective study skills. Emotional "incompetencies" are deeply ingrained, learned habits. As people acquire their emotional repertoire of thoughts, feelings, and actions/reactions throughout life, the neural connections that support these become dominant pathways for nerve impulses.

So for emotional learning to take place we first have to identify, then loosen, and ideally, disconnect those old connections before establishing new ones. Emotional learning often involves ways of thinking and acting that are central to a person's identity. Changing a habit—such as learning to approach new people positively instead of avoiding them, or to become a listener instead of doing all the talking—is a challenging undertaking that requires a lot of firm but gentle support. One-day seminars just won't do it.

A long list of survey instruments have been developed and tested to measure a person's EI. Some of them are self-scoring, while others are 360-degree surveys

Table 9.6 Competencies of Emotional Intelligence

Cluster Competencies	Characterization	Features
Self-Awareness Emotional awareness Accurate self-assessment Self-confidence	The ability to read one's emotions and recognize their impact. Understanding one's internal states, preferences, resources, intuitions, and the link between emotions, thought, and action. This competence refers to how much we understand ourselves and have confidence in our feelings and abilities. Equipped with this awareness, an individual is able to manage his or her own emotions and behaviors, as well as relate to other individuals and groups.	Realistic self-assessment, Self-deprecating sense of humor; trusting one's "gut" feelings to guide decisions
Self-Management Self-control Trustworthiness Conscientiousness Adaptability Achievement orientation Initiative	Involves controlling one's emotions and impulses and adapting to changing circumstances, to control emotions or to shift undesirable emotional states to more adequate ones. Self-management competence addresses our capability to work under stress and how we are able to handle our emotions to achieve ends without harming ourselves or others.	Management of one's disruptive emotions and impulses; reliance on honesty/integrity; ready to act on opportunities
Social Awareness Empathy Organizational awareness Service orientation	Ability to sense, understand, and react to others' emotions while comprehending social networks. Before we can direct others, we need to be able to see the world through other people's eyes, appreciate diversity, and be ready to listen actively and understand others.	Sensing others' feelings; reading a group's emotional currents and power relationships; recognizing and meeting people's needs.
Relationship Management Developing others Leadership Influence Communication Change catalyst Conflict management Building bonds Teamwork and collaboration	Capability to inspire, influence, and develop. The individual competencies are related to communication, influence, conflict management, leadership attitude, being a change agent, building bonds, collaboration, team synergy . . . as they allow entering and sustaining satisfactory interpersonal relationships.	Expertise in building and leading teams; inspiring and guiding individuals and groups; listening openly, and sending convincing messages; negotiating conflicts; creating group synergy.

done by others. The one we use as an example here is the Trait Emotional Intelligence Questionnaire (TEIQue), which consists of 153 questions with 7 possible responses, on a Likert scale, ranging from "Disagree completely" to "Agree completely." A partial list of these questions is given in Table 9.7.

Table 9.7 Partial List of Questions from the TEIQue

		Disagree						Agree
1.	I'm usually able to control other people.	1	2	3	4	5	6	7
2.	Generally, I don't take notice of other people's emotions.	1	2	3	4	5	6	7
3.	When I receive wonderful news, I find it difficult to calm down quickly.	1	2	3	4	5	6	7
4.	I tend to see difficulties in every opportunity rather than opportunities in every difficulty,	1	2	3	4	5	6	7
5.	On the whole, I have a gloomy perspective on most things.	1	2	3	4	5	6	7
6.	I don't have a lot of happy memories.	1	2	3	4	5	6	7
7.	Understanding the needs and desires of others is not a problem for me.	1	2	3	4	5	6	7
8.	I generally believe that things will work out fine in my life.	1	2	3	4	5	6	7
9.	I often find it difficult to recognize what emotion I'm feeling.	1	2	3	4	5	6	7
10.	I'm not socially skilled.	1	2	3	4	5	6	7
11.	I find it difficult to tell others that I love them even when I want to.	1	2	3	4	5	6	7
12.	Others admire me for being relaxed.	1	2	3	4	5	6	7
13.	I rarely think about old friends from the past.	1	2	3	4	5	6	7
14.	Generally, I find it easy to tell others how much they really mean to me.	1	2	3	4	5	6	7
15.	Generally, I must be under pressure to really work hard.	1	2	3	4	5	6	7
16.	I tend to get involved in things I later wish I could get out of.	1	2	3	4	5	6	7
17.	I'm able to "read" most people's feelings like an open book.	1	2	3	4	5	6	7
18.	I'm usually able to influence the way other people feel.	1	2	3	4	5	6	7
19.	I normally find it difficult to calm angry people down.	1	2	3	4	5	6	7
20.	I find it difficult to take control of situations at home.	1	2	3	4	5	6	7

Source: www.psychometriclab.com.

The scoring offers insights, along with the competencies listed in the table, such as adaptability, social awareness, and so on. For example, a high score on info emotional management means that the scorer is able to make others feel better when they need it, to calm them down, and motivate them to move on. A low score, on the other hand, means that the person will be overwhelmed when he or she has to deal with other people's emotional problems.

Studies using some of the available instruments showed that women are much more adaptable than men; they show more empathy, are more skilled at interpersonal relationships, have greater self-awareness, and are more service-oriented than males. Some studies also assert that high-performing managers possess significantly higher levels of self-awareness, self-management capability, social skills, and organizational savvy, all considered aspects of the emotional intelligence domain.

9.3 POWER AND PROBLEMS OF TEAMWORK

Using teams is considered a simple way to enhance productivity. But as we learned in Chapter 4 when we discussed the performance of a carpentry crew, this is not necessarily true. This section investigates what it takes to build a successful team and what dangers are lurking that could easily waylay its efforts.

Cohen and Bailey (1997) defined a team as:

> [A] collection of individuals who are interdependent in their tasks, who share responsibility for outcomes, who see themselves and who are seen by others as an intact social entity embedded in one or more larger social systems and who manage their relationships across organizational boundaries.

This definition applies in particular to teams consisting of members from different areas of the organization, who are brought together to work on a problem that cannot effectively be solved by individuals working on their own. Hackman (1987) stressed that team effectiveness is a multidimensional value that includes both its output productivity and its internal working relationship. As might be expected, the mathematical formulation for what produces team effectiveness includes a variety of factors, such as: (1) performance, (2) behavior, (3) attitude, and (4) team member style and corporate culture. Which of these are most important is still being researched. Nevertheless, psychologists have discerned a series of prerequisites that good working teams need. Table 9.8 lists those qualitative factors that have been found to be linked to successful teams.

Based on what we learned in section 9.2.6, it is not surprising that both the leader and the team members should possess sufficient emotional intelligence (EI), as the team may experience stress and possibly setbacks, which it will have to overcome. Research has also found that the person most in need of EI is the leader, as he or she will be responsible early on for establishing emotionally healthy group norms.

9.3.1 Team Dynamics

Before the beginning of agriculture, 10,000 years ago, early man depended on hunting and gathering for survival. Moving from place to place, nomadic people had to find ways to protect themselves. This was accomplished by building shelter and depending on "safety in numbers," as in a band or tribe, which was not so large that they would run out of food sources too quickly. Societies, composed of as few as 10 to 20 people, needed members with various skills, such as hunting and fishing and, of course, gathering.

Table 9.8 Prerequisites for Successful Teamwork

Requirements for a Successful Team	Conditions Where Teams Out-Perform Individuals
Members that respect, support, and trust each other	The solution path is not apparent
Defined goals, common purpose, role clarity, appraisal system	Task is complex; needs cross-functional expertise
Strong interpersonal skills, open communication, productive conflict resolution, shared values and beliefs	The implementation solution will require the cooperation of multiple divisions or departments
Members subordinate their own objectives to those of the team	Problem requires innovation and experimentation
Resilience to setbacks; see problems as opportunities; creative climate	Fast learning is required
Autonomy and active support by senior management	Emotionally competent team members
Team training emotional competence	Emotionally competent leader

A destructive use of teams is to fight wars. Nevertheless, "war is the mother of all things," even the effective use of teams and teamwork. For example, the Egyptians learned through warfare how to organize large forces (e.g., directing and supplying resources), a skill they also needed to construct the pyramids.

Even with this long history of using teams, still today we know very little about how to sustain high-performance working teams; in fact, only 50 percent of work teams actually succeed. But when they do, well-organized teams deliver a company important advantages, as illustrated in Figure 9.6.

Self-managed work teams have the authority to control their work and to develop group norms to regulate activities. They plan, organize, coordinate, and take corrective actions. In short, self-managed work teams are given responsibilities usually held by managers; but control comes from the team, rather than from an authority figure. In this way, a manager's supervisory tasks are minimized, and group norms are maximized. Self-managed work teams are not for all organizations, however; the characteristics needed for success include:

- **Cross-training**, which allows team members to move from job to job within the team, is essential. Thus, team members should receive training in the specific skills that will broaden their personal contributions to the overall effort.
- **Interpersonal and problem solving skills**, which are needed for decision making within and among teams and for conflict resolution. Teams need to

Figure 9.6 Creative Teamwork

learn how to analyze problems, gather relevant data, generate alternatives, and select best solutions.

In the next sections, we will review the many issues that a manager should be aware of when building or participating in a team. Thereafter, we will discuss ways to assess team performance.

9.3.2 Important Team Characteristics

Construction sites are made up of crews that are, in fact, work teams. Construction teams also are used to manage projects, generate a large bid, or investigate an accident. Each of these teams has its unique form and social structure, made up of members with different levels of expertise, emotional intelligence, and motivation, but joined to pursue a common goal. Based on the assignment, a leader may be assigned or self-directed, and the team members may work together at one place or be geographically dispersed; but they all need to collaborate, coordinate, and share information to accomplish their mission. The team might be organized into

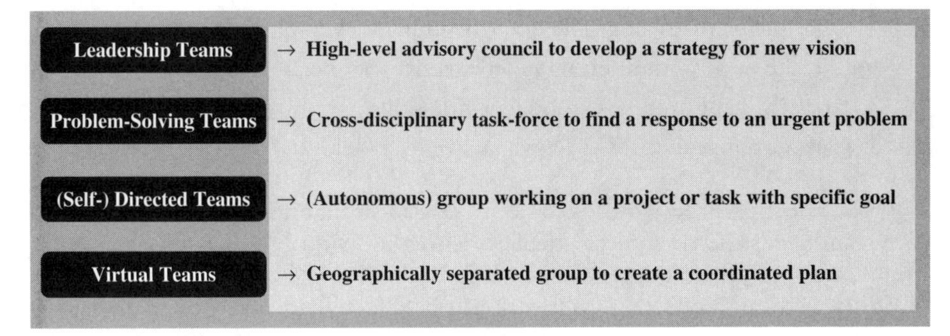

Figure 9.7 Types of Teams Found in Construction

subgroups or task groups with parallel assignments, to speed up the progress, while joining intermittently to share and refocus the mission, if necessary.

Figure 9.7 introduces four common types of teams, with their respective assignments. While leadership teams, also called *advisory councils*, are generally populated by key management, teams consist of problem-solving members who bring with them key competencies related to the problem at hand. Thus, such task forces may include professional specialists from all levels of the organization. The traditional or directed teams and the self-directed teams are mostly found in the manufacturing area or at the construction site. For example, steel crews are hired as teams rather than as individuals. Their members have developed the trust and the work ethic that makes them highly efficient.

Information technology offers ever-more convenient tools for working as a team over the Internet. As Chapter 10 on communication will reveal, virtual teams have to meet a series of special prerequisites in order to be successful. Nevertheless, the globalization of business will inevitably push the number of virtual teams upward.

9.3.3 Life Cycle of a Working Team

As we can deduce from Figure 9.7, the impetus for team building is commonly a problem to be solved, a complex plan to be put together, and so on. It is apparent that the nature of the targeted goal will dictate the technical and/or managerial competencies needed on the team, the number of people, the resources to assign, and autonomy of the members. And, as noted earlier, there are emotional competencies and motivational expectancies that are equally important factors when selecting the team members. Furthermore, potential problems with intra- and extra-group relationships between members need to be checked. When interviewing team applicants, these are some of the topics to address:

- Which strengths of the applicant meet the needed competencies?
- Which competencies is she or he willing to work on?

- What problem-solving style does the individual employ?
- Can she or he communicate in an effective manner?
- Does the applicant have good listening skills?
- Can the applicant provide constructive feedback?

The selection process needs to be structured so that it is not biased. An effective team needs the thoughtful, detail-oriented individual, the creative mind, the technical expert, the managerial all-rounder, as well as the outgoing, insightful individual. Although effective teamwork can be affected adversely by many factors (e.g., uncooperative behaviors, inadequate process skills, poor technical skills), evidence to date supports theoretical expectations that teams with greater autonomy perform better and experience less stress and strain.

Group dynamics as related to development concerns not only why groups form but also how. The most common framework for examining the how of group formation was developed by Bruce Tuckman in the 1960s. In essence, the steps of group formation imply that groups do not usually perform at maximum effectiveness when they are first established. They go through several stages of development as they strive to become productive and effective. Most groups pass through the same developmental stages, with similar conflicts and resolutions.

According to Tuckman's theory, there are five stages of group development: forming, storming, norming, performing, and adjourning. During these stages, group members must address several issues; and the way they resolve these issues determines whether the group will succeed in accomplishing its tasks.

Forming

This phase commonly begins with an initial get-together, during which some members will meet for the first time. At this meeting, they learn the goal(s) of the group and its rules, and possibly share their expectations and motivations. Usually, at this time, everyone is very friendly, helpful, open, and patient.

A wide variety of exercises and even games are recommended to encourage camaraderie, such as a wilderness camp outing. The exercises all are designed to develop trust and offer fun experiences that bind the team members together. Experience has shown that this phase is critical, in that it builds relationships that are essential later in the process.

Storming

As the name of this phase indicates, the team advances from the gentle "breeze" of the forming phase to experiencing the first internal storms. They are natural occurrences on teams whose members have different competencies,

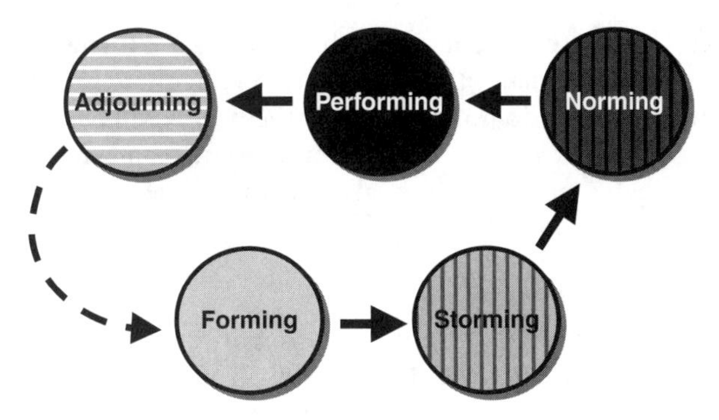

Figure 9.8 Team Life Cycle

and so view the same problem from totally different perspectives, and want to address it from his or her realm of expertise, sometimes causing conflict. A common source of team failure is caused by team members interpreting the same task in different ways. The point is, disagreement and conflicts are normal during this phase.

In order to overcome this hurdle, the members need to be willing to listen to each other and begin work on finding a mutually acceptable way to approach the problem. This helps to strengthen the cohesion of the group, leaving them united and ready to take on the assigned task as a unit.

Norming

This phase, too, is necessary to solidify a team, whose members are able to think and act as a team, rather than as individuals coming from different parts of the company, each with its own rules and performance standards. Group norms define acceptable standards of behavior within a group, which are shared by its members. Norms, which define the boundaries of acceptable and unacceptable behavior, are typically established to facilitate group survival, make behavior more predictable, avoid embarrassing situations, and express the values of the group. Norms answer such questions as: What do deadlines really mean? How high a level of quality is necessary? Does every member have to be at every meeting? What about developing subteams?

If the team can establish harmonious relationships at this stage, they are ready to move on to the performing stage. To that end, the majority of the group must agree that the norms are appropriate in order for them to be accepted. There must also be a shared understanding that the group supports the norms. Eventually, each group will establish its own set of norms, which might determine anything from appropriate dress to how many comments to make in a meeting. The norms often reflect the level of commitment, motivation, and performance of the group.

Note, however, that there are situations where norms are thrown overboard. One such example is when the majority of members fail to adhere to the norms. Otherwise, group members who do not conform to the norms will be penalized by being excluded, ignored, or asked to leave the group. Some teams, however, disband at this stage, primarily because of emotional or attitude issues.

Performing

A team entering this phase of the cycle begins work in earnest. Each member by now understands what needs to be done, the problems to be solved, and the process necessary to accomplish the goal(s) of the group. Members of the group make decisions through a rational process that is focused on relevant goals.

As we just learned, the norming phase establishes performance levels for each individual. In fact, some members of a group may have the ability to perform at higher levels than the group's average, yet they are expected to help meet a group norm. For example, a mason may stop working 20 minutes before quitting time in order to clean his or her own or the others' workspace so that everybody can leave on time.

Adjourning

Every construction project eventually comes to an end, and the project team is disbanded, or "adjourned." The same holds true at a lower level, with teams or crews. On the one hand, this can be a happy stage, with members congratulating one another on a job well done. On the other hand, adjournment means the disruption of working arrangements that may have become comfortable and efficient.

Work teams sometimes adjourn before reaching their goal, such as when an important member has to leave the team and can't be replaced, the task for the team is cancelled, or needed information or other resources are not provided by the upper management.

9.3.4 Group Dynamics

Because group members also interact on an emotional level while they're together, groups develop a number of processes that separate them from a random collection of individuals. Kurt Lewin studied how the many forces come into play in the way groups work and make decisions. According to his *force field theory*, Lewin noticed that individuals' habits, customs, and attitudes both drive and restrain team activities. So when assessing the performance of a team, we have to study the roles that individual members play and their relationship to each other (e.g., communicate); we also have to pay attention to an

Kurt Zadek Lewin

Kurt Lewin was an American social psychologist commonly considered to be the founder of the movement to study groups individual's need to belong; the differences in task and social influences; and the behavior of individuals within the group norms (e.g., conflict resolution).

One interesting and important issue is the changing roles that different members might assume within a team. Lewin found that the roles assigned to persons at the beginning of a project over time be replaced by others, which he categorized in three groups: (1) work, (2) blocking, and (3) maintenance.

scientifically. He coined the term *group dynamics* to describe the way groups and individuals act and react to changing circumstances. "An issue is held in balance by the interaction of two opposing sets of forces—those seeking to promote change (driving forces) and those attempting to maintain the status quo (restraining forces)."

Work Roles

Figure 9.9 lists and describes nine common roles that support the mission of the team, referred to as *work roles*. The *initiator*'s role, at the top, is to propose actions and methods to move the team forward; next in the figure is the *information seeker*, whose role it is to remind the group of the need for more and better information, before making important decisions. The member who has the most experience and the most relevant information is the *information provider*, who provides not only information but also his or her opinion and advice. Critically important is the *devil's advocate*, who questions and tests the thinking of the group to assure it will withstand any attack by critics who might otherwise derail their work. This member is able to anticipate the reactions of key

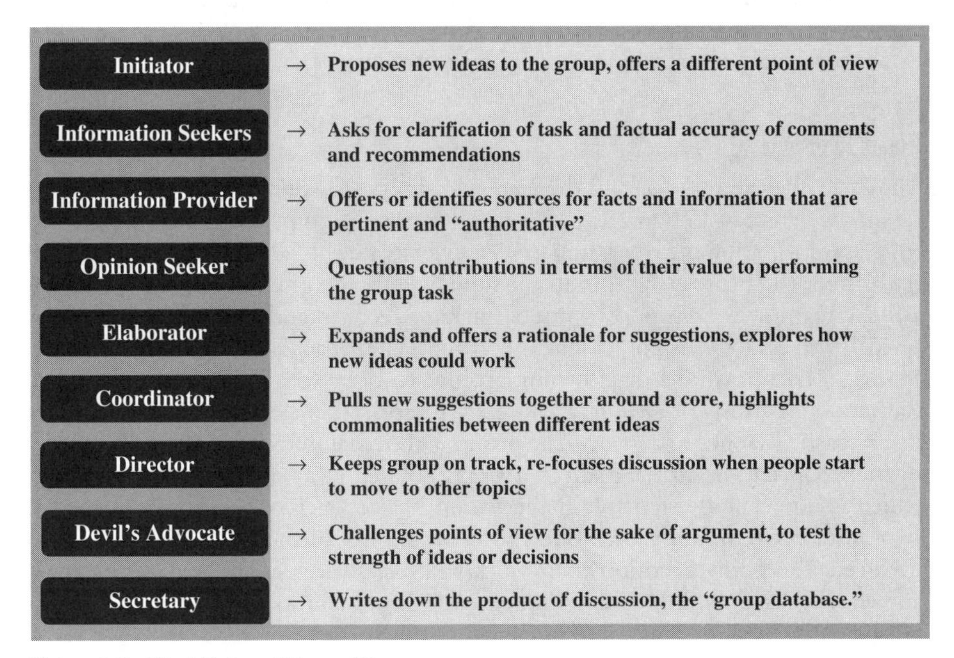

Figure 9.9 Work Roles of Formal Groups

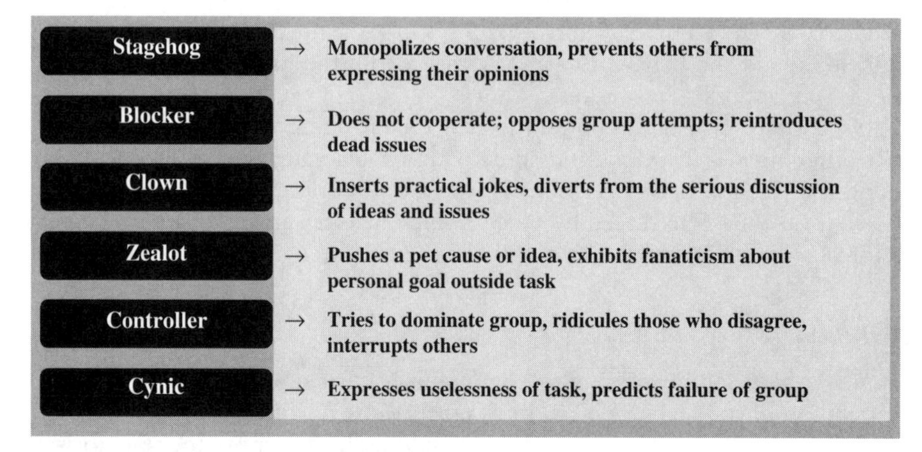

Figure 9.10 Blocking Roles in Work Groups

members of the company who will be impacted by the work of the team. The devil's advocate gives voice to such critics, requiring the other members to constantly test and defend their choices and decisions.

All these roles contribute positively to the success of the group's work; but inevitably—and especially on less carefully filtered or self-selected, more informal teams—there will be those members who pursue their personal goals and do not agree with or even oppose the mission of the group. As Figure 9.10 highlights, these individuals, who take on *blocking roles*, can be very interesting characters.

Blocking Roles

The first two, the *stagehog* and the *blocker*, can seriously hamper the group's progress if they are not recognized early on. They pretend to perform one of the work roles as information provider, initiator, or even as a devils advocate, but, in fact will try to steer the team away from its mission. Specifically, the *blocker*'s goal is to derail the team at every opportunity by making recommendations that are "poisoned" with issues that could split the group. The *clown* is not a member simply sharing a harmless joke or clowning around to break the tension of the group. Rather, an aggressive *clown* uses "humor" to criticize members. The clown jokes in a sarcastic manner or distracts the group with trivial information or unnecessary or inappropriate humor. Finally, a *cynic* is able to wear down the group spirit, which is especially destructive if the group has to overcome setbacks early on. By constantly reminding everyone about the futility of their task, the huge risks involved, the poor reception their work will likely receive, and so on, the cynic succeeds in discouraging other members. An "I told you so" attitude will immediately identify a *cynic*.

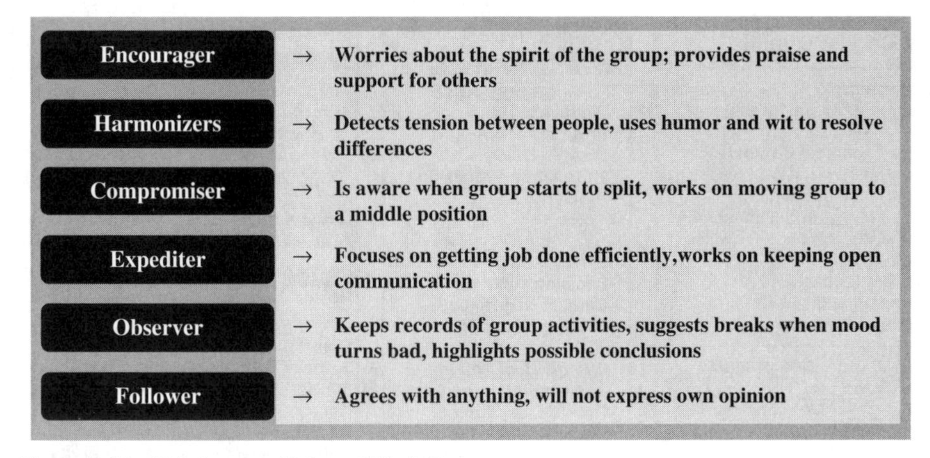

Figure 9.11 Maintenance Roles of Work Groups

Maintenance Roles

Maintenance roles are taken on by members who already fill a working role. These people provide social/emotional support to members who are prone to dissatisfaction and disappointment during the project. They also help to reduce tension and work out differences when members experience fatigue or other stresses.

As shown in Figure 9.11, both the *encourager* and *harmonizer* should be members who have a high level of emotional intelligence. The *encourager* should be respected by everyone and have a friendly, warm, and responsive demeanor. The *harmonizer* detects tension before anyone else. He or she is able to identify the source of the tension, keep communication channels open, and reconcile differences early on, while pointing our common opportunities to improve the situation.

9.3.5 Team Building

A team can be considered a production system within an organizational structure, one that consists mainly of people who have to interact and collaborate. Typical output of engineering and construction teams are design drawings and construction methods, schedules, and estimates for large projects (bid package), such as building a bridge, installing a new computer system, or erecting a radio tower. Teams are generally effective for solving complex problems whose solutions require different yet complementary expertise. Needless to say, however, not every team is efficient or effective, so it's important to figure out how to form teams that work well together and succeed at their assigned tasks.

The first issue we have to address is to agree on what effectiveness isis and how to measure it. When we looked at production teams or crews, such as masonry or

Input	Critical Qualities
Team Leadership	Respected, No Hidden Agenda, Coach, Emotional Intelligence
Individual Team Members	Cross-Technical Competence Cross-Managerial Competence Emotional Intelligence Shared Motivation-Expectancy Shared Vision Creative Skills Learning Skills Communication Skills Trust and Confidence
Rewards	Clear Metrics, Team-Based
Objectives	Clear, Important, Urgent
Time	Sufficient for Meetings-Work
Training	Appropriate for Leader-Team
Information	Shared, Sufficient, Timely

Teamwork Process
Process Tools -Planning -Progress Tracking -Brainstorming -Information Processing -Problem Solving
Team Actions -Synergistic Collaboration -Decision Making -Conflict Negotiations -Creative Thinking
Individual Actions -Contributing Expertise -Critical Questioning - Accepting Contributions -Obstructing-Retracting

Team Output	Quality
Recommendations Plans, Visions Products Innovations Cost Estimates Cost-stress-accid.	Completeness Pragmatism Acceptability Timeliness Cost Creativity Aligned w. Goal

Measures	
Process	***Lagged***
Contribution Range	Higher Profits
Setback Resilience	Incr. Productivity
"I" vs. "We" Attitude	Less Accidents
Acceptance of Ideas	Less Waste
Sharing of Success	Awards
Adherence to Plan	Recognitions

Figure 9.12 I/O Model of Working in Teams

concrete placement, we learned about Taylor's Scientific Management, Gilbreth's motion studies, and Deming's statistical control. When we looked at managers, we learned about the 360-degree evaluation and Kaplan and Norton's Balanced Scorecard for assessing the performance of entire departments and companies. The key to all these methods was the establishment of metrics that could be used as units of measure, similar to meters and miles. Here again, we will model teamwork to help us to better design an approach for managing teams successfully.

As in a production process model, each input resource comes with qualities that will impact the success of the teamwork. The alignment of those qualities with the goal(s) of the team will dictate its expected performance. The process itself is shown in Figure 9.12 as consisting of three components: (a) tools, (b) team actions, and (c) individual actions. The output of the team is not as important as its quality, in terms of its value in meeting the objectives of the overall organization. In other words, the number of recommendations and plans is less important than their alignment with the vision of the company or their acceptability by the other members of the company. Possible measures to monitor the success of the team are varied, and include team output, quality of the output, and various factors to be observed while the work is ongoing (e.g., adherence to own plan); other success measures are factored in long after the team is disbanded, such as higher profit.

Introducing a teamwork culture in an organization that has traditionally valued and rewarded only the contributions of individuals has to be carefully planned. For one, members have to be trained to function in and as a team. Second, new appraisal systems for performance evaluation and promotion have to be implemented and aligned with the motivational expectancies of team members. Finally, the organization itself has to adapt, as the output of teamwork may require companywide change. Clinging to ingrained habits and the "we have always done it this way" attitude has to be replaced with a *culture of change*, and that will not happen quickly.

Because of its importance, in the next section, we will address critical planning steps for each of the main teamwork I/O modules: input process/output measures.

9.3.6 Planning an Effective Team

Are you sure you need a team for the job? Teams are not always the best solution. As mentioned earlier, the problem at hand needs to be sufficiently complex and important to the company, and so requires a variety of technical, managerial, and operational skills to solve, develop a plan to carry out the solution, manage a long-term project, and so on. A team by its nature consists of experts from various areas.

Before considering who will be on the team, some key parameters have to be established and key questions answered:

What's the goal? Given that the job at hand is important for the company, it should be easy to define the objectives for the team and its relevance to the strategic goals of the company. Some questions to ask, with the answers later delivered in a memorandum to the team, are:

1. How do the objectives of the team fit into the business strategy?
2. What are the desired results of the team in terms of quantity and quality?
3. How will the performance of the team and its members be assessed?
4. What is the "life expectancy" of the team?
5. What is management committed to do in support of the team?

What kind of leadership? Should the team self-select a leader or should a leader be chosen by management up front? As we learned earlier, the different members of an autonomous team will take over the key roles necessary for its operation. They include the initiator, coordinator, director, or the secretary. However, an "official" leader could emphasize the importance of the team. Therefore, the question is: What are the competencies of the team leader? Should it be a line manager, or a person with a nonsupervisory assignment?

Let's remind ourselves about the purpose and the nature of work teams. Teams are formed in order to solve difficult problems, with the help of a group of multiskilled people charged with synergizing their talents. Teams depend on open communication and an innovative spirit, whose members are not afraid of failure; rather, they treat failure as part of the process to break through to new horizons. The traditional role of line managers is to implement top-down directives through appropriate instructions to subordinates. It is assumed in this model, rightly or wrongly, that the line manager knows how to execute the directives in the optimal manner. It is apparent, then, that having a line manager serve as a team leader will raise insurmountable barriers to efficient teamwork.

In view of the needs of teams, we can conclude that the function of a leader is more that of an enabler or facilitator, who develops the necessary environment in which creative minds can work without interference from an "all-knowing" supervisor. The team leader is responsible for providing the resources and encouraging the members to take control of and responsibility for their own efforts and achievements. The role of the facilitator requires formal training and development, in particular, in: teamwork tools, group dynamics, creative problem solving, and conflict resolution. So if a company is considering using its line managers as team leaders, it should ensure that they, first, "unlearn" their former modus operandi, and second, pass a formal training course.

It is evident that having a supervisor act as a leader of a team that includes his or her subordinates could create conflict within the team, especially if insufficient training has been provided, not only for the leader but all its members. Training should focus on soft process skills such as communication, creativity, negotiating, and so on. Other dangers emerge when line managers lead teams, such as difficulty remaining neutral when dealing with team members, not allowing sufficient time to complete a heavy workload, or failing to communicate clearly the priorities of upper management.

Who to put on the team? Once the leader has been selected, an equally careful evaluation of team members ensues. While the nature of the team objective provides the basis for defining the necessary and complementary technical and managerial competencies, equally important, and critical to success, is choosing members with compatible personal traits. Table 9.9 expands on the model in Figure 9.12, where critical quality issues are listed.

It is apparent that no perfect alignment of all personal factors is possible. As a consequence, the eventual selection will be based on expectations of the behavior of team members over time. Notably, many young and ambitious employees might regard working on a team as a once-in-a-lifetime opportunity to "shine" and make significant contributions to the team effort while at the same time gaining important insights by learning from other more senior members.

Table 9.9 Critical Qualities of Successful Teams

1.	Emotional intelligence	Self-efficacy, ego, temperament, personality
2.	Shared motivation/ expectancy	Comparable personal motivators and expectancies
3.	Shared vision	Individual vision aligns with that of the company
4.	Creative skills	Ability to think outside the box, and "dream"
5.	Learning skills	Capability of inquiry-based and self-guided learning
6.	Communication skills	Public, face-to-face, oral, written, graphical, electronic
7.	Trust and confidence	Degree of respect for, and reliance on, the honesty of others

An optimal team size probably does not exist, as the number of members required depends on many factors, such as the complexity of the task, the skill of the team leader, the degree of globalization, and, finally, the breadth of the competencies of its members.

How to reward the team? Teams are made up of individuals who have needs and expectations that motivate their performance. Similar to attempting to apply traditional supervisory techniques to the role of team leader, individually-based performance appraisals as a basis for promotion and reward are incompatible with the needs of a functioning team. Instead, a team-based reward system must be implemented, which recognizes the motivational factors, as well as the expectancies, of every team member. But this is not enough. The reward system also has to be based on measures that are clearly linked to the team's overall goal, since people tend to focus on measures that are linked to rewards. For example, by using the number of reports or plans produced by the team as the key measure of success, the usefulness of those products in achieving the company's strategic goals gives way to the amount "pushed out".

Figure 9.12 identifies several measures and indicators that possibly underpin an effective reward system. First, as in any dynamic system, teams should be encouraged to grow and learn over time, resulting in ever higher-quality output. A reward system also needs to be flexible enough to consider the length of time a team has been together, noting that longer-term teams have been shown to be significantly more productive than shorter-term teams.

Indicators: The way the team works, and how the members interact, offers several indicators linked to meeting goals over long periods of time. Table 9.10 expands on some of the process-related measures shown in Figure 9.12, which are considered indicators of success.

Measures: Different from indicators, measures can be linked to actual team outputs. Care needs to be taken in measuring factors that are relevant to the

Table 9.10 Internal Group Behavior Measures

1.	Contribution range	Displays the balance and depth of the team in working on goal
2.	Setback resilience	Reveals the toughness of the team; ability to learn through failure
3.	"I" versus "we" attitude	The use of "I" or "me" instead of "we" and "us" indicates that the goals of a group are not shared by an individual.
4.	Acceptance of ideas	Readiness to accept and adopt ideas from others on the team
5.	Sharing of success/failure	Demonstrates whether it is a team or a group of individuals
6.	Adherence to own plan	Shows how the group is managing their own progress

Table 9.11 External Group Performance Measures

1.	Completeness of output	Assesses the comprehensiveness of the team's production
2.	Pragmatism of output	Evaluates the attainability or effectiveness of recommended solutions
3.	Acceptability of output	Appraises how other company employees will accept the plans
4.	Timeliness of output	Values whether the plans/recommendations are submitted in time to be effective
5.	Cost of output	Considers the extra costs associated with the outcome
6.	Output creativity	Judges the originality of the team results
7.	Output goal alignment	Reflects on how the output is aligned with the goals
10.	Higher profits	Difficult to assess from the results of one team alone
11.	Higher productivity	Requires pre- and postanalysis of production (Hawthorne effect)
12.	Fewer accidents	Care needs to be taken to separate "acts of God"
13.	Less waste	Pre-/postcomparison of waste production necessary
14.	Business awards	Special awards given to the company as a result of team effort
15.	Company recognition	Acknowledgment of company by public/media

achievement of the goals predetermined by the upper management. The model in Figure 9.12 lists several success factors, some of them linked to the quality of the output and others related to possible long-term goals. Table 9.11 expands on those factors.

Of course, the type and size of the reward will depend not only on the importance and term length of the team, but also on the valences of its members. It might be appropriate to compile a team-based package that recognizes the motivational differences by including not only financial rewards but also increased responsibility, recognition, job security, and other "perks" (e.g., company car).

One final but important factor needs to be highlighted.

Does the team need training? The simple answer is an emphatic *yes*. It should be apparent that teamwork is, in itself, a competency, one that requires both knowledge and skills. In particular, all members need to understand the life-cycle phases that they will pass through, the different roles that are necessary for the team to function, team dynamics, and the various tools and techniques they will need to innovate, make decisions, and so on.

Figure 9.13 Synergy with Trust

To build an effective team, it is recommended to take advantage of independent experts, who specialize in coaching teams and who can offer a rich expertise gained from working with teams in various industries.

9.3.7 Launching the Team

Assuming that the team has been put together in optimal fashion, and its members trained (i.e., needed competencies, appropriate size, goals clearly defined, expectancies aligned; leader and members trained; reward system approved; sufficient time, space, and information technology available), their first meeting will lay the foundation for the forming and norming phases, especially when the members are not yet familiar with each other (see section 9.3.3). Leaders can take advantage of a treasure chest of techniques to advance a new team through the stages necessary to build up trust in one another and discover their respective strengths and weaknesses.

Figure 9.13 symbolizes the state of a team that is ready to perform at a high level, with the trust that enables them to overcome natural barriers. As shown, the alligator allows a bird to forage in its open snout, to clean between its teeth for bits of food. This is an illustration of a win-win situation: the alligator gets rid of annoying insects, and the bird gets a nourishing meal.

A wide variety of group games and so called ice-breaker activities have been developed to help team members get to know each other as quickly as possible. Such activities can target different aspects of team building, such as individual development, self-confidence, emotional intelligence, communication, or background sharing. The sidebar on team-building exercises briefly describes three examples, each with a different level of intensity and purpose.

The Power of Synergy

Synergy, from the Greek *syn,* "together," and *erg,* "to work," describes a system of individual parts, factors, or effects working together to produce results that are far advanced from what each could achieve alone—the whole is greater than the sum. A newly developed synergistic system might produce totally unpredicted results.

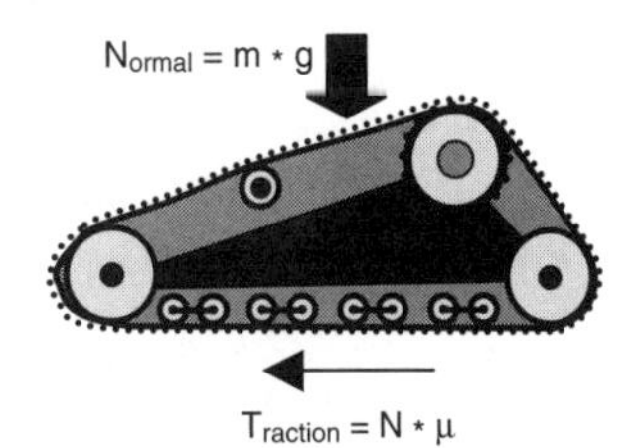

Synergy in Construction: Dozer Tracks

Tracks synergize several physical and mechanical phenomena:

1. Gravity
2. Density of steel
3. Flexibility of chains
4. Shear resistance of soil
5. Distribution of force by track plates, and
6. Mechanical advantage of sprocket wheels.

Team-Building Exercises to Get to Know Each Other

Cellphone Ring Tone

This could be the first team-building activity to be carried out, when the different members introduce themselves. Each person is asked to:

1. Reveal (as a sound) **what** ring tone they have chosen.
2. Explain **why**, and
3. **What** they would install if it were available.

Ring tones are for many an expression of their personality. The facilitator could easily encourage add-on anecdotes beginning, "Once my cellphone . . . ".

Dream Island Development

The team is divided into pairs for this exercise. They gather around a large table with a large sheet of paper in front on them—the bigger the group, the bigger the paper. Each pair is assigned a section of the paper onto which they are to draw a section of their dream island in the Pacific where they would like to retire. They are given a set of color markers and asked to first draw the coastline, so that they connect to the adjacent two groups.

Next, each team fills its area with whatever they desire, to develop a *self-sustaining* community on their part of the island (mountains, beaches, cliffs, roads, buildings, garden spots, agriculture, fishing, shopping malls, golf courses, restaurants, utilities, harbor, airport, wind farms, solar energy sources, etc.). The facilitator might call the teams' attention to the need to develop a common scale for buildings, roads, and other infrastructure, and connect highways and utilities at the boarder of the territory.

Cooperation will be important, since the energy production on the island will probably be concentrated and traded. A variety of issues and situations will be discussed, negotiated, agreed on, or argued. After the teams have reached some basic agreements, each is asked to describe three food dishes they could live on for an entire year.

The final task is for the teams to agree on where to grow, catch, process, distribute, market, and cook all the different foods selected by the teams (corn, vegetables, fruits, meat, fish, etc.). Remember, the island community has to be self-sufficient, but bartering is permitted between the areas.

What Do You Do?

This activity is designed to improve team members' understanding of each other's roles and responsibilities within the company. Each team member, or a pair, is asked to prepare a short presentation of the job they are doing, and where (department, office, construction site, etc.), which they will give to the rest of the group. The presentations can be informal (flipchart or discussion style) or more formal (PowerPoint), depending on the capability or time available.

- In the first round, the presentation should address following subjects: (1) What is the mission of the area where the member works? (2) Why is the area important for achieving the vision of the company? (3) What have been the member's most difficult challenges in the last half year? (4) What other area(s) in the company could help most to improve the member's performance, long term?
- The second round is more impromptu and requires a rapid response to a randomly picked question, some humorous and some more serious. Here is a list of possible questions. The person whose turn it is picks a number:
    1. A genie allows you one wish to help your work. What would it be? (Sorry, the genie has no money.)
    2. What is the number-one job-related stress in your area of work?
    3. What should be done for this year's New Year holiday party to make it more fun?
    4. What should be done to make sure that every job is completed on time?
    5. The company wants a new logo. What should it look like?
    6. What should be done to ensure every job comes in under budget?
    7. What would be the top performance motivator in your area of work?
    8. What is your best idea for the company to save cost?
    9. What is your best idea for the company to improve quality?
    10. Your company plans a 30-second commercial for TV. What should it say and show?

9.3.8 Tools Used by High-Performance Teams

By definition, teams are charged with solving/achieving a complex and multifaceted problem/goal. That is why its members come from different areas of the company, bringing with them different sets of competencies. As a result, the different team members will look at the issue to be addressed from different perspectives and pave different paths leading to potential solutions. In order to avoid the potential for chaos during this process, successful teams follow a structured approach, a series of problem-solving steps, which are shown in Figure 9.14. Table 9.12 provides a short description of each step.

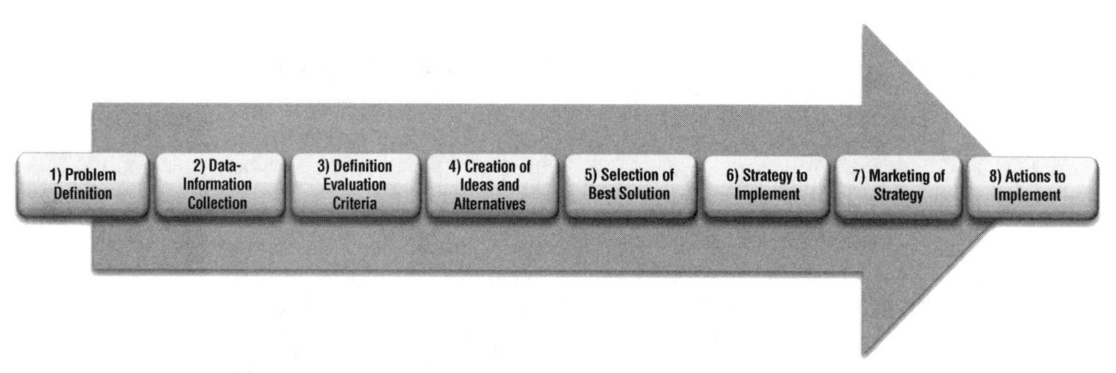

Figure 9.14 Sequence of Problem-Solving Phase

Table 9.12 Stepwise Approach to Problem Solving

Step	Activities and Questions for Team	Result(s)
1. Problem definition	A team that can define the problem at hand is halfway to solving it. What are all the factors linked to the problem? What are the cause/effect relationships that are related to each factor? What secondary factors impact the identified causes? Which relationships are not clearly understood? What future developments could have significant impact on the relationships? What sub- or sub-subproblems can be defined? What is going right, and why? What data, information, forecasts, etc. are missing?	Fishbone diagram that shows the main cause/effect relationships A succinct problem statement identifying the key issues List of information or data needed
2. Data/ information collection	Gathering, processing, dissemination, and discussion of missing data and information. The validity and accuracy of critical facts and forecasts should be verified through a second source.	Shared understanding of factor relationships quantitative/ qualitatively
3. Definition/ evaluation of criteria	This is a very critical stage of the process. Before searching for solutions, the team has to clearly define the desirable features of a solution or plan. Finally, the members have to prioritize them according to importance.	List of weighted criteria to evaluate final outcome(s)
4. Creation of ideas and alternative solutions	This is the creative segment of the process, when the synergistic capability and competence of the team should come to fruition. The most common tool used in this phase is called brainstorming, used to generate ideas. Brainstorming may look unstructured but it follows some ground rules. In a second step, the creative ideas have to be aggregated into three to four complete alternative solutions.	List of creative ideas underpinning three or four worked-out alternative solutions

5. Selection of best solution	This stage requires the use of weighted evaluation criteria, which will provide the bases for selecting the best solution. With the help of a point scale and quantitative analysis, the alternative solutions can be compared. Best solutions need to be finalized in the level of detail necessary to achieve the given goal.	Description and justification for selecting the preferred solution Details necessary for implementation.
6. Strategy to implement	Most new solutions or plans require investments, changes in procedures, and the acceptance by departments or individuals. There will be weaknesses and risks of failure. Thus, a strategy must address the financial, technical, organizational, and social aspects. What steps are necessary to mitigate the risks? What progress controls should be implemented?	Strategy for realization of best solution that addresses every aspect critical to its success. Recommendation for progress monitoring.
7. Marketing of strategy	The proposed solution will most often require changes, so goodwill from company staff is essential. To ensure that, first-hand information of the exact nature, the benefits, and the impacts of the project should be provided. The team members would be most appropriate to deliver this information, as they are knowledgeable and work in different areas of the company.	Information campaign that creates goodwill in the company prior to implementation.
8. Actions to implement	Initiation of the first steps of the carefully developed solutions. At the same time, the procedures to monitor progress and performance have to be put in place.	Implementation of action plans and monitoring system.

Worked-Out Problem 9.1: Brainstorming Rebar Placement Improvement

RodBuster Inc. is a highly specialized company in the area of rebar placement for large concrete structures such as bridges, hospitals, high-rise buildings, and large military facilities. The company president has charged a special team, consisting of a rebar detailer, purchase agent, trucker, crew foreman, and a long-time rodman, to propose a strategy to boost the productivity of the company's crews by a minimum of 30 percent. We were hired to help them to prepare their brainstorming session planned for tomorrow at 11:00 A.M. to run until 1:00 P.M.

Assumptions and Models

We assume that all the team members know each other and have gone through a training session preparing them to work efficiently as a team and

(continued)

(continued)
where they learned the basics of brainstorming. Each person brings with him or her competencies related to one or two aspects of placing rebar for structures, be it walls, column heads, bents, beams, elevated slabs, and so on. We also assume that they are familiar with the material presented in section 3.5.2, where the supply chain for rebar delivery was discussed.

Preparation for Meeting

The meeting should be organized into four separate phases: (1) welcome and review of the day's agenda and restatement of goal; (2) brainstorming session; (3) reverse brainstorming; (4) development of fishbone diagram.

Review of Brainstorming

Key Points of Brainstorming Steps for Improving Productivity

1. Goal is to generate as many ideas as possible; ''dreaming'' is encouraged—''what if,'' ''how about?''
2. For five minutes, everybody writes down ideas.
3. Each member calls for one idea to be presented for everybody to see (on a flipchart, tablet, etc.).
4. Before the next round, members are invited to expand on other members' ideas—''piggybacking.''
5. No criticism, no buts, no whys are allowed at this stage.
6. Keep going until no more ideas or questions emerge.
7. Maximum time is XX minutes, followed by a five-minute break.

Key Points of Reverse Brainstorming: Causes of Low Productivity

1. Generate as many causes/obstacles as possible.
2. Link causes and obstacle to the list of new ideas.
3. Maximum time is XX minutes, followed by a five-minute break.
4. No criticism, no buts, no whys are permitted.
5. All links are recorded for everyone to see (e.g., with colored arrows).

Reminders for the Leader of the Brainstorming Event

1. Promote synergism between the members by stimulating expansions on each promising idea.
2. Get every group member involved. Encourage a member without ideas to ask a key question.
3. Don't let group members get sidetracked analyzing one idea at length.
4. Prepare provocative statements to reenergize the team that is stuck.

5. Counter "killer statements" by reminding the group of the charge by the company president to come up with new and so-far untried methods. Sample "killer" statements to respond to immediately include:
   - This will never work.
   - Great idea, but . . .
   - Let's stick with what works.
   - We've done all right so far.
   - If it ain't broke, don't fix it.
   - You can't argue with success.
   - This place is different.
   - Are you working for the competition?
   - We're not ready for that yet.

Review of Fishbone Diagrams

Phase 4 calls for the drawing of a fishbone diagram, a tool to help the team systematically flush out the key relationships attached to the goal state. Its graphical format is excellent for prompting a team to discuss and display various options, while reaching successively higher levels of detail.

The leader should begin this phase with a quick review, using an incomplete example related to the reverse brainstorming session earlier. Then, after a short discussion of the process, start drawing the diagram for the final goal: *improvement of productivity by 30 percent*.

Step 1: Define the problem or goal in three to five words; write them in a box on the right side of an empty page and draw a long horizontal arrow pointing at the box.

As indicated in Figure 9.15, the problem or the goal is represented by the head of a fish skeleton, where the backbone connects to its fins. The backbone serves as a hub for all the causes that are linked to the problem or the goal—the fish head.

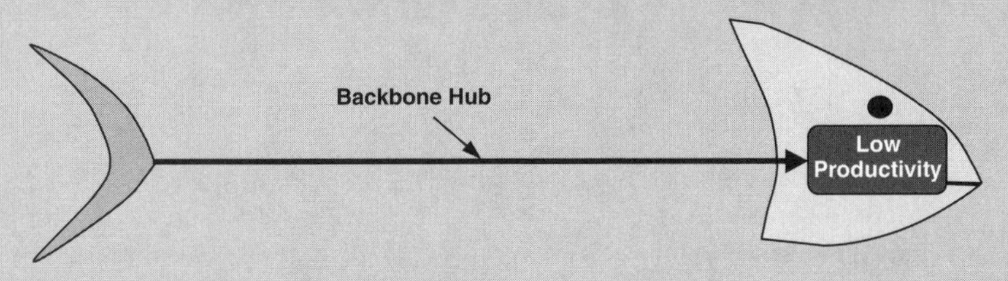

Figure 9.15 Setup of Fishbone Diagram

(*continued*)

(*continued*)

Step 2: Agree on the four to six main causes or factors that are most pertinently related to the problem or that contribute to achieving the desired goal. Add them to the backbone as secondary bones or fins.

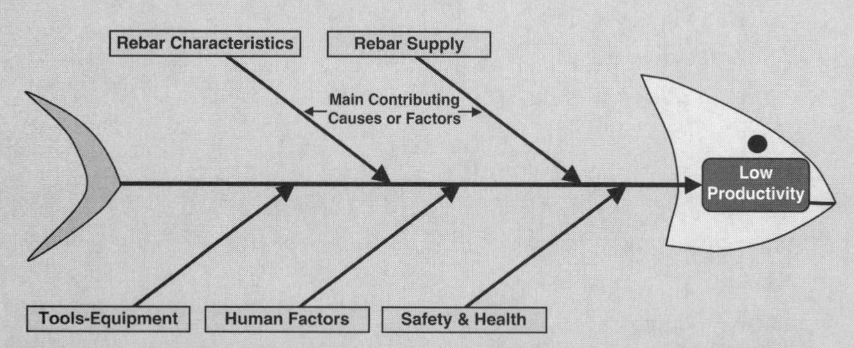

Figure 9.16 Fishbone Diagram with Main Bones Showing Top-Level Causes or Factors

Step 3: Identify the second-level causes and factors by searching for more detailed explanations for each of the major cause categories (Figure 9.17).

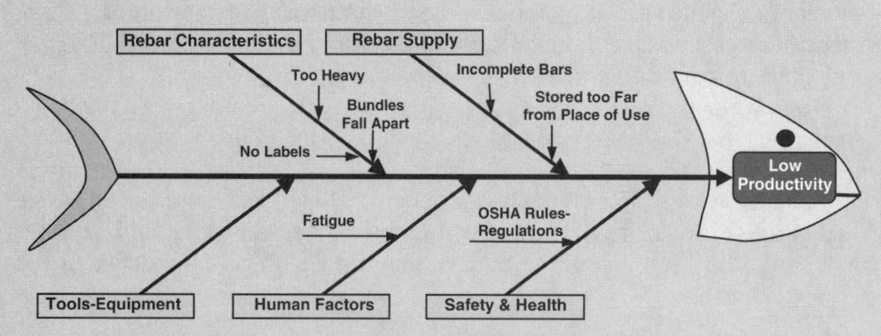

Figure 9.17 Fishbone Diagram with Second-Level Causes

Step 4: Try to break down the second-level causes and factors into a third level.

Now the diagram is getting overcrowded, with three levels of causes (see Figure 9.18). However, the flexibility of the cause/effect concept offers the opportunity to begin a new fishbone diagram on another page, putting the higher-level causes or factors in the box at the fish head.

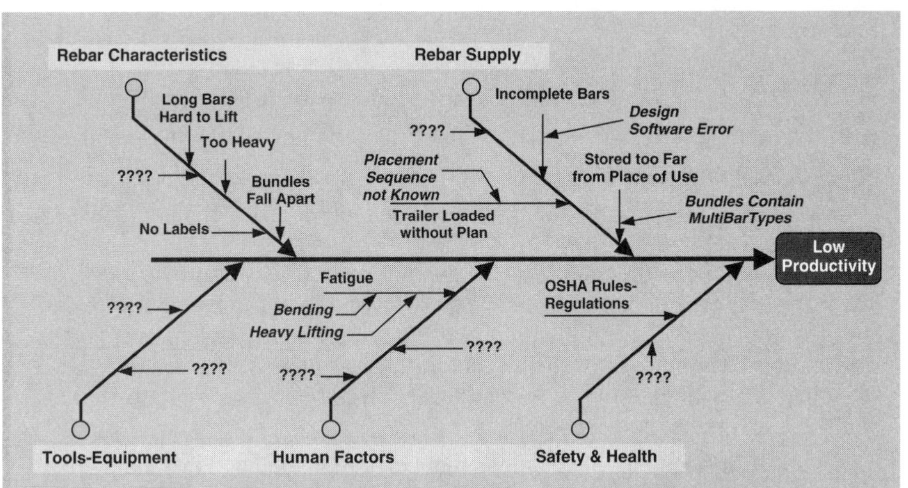

Figure 9.18 Partially Completed Fishbone Diagram with Third-Level Causes or Factors

Agenda for Brainstorming Meeting

11:00 A.M. Welcome and review of team goal, defined in previous meetings

11:10 A.M. Review of brainstorming rules, key points, election of recorder

11:20 A.M. Ask everyone to write their ideas and questions on a piece of paper:

What can I do to push the productivity of rebar placement up 30 percent?

What could others do that would help me?

What would be a dream come true?

11:30 A.M.Begin first round, soliciting one idea per person; no negative comments allowed.

Recorder writes all ideas on a place clearly visible to all.

Members are asked to review their written thoughts and try to expand on ideas that have been recorded, and write down questions.

11:45 A.M. Complete recording of ideas and questions so that everyone can contribute.

12:00 P.M. Five-minute break

12:05 P.M. Begin reverse brainstorming. Again, everybody writes their ideas down:

What are causes for low productivity?

(*continued*)

(*continued*)

What could cause even lower productivity?

What are obstacles to improving productivity by 30 percent?

Begin recording after 10 minutes. Follow brainstorming pattern from above.

12:25 P.M. Five-minute break

12:30 P.M. Review the fishbone diagram method using the partial example.

Begin new fishbone diagram with: 30 percent productivity improvement
With the brainstorming result visible to everyone, ask the members to identify the *top-level* areas that would either drive productivity improvement or hinder improvements. Use different colors to draw drivers and blockers.
Continue fishbone diagram with second- and even third-layer relationships.
Add new diagrams, if necessary.

1:00 P.M. Close meeting unless team is not finished and members are willing to stay.

Promise that results of the meeting will be summarized and sent to everyone within two days.
Set next meeting time and preliminary agenda.

9.4 BASICS ABOUT CREATIVITY

Arguably the most creative scientist of the twentieth century, Albert Einstein, wrote: "I have no special talent. I am only passionately curious." Later, he added, "I am able to dream." Obviously he did not dream all the time—in particular, when he passed his physics exam at the Swiss National Institute of Technology at the top of his class.

What makes someone a creative thinker? One of the greatest inventors, Thomas Edison, may have given us a clue to this mystery when he said:"*Genius is one percent inspiration and ninety-nine percent perspiration*," and "*Opportunity is missed by most people because it is dressed in overalls and looks like work.*"

Psychologists, neurologists, and philosophers,. have long tried to understand how our brain works, but we still have only a limited understanding how people produce creative solutions or devise innovative ideas. One model that many seem to have agreed on is one that highlights the importance of the two sides of the brain: left brain and right brain. Each side "specializes" in a different way of thinking, referred to as *convergent* and *divergent thinking.*

9.4.1 Convergent Thinking

Convergent thinkers tend to seek out a single, correct solution to a problem. Traditional education is geared toward the convergence of a lot of information—for example algebra—to solve variations of set problems during tests. As Figure 9.19 illustrates, a convergent-thinking engineer will try to draw on his or her knowledge

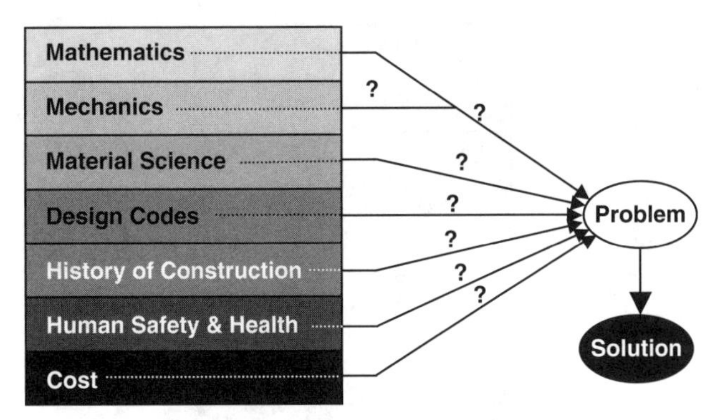

Figure 9.19 Convergent Problem Solving

of various subjects, which usually works to solve a problem in a linear form. The problem pattern will provide the clues to search the knowledge base, a search that continues until a solution is found. Thus, a solid grasp of established facts and professional practice is the base from which convergent thinkers operate.

Even creative people need to be able to apply convergent thinking. For example, Einstein needed to learn to operate with tensors as well as physics, to model and solve some of the complex problems he was facing.

9.4.2 Divergent Thinking

One of the first to define divergent thinking was William James (1890). He described it this way:

> Instead of thoughts of concrete things patiently following one another in a beaten track of habitual suggestion, we have the most abrupt cross-cuts and transitions from one idea to another . . . unheard of combination of elements . . . we seem suddenly introduced into a seething caldron of ideas . . . where treadmill routine is unknown and the unexpected is the only law.

Young children don't keep to the beaten track of habits when they play. They use their imagination and "specialized" tools unencumbered by tradition. The right hemisphere of the human brain supports all the thinking traits they need to be creative. Everybody has right brain; every person should be able to produce novel ideas through activities that not only allow but actually encourage creative thinking, such as fiction play, experimentation, sketching, debate, and others. Creativity trainers like to tell clients, "If you keep thinking the same old way, you'll always get the same old ideas." However, that is not to imply that one should reject knowledge gained through convergent learning; on the contrary. As Figure 9.20 shows, divergent thinkers very much depend on the raw knowledge they have learned from diverse sources and that lay outside the main domain (e.g., Einstein was also a philosopher and violinist).

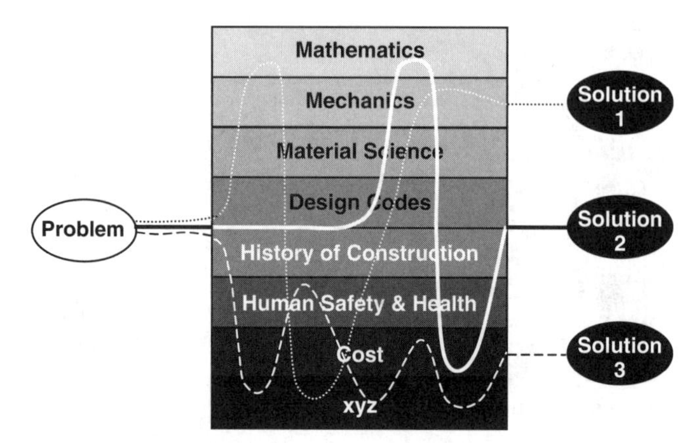

Figure 9.20 Divergent Problem Solving

For most people, creative ideas come when their minds are involved in an unrelated activity, such drawing, or showering in the morning. Even when we sleep we are working on problems, which can be thought of as a fermentation process. Simply, when we are relaxed, we are better able to connect the conscious with the unconscious mind, for "downloading." We must keep in mind, though, that the brain must be able to access all the up-to-date knowledge it needs to "play" with an infinite number of combinations of thoughts. The brain's capability to dream allows us to dispel learned associations, to find the underlying "truths" and remove invisible thought barriers. At some point, new combinations float up and break through into our consciousness, which we experience as an "ah hah" moment, or describe as a "bolt out of the blue."

One prerequisite to creative processing is to be confronted with a complex problem that cannot be solved using linear or convergent approaches. A person who is able to "cross-cut," as William James put it, will jump laterally from one field of knowledge into another, possibly using many different starting points in order to change perspective. In this way, many different solutions may result, some better than others, of course.

9.4.3 Left-Brain/Right-Brain Collaboration

As we just learned, a creative person requires both convergent and divergent thinking capabilities, each associated with one side of our brain. Figure 9.21 illustrates the characteristics of the two sides, as we know it today.

Immediately apparent is that the two lists align with the description of convergent and divergent thinking, and Figure 9.21 indicates that a person will "visit" both sides while thinking. Solving an engineering problem will usually require a look at the detail, some math, and the use of standards and codes, and involve some common sense and a big-picture view of the solution to be correct.

Left Brain Features

Uses logic; rational
Detail-oriented
One-at-a-time processing
Looks at a detail
Knows "how"
Facts "rule"
Focus on present and past
Loves math and science
Acts systematically; linear
Remembers numbers, names
Follows important rules, laws
Relies on words, facts
Forms strategies
Thinks convergently
Operates cautiously
Sees cause and effect
Objective

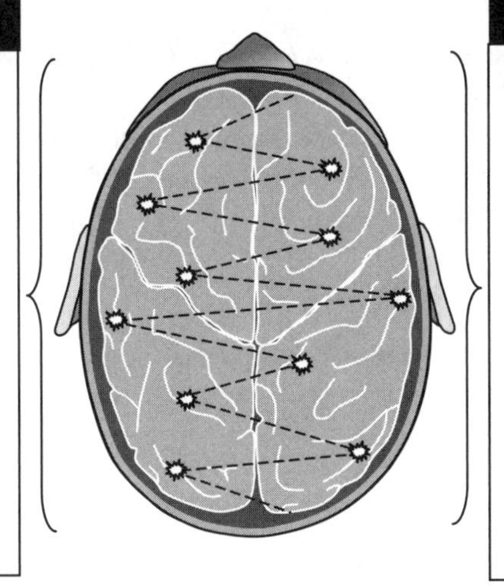

Right Brain Features

Uses feelings, intuition
Big-picture oriented
All-at-once processing
Looks at the whole
Discovers "why"
Imagination "rules"
Focus on present and future
Loves philosophy
Acts casually; random
Remembers places, faces
Follows common sense
Relies on images, emotions
Forms possibilities
Thinks divergently
Operates adventurously
Sees resemblances
Willing to be subjective

Figure 9.21 Characteristics of Left- and Right-Brain Capabilities

Creativity may not be needed to solve even a complex problem, as long as the problem can be broken down into smaller, easier-to-solve ones. Creative thinking is only needed when the problem is "nontraditional," and possibly is too complex for one person to solve alone.

Brainstorming is the tool, presented earlier, designed to synergize multiple brains with complementary features. Two divergent thinkers can feed off one another, in that one will pick up an idea from another and move laterally through his or her knowledge base, dramatically extending the reach to find possible combinatory solutions. In this way, the chances for achieving a valuable and unexpected outcome also increases.

9.4.4 Attributes of a Creative Individual

While no creative person is exactly like another, there are some basic features that have been found to apply to all of them. Here are some examples:

- ***Inquisitive*** in many unrelated areas, pursues divergent interests
- ***Expresses radical and spirited opinions***, tenacious or persistent defending a unique idea
- ***Willing to make a "leap of faith,"*** high risk taker, adventurous, always ***prepared to risk failure*** in order to learn, shows perseverance to make it work
- ***Playful, a daydreamer***, imaginative, pursues "what-if" questions.
- ***Unconventional sense of humor*** that may appear inappropriate and irreverent to others.

- ***Nonconforming*** individualist, classified as "different", sees the good in the bad
- ***Critiques constructively*** without inhibition with the goal to improve, seeks out other people to critique his or her own new ideas or projects.
- ***Never satisfied with the status quo***, constantly looking to improve constructively
- ***Plays "devil's advocate"*** in order to explore the other side of an argument
- ***Receives satisfaction from "work"*** challenging his/her creative side, curiosity regardless of school grade or paycheck
- ***Believes all problems are opportunities,*** solvable with commitment of time and energy

Signs of Blocked Creative Thinking

As stated, all people have the ability to be creative, to access the right side of their brains, given the necessary knowledge available on the left side. But innovative thinking can be buried or blocked, by, for example, strict adherence to learned conventions and anxieties such as fear of failure or ridicule. The following statements indicate blocks to fully tapping into the creative side:

- I am ***not creative***, ***I can't*** do this (insecurity)
- It ***can't be done*** (fear of failure, negativism)
- We always have done it this way, ***why change*** (fear of change)
- We worked on this so long now, it will ***never work*** (lack of perseverance)
- Why do you want me involved in this? ***I don't know anything*** about this (insecurity)
- This smells like more ***work for me*** (laziness, apathy)
- ***Why should I help*** these other guys (have a safer workplace)? (jealousy, toxic behavior)
- ***The others*** on the brainstorming team ***don't know anything*** (prejudice, conceit)
- I don't want to ***embarrass*** myself with some ***silly pipe dreams*** (pear of ridicule)
- Management is not really supporting this; they're only testing us to see who is ***stupid enough to try*** (cynicism, mistrust)

9.4.5 Exercises to Break Down Blocks to Creative Thinking

As the list of blocks to creative thinking indicates, most if not all are related to problems of mental hygiene. Blocks are associated with poor self-esteem, anxieties, and fear. Unfortunately, there is no easy "cleaning" process for these hygiene issues; it takes *affective* activities to slowly improve the state of mental hygiene and generate positive emotions over time.

Here are things a manager or brainstorming facilitator can do to help unblock creative thinking:

Tools to Promote Creativity

- Suggest carrying a small notebook ("idea journal") and pen/pencil to write down ideas during the day or when just waking up.
- Hold a coaching session on sketching, using Leonardo da Vinci's journal as an example. Hand out napkins.
- Organize an exercise to design a construction method to install a soil anchor on the moon.
- Practice visualization; demonstrate how to draw concept maps, illustrative schema, or big-picture maps that show how everything is connected across different "worlds."
- Role-play by having team members take on unusual personas (e.g., a customer buying a big chicken breast; a chicken in a farm falling over from overweight; human immune system after eating chicken with lots of antibiotics). Suggest a hair-raising scenario to start the play (e.g., after eating so much chicken, the person is attacked by bacteria and does not have any immune protection and is quickly dying). Then ask: Who are the ones who suffer, and who benefits most, and why?)

Supportive Behaviors During a Brainstorming Meeting

- Stress expertise of individuals.
- Model creative behavior, and reinforce through verbal commentary.
- Encourage critical questions based on facts, not assumptions.
- Highlight the human spirit to explore new horizons (history).
- Applaud simple/first contributions.
- Appeal to human nature to help others in new ways.
- Help to get out of an invisible rut by encouraging what-if escapes
- End the meeting leaving major issues unanswered, and encourage everyone to "sleep on them."

CHAPTER REVIEW

Journaling Questions

1. This chapter argued that a leader needs *cognitive* and *emotional* intelligence, as well as *interpersonal* skills. Reviewing Table 9.2, "Differences between Managers and Leaders," assign the listed items to these three

competencies. Recognize that you cannot distribute them all. Propose a fourth and possibly a fifth competency that a leader possesses.

2. Table 9.7 contains 20 survey questions. Assign each of the questions to one of the four clusters of emotional intelligence. Describe which competency it addresses and what it intends to measure.
3. As a student, you probably have been a member of several teams. Discuss the life cycle of the best and the worst teams you have worked on. Describe some of the dynamics that you remember that made it good or bad. Identify some of the roles in Figures 9.9, 9.10, and 9.11 that students play on a team you are in right now (note: one person may play several roles).

Traditional Homework

1. Return to Table 9.1 and review your answers to the three questions. Discuss whether you will change your answers now, after completing the chapter.
2. Read Header Problem 9.1 again, then answer these questions:
   a. What common mistake did John make when starting his new position?
   b. What skill is John missing in the way he reacts to the team's performance?
   c. What competencies should John acquire that will allow him to overcome that problem?
   d. What would you have done differently if you had been given John's assignment?
3. We learned that a leader needs *cognitive* and *emotional intelligence,* as well as *interpersonal* skills.
   a. Assign the 21 attributes that are listed in Table 9.13 to one of the three competencies.
   b. Add five additional attributes that a leader needs to have but that are not on the list. Explain why they should be included.
4. Imagine that you are living in Leonardo da Vinci's time and are thinking about building a tidal power plant. Sketch the main components that have to be constructed. Detail how the water will power the waterwheel during low and high tide. Draft a mechanism that will enable the millstones on shore to always rotate in the same direction.

Open-Ended Question

1. Engineers are considered poor communicators. After working through this chapter and reviewing the literature, you might realize that the failure of college education to prepare young engineers to become leaders is far more systemic.

Table 9.13 Leadership Attributes

Attributes of a Leader
1. Engages in life-long-learning.
2. Rewards excellent contributions by others.
3. Articulates vision and goals convincingly.
4. Ensures the health and welfare of everyone.
5. Empowers subordinates to act on visions and goals.
6. Is aware of basic human needs.
7. Works and thinks across boundaries.
8. Listens to others with an open mind.
9. Emphasizes the importance of intrinsic values.
10. Accepts critical feedback.
11. Has strong analytical capabilities.
12. Stays honest and fights corruption.
13. Forbids bullying by any group or person.
14. Attracts and coaches talented people.
15. Is an innovative and original thinker; learns from mistakes.
16. Shares information freely.
17. Is a proactive problem solver and effective planner.
18. Inspires trust.
19. Is able to ask the critical questions.
20. Has long-term and global perspective.
21. Is unafraid of making mistakes in pursuit of the vision.

a. What basic principles in the way engineers are educated (four-year program) do you think should be changed (e.g., lecture versus problem-based Socratic teaching, internships, humanities)?

b. What specifics have to be modified in order to prepare engineers to become leaders? (Note: You may not add a new course unless you eliminate one).

c. Outline a four- or four-and-a-half- year plan.

BIBLIOGRAPHY

Alasoini, T., A. Heikkila, and E. Ramstad, E. High-Involvement Innovation Practices at Finnish Workplaces. *Inter. J. Prod. Perform. Mgmt.*, vol. 57, no. 6, 2008.

Balkema, A., and E. Molleman. Barriers to the Development of Self-Organizing Teams. *J. Managerial Psych.,* vol. 1(2), 1999.

Bass, B. M. *Leadership and Performance Beyond Expectations* (New York: Free Press), 1985.

Beech, N., and O. Crane. High-Performance Teams and a Climate of Community. *Team Perform. Mgmt.*, vol. 5, no. 3, 1999.

Bennis, W. G. *On Becoming a Leader*, Addison-Wesley, 1989.

Bonasso, S. G. Engineering, Leadership, and Integral Philosophy, *J. Prof. Issues Eng. Ed. Pract.*, vol. 127, no. 1, January 2001.

Bowie, N. A. Kantian Theory of Leadership. *Leadership Org. Dev. J.*, vol. 21/4, 2000.

Bowman, B. A., and J. V. Farr. Embedding Leadership in Civil Engineering Education. *J. Issues Eng. Ed. Pract.*, vol. 126, no. 1, January 2000.

Bradberry, T. R., and L. D. Su. Ability versus Skill-Based Assessment of Emotional Intelligence. *Psicothema*, vol. 18, 2006. www.psicothema.com.

Bryde, D. J. Modelling Project Management Performance. *Inter. J. Quality Reliability Mgmt.*, vol. 20, no. 2, 2003.

Burns, J. M. *Leadership*. New York: Harper & Row, 1978.

Butler, C. J., and P. S. Chinowsky. Emotional Intelligence and Leadership Behavior in Construction Executives. *J. Mgmt. Eng.*, vol. 22, no. 3, 2006.

Chan, A. T., and E. H. Chan. Impact of Perceived Leadership Styles on Work Outcomes: Case of Building Professionals. *J. Constr. Eng. Mgmt.*, vol. 131, no. 4, April, 2005.

Cohen, S. G., and D. W. Bailey. What Makes Teams Work: Group Effectiveness Research from the Shop Floor to the Executive Suite. *J. Mgmt.*, vol. 23, no. 3, 1997.

Crosbie, R. Learning the Soft Skills of Leadership. *Ind. Com. Train.*, vol. 37, no. 1, 2005.

Denton, D. K. Developing a Performance Measurement System for Effective Teamwork. *Inter. J. Quality Prod. Mgmt.*, vol. 07, no. 01, 2007.

Drexler, J. A., T. A. Beehr, and T. A. Stetz. Peer Appraisals: Differentiation of Individual Performance on Group Tasks. *Human Res. Mgmt.*, vol. 40, no. 4, 2001.

Farr, J. V., S. G. Walesh, and G. B. Forsythe. Leadership Development for Engineering Managers. *J. Mgmt. Eng.*, vol. 13, no. 4, 1997.

Fiedler, F. E. and J. L. Macaulay. The Leadership Situation: A Missing Factor in Selecting and Training Managers. *Human Res. Mgmt. Rev.*, vol. 8, no. 4, 1998.

Goleman, D. *Emotional Intelligence.* New York: Bantam Books, 1995.

Goyal, A., and K. B. Akhilesh. Interplay among Innovativeness, Cognitive Intelligence, Emotional Intelligence, and Social Capital of Work Teams. *Team Perform. Mgmt.*, vol. 13, no. 7/8, 2007.

Grant, A. M. Enhancing Coaching Skills and Emotional Intelligence Through Training. *Ind. Commercial Train.*, vol. 39, no. 5, 2007.

Gratton, L. Managing Integration Through Cooperation. *Human Res. Mgmt.*, vol. 44, no. 2, 2005.

Groves, K. S., M. P. McEnrue, and W. Shen. Developing and Measuring the Emotional Intelligence of Leaders., *J. Mgmt. Develop.*, vol. 27, no. 2, 2008.

Hackman, J. R.The Design of Work Teams. In *Handbook of Organizational Behavior* (J. Lorsch, ed.). Englewood Cliffs, NJ: Prentice-Hall, 1987.

Hay, A., and M. Hodgkinson. Rethinking Leadership: A Way Forward for Teaching Leadership. *Leadership Org. Develop. J.*, vol. 27, no. 2, 2006.

Heilman, K. M., S. E. Nadeau, and D. O. Beversdorf. Creative Innovation: Possible Brain Mechanisms. *Neurocase*, vol. 9, no. 5, 2003.

Houldsworth, E. Leadership Team Performance Management: The Case of BELRON. *Team Perform. Mgmt.*, vol. 14, no. 3/4, 2008.

Ingram, H. Linking Teamwork with Performance. *Inter. J. Team Perform. Mgmt.*, vol. 2, no. 4, 1996.

James, W. *The Principles of Psychology.* New York, Henry Holt and Co., 1890.

Jeffery, A. B., J. D. Maes, and M. F. Bratton-Jeffery. Improving Team Decision-Making Performance with Collaborative Modeling. *Team Perform. Mgmt.*, vol. 11, no. 1/2, 2005.

Kerr, R., J. Garvin, N. Heato, and E. Boyle. Emotional Intelligence and Leadership Effectiveness. *Leadership Org. Develop. J.*, vol. 27, no. 4, 2006.

Koman, E. S., and S. B. Wolff. Emotional Intelligence Competencies in the Team and Team Leader. *J. Mgmt. Develop.*, vol. 27, no. 1, 2008.

Kouzes, J. M., and B. Z. Posner. *The Leadership Practices Inventory:Theory and Evidence Behind the Five Practices of Exemplary Leaders.* Hoboken, NJ: John Wiley & Sons, 2002. http://media.wiley.com/asests/61/06/lc_jb_appendix.pdf.

Leach, D. J., T. D. Wall, S. G. Rogelberg, and P. R. Jackson. Team Autonomy, Performance, and Member Job Strain: Uncovering the Teamwork KSA Link. *Appl. Psych. Inter. Rev.*, vol. 54, no. 1, 2005.

Macneil, C. The Supervisor as a Facilitator of Informal Learning in Work Teams. *J. Workplace Learn.*, vol. 13, no. 6, 2001.

Mastrangelo, A., E. R. Eddy, and S. J. Lorenzet. The Importance of Personal and Professional Leadership. *Leadership Org. Develop. J.*, vol. 25, no. 5, 2004.

McCallum, S., and D. O'Connell. Social Capital and Leadership Development: Building Stronger Leadership Through Enhanced Relational Skills. *Leadership Org. Develop. J.*, vol. 30, no.2, 2009.

McCuen, R. H. A Course on Engineering Leadership. *J. Prof. Issues Eng. Ed. Pract.*, vol. 125, no. 3, July 1999.

McEnrue, M. P., K. S. Groves, and W. Shen. Emotional Intelligence Development: Leveraging Individual Characteristics. *J. Mgmt. Develop*, vol. 28, no.2, 2009.

McGrath-Champ, S., and S. Carter. The Art of Selling Corporate Culture: Management and Human Resources in Australian Construction Companies Operating in Malaysia. *Inter. J. Manpower*, vol. 22, no. 4, 2001.

Mendibil, K., and J. MacBryde. Factors That Affect the Design and Implementation of Team-Based Performance Measurement Systems. *Inter. J. Prod. Perform. Mgmt.*, vol. 55, no. 2, 2006.

Miller, D. M., R. Fields, A. Kumar, and R. Ortiz. Leadership and Organizational Vision in Managing a Multiethnic and Multicultural Project Team. *J. Mgmt. Eng.*, vol. 16, no. 6, 2000.

Mumford, M. D., S. T. Hunter, D. L. Eubanks, K. E. Bedell, and S. T. Murphy. Developing Leaders for Creative Efforts: A Domain-Based Approach to Leadership Development. *Human Res. Mgmt. Rev.*, vol. 17, 2007.

Oliver, R. L. Expectancy Theory Predictions of Salesmen's Performance. *J. Marketing Res.*, vol. 11, August 1974.

Parolini, J., K. Patterson, and B. Winston. Distinguishing Between Transformational and Servant Leadership. *Leadership Org. Develop. J.*, vol. 30, no. 3, 2009.

Politis, J. D. Self-Leadership Behavioural-Focused Strategies and Team Performance: The Mediating Influence of Job Satisfaction. *Leadership Org. Develop. J.*, vol. 27, no. 3, 2006.

Ross, T. M., E. C. Jones, and S. G. Adams. Can Team Effectiveness Be Predicted? *Team Perform. Mgmt.*, vol. 14, no. 5/6, 2008.

Salas, E., D. D. Granados, C. Klein, C. S. Burke, K. C. Stagl, G. F. Goodwin, and S. M. Halpin. Does Team Training Improve Team Performance? A Meta-Analysis. *Human Factors*, vol. 50, no. 6, December 2008.

Salovey, P., and J. Mayer. Emotional Intelligence. *Imag. Cogn. Personality*, vol. 9, no. 3, 1990.

Shah, J., and E. T. Higgins. Expectancy × Value Effects: Regulatory Focus as Determinant of Magnitude *and* Direction. *J. Personality Soc. Psych.*, vol. 73, no. 3, 1997.

Shahin, A., and P. L. Wright. Leadership in the Context of Culture: An Egyptian Perspective. *Leadership Org. Develop. J.*, vol. 25, no. 6, 2004.

Skipper, C. O., and L. C. Bell. Influences Impacting Leadership Development. *J. Mgmt. Eng.*, vol. 22, no. 2, 2006.

____________. Assessment with 360° Evaluations of Leadership Behavior in Construction Project Managers, *J. Mgmt. Eng.*, vol. 22, no. 2, 2006.

Spatz, D. M. Leadership in the Construction Industry. *Pract. Period. Structural Design Constr.*, vol. 4, no. 2, May, 1999.

Stansfield, T., and C. O. Longenecker. The Effects of Goal Setting and Feedback on Manufacturing Production: A Field Experiment, *Int. J. Prod. Perform. Mgmt.*, vol. 55, no. 3/4, 2006.

Sunindijo, R. Y., B. H. Hadikusumo, and S. Ogunlana. Emotional Intelligence and Leadership Styles in Construction Project Management. *J. Mgmt. Eng.*, vol. 23, no. 4, 2007.

Thompson, P., and R. Wallace. Redesigning Production Through Teamworking: Case Studies from the Volvo Truck Corporation. *Inter. J. Operations Prod. Mgmt.*, vol. 16, no. 2, 1996.

Toor, S. R., and G. Ofori. Developing Construction Professionals of the 21st Century: Renewed Vision for Leadership. *J. Prof. Issues Eng. Ed. Pract.*, vol. 134, no. 3, July, 2008.

Waal, A. A. Stimulating Performance-Driven Behaviour to Obtain Better Results. *Inter. J. Prod. Perform. Mgmt.*, vol. 53, no. 4, 2004.

Chapter **10**

Communication: The Nerve System of Construction

One of the basic causes for all the trouble in the world today is that people talk too much and think too little. They act impulsively without thinking.

—Margaret Chase Smith

I know you believe you understand what you think I said, but I am not sure you realize that what you heard is not what I meant.

—Robert McCloskey (attributed)

A miscommunication between workers at a Cobb County warehouse led to the accidental death of a forklift truck driver Thursday morning, police said.

—"Miscommunication Leads to Forklift Driver Death,"
The Atlanta Journal-Constitution, October 4, 2007

Every dynamic system, from the smallest sentient creature to the largest business conglomerate, depends on effective communication in order to sustain itself. Human history is full of examples of how the speed at which good, or bad, news was communicated created heroes, decided battles, or, if arriving too late, led to the death of innocent people. Probably one of the most well-known newsbearers was the Athenian Pheidippides, who ran 42 kilometers (26.2 miles) from Marathon to Athens with the unexpected news of the victory of the Greeks over the invading Persians. On his previous "news run," Pheidippides carried a message from Athens to Sparta, a 150-mile journey that took him two days. The Romans extended the Assyrian postal system (1700 BC) by establishing the Imperial Post, which comprised 1000 stations equipped with fresh horses along a postal route that could only be used for official business. The American equivalent was the short-lived Pony Express, which connected St. Louis with San Francisco, and took 10 days, or approximately 220 miles/day, to traverse. Different from the Roman messengers, the Pony Express riders delivered only envelopes or small packages. The Roman emperor, in contrast, received his messages more slowly but was able to question the messenger, to learn more than what was contained in the "official" message.

Long-distance communication by messenger became obsolete when the Pony Express was "overtaken" by the speed of electromagnetic current carried by wire transmitting coded signals, with the transmission of the first telegraph in Morse code in 1844. Thus began a new era of high-speed mass communication that eventually led to today's information technology (IT).

10.1 ENGINEERING DRAWINGS: THE ANCIENT COMMUNICATIONS MEDIUM

Communication is the process of sharing ideas, information, and messages with others in a particular time and place. Communication includes writing and talking, as well as nonverbal messages, such as facial expressions, body language, or gestures; visual information exchange through the use of images or pictures (sketches, paintings, photographs, video, film); and electronic communication, via telephone, electronic mail, cable television, or satellite broadcasts.

Without doubt, the image that has come to symbolize civil engineering and construction, second only to the image of the hardhat, is the blueprint drawing. Interestingly, both are fairly recent innovations when compared to other symbols of other professions. The blueprint reproduction procedure was discovered in 1842, and the first "hardhat zone" was established in 1933, during the construction of the Golden Gate Bridge. Whereas the rod of Asclepius used to symbolize the medical profession is an ancient Greek symbol associated with healing the sick; and the blindfolded Justitia, who represents the justice system with her sword and a scale, was the Roman Goddess of justice.

10.1.1 The Evolution of Models and Blueprints

Historical artifacts from the time of the Chaldeans, who preceded the Sumerians, prove that the genesis of drafting and technical drawing on a flat surface to represent plans, cross sections, and elevations of a building can be traced to even earlier than the Greek gods. A statue of Gudea, the ruler of Lagash in southern Mesopotamia around 2100 BC, indicates that technical drafting must have already been commonplace at that time. Exhibited in the Louvre in Paris, and shown in Figure 10.1, the

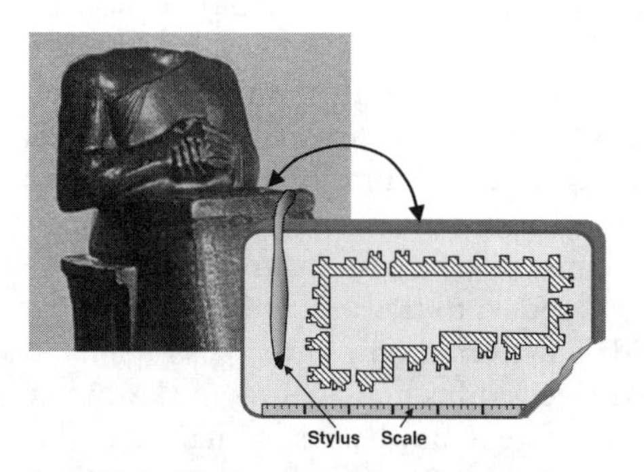

Figure 10.1 Gudea the Architect with drawing of temple complex

statue depicts a prince with a clay tablet on his knees, alongside a scaled ruler and a stylus for drawing in wet clay, implying that the tablet is a drawing board. Through archeological excavation it was confirmed that the drawing depicts the foundation, towers, and six gates of the temple complex at Eninnu of Ningirs.

Later, a letter from the prior of Canterbury Cathedral to the archbishop around 1494 also talks about early architectural drawing of plans.

> . . . I have communed with John Wastell your mason, berer hereof, to perceive of hym what forme and shape he will kepe in reysyng up of the pynacyls of your new towre here. He drew unto us ij patrons of hem. The on was with doble fineall wihoute croketts and the other was with croketts and sengle fineall. . . . shew hym your advise aqnd pleasure whyche of them ij, or of any other to be divised, shall contente you gode Lordshyp to be appointed."
>
> —***Sheppard, J.B. 1877, "Christ Church Letters," Camden Society, N.S. XIX, London, England.***

While drawing and drafting were the preferred tools of architects to express form and symmetry, the resulting designs were not very helpful to the masons and stonecutters who actually had to execute the designs. The drawings provided no guidance regarding, for example, dimensioning stones for a medieval church or cathedral. Shelby (1964) reported that because in medieval England, " . . . the drawings lacked dimensions and jointing of the stones, they could not be used as 'working drawings' or 'blueprints' in the modern sense of those terms—for they required too much interpretation before the necessary technical information could be made available to the masons for the actual cutting and placing of the stones." To solve the problem, the English master mason used another common communication tool: scaled models. According to Shelby, "In this tracing house the master mason draws plans, elevations, or sketches on parchment, and there he traces on board the molds that directed the stonecutters on their work. Thus from the master mason's drawing office there emanated both the design of the building and the means of supervising the execution of that design."

The emergence of cast iron as building material in the construction industry with the construction of the bridge at Coalbrookdale, England, in 1779 was followed by the introduction of even stronger puddle iron and carbon steel in 1855. These materials required a different approach to communication among architects, engineers, and the "ironmen." Good examples may be found in the drawings for the two largest buildings of their time, both built for the 1889 World's Fair in Paris. Figure 10.2(a) presents the design detail of the crowns of the two hinged arch elements, which were made of riveted double T-flanges, 9 feet (3 meters) apart. The 900,000-square-foot Galeries des Machines was, with 364 feet between its two bases, the largest wide-span, iron-framed structure ever built. Right next door, at 984 feet (300 meters), stood the Eiffel Tower, the tallest structure ever built at the time. Named after its designer and chief engineer, Gustave Eiffel, the Eiffel Tower retained that distinction until 1930, when New York's Chrysler Building was completed, reaching 1,046 feet (319 meters).

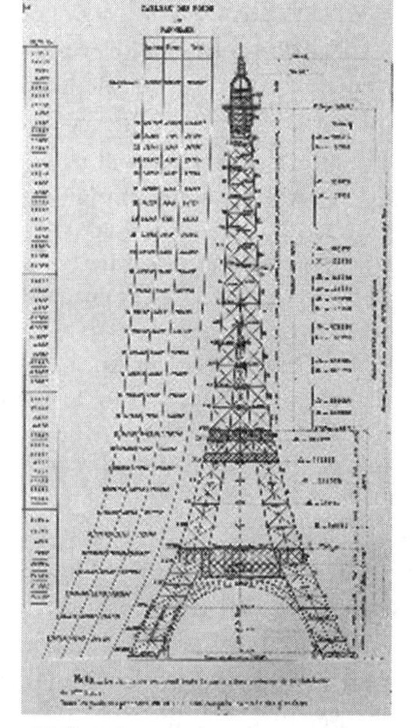

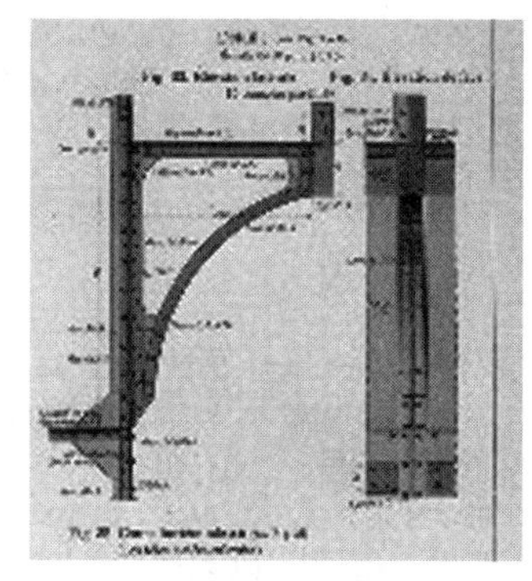

a) Hinged crown of two-member arch b) Elevations and section overview c) Detail of cantilevered balcony at the lantern

Figure 10.2 Technical Drawings for the Eiffel Tower, around 1880. (Source: www.tour-eiffel.fr/teiffel/uk/)

Immediately recognizable is the degree of detail each draftsman included to indicate curvatures, cross-sectional views, rows of rivets, readable labels (type, size, elevations), and renderings to maximize visual understanding. As shown, the level of detail used in the visualizations is matched by the scale. For example, every rivet-head stands out because of its shaded rendering (see Figure 10.2(c)); they are drawn as small circles or dots in Figure 10.2(a). Shading is effectively used to depict curved elements and to indicate different pieces sandwiched against each other.

With the increased use of steel and its ability to sustain large stresses encouraging the design of larger and larger projects, along with some spectacular failures of iron-bridges, came the necessity to apply the principles of physics to predict the action of forces on a structure. These principles then had to be represented visually. Not surprisingly, the need to preserve transparency asked for a method that allowed the graphical integration of structural drawings with internal/external forces, moments, stresses, etc. Early examples of how this was accomplished are design drawings that were produced in conjunction with the construction of the Panama Canal. The examples in Figure 10.3 show the quality of the design drawings of a lock gate for the Third Locks Project in 1943.

It is remarkable to note how much effort went into the details of the cutaway 3-D drawings showing the sunlight on the gate, drawn and rendered entirely by hand, before the sophisticated design software of today. The effectiveness of the visual display of the two options, miter and arch, is undeniable. The other two graphs combine the cross-sectional representation of the steel-structure of one gate, both

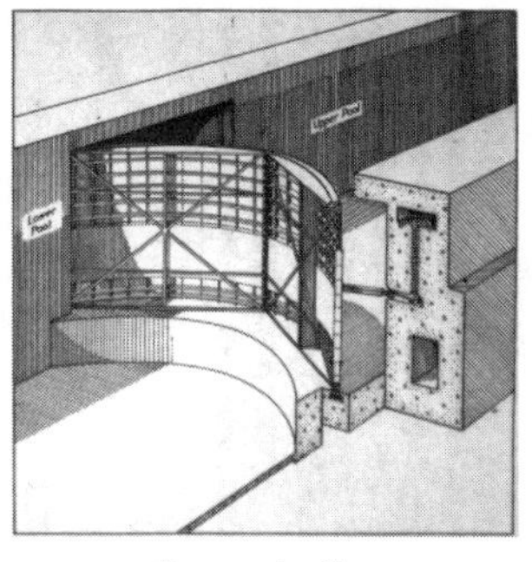

a) Cut-away 3-D designs for comparative evaluations

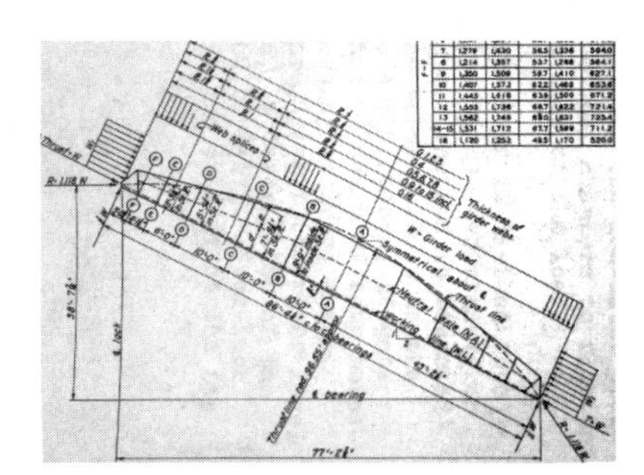

b) Graphical representation of load distribution

Figure 10.3 Design Options and Load Model for Arched Gate 1943. (Source: Civilian Records, National Archives at College Park, Records of Panama Canal RG 1851880-1979.)

horizontal and vertical, with the invisible distributed unit-loads that "lean" on the gate. We can recognize without difficulty the location of the horizontal girders, G1–G16, and the stiffeners supporting the double walled gate. Figure 10.3(b) shows the intricacy of the gate design, including the forces from the water, the reaction forces from the bearings and the second gate. Using vectors to represent the forces, we can identify the directions that were used to calculate the resulting stresses in the girders.

In the early 1970s, design professionals started to use computers to calculate forces and moments in large structures. FORTRAN (for Mathematical **For**mula **Tran**slating System), developed by IBM, was a popular programming language of the time used by engineers. It involved the use of punched cards, one for each line of code. The printed output consisted of long data columns, which could be drawn by hand. About the same time, the finite element method came into mainstream use; it depended heavily on the speed of calculations provided by mainframe and mini-computers.

The emergence of interactive graphics workstations and object-oriented computer languages created a new reality for the design engineer. With the capability

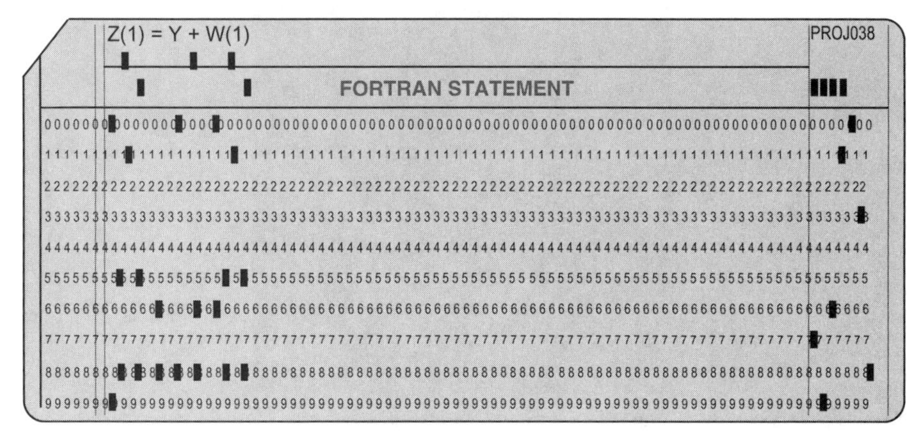

Figure 10.4 Single-Command Fortran Punch Card

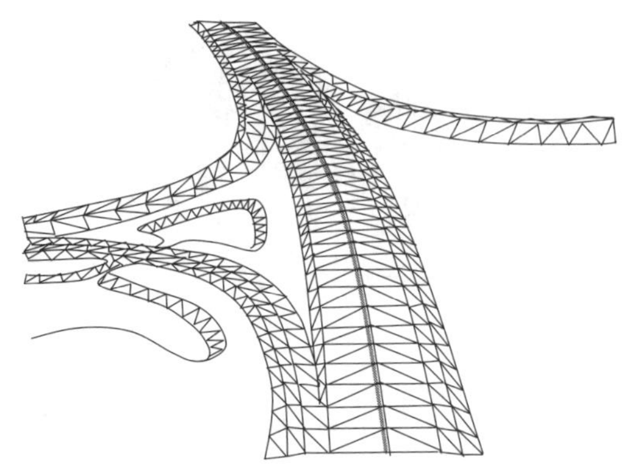

Figure 10.5 Wire-mesh CAD Design

to run multiuser/multitasking operating systems on minicomputers, they also could network to larger computers for engineering analysis and design visualization. Not surprisingly, the advances in graphic displays fostered the emergence of many computer-aided drafting (CAD) packages that offered both 2- and 3-D capabilities. Wire-frame modeling allowed the representation of objects via a mesh of edges, which can be made to appear solid by rendering the areas described by the mesh. Figure 10.5 shows a wire mesh representation of a highway connector, which approximates the form of each subbase and road base and applies color to represent different materials (not visible in the figure). The observant reader will notice that an important object is missing: namely, the support for the bridge, from left to right across the main road in the center of the Figure 10.5. It is, in fact, the capability to turn objects on and off that gives designers the opportunity to work through many what-if scenarios, and to simulate and rehearse the sequences for assembling large structures.

Solid-modeling software has become so sophisticated that users can view, by means of special goggles coupled with a head movement sensor or other interface methods, stereoscopic imaging displays (see Figure 10.6). By creating two slightly

Figure 10.6 Joystick Control Virtual Reality in Stereo

offset images of the same object, and displaying them on two small video screens, one for each eye, the viewer becomes immersed in a virtual world, a 3D space populated with CAD models, which simulate the drawings as if they already existed. To track the movement of the head, which indicates the direction the eyes are pointing, triangulating sensors "feed" the software to calculate appropriate views, which are immediately displayed on the video screens. In this way, interactive software packages permit users to "walk through" buildings composed of thousands of parts and millions of images.

10.1.2 Will It Work?

Construction has been compared to an assembly-line operation, where numerous pieces have to be joined together to make an object. But it is different from the assembly line in, say, the automotive industry where workers at fixed stations each add a different piece to a chassis as it moves down the line, until it is a complete car. In contrast, construction workers have to operate within an ever-changing environment and handle many different materials in various shapes, sizes, consistencies, and weights, with the help of a variety of different tools and equipment. As the object (i.e., a structure) is put together, spatial obstructions (e.g., steel columns and girders) are erected that eliminate the open space available at the beginning of the project. A serious problem may develop when two systems (e.g., mechanical and structural) need to use the same space—for example, during handling, transport, or construction, or in their final location.

To avoid such serious problems, many contractors spend large sums of money to build scaled models, to test whether there is any interference at the site, an activity referred to as *interference checking*. In the 1980s, wood and plastic scaled models began to be replaced by computational models in 2D and 3D, using computer graphics to display the spatial information of the designed structure or the movements of large equipment, such as cranes. Figure 10.7 exemplifies the problem of

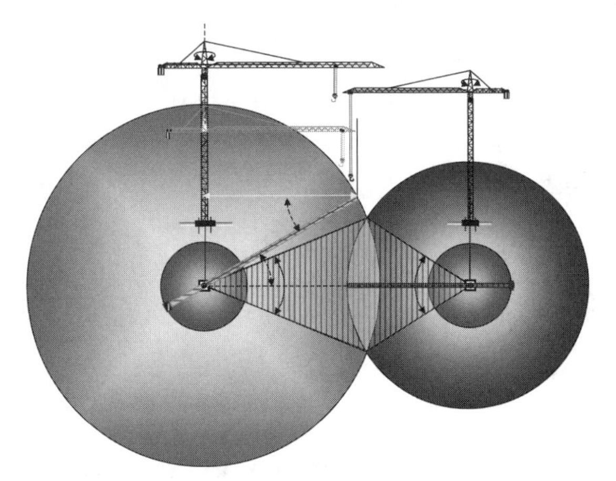

Figure 10.7 Computer Graphics to Interactively Check for Crane Interference

interference, and how computers can provide quick answers to what-if questions. Shown are two different-size tower cranes that have overlapping service areas, with the two jibs sharing the most space when aligned center to center at the two circles. The danger of a direct collision is mitigated because the jibs operate at different heights. Thus, a collision can occur only if the load line of the higher crane is positioned to interfere with the jib of the lower. The graph highlights the danger area in white, with lines crossing in two directions. The radial lines enclose the two angles that constitute the most critical zones. Moving one of the cranes may not be desirable, however, as the area covered by both allows it to be serviced by either one, or to "relay" a load from one crane to the other. Furthermore, moving the circles, the "cranes," may be prohibited by other obstacles, such as adjacent buildings or power lines. Interactive CAD software provides the means to move the center of one crane and immediately see the effect of the move.

As computer and software technologies continue to advance, the ability to render and visually simulate reality expands exponentially. What used to take a computer all night to produce can now be done in minutes, or even seconds. Figure 10.8 offers three examples of the exciting opportunities that designers and construction engineers now have to maximize their output.

Figure 10.8(a) depicts a simple cross-sectional view of a tunneling operation, which includes the configuration of soil layers, symbols for people and equipment positioned at pertinent locations, and informative labels. This graph was produced with the drafting tools on a word processor. Figure 10.8(b) is an example from an object-oriented solid modeling software program, in which each component is modeled as an object that can be individually manipulated—the size of the pipe can be changed, removed, or pushed down the track, together with the cart, for example; likewise, the height of the person can be easily changed. An automatic interference checker would alert the user in real time, if the person's hand touched the tunnel rib, for instance. Figure 10.8(c) is the output from a solid modeling software program, which allows the user to set chosen objects in motion. Here, the cutter head could be set to turn clockwise or counterclockwise; or the hydraulic pistons at the back of the shield could be retracted and pushed out. The use of 3D head displays and a head-motion monitor allow a user to virtually step into the CAD and look around.

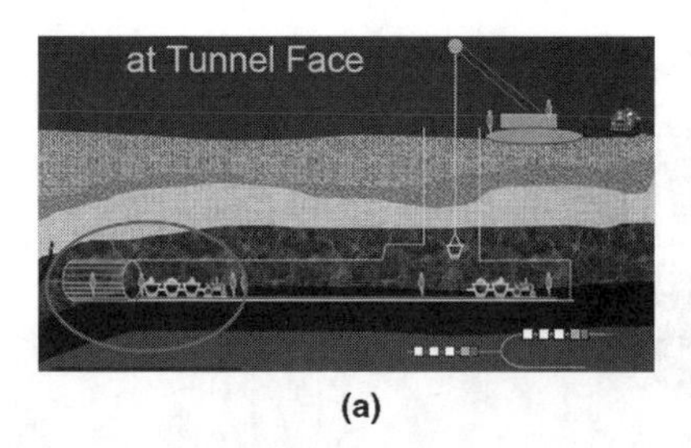

(a)

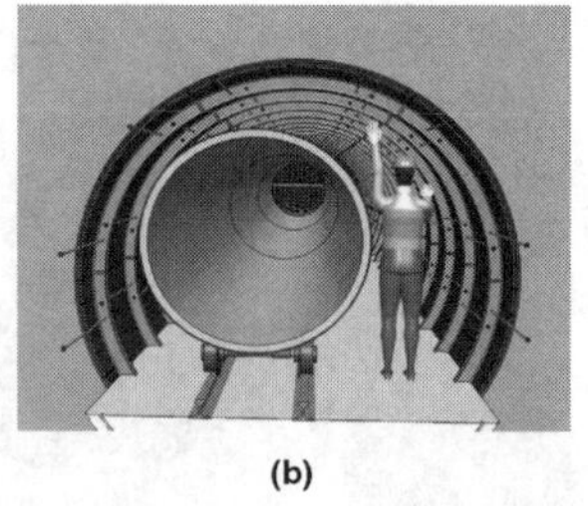

(b)

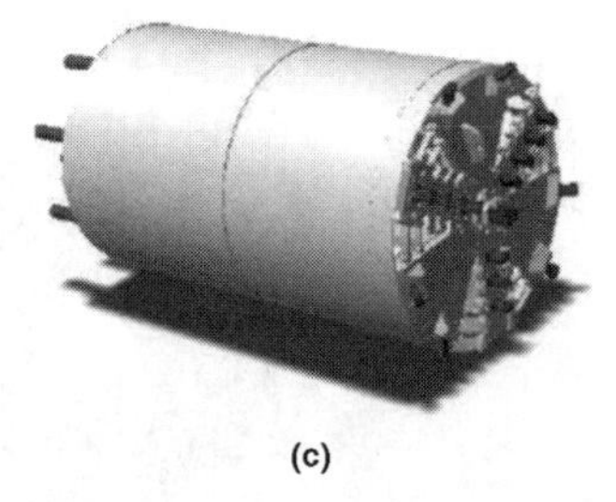

(c)

Figure 10.8 Visualization schemes for a Tunneling Project: (a) Conceptual Sketch with Lines and Generic Symbols; (b) Scaled 3D View of Tunnel with Pipe and Worker; (c) 3D View of Cutter Head from Afar

10.1.3 The Internet Revolution

During the 1990s, the Internet became the new communications medium over which to exchange digital information easily. Parallel developments were the rise of digital audio, video, and still-image cameras that had the capability to link to a computer where images could be downloaded and manipulated. Now, text, audio, pictures, videos, and CAD drawings could all be sent conveniently over the Internet. Initially, however, commercial connections were slow and the files were large. But driven by demand, fiberoptic cables and switches improved rapidly; and soon, innovative systems were developed that could transmit large numbers of digital documents and graphics files at high speeds.

Global use of the World Wide Web, which could run on multiple operating systems (Unix, VMS, Macintosh, NextStep, and Windows), followed by so-called Web-hosting platforms made other important services widely available, such as remote management of computer files, images, audio files, and electronic documents. Web hosts make files accessible to authorized personnel internally as well as externally. The concept of "hypertext," which made it possible to store and find information in any format, quickly, was combined with the idea of browsing and serving information over the Web. The HyperText Markup Language (HTML) was created to define the structure of Web "pages." Using HTML it became seamless to handle a wide range of multimedia documents. The structure of modern HTML was agreed upon at the first World Wide Web Conference, May 25, 1994, to include tables, graphics, and mathematical symbols. Referred to as *content management systems* (CMS) these independent services manage information without regard to the source or the required use, thus providing a uniform repository, or "warehouse," for all types of information using a uniform structure. Expensive redundancies and associated problems with information consistency were henceforth eliminated. Subsequently, browsers and servers all came to use HTML and so could communicate seamlessly across the Web, where files could be viewed, uploaded/downloaded, and, most importantly, shared.

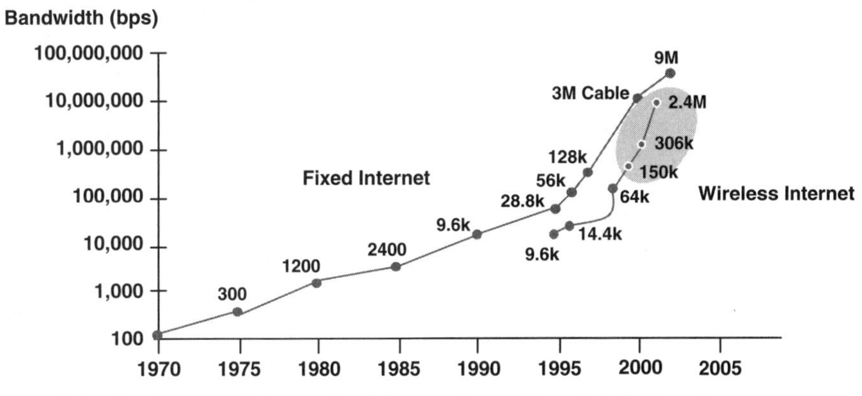

Figure 10.9 The Exponential Growth of Bits-per-Second Bandwidth Over Time

The client/server model, another important service to emerge, was based on Web servers (database, email); these models accepted requests from different client Web browsers, processed them, and returned responses to the clients. For example, instead of maintaining a CAD drawing on a personal or mainframe computer, it could be managed by a Web server, which would make it, or any part of it, accessible to anyone granted access. Finally, Web-based technology offered an easy and real-time concept for sharing of information among project teams, referred to as *project portals*. The client/server model gained popularity in the project management arena, and project portals served as central repositories for information and provided Internet connectivity across multiple platforms and team members, known as *virtual teams*, who could reside in locations all over world.

10.1.4 Costly Gaps in Electronic Communications

In 2004, FIATECH from Austin, Texas, published a report titled "Capital Projects Technology Roadmap" which expressed the thoughts of experts who had looked at a variety of key issues facing the industry. FIATECH was founded in 1999 as a consortium of companies interested in advancing breakthrough technologies in construction. The text box presents some key statements from three different sections in the report.

"One of the largest problems facing any capital project or capital facility operation is timely access to accurate and complete information. Companies spend millions of dollars each year searching for data that exists but is not readily accessible.

The capital facilities industry is not only fragmented and based on relatively transient projects, but it also suffers because its information technology systems offerings aren't well aligned with what people do all day, so that the interfaces between the systems used aren't well aligned with the interfaces between work processes.

The same information is typically recreated at great expense from project to project, and useful knowledge captured by the experience of key individuals is lost over time.

"(Element 9: Lifecycle Data Management & Information Integration, http://fiatech.org/lifecycle-data-management-information-integration.html)

"Tools for automated information authoring, transfer, and collection have propagated into design, engineering, and purchasing areas, but not into construction field functions.

The problem for the construction industry is that the information supplied and required at the field level has multiplied, but traditional manual processes are still in effect."

(Element 4: Intelligent & Automated Construction Job Site http://fiatech.org/construction.html)

"Some specific problems include:

- Poor access to accurate data, information, and knowledge in every phase and function.
- Tools for project planning and enterprise management are maturing, but an integrated and scalable solution that delivers all needed functionality for any kind of project is not available.

The missing link is the ability to integrate all functions of a facility's planning and management system and all required information in a unified project/facility management environment."

(Element 6: Real-time Project and Facility Management, Coordination and Control, http://fiatech.org/project-mgmt-controls.html)

The direction of the envisioned road map becomes immediately clear when we focus on the words repeated again and again: *information, data, and knowledge.* The map promotes the total integration of all project stakeholders under one project network umbrella.

At about the same time, the National Institute of Standards and Technology (NIST) published a study on the estimated cost of the lack of integration between design and management software in the construction of capital facilities. The study group estimated the costs associated with the resulting delays and added man-hours and interviewed a large number of experts from the industry, to pinpoint the reasons behind, and result of not interfacing the various systems.

Here are some statements that highlight their findings:

Many parties, each with expert knowledge in different disciplines, often operate in isolation and do not effectively communicate knowledge and information with teaming partners, both internally and externally.

Many projects require a significant number of requests for information questions and drawings between owners, architects, contractors, and subcontractors, often as many as 300 to 500 on a typical project.

Design, engineering, and operations systems are typically not integrated. In some firms, an estimated 40 percent of engineering time is dedicated to locating and validating information gathered from disparate systems.

Many firms use both automated and paper-based systems to manage data and information. In many cases, the hard-copy construction documents

(*continued*)

(*continued*)
are used on the jobsite in lieu of electronic copies. Electronic versions, therefore, often do not reflect facilities' as-built specifications.

Many smaller construction firms and some government agencies do not employ, or have limited use of, technology in managing their business processes and information.

Interoperability is the ability of systems, units, or forces to connect to each other in order to operate collaboratively and exchange information, services, etc.

General contractors as a whole had the lowest level of technology adoption of the interviewees. . . . most of their work is performed using paper copies of design and engineering files. Respondents said this is because the work environment at the construction site precludes widespread usage of computing technologies. General contractors also believe that the return on investment for construction equipment is greater than the return on information technology.

Interviewees stated that the majority of their work is done on paper and that paper is passed off to processing teams, including those for information requests, materials management and procurement, and inspection and certification. Respondents also reiterated the same coordination issues expressed by architects and engineers, in that collaborations require a significant amount of double entry into and among paper and electronic systems. The key areas in which opportunities exist to reduce interoperability costs are

- information request processing,
- document management
- project management,
- procurement, and
- facility planning and scheduling

NIST, 2004

The cost of the many interoperability issues among the participants of capital facilities projects in the United States in 2002 was estimated as follows:

Cost to architects and engineers	\$1.2 billion
Cost to general contractors	\$1.8 billion
Cost to specialty fabricators and suppliers	\$2.2 billion

It is easy to recognize that the vast majority of the cost accrues to the owner and operator, mainly in delays in operation start-up for the new facility. These delays then result in lost revenue for renting or production, plus added interest payments on construction loans.

In Chapter 3, section 3.1.3, we discussed the Building Information Modeling (BIM) approach to integrating design and construction, similar to an information supply chain. NIST's report on the high cost of interoperability provides an economical reason for the construction industry to embrace BIM as the basis from which to close the large gaps.

10.1.5 On the Communication Trail

Miscommunication between people is responsible for many deadly accidents in the workplace. Likewise, misunderstanding the meaning of a message between sender and recipient is responsible for many quarrels, and even physical fights, on the job. Finally, the lack of communication can lead to the errors on the job, job stress, and/or depression.

Communication is a complex process, fraught with built-in potential for unintentional errors, which may go unnoticed by either the sender or the recipient of a message. The following short discussion highlights some of the main processes involved in supporting human communication.

Figure 10.10 models a two-way communication path between two members of a construction company, Jane and George. As indicated, both Jane and George are sender and recipient of messages transmitted between the two. The dotted vertical line between them illustrates their physical separation, which is breached by arrows representing message transmittals from sender(s) to receiver(s). The communication between Jane and George is modeled as bidirectional, indicating

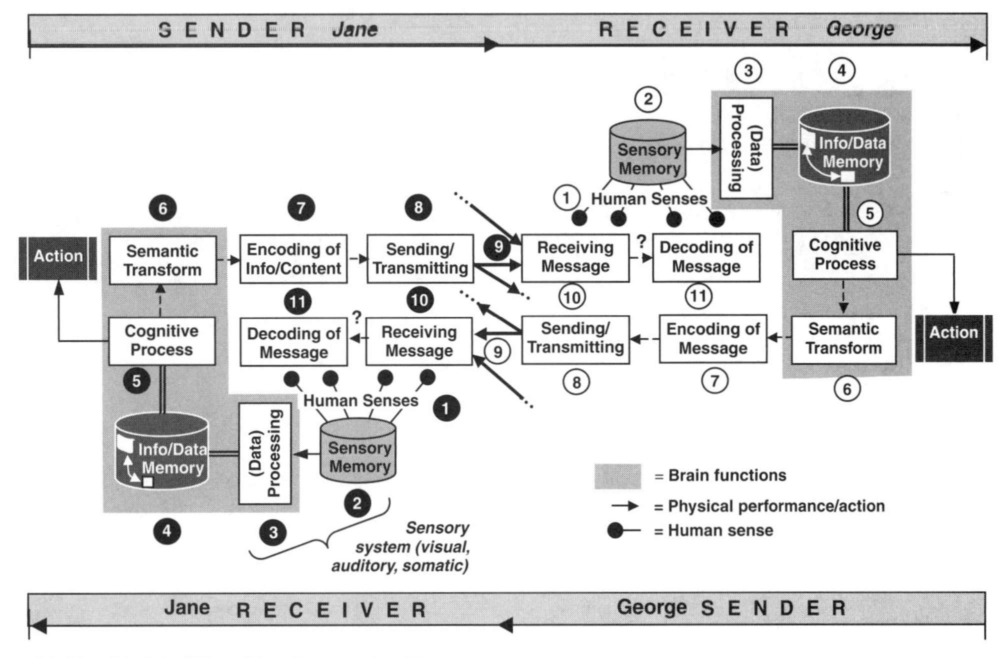

Figure 10.10 Model of Two-Way Communication

an exchange or question/answer-type interaction. In hierarchical organizations, this is not always the case, since higher-level units transmit directives or instructions to lower-level units. In such cases, the communication goes in only one direction and concludes with actions at the receiving end.

Let's assume that Jane is the safety manager and she is visiting a construction site managed by George. Jane is walking through the site to check whether everything is safe. She notes that two sets of slings are being used by the crane, one to move rebar and one to install wall forms. As she gets closer to the two slings used to fasten the long rebar to the hook, she thinks she spots a problem. From here, we'll begin the course of decision making and communication between her and George, shown in sequence in Table 10.1.

Table 10.1 Steps on the Communication Path

Step	Description	Jane's Actions	George's Actions
1	Observation by one of the five human senses. The photosensitive membrane of retina detects the light, dark, and color of written text, pictures, signs, and so on. The ear identifies changes in air pressure created by sound waves. The skin senses pressure; the nose detects odors, and the body feels vibrations.	Jane's eyes detect two yellow fasteners attached to a set of long, steel bars. She sees stitches sticking up and fibers cut.	While George is listening on the phone, he is also looking out the window from his desk, where he sees Jane standing at the rebar storage area holding two slings.
2	Analogue sensory inputs are compressed and stored for 20 to 50 milliseconds.		
3	Stored sensations are processed into nerve impulses and sent to the cortex of the brain.		
4	The cortex, a section of the human brain responsible for higher-level processing of nerve impulses, constructs the meaning of the sensory signals and stores it in short-term memory.		
5	The cognitive sections of the brain comprehend, learn, predict, analyze, and make decisions using information and knowledge from long- and short-term memory.	Jane realizes that she is looking at synthetic web slings. She remembers that they are covered under OSHA 1926.251, Rigging Equipment. Cut fibers and broken stitches indicate damage. Damaged slings need to be replaced not reused.	George realizes that Jane is holding the same slings that he found damaged two days ago. He told the superintendent to dispose of them and to buy new ones, as he did not have time to do it himself.

Table 10.1 (*Continued*)

Step	Description	Jane's Actions	George's Actions
6	The outcome of the cognitive process results in an action or a communicative response to the observation in step 1.	1. Prohibit laborers from using slings. 2. Talk to George about how to replace slings. 4. Plan training on rigging safety.	1. Thank Jane. 2. Tell her what happened before. 3. Call the superintendent about the new slings.
7	A response or message content is transformed into a desired format (e.g., letter, speech, drawing, email).	1. Oral communication 2. Call George on the phone and talk about replacement. 3. Request that George schedule a toolbox talk.	1. Oral communication by phone with Jane 2. Oral communication 3. New cellphone call
8	The response or message is encoded into a form that matches the selected format (e.g., writing with pen/pencil, using ASCII code for digital transmission, speaking, or using hand signals).	1. Verbal instruction 2. Two-way cellphone communication	1. Verbal appreciation 2. Verbal explanation 3. Verbal request
9	The message is transmitted (an addressed and stamped letter is deposited into a mailbox; the Send button is pushed to transmit email).	Oral: "Hi George. I just found two slings severely damaged. We need to replace them. Do you have another set?"	Oral: "I can't believe it. Thank you very much for finding this safety problem. As a matter of fact, just two days ago I had asked my superintendent . . . "
10	The transmitted message is accepted (the letter is opened; the email message is clicked on, etc.)		"What are you saying?" (Continue at step 1.)
11	If need be, the is decoded message (e.g., email transmitted in ASCII and transform into text or visuals to be viewed).		

Header Problem 10.1: Transformations on the Communication Path

Beth reads an email from Indiana Steel Inc. (ISI) stating that its delivery and erection process could start two weeks earlier than planned. ISI would like to know if its workers can start erection two weeks ahead of schedule. It is Beth's job to determine whether the anchor bolts will be finished and the shakeout area available.

(*continued*)

(*continued*)

Phase A: Confirming the Written Request

To: Beth Bertsch <JB@L&M.rr.com>

From: George Friday<Gfriday@Indisteel.com>

Subject: Moving schedule up

Date: May 15, 2009

Dear Beth

We are doing extremely well in our shop thanks to the Web-based project management system that speeds up our submittal and approval process. We could begin two weeks ahead of the present start date of August 31, 2009. Can you confirm that this is possible by ensuring that a laydown area would be available and the anchor bolts for column rows A through C inspected?

Thanks.

George

Phase B: Development of Response

This phases involves Beth making a decision about what to do in response to the message from George, by assessing its meaning relative to the relevant circumstances, recovering stored heuristic rules that could be used, and using logical reasoning or simple guessing (trial and error).

Beth's thoughts

"This would be so great. Walt Steiger from JC Footing Inc. told me yesterday that they are on schedule, too. I need to call him. Oh no! I am not going to talk to him; he hates me and will treat me like one of his laborers. What should I do so he will collaborate rather than throw his arms up as if to sacrifice his own son? I need to find a win-win solution, because steel is on the critical path. John would surely recognize my work, like last time. What should I do?"

It is apparent from "reading Beth's mind" that she has to solve several problems, even as she is experiencing real or imaginary barriers to effective communication.

Main problem: Getting Walt to prepare the site so that George can start steel erection two weeks ahead of schedule.

Barriers to Communication

Emotional: Beth expects Walt to push her around and talk down to her. This could make her angry if she talks to him on the phone. She knows that she would become defensive and accuse him of being a sexist, something he would deny.

Misrepresentation: Beth expects Walt to misrepresent the situation in order to declare that it would cause him a major problem. He would do this despite yesterday's declaration to her that they were on time. He will try go get a change order and, with it, extra money.

Beth realizes that she could use writing as a form of communication, which would eliminate the emotional barrier. She knows she writes well, whereas Walt has a problem with spelling. He might call, though! The potential for misrepresenting the actual situation and problem still exists, as long as Walt is not willing to collaborate. How can Beth find something that he wants but would not cost her money?

Option 1: Call a brainstorming session with Fred Thomson (the head of planning) and Marge Baer (head of quality control).

Option 2: Meeting with John Price (her boss) to discuss the opportunity.

Option 3: Find her own solution by using creative/lateral thinking, one of her strengths.

Assessing the costs and benefits in her head, Beth decides that option 3 has the greatest utility for her. What does Walt want and what would not cost her any money? She starts to draw a cause/effect mind-map on the piece of paper in front of her, starting with the answer to the question (see Figure 10.11).

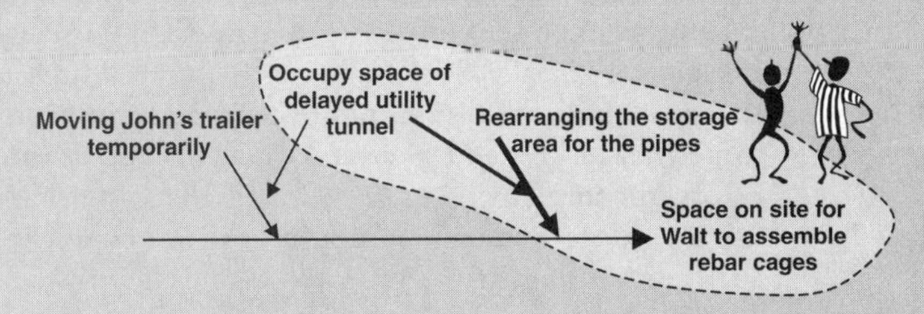

Figure 10.11 Beth's Mind-Map

Beth realizes that she has the winning solution, one that takes advantage of the delay in the utility tunnel, which runs along the pipe storage area. All the new deliveries could be stored there, to make room in the front for Walt to start a preassembly area. She plans to confirm this possibility with Fred and make sure that he has not given the space to somebody else.

1. What technology should Jane use to transform her plan into information she can communicate to Walt in such a way that he does not misunderstand?
2. Draw a decision tree (cause/effect diagram) showing possible reactions from Walt, followed by possible responses from Beth. Devise a zero-sum game, as well as a win-win/no-zero-sum game.

(To be continued)

10.2 COMMUNICATION STRATEGIES EMPLOYED BY ORGANIZATIONS

Construction companies are organizations or production systems that depend on managers and workers to achieve their goals. Achieving a strategic goal with a specialized workforce requires organizing people into departments and forming an information network that lets people know what to do and how to carry out the mission of the company. This top-down bureaucratic method of disseminating information is the traditional view of organizational communication, a view that, more recently, has lost its importance due primarily to the availability of highly efficient modern tools and technologies.

10.2.1 Main Functions of Communication

The management of projects has always depended on some lateral communication to solve difficult problems, communication that involves more than one participating entity. The complexities of today's projects typically require that designers, engineers, construction managers, and owners all participate, along with lower-level craftspeople.

Concepts of Communication

1. Present professional material (traditional)
2. Transmit messages (traditional)
3. Socialize (internally and externally)
4. Manage information
5. Formulate knowledge (teach, train)
6. Discussion (debate, negotiation)

The role and meaning of communication went through major transitions in the twentieth century. As the sidebar shows, at least six concepts have been identified and studied by sociologists, behavioral and computer scientists, and many others. The engineering profession treats communication as the means to present professional material (1), usually in writing and/or verbally. Sociologists and, most recently, computer scientists modeled communication as a transmission process, using transmitters and receivers, which can, however, produce errors in communicating (2). Now it has also been generally acknowledged that communication plays a key role in creating social networks of trusting people within their own and external organizations (3). Managing information (4), reemphasizes the fact that communication and information sharing are not the same. Communication is more than just a transmission process; information is associated with an extensive set of related issues (e.g., processing, storing, securing, retrieving). Knowledge management, teaching, and training (5) comprise another area that has gained importance of late, and is being studied. Finally, discussing and negotiating (6) are critical traits of teamwork that also are necessary for consensus building within an organization.

10.2.2 Communication as the Enabler of Managerial Functions

The overall objective of management in a for-profit organization is to create value for its shareholders. As businesses grow from "mom-and-pop" operations to larger enterprises, management is distributed among specialized individuals or teams, which belong either to the *top management* (strategic), *middle management* (support), or *low-level management* (operational). Over the years, some general functions that are part of every management level have been defined. The list of generally accepted management functions includes:

- Setting goals
- Planning actions, to achieve goals
- Acquiring and coordinating the resources needed to carry out the actions
- Leading and motivating the personnel who will execute the actions
- Monitoring and controlling progress against plans, which may need revision

Two other threads running through all management levels and all functions are the key management tasks of *problem solving* and *decision making,* both of which depend heavily on information of different kind. Strategic decisions require forecasts about the future of the economy, political environment, and so on. Problem solving, done in support of the operation, may need data about, for instance, expected maximum flood level, selection of a foreign supplier based on currency expectations, or the expected production rate of the concrete crew during December. Operational decisions have to be made on an hour-to-hour or even moment-to-moment basis, decisions that require real-time information, such as expected wind speed, concrete strengths, employee shortages, and others.

As we learned earlier, information management is concerned with processing and storing information so that it is available when needed; this involves channels and many additional efforts that rely on other aspects of communication. To understand this better, we'll model the functions of project or construction management from the perspective of success as measured by contributing to the strategic goals. In a second step, we will add the information needs and required communication practices.

Figure 10.12 organizes the managerial activities as a sequence from planning to controlling, by subject area. Clearly indicated is that managerial processes and

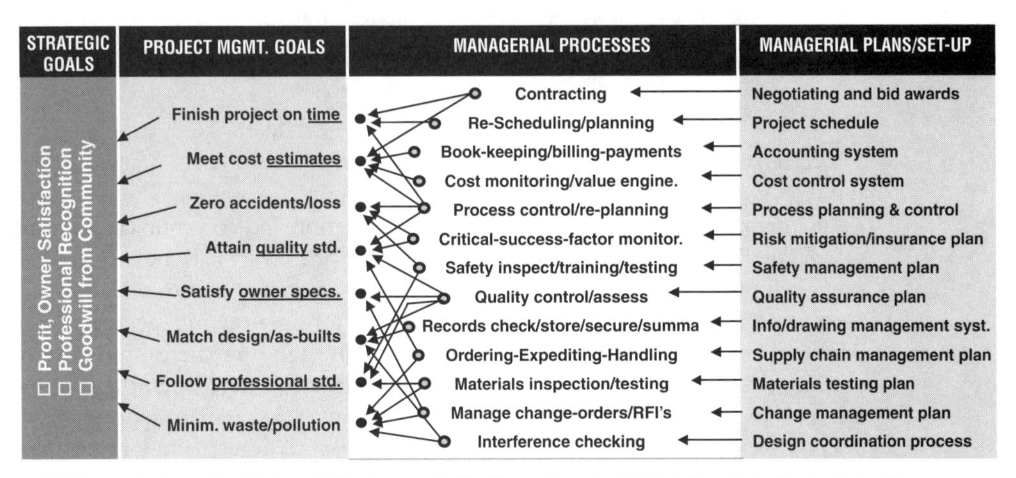

Figure 10.12 Relationship Model of Goals and Activities of Project/Construction Management

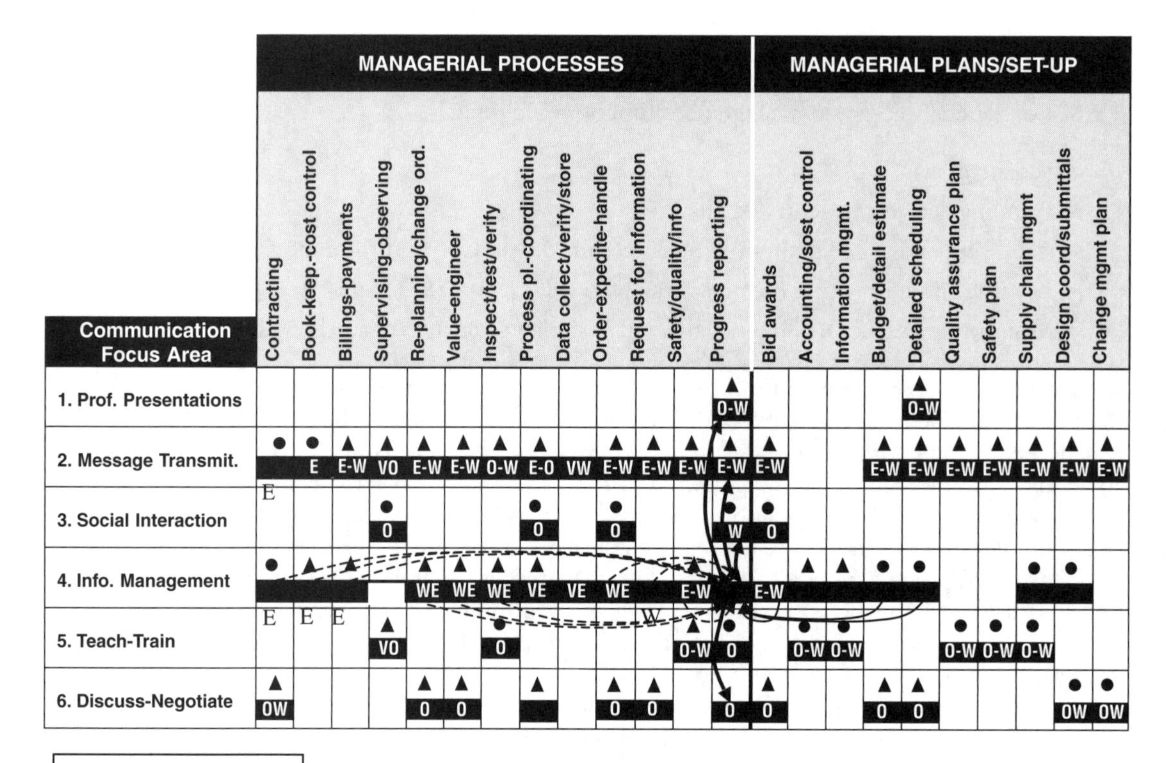

▲ = Plays crucial role
● = Plays significant role
→ = Supply of plan data
- → = Supply of actual data

O = Oral communication face-to-face, person-to-person, video/teleconference,
V = Visual info gathering, reading-observing-video-photo-drawings-
W = Written documents-reports-letters-receipts-shipping lists
E = Electr./computer data entry/processing/storing/securing-CAD-WP- email-simu-PPoint
OW = Oral communication resulting in written document
VW = Visual observations/inspection resulting in written document
WE = Data entry from written documents with subsequent analysis/storage
VE = Visual observations followed by modeling and computer simulation
VO = Visual observations and oral instruction/commands

Figure 10.13 Relating Common Management with Traditional Communication Activities

project goals are not only precisely aligned but that one process may contribute to multiple goals, and one goal may depend on multiple processes. The complexity of these interrelations is overlaid by a web of information and communication activities.

Figure 10.13 highlights how some of these procedures are networked to support the project management functions. It is apparent that communication areas 4. Information Management (data entry, verification, processing, security, storage, printing, drawing, etc.), and 2, Message Transmittals, are the dominant areas of communication. Oral communications carried out when operating in area 6, Discuss and Negotiate, is associated with many managerial functions. With the purpose of showing that most of these areas are tightly intertwined, the managerial task of presenting progress reports is utilized. The center pin of reporting is the capability provided by area 4, the information management system, as it holds the data, provides access, and offers the tools to organize it into written and/or multimedia documents. In addition, the oral presentation and discussion of project status and forecast with the owner and/or architect is linked to focus areas 1, Professional Presentations, and 6, Discussion and Negotiations. Both are crucial for the project team, as its skillful execution will create goodwill and encourage cooperation from those who "pay the bills." As shown in Figure 10.13, the importance of this aspect of managing a project is underlined by marking communication area *3*, as Social Activity, significant in submitting a report.

10.2.3 Strengths and Weaknesses of Various Communication Media

As we observed, the four main means by which human beings communicate in organizations are: (1) written, (2) images, (3) verbal, and (4) nonverbal. Each of them can be exchanged along many different channels, and each with its unique strengths and weaknesses. One commonly used approach to distinguish between the various communication media is the number of channels that can be used in parallel. A face-to-face (FTF) meeting is referred to as *rich,* whereas a typed letter is an example of *lean*. Other aspects of communications media, besides the visual and audio aspect of "face time," include simultaneity, the social atmosphere, and the relative permanence of the communication (e.g., written records versus electronic messages). The next sections provide short descriptions of the most common media.

Face-to-Face (FTF) Discussion

This communication medium is rich in that it offers the "human touch," needed to build trust, provide an assessment of a person's knowledge, and clarify the common goal that brings and keeps a group together. Nonverbal messages, for example, can provide immediate feedback to a speaker in the form of wide-open eyes or blank stares, indicating that something he or she said was not understood. Shaking one's head may serve to indicate silent disagreement with a statement.

Similarly, a leader who is "tuned in" will be able to pick up such "statements" in a meeting, and open a path for clarification or discussion. Trust and confidence are extremely important when the task of a meeting is to brainstorm to solve a complex problem, as it requires everyone to share their ideas and knowledge. Trust is obviously also helpful in negotiations, where all parties need to compromise for the good of the whole. This will be more difficult to accomplish when there is conflict between certain of the parties, or when the personal camaraderie of the group sidetracks it from its professional goal.

Person-to-Person (PTP) Phone Conversation

Conversing with a supplier or contractor precludes visual clues, and so can be difficult in situations where the two parties have not met in person previously, to establish a rapport. This is especially true when a language barrier also exists. A person-to-person phone call, although not as rich a format as videoconferencing (described below), is very effective when the parties already know one another and speak a common language, yet still need or prefer being able to hear voice intonation or require an immediate response.

Teleconferencing (Voiceconferencing)

This simple-to-use medium enables people at different locations to meet around a virtual table and converse with each other in real time. However, being able to only hear the voices of the others can cause problems when the distances between the locations are significant and when all the parties do not know one another, or well, and cannot recognize voices, and so cannot tell who is speaking, sometimes leading to embarrassing exchanges. Except in cases where—as in a webinar—one person is presenting material to the rest of the group, teleconferencing is only recommended between people who have met before and can match phone voices with the appropriate individuals.

Videoconferencing

The use of videoconferencing requires balancing the effort that goes into its preparation against its benefits. If the quality of the video is bad, or the sound and lighting are not well handled, videoconferencing can distract from the goal of the conference and may even diminish its results. It has been found that this form of communication is most beneficial for bringing together a large group who have never met before. Table tags can be used to identify the participants, and overhead cameras can focus on drawings and other materials referred to by the presenter.

Electronic/Instant Messaging (IM)

Akin to a teleconference, this near-real-time Internet-based communication medium connects two or more people who are online. It is sometimes referred to as online chatting. IM users are not compelled to take or reply immediately to an incoming "call," but typically the communication happens quickly, with the sender getting an acknowledgment or reply very quickly—hence the term "instant." Messages may be accompanied by photos or video sent via Web cameras. IMs are logged, meaning a history of exchanges can be kept, and URLs or document snippets can be pasted into subsequent messages.

Electronic Mail (Email)

Almost 80 percent of a manager's communications are done via email, whether in the form of long letters, brief sentences, or one or two words in response to a simple question or a new idea. Email, an asynchronous medium, is very convenient and effective, and can be sent sitting at a desk in front of a computer or walking in a field carrying a handheld device such as a BlackBerry or cellphone. It is possible to write/type a new thought spontaneously and send it to as many people as desired, when in range of a network.

Email is an excellent substitute for a PTP phone conversation when a language barrier exists, as online translators are available to translate unknown words. Users also can check what the receiver wrote exactly two days before. The response can be edited after the sender sleeps on it overnight, to sharpen its meaning.

But email lacks "warmth," and it is difficult to interpret a correspondent's true feelings or intent, sometimes causing misunderstandings. In an attempt to address that problem, so-called emoticons have emerged, symbols of emotional cues. Email is also prey to viruses, worms (on Windows-based systems), spam, and other unsolicited and unwelcome mail from unknown sources.

10.2.4 The Effect of Distance on Media Selection

The number of people attending a company-sponsored conference meeting may vary from two to hundreds. While the more informal FTF meeting works well for small groups, bring together a large number of active participants is a logistical challenge, one that requires the careful selection of an effective communication medium. Figure 10.14 takes a matrix approach to assess the various options according to distance, group size, and immediacy.

This simplified matrix is intended to provide general guidelines, with the understanding that many different scenarios are possible. For example, it is not uncommon for two people who work in the same office to prefer to discuss an issue using email; first, because it is more convenient (they can respond during spare time); and, second, it is more effective, in that they can take time before responding to contemplate the issue at hand. For other reasons, too, it is not always desirable, nor is it necessary or even always possible, to meet face to face in the same room—especially as the distance between the parties grows larger.

Asynchronous communication offers many benefits that are especially attractive to time-starved executives, which is the reason that the many Web-based tools now available are of great interest to the management community. Still, the value of the social aspects of personal interactions to the establishment of trust cannot be overstated, and are essential to using virtual meeting tools successfully.

10.2.5 Matching Media to Meeting Objectives

Brainstorming an innovative engineering solution to a complex construction problem requires a different meeting strategy than, say, negotiating a deal with a supplier or subcontractor. It follows then that selecting the communication medium for a meeting depends not only on distance and number of attendees but also on the primary objective(s) of the meeting.

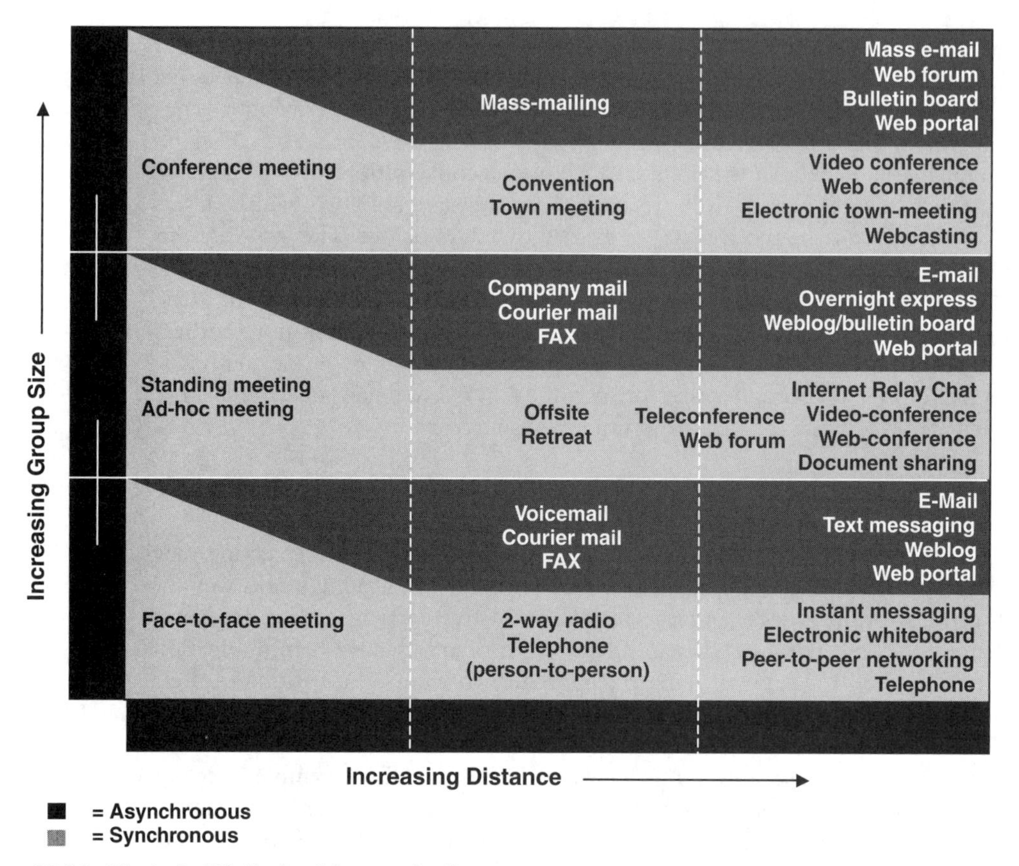

Figure 10.14 Clustering Methods of Communication

To study this issue, we distinguish four different meeting goals:

1. *Brainstorming to find solution:* The group applies creativity tools to search for innovative ideas.
2. *Choosing an action strategy:* The group discusses pros and cons, weighs and ranks options.
3. *Developing an action plan:* The group agrees on subtasks, task sequence, and duration probabilities.
4. *Negotiating:* The group discusses how a deal measures up against a set of weighted selection criteria and past performances of the parties involved. The deeper the trust, the better the deal.

Whether a meeting takes place in real time will have a major influence on the selection of the appropriate communication medium. Figure 10.15 presents a

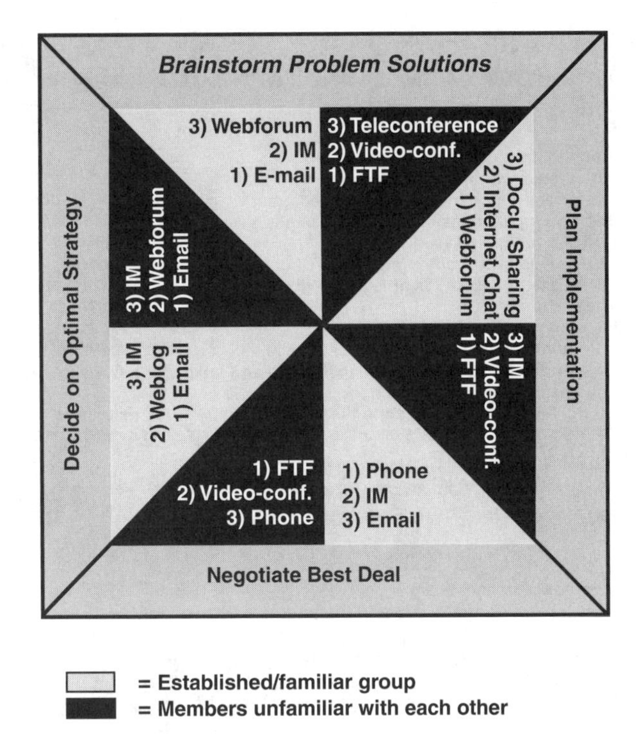

Figure 10.15 Communication Media Tailored to Key Objectives

graph that clusters the various IT technologies according to the four main meeting objectives and another relevant variable—personal relationships. Recall from earlier that trust between team members, which is imperative for teamwork, cannot be built using lean communication media. Conversely, too much media richness can become a hindrance; and large group size will diminish the contributions of individuals.

Much has been written about each of the different meeting purposes, although brainstorming has received by far most of the attention. There are a variety of do's and don'ts for these meeting types, and even more variations to the basic concepts are being promoted by managerial consultants.

With the exception of decision making, the best medium for teams who are not familiar with each other is the face-to--face (FTF); second is videoconferencing. And, again, email can be an excellent tool for those situations that do not require real-time message exchanges.

Earlier chapters described the differences in the way people learn—how their personalities, needs, fears, and emotional

Brainstorming

Brainstorming is a powerful technique for creative problem solving, out-of-the-box thinking, innovative planning, and team building. Brainstorming forces lateral cross-disciplinary views, motivates, and builds team spirit, getting participants working together. Brainstorming is not a random activity; it follows a process.

Dos, Don'ts and Death Nails to Running FTF Meetings

Preparation as the Leader

Dos:

- Write down the questions that you can't answer yourself.
- Decide on the communication medium (FTF, web conference....)
- Prepare the activities that best match the questions needing to be addressed (brainstorming, discussion, work in groups,..)
- Include some cutting edge educational stuff.
- Design an agenda that involves everybody right away. Check the titles and content with one or two key people who will attend.
- Organize appropriate refreshments.
- Have visual aids and learning aids on hand.
- Distribute the agenda so participants can prepare in advance.
- Have a bar chart that shows the approximate time allotments and space for new items. Reserve time at the end for summary and assignments of tasks.
- Assign a recorder who will take minutes and send them out promptly (even real time.)

Items in the Minutes:

1. Attendance
2. Key questions, problems addressed
3. Main decisions made
4. Assignments of follow-up tasks
5. Next meeting

Don'ts:

- Don't call meetings that are unnecessary
- Don't call a meeting when some important members have to be late or can't attend at all
- Don't call a meeting with no time to prepare

Preparation as a Participant

Dos:

- Understand why your presence is necessary.
- Think about the role that you are going to play and the path that you will follow.
- Plan so you can be on time.

Don'ts:

- Don't attend meetings without preparing yourself
- Don't be late and without the material needed to work

Opening of Meeting

Dos:

- Start on time in respect to those who arrived on time.
- Take time to tell and hear stories.
- Introduce members that are new or guests. Let them present themselves if you have given them advanced notice.
- Welcome and thank attendees. Emphasize that you value their ideas, opinions and questions.
- Paraphrase the agenda make certain that everybody understands and agrees with the core questions that gave rise to the meeting. You need to be ready to change them.
- Go through the different activities that you had planned and explain the role(s) each participant is playing. Verify that the procedures are understood by everybody.
- Start late to accommodate stragglers.

Don'ts:

- "I called you together so we can talk a little about what to do about our department"
- "I thought we needed to have a meeting since we haven't have one for a long time"

Conduct of Meeting:

Dos:

- Start with a smile and keep in mind "*Look who is coming to the meeting*!"
- Stay on track but leave the time to look at all aspects of the problem.
- Let the people carry the content; you guide the process and ask the right questions.
- Test from time to time how the participants feel about the progress by asking.
- Acknowledge and reinforce constructive contributions.
- Recognize conflict early and take counteractive measures.
- Have outcomes shared if more than one sub-group is working in parallel.
- Show the progress on the barc hart (cross out what is completed).
- Close meeting on time but not before summarizing what was accomplished and assigning followup tasks, etc. If needed, agree on time for next meeting.

Figure 10.16 Do's and Don'ts for Running FTF Meetings

Don'ts:

- Constantly interrupt the work of the groups or a presentation
- Dominate the discussion
- Get personal
- Embarrass yourself
- Force people to participate
- Let the meeting be taken over by somebody else

Sure Paths to Disaster:

1. Always show people that you know more than they.
2. Use meetings only to let the participants know what you think.
3. Usurp every good idea without giving credit to the originator.
4. Ridicule every idea that does not fit your pre-conceptions.
5. Ensure that highly qualified people are not invited to the meetings.
6. Choose a meeting room that is small and without temp. controls.
7. Don't waste money on refreshments.
8. Don't speak personally to the participants before and after.
9. Ask for a 200-page follow-up document from every member.
10. Call lots of meetinas.

Figure 10.16 ***(Continued)***

intelligence, affect the learning process. People attending a meeting bring with them all of those individual characteristics, augmented by their expectations of the meeting. Over the years, psychologists have characterized the behaviors of some of the most common personality types attending official meetings. Figure 10.17 summarizes some of the most easily recognizable characters.

10.2.6 Local Communication Patterns

In 2006 and 2007, the lead author of this book conducted three surveys in three countries: the United States, South Korea, and Finland for the purpose of investigating local preferences in the way construction managers and engineers communicate. Figure 10.18 gives the comparative results in percentage of the respondents who use a particular channel.

It is striking to note the heavy dependence on the cellphone, and an equally prevalent use of email; followed by, to a slightly lesser degree, the land-based telephone. Very interesting are also the recognizable differences in the use of express mail services and face-to-face conversations outside meetings. Construction managers in Finland prefer to talk things over, face to face, whereas their colleagues in South Korea do this 40 percent less often. The other big gap is between the use of express mail, which is popular in the United States virtually unused in Finland.

Though this three-country survey represents a small sampling, it does serve to highlight that the choice of communication channels is as much a matter of technology as it is of cultural background and legal structure.

A The **Monopolizer** talks on and on as though their ideas or beliefs are inherently more important than those of others. When the leader allows someone to monopolize a meeting, it sends the message that rudeness is sanctioned. Someone needs to indicate an interest in hearing from others in the meeting, to remind the Monopolizer that others can speak as well as listen.

B The **Space Cadet** seizes the discussion by twisting the topic to a remotely related matter. One minute you're on topic and the next you're in "left field." The leader's ability to recognize the tangent and to refocus is essential to a productive meeting. "Let's remember to confine ourselves to the topic at hand" is a good way to get back on track. As well, one can "facilitate" such "space items" by noting them on an "other issues" list to be addressed later.

C A **Devil's Advocate** is in most business meetings. This person seems to relish taking the opposite tack, in taking an opposing view. Often this employee begins by saying "just for the sake of argument—I believe the opposite is true." While there is value in looking at issues from multiple points of view and to avoid group think, the Devil's Advocate makes a sport out of it. A good leader will praise the person's contribution to have the group look at all the issues but redirect the discourse "Let's keep this thought in mind when we move on to…."

D The **No Way** person, a cousin of the devil's advocate, is able to take the air out of any tire that is starting to get traction. "It won't float." "We tried it once and it was a failure." They don't believe that there is a better way of doing things. The best way is to re-program their destructive attitude into that of its "cousin." "Suppose for a minute that…." "What would it take to…?" This will force the person to use arguments that can be dealt with.

E **I Just Don't Know** is unable to make decisions. Despite being in a deliberative body, they are conflicted by multiple arguments, and can't evaluate all the options too afraid of being wrong or to offend somebody important. They are embraced by the Devils' Advocate and the No Way members since they are able to block progress. The leader needs to reassure the wary I Don't.

F The **Bombers** have matches to ignite the dynamite. They are able to push "hot buttons" that are sure to strike a nerve, to provoke the emotional reaction that moves the group into a quagmire. The leader who sees the firing cord burning can put it out by affirming "let's not go there." Other phrases like "let's cross that bridge when we get there" or "that's a hornet's nest we don't need to disturb" are also "fire extinguishers" that stop bombers from blowing up a business meeting.

G There's likely a **Pleasing Beaver** in a business meetings. This employee is incessantly carrying sticks for the boss in order to please whoever is in power. They are able subvert themselves to whatever the boss says. They are seen as "willing slaves" and are not trusted. The meeting leader should elicit their ideas and preferences before asking others as a way of drawing them out.

H The **Fighter** uses a confrontational style to object to others' ideas with vigor. Without regard to hurting others' feelings, the Fighters don't even realize when they're attacking on a personal level, the meeting leader needs to intervene sharply since attacking a person is poisonous and unacceptable.

I And there will be the **Clown(s)**. There is a time and place for joking. While we all like a good laugh, constant joking disrupts a meeting and distracts attention from where it should be. Their constant infusion of humor can belittle others' motions and make it difficult for some to be taken seriously. A business meeting leader can designate several minutes at the start or middle of a business meeting specifically for humor. When it crops up elsewhere and is deemed disruptive, the leader can remind people that the time for humor is passed or forthcoming, so as to control

Figure 10.17 What Personalities Are at the Meeting?

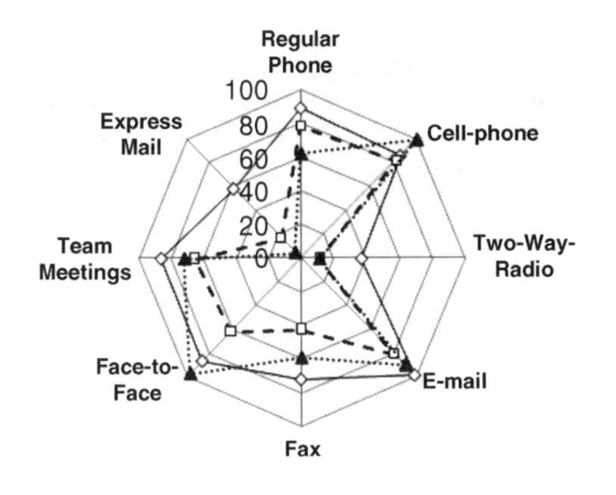

a) Channels Used to Request Information

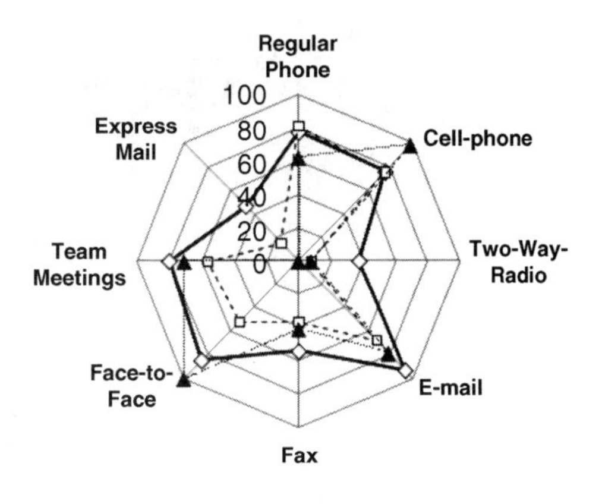

b) Channels Used to Distribute Information

Figure 10.18 Three-Country Comparison of Communication Channel Preferences in the Construction Industry

Worked-Out Problem 10.1: Patterns of Communication

Provide four construction examples for the two most common patterns of communication:

- One-to-one (Table 10.2)
- One-to-many (Table 10.3)

Assumptions

The two most basic types of communication are:

- Request for information
- Automatic messaging

Definition of Terms

Data: A single fact, statistic, code, word, or picture (e.g., address, number, drawing).

Message: A statement containing data, news, advice, request, or the like, initiated by a sender.

Information: A message with meaning for receiver. Data that has been processed/interpreted.

(continued)

(*continued*)

Table 10.2 One-to-One Communication

Source/ Destination	Request	Response	Automatic Message
Crane—truck with steel	Number of >.5 and >1.0 ton steel pieces	5 >.5 ton; 8 >1.0 ton	Tag number and weight of piece picked up
Superintendent: Steel/truck	List of all steel pieces loaded	List of tag numbers, weights	Unloading location upon arrival
Superintendent: steel/ structural engineer	Clarify orientation of steel piece with tag number	Steel piece with tag number connection details; e.g., A.34.87 (1,2)	Daily progress
Foreman: steel/ safety manager	Request test material for new hire	Safety material and test booklet	Name of new hire passing safety test

Knowledge: An organized set of information and rules about a topic that can be learned.

Wisdom: Human capacity to use knowledge and facts with insight, based on long years of experience.

Table 10.3 One-to-Many Communication

Source/Destination	Request	Response	Automatic Message
Superintendent steel: Not all steel pieces installed	Update location	Steel 1 is at coordinate x, y, Steel 2 is at coordinate x, z	Time stamp when entering construction site
Construction manager: Equipment on diesel	Provide diesel level	Equipment 1: 5.2 gal, Equipment 3: 8.2 gal	Snowfall expected in next 20 minutes
Equipment rental: All equipment on-site	Provide hours operated today	Equipment 2: 8.2 hours Equipment 3: 6.5 hours	Send security code after linking to GPS security system
Owner: Design team	Availability for meeting on August 1	Designer 1: all day, Designer 2: all morning	Submit hours spent on project during the month of June

Traditionally, information flow has been modeled along organizational hierarchies as top-down or bottom-up, to mirror levels of authority and responsibility. Recent developments in information logistics suggests, however, that those channels have become too long and cumbersome, especially since electronic networks now offer tools to securely access, protect, and authenticate information. Also note that the integrated communication model in Figure 10.15 does not contain arrows, nor does it reflect the many different forms used to send messages. In many ways, the figure represents an ideal world of data and information sharing among all the possible sources and destinations.

In the next section we delve into the reality of construction and its relationship to information logistics and communication.

10.3 LOGISTICS OF PROJECT INFORMATION

Originally, logistics functions were defined in terms relating to a military's need to resupply its troops when away from its home base. As military history shows, mastering supply logistics often meant the difference between the success or failure of entire armies. A more business-oriented definition of logistics is provided by the Council of Supply Chain Management Professionals (CSCMP):

> Logistics management . . . plans, implements, and controls the efficient, effective forward and reverse flow and storage of goods, services, and related information between the point of origin and the point of consumption in order to meet customers' requirements (http://cscmp.org/Resources/Terms.asp, 2007).

From this definition, information logistics is concerned with the flow and storage of information between the source and the user. While it does not include its creation, the information supplied should meet the users' needs, fit the selected communication media, and users' preferences. Thus, content, format, and timeliness are of importance.

Similar to manufacturing, resources and information logistics are closely related. Knowing the status of an engineering design, location of a pipe-valve, or the exact length of a precast panel can be critical in avoiding costly mistakes. As was discussed earlier, construction focuses increasingly on creating networks that enable authorized personnel to track materials, while the maintenance shop updates records on each tool after it has received scheduled maintenance.

The next section discusses the concept of information logistics for a basic construction project involving subcontractors, suppliers, and design services. As a first step, we will look at the interorganizational data flow.

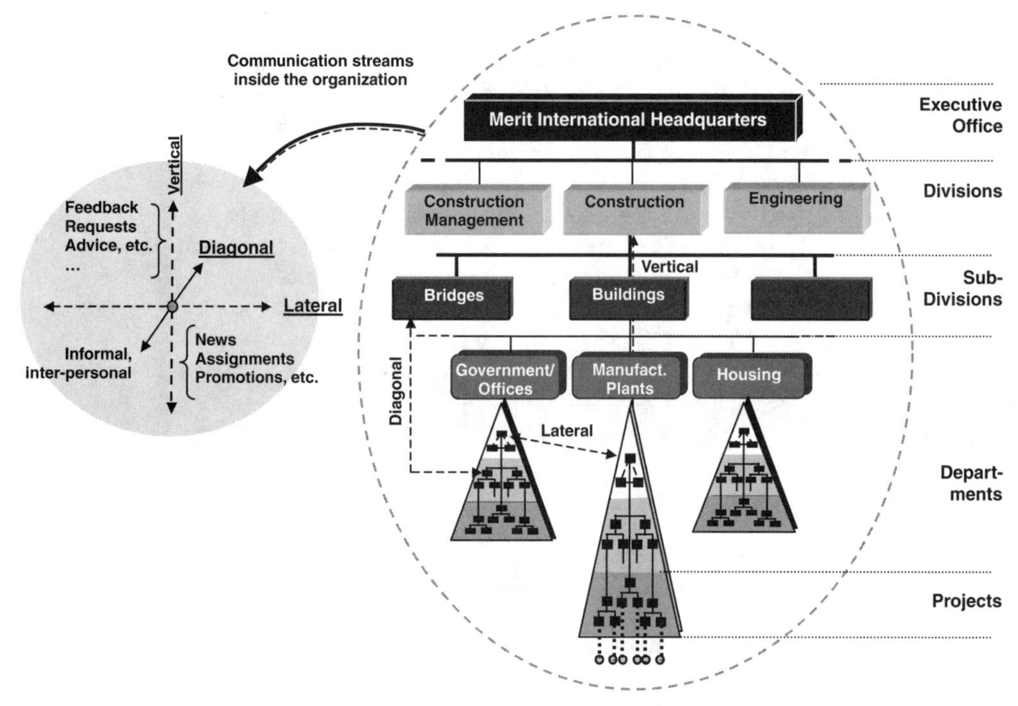

Figure 10.19 Modeling Communication Channels as an Overlay to an Organization

10.3.1 Interorganizational Information Flow

In Chapter 3, we learned about the Work Breakdown Structure (WBS) and the Organizational Breakdown Structure (OBS) that led to an effective coding structure useful for both planning and control. The same hierarchical OBS can also be used as a basis to design the information flow vertically, from top to bottom (e.g., request for response) and from bottom to top (e.g., feedback to request).

The media for this formal kind of "transmissions" include standard memos, written follow-ups to verbal exchanges, announcements, and Web postings (the MEMORANDUM insert presents some basic rules). While lateral "information exchange" has long been thought of as superfluous and harmful gossip between colleagues, it does fulfill a list of important functions—not the least of which is getting help from colleagues to find a solution to a problem. It is well known that departments in large construction companies, as in universities, are reinventing the wheel that has already been designed by others in the same organization. This is an obvious sign of a lack of lateral or diagonal communication. As indicated, vertical communication follows the hierarchical channels, while horizontal and diagonal communications create their independent networks. Media selection for the "informal" contact follows the path of "least resistance." A meeting around the water cooler, a quick phone call, or an email is all it takes.

How to Write a Memo

MEMORANDUM

TO: The person receiving the memorandum

FROM: The person writing the memorandum

DATE: April 20, 2010

SUBJECT: Short title descriptive of the topic

CC: Copies to . . .

- *I am writing because . . .*
- *The facts are . . .*
- *Based on . . . I will, or I propose that you . . .*
- Use block-style paragraphs.
- Single-space your memo; add six points above paragraphs—double-space a very short memo.
- Use 12-point Times Roman as the default font.
- Do not add a complementary close (e.g., "Sincerely"). Do not sign memos at the bottom.
- In a footer, add the date and number the pages of the memo, for example "2/23/96, page 2 of 4."
- Begin long memos with an executive summary. Your reader determines what "long" is.
- Use the spell-check function.
- Break long lists of bullets into several clusters.

Without a doubt, management should use the climate of its communication network as a key indictor for measuring the health of the entire organization. Positive and open channels are free of degrading jokes, inaccurate "facts" about others; staff is motivated and willing to meet the challenges set by supervisors. In the same vein, supervisors are willing to listen and are open to adopting new ideas (from others) or to willingly provide assistance.

The backbone for exchanging large volumes of digital communication within the office building is provided by a local area network (LAN) or a wireless local area network (WLAN). While the first uses Ethernet networking technology, the second operates the wireless system using radio signals based on the IEEE 802.11 standards. WLANs offer users the mobility to move around within a building and take their laptops to meetings or other offices and never lose the contact to an online database, CAD, or any Web-based application software.

The security for LANs and WLANs from unknown intruders can be achieved by installing a firewall, which creates a demilitarized zone (DMZ) a

Wi-Fi—IEEE 802.11b/g

Wi-Fi has several meanings. First, it denotes a global industry association of more than 300 member companies. Second, it is a "wireless Ethernet" using the IEEE 802.11 set of protocol standards using 2450 MHz and 5800 MHz. Wi-Fi networks are able to link together electronic equipment, such as computers and printers, and transmit files or set up audio links. Wi-Fi uses the same radio frequencies as Bluetooth, but with higher power, resulting in a stronger connection.

protected subnetwork. Servers are used to serve clusters of applications that need standardized software and hardware interfaces, to include file servers, database servers, backup servers, print servers, mail servers, or web servers. Hubs or switches work similarly to power strips, in that one connection to the Ethernet is split into multiple lines that connect to client devices. The hub acts like a switch, opening only one "gate" at a time to send a packet of data.

Wireless communication with a WLAN begins with an access point serving as a transceiver, a device capable of transmitting and receiving radio waves and serving as a switch and a security gate between the WLAN and the LAN or the Internet. Business applications requiring longer reach to cover multiple stories or a large maintenance facility rely on repeaters that can extend the extent of the access point. As a consequence, employees within the covered areas can connect to the LAN or to any other device within the WLAN equipped with the proper interface system. For example, a user might be able to connect to a wireless camera, overlooking a laydown area outside the building, from a wireless laptop or PDA at the office. Barcode scanners or RFID readers with wireless links allow updating the status of a CAD drawing, or record automatically the arrival of delivery trucks in real time.

In summary, whichever information network is being implemented within a company structure, its main objective is to support the mission of the organization. Besides supporting decision making by teams, or basic internal communication, specific missions include:

- Make access to information for authorized persons as convenient as possible.
- Enhance the rapid adaptation to changes in the technological, industrial, economical, or political arenas.
- Foster continuing education within the company.
- Create a community of highly qualified individuals, respectful of each other, working together ethically, willing to share successes and to accept responsibility for failures.

A positive atmosphere with open communication channels within the construction company provides a robust platform for connecting safely to outside groups/companies to create an information network dedicated to a project. Referred to as an *extranet*, such a project net is able to link the project participants for the purpose of exchanging digital information more securely and efficiently. As on the Internet, there is no central server; every computer on the extranet is connected with every other computer using HTTP, FTP, or other

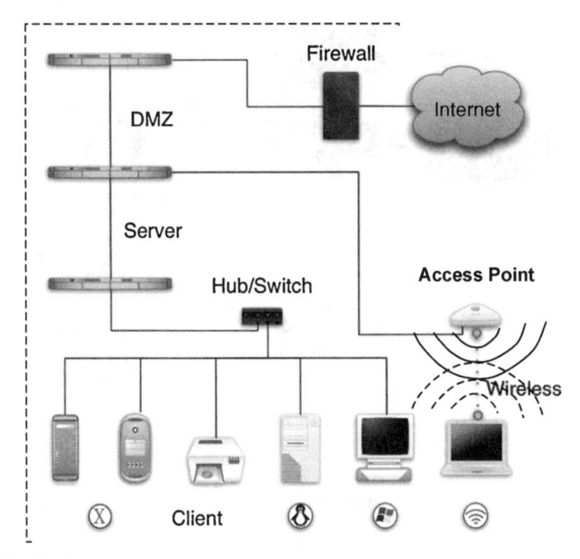

Figure 10.20 LAN Architecture

protocols. The following section presents the basic concept of a project net that is using the extranet configuration.

10.3.2 Information Flow on the Project Net

The objective of an extranet, like the project net, is to maximize information exchange between a finite number of participating partners. By nature, these partners have to be willing to let others to have access to some of their own data, which is mainly motivated by a common goal. In other words, the members of a project net should behave like a business alliance that is established to cut costs and improve service for a customer. Since alliances are uncommon in the lowest-bid-oriented construction industry, the free sharing of data among all members on a project is unlikely. Nevertheless, the flexibility of the extranet protocol allows the creation of dedicated networks between interested entities for the duration of a project.

The resulting project information scheme for exchanging messages, shown in Figure 10.20, contains the traditional, and an extranet in parallel. As the drawing depicts, three project participants have joined in an information-sharing alliance that is created by an extranet between the three companies, which is also tied into the construction site. In other words, the general contractor, the steel designer, and the steel supplier are all willing to share their project-related data of interest to the others in the alliance, while all of them also have access to data and information that is created or available from the construction site (see Figure 10.21). The latter capability could be extremely useful, in that project webcams could be accessed in order to verify that safety measures are being implemented, that delivery trucks are arriving on time, and so on.

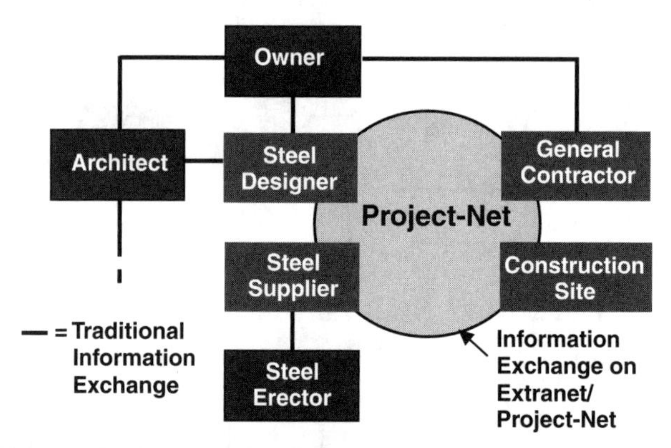

Figure 10.21 Extranet for Select Project Members

10.3.3 Traditional Communication Patterns

Because of the importance of the construction site as the source and the destination of large data and information streams it seems appropriate to begin this review in the field. A study conducted by two researchers, Igal Shohet and Shay Frydman, focused on reporting on communication patterns in exactly this area. The following highlights some of their findings, published in the article "Communication Patterns in Construction at Construction Manager Level," in ASCE's *CEM Journal* (129(5), 2003, pp. 570–577).

One of the goals of the study was to identify the communication flow and its impact on project performance. After surveying managers of 30 construction projects with an average of 20 subcontractors and a projected average cost of $18 million, the authors found that on average: (1) 45 percent of all drawings required extensive modifications because of errors and constructability problems; (2) original blueprints were supplied two days before construction had to begin; (3) revised drawings were delivered 5.4 days behind schedule; and, most crucial (4) 1.9 days were lost per month because of late designs. It was felt that the poor supply chain of design drawings was due to lack of poor communication between the design office and construction management.

In focusing on the communication by the construction manager, Shohet and Frydman found that 52 percent of it was written, 28 percent verbal during meetings, and 20 percent verbal using telecommunication. Considered as written were design drawings, letters, specifications, and emails. It would have been interesting to know more specifics of the flow of the written information—in particular, the specific content, their sources, and the destination. SuChart Nuntasunti investigated the content of the information exchange during weekly and monthly meetings for a new $25 million classroom and laboratory building for chemistry/physics at North Caroline State University campus ("The Effects Of Visual-Based Information Logistics in Construction," PhD thesis, NC State University, 2003). The general contractor called and chaired the weekly meetings, lasting usually 70 minutes, every Monday,

Table 10.4 Time-Based Distribution of Communication Content During Construction Meetings

Content of Messages	Weekly (%)	Monthly (%)
Reporting on progress	22.5	13.5
Informing about short-term plans	24.7	10.8
Warning about upcoming disturbances	2.0	9.5
Notifying about unforeseen changes	1.7	2.7
Alerting about change orders	0.3	4.0
Clarifying conflicts in drawings/specs	2.3	20.7
Confirming previous informal agreements	1.3	11.3
Requesting (more) information	4.6	4.5
Reviewing safety standards, milestones, . . .	23.4	11.7
Recording verbal agreements into minutes	0.0	2.3
Miscellaneous information	17.2	9.0
Total	100.0	100.0

with an average of six subcontractors in attendance. Once a month, on the second Thursday, the architect chaired a meeting that lasted about two hours with 20 people in attendance. Nuntasunti participated and analyzed the minutes of 36 weekly and 5 monthly meetings in 2002, and developed detailed statistics of the information flow.

Table 10.4 highlights the different emphases of the two meeting types. While the monthly meeting could be considered a proactive problem solving session among all 20 stakeholders of the project, in the weekly meeting, the general contractor looks ahead to the next week and underlines the importance of the pertinent safety rules and regulations.

Nuntasunti categorized the messages that were exchanged further, according to three functional categories: (1) dissemination of information, (2) question or request for more information, and (3) answer to questions.

Again, the data in Table 10.5 reflects the purpose of the two meetings, information dissemination and reinforcement of important standards (e.g., safety) by the general contractor during the weekly meeting, while a greater emphasis is given to sharing information and addressing potential problems by all the participants. In effect, the weekly meetings could be called *operational* and the monthly *strategic*.

Table 10.5 Messages classified by function

Purpose/Function of Messages	Weekly (%)	Monthly (%)
Dissemination of Information	88.7	63.1
Questions	7.3	19.4
Answers/response to questions	4.0	17.5

Shohet and Frydman also grouped the surveys into two clusters: effective and ineffective construction projects. Effectiveness of a project was measured in terms of: (1) timeliness (adherence to schedule), (2) productivity (efficiency of labor), and (3) safety. Table 10.6 presents a comparison of the six message categories of the communication stream for the two clusters.

What is immediately striking is that half of the message categories are equally strongly represented while, at the other end of the spectrum, communication about quality control shows a discrepancy of 14 percent. As the arrows intend to indicate, there seems to be a relationship between effort spent on quality management and the two remaining—instructions and cost control. The data gives the impression that increased quality management reduces the need to send out instructions and messages related to cost control. This phenomenon might be based on the proactive problem-solving nature of effective quality control, which is based on "doing it right" instead of "fixing it when needed."

Table 10.6 Comparison of Message Contents

Content of Messages	Percent by Effective Projects	Percent by Ineffective Projects	Importance to Effective Management (%)
Instructions	28	35	–7
Materials and equipment management	12	11	1
Quality management	18	4	14
Scheduling and coordination	23	23	0
Cost control	17	25	–8
Other	2	2	0
Total	100	100	0

Table 10.7 Comparing Distribution of Communication Partners

Communication Partner	Percent by Effective Projects	Percent by Ineffective Projects	Importance to Effective Management (%)
Owner	7	6	1
Users	12	12	0
Designers	19	32	−13
Project engineer	16	16	0
Suppliers	9	3	7
Superintendent	19	11	8
Subcontractor	17	20	-3
Total	100	100	0

Another interesting analysis focused on the distribution of communication between the construction manager and project participants. The result of this comparison, depicted in Table 10.7, hints at the fact that a higher-quality communication between construction manager and the designer reduces the quantity of communication (−13%), which was used to work/communicate more extensively with suppliers (+7%) and the superintendent (+8%).

While the two studies discussed here can't tell us the entire story about communication on a construction site, the findings reinforce the importance of open communication channels, or in other words, the result of poor communication on the effectiveness of project management. The following section presents the on-site communication network that is being worked on by various agencies worldwide.

LAN with Ethernet

A LAN (local area network) is a communication network covering a limited area, such as a construction site. Ethernet, introduced in 1980, is the dominant technology using a combination of twisted pairs of copper wires, along with the fiber optic versions for site backbones. Wired Ethernet cables run at 10 Mbit/s, 100 Mbit/s), and 1 Gbit/s (mega- or gigabits per second). Wireless Ethernet IEEE 802.11.g operates at approximately 22 Mbit/s.

10.3.4 Agent-Based, Ubiquitous On-Site Communication

The BIM model will open new opportunities to substantially improve the design process, especially in addressing interoperability issues. But what happens when the designs, schedules, and budgets arrive on-site? The answer to this question should be easy to answer by now. The construction site has been mostly left out of the information-driven revolution, thus strangling the possibilities to improve productivity that were possible in manufacturing and even agriculture.

FIATECH's "Capital Projects Technology Roadmap" put the finger on the spot when it points to the still mainly

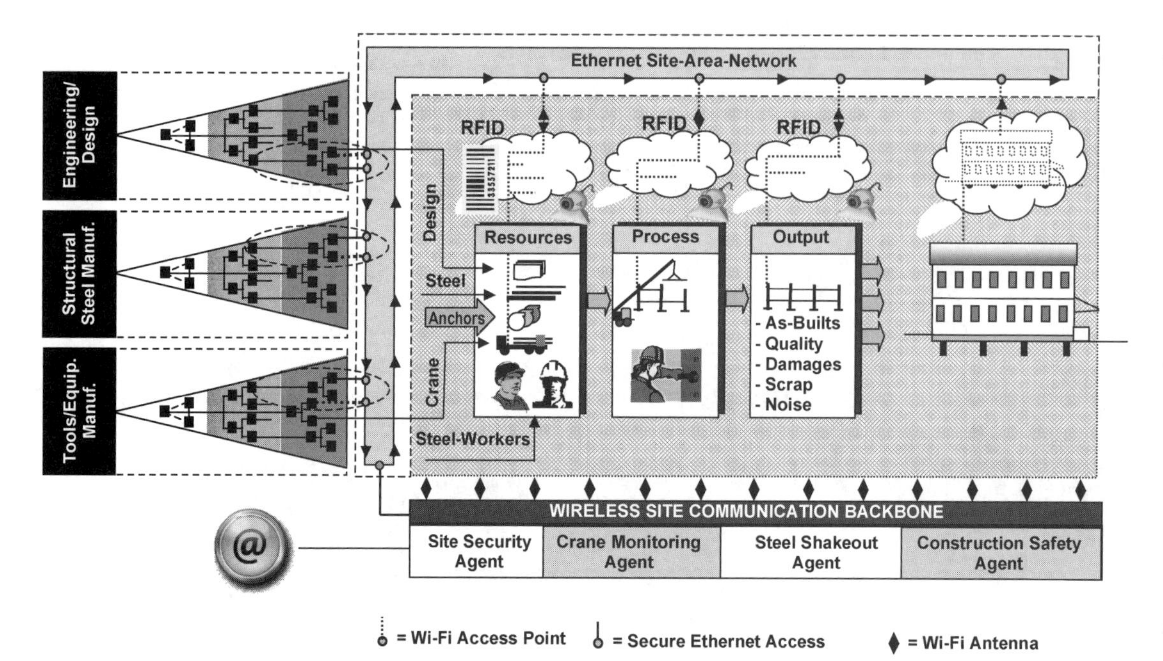

Figure 10.22 Cradle-to-Grave Information Network with Real-Time Communication with Work Front

paper-based construction sites, lack of information technology at the work front, and the total lack of an integrated information system that links the owner, architect, engineers, contractors, and suppliers with site-based equipment, automated data collection tools, and computerized information agents.

Figure 10.22 depicts an integrated communication system consisting of two LANs, one high speed cable and one wireless Ethernet, connected to the Internet. The modeled system implies an uninterrupted and mostly automated supply chain of information: top-down, bottom-up, as well as on demand. In other words, information is being carried electronically by the system in a cradle-to-grave manner where anybody allowed to enter the system has access. Since the amount of new information will be excessive, computerized agents are "plugged" into the system, enabled to monitor data flows and intelligent enough to make decision within their specialty area. The following presents several technologies that the construction is getting ready or has already adopted.

Connecting Islands of Information

Only 25 years ago, the concept of public cellphones was a fiction, but it is a global phenomenon today. Automobiles and trucks are now hooked to gigantic telematics systems that provide real-time information about the nearest restaurant or upcoming traffic problems. With recent advances in computing, smart devices are becoming progressively smaller, and communication more powerful. This trend opens the door to finally address the many problems that are related to "islands of information" in construction.

In 2003, Cisco Systems presented a result of a project intended to measure the benefits of wireless local area networks (WLANs) at work. According to the study, over 12 percent of 603 U.S. organizations (with 100-plus employees) deployed WLAN infrastructure in their offices. The report estimated that WLAN increased work efficiency by approximately 1.29 hours per day.

Wi-Fi antennas, connected to an Ethernet access point, come in various sizes and provide signal coverage of up to 100 to ~500 feet indoors and 1000 feet outdoors. Clients within the shaped 3D space that is reached by the radio signals are able to send and receive data from an access point connected to the Wi-Fi antenna. Due to other electrical signal energy, called background noise, as well as obstacles between a client and the antenna, the signals attenuate, losing transmission strength and stability. Such losses can be anticipated and signal loss avoided by installing repeaters at strategic points. All a repeater does is to invigorate the low radio signal and send it out at much higher strength. As a result, Wi-Fi-enabled devices such as notebooks, tabled-PC, IP-video cameras, PDAs, and electronic sensors are able to connect to the Internet. The appropriateness of this wireless concept for the construction site is enhanced by the data transmission rate of 802.11g 50, sufficient to handle an average of 50 wireless cameras at a time.

Which islands will be connected? The answer is, every island that has a Wi-Fi transceiver. Imagine the crane operator. He or she can access that the Internet, to see each piece that is on a trailer-truck that just arrived, along with coding, weight, and the location where it should be unloaded, or installed in the building. A barcode or RFID reader mounted on the crane hook is able to read the code to verify it before the crane lifts it to its required location, automatically updating the progress of steel erection. Or the quality control engineer for concrete, working on-site, can immediately send the result of each test to a website that automatically monitors the concrete quality and alerts the construction manager sending a cellphone message. The next, more specialized, cases provide a few examples from an almost endless list.

Software Agents to Manage the Site Network

Creating data-intensive networks raises the risk of data and information overflow. A software agent is a stand-alone node built into the network, capable of autonomous action on behalf of a user. In one application, such an agent will take over routine activities for the user, such as automatically checking the availabilities of desired project participants to participate in a conference call and book a time on everybody's calendar. Another very critical area is the management of information networks able to capture, store, analyze, massage, and transmit field data/information quickly and in vast quantities. Such agents are able to discover problems (e.g., indicated by sensory data), respond and recover from routine and extraordinary events that might cause failure (e.g., security breach), and coordinate with other systems (e.g., local police), all without direct human participation.

Wireless Environmental Monitoring and Recording

Information about wind, temperature, rain, and humidity inside a construction site are excellent indicators of what is going on or what will happen, if looked at comparatively over several days or as differentials. For example, the humidity measured inside the building will tell if its dry enough to install flooring. Changing wind speeds warn about an upcoming storm. Wireless battery-powered sensor units will supply all these data in real time to a weather station programmed to monitor, store data onto a website, and send warning messages when certain threshold values are met. Due to the flexibility provided by the wireless link, the sensors can be quickly moved to different areas, as needed. Other wireless environmental sensors that provide invaluable data to a monitoring agent measure noise, fumes, dust, and even security cameras.

Virtual Site Meeting for Real-Time Problem Solving

Designing a construction site with a continuous Wi-Fi signal coverage allows the direct communication link between any point on-site with any point on the Internet. In particular, an inspector equipped with a personal area network (PAN) to integrate a wireless video camera, wireless tablet PC, and Internet phone is able to communicate live with a design engineer anywhere online. What results is virtual human-to-human interaction supported by Web-based tools, allowing both to look at the same drawing, image, and other documentation while discussing or brainstorming a problem. The possible savings are significant, stemming mainly from cutting out the time to send a request for information (RFI) and getting a response from the engineer, architect, supplier, or other involved party.

PAN with Bluetooth

A PAN (personal area network) establishes communication among electronic devices in the close vicinity of a person (10 meters, or 33 feet). PANs connect personal devices themselves (intrapersonal communication) or to a higher local area network (LAN) and the Internet. A Bluetooth uses 2450 MHz and handles up to eight active devices.

Equipment Alert Agents

One of the environmental threats to cranes is the sudden appearance of a strong wind at speeds above 20 mph (32 km/h). Having a wind sensor directly mounted on the crane does not suffice, since it provides information that might be too late to avoid a disaster. A crane alert agent manages several sources of weather information with data coming from the site, based on weather sensors, but also from the National Oceanic and Atmospheric Administration (NOAA) and the closest airport. The latter will provide weather forecasts for the region that are calibrated with the actual reading. The goal is to alert the crane operator using different alarms when specified threshold values reached.

Smart RFIDs that Interact with the Surrounding

RFIDs have had major positive impacts on the productivity in the supply chain of many industries, including shops all over the world that have installed electronic sensors at the exit in order to stop thefts of merchandise equipped with small tags. Who has not marveled at the automatic toll payment collection systems? When a car approaches a tollbooth, a reader at the booth sends out a signal that "wakes up" the transponder on the car windshield. The transponder then broadcasts its unique ID to the reader, which in turn charges the account and opens the gate for the car. All that while the car is moving.

So far, the majority application of RFID tags have been used to monitor the location of equipment and materials, but here are other specific uses:

- *Component tracking:* Precast elements or pipe elements as they progress through the production cycle, where they are stored, and when they are being loaded and delivered to the site
- *Access control:* Securing personnel entrance to the construction site
- *Equipment management:* Automatic recording of equipment use and service scheduling.
- *Security:* Monitoring of equipment location and movements at unusual hours
- *Asset management:* Location of assets at construction sites or in transfer

While the majority of applications today employ passive RFID tags, requiring the energy of a magnetic field to be recognized, active RFIDs send out signals on their own. They are becoming not only cheaper but also more powerful, usually operating at 433 MHz, 2.45 GHz, or 5.8 GHz, with a signal range of 20 to 100 meters (60 to 300 feet). Active RFIDs may include three main components: (1) active battery-powered tag, (2) active transponders, and (3) beacons.

Beacons are elements of a real-time locating system (RTLS), able to detect the precise location of an active tag attached to a piece of equipment or material. A beacon emits a signal, with its unique identifier at preset intervals. The beacon's signal is picked up by at least three reader antennas positioned around the perimeter of the area where assets are being tracked.

Because of their power source, active RFID tags can be expanded and adapted to include additional memory and local processing. They can read, write, and store significant amounts of data, and can be attached to sensors to store and communicate data to and from other devices.

10.3.5 Improvements in and Opportunities for Wireless Communication

In the survey of construction managers mentioned earlier, two other questions were asked. Figure 10.23 plots the answers, in percentage of total respondents, to

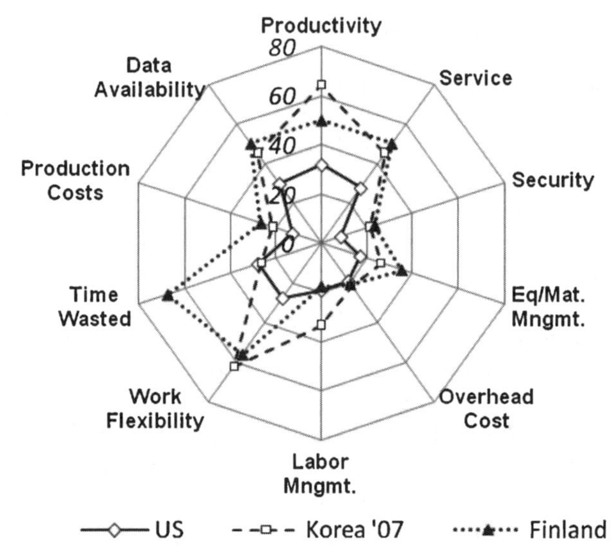

Figure 10.23 Improvements Due to Wireless Technology Use

the first question, which queried respondent whether they had observed improvements on 10 different scales that were the result of introducing wireless communication (including cellphones).

The noticeably smaller percentage of survey participants from the United States reported improvements due to wireless technologies. The absolute largest improvement was observed by the Finnish managers in reducing time wasted (70 percent) and greater work flexibility (58 percent), topped only by the Koreans (62 percent). Similarly impressive was the 65 percent of Koreans who saw improvements in productivity. The discrepancies are interesting, as the Finnish contractors reported a 20 percent higher use of cellphone, which produced significant improvements on several scales.

Another question asked the respondents to identify which area or task in construction might be an opportunity for Web-based wireless communication. Also plotted in percentages, Figure 10.24 provides a contrasting picture, as U.S. managers by far have the highest expectations. Is it because the respondents were younger, or had higher information literacy? The winner for everybody is the exchange of e-drawings, which is clearly related to the emergence of BIM. Similarly, the approval of designs is a sign of BIM's influence. A 50 percent "enthusiasm" created the automatic data collection, site monitoring, and data sharing with project network.

It will be interesting to observe where the industry will actually be in 5 to 10 years.

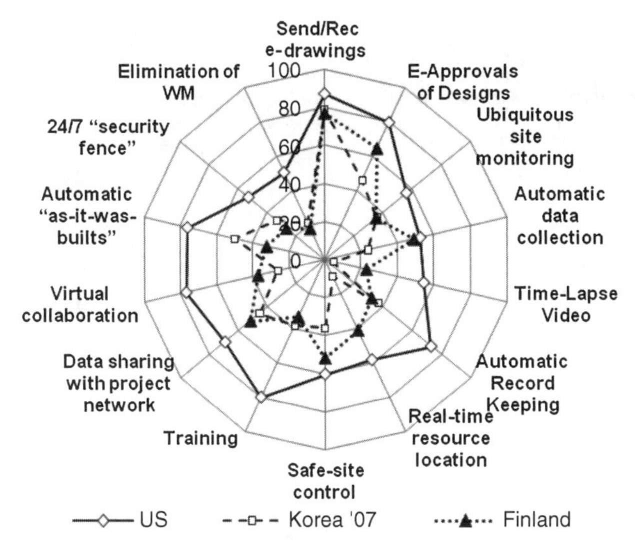

Figure 10.24 Opportunities for Wireless Communication

Header Problem 10.1: Transformations on the Communication Path (continued)

Phase C: Communication of Message

Beth decides to use email to compose a letter to be sent electronically. She laid out the letter into two sections: (1) offering space for right now, and (2) asking for space for July.

To: Walt Steiger <Walt@JCFooting.com>
From: Beth Bertsch <JB@L&M.rr.com>
Subject: Opportunity for more space
Date: May 15, 2009
Hi Walt,

I have been thinking about how I could help you after you pointed out several times that the lack of space during this phase of your work creates problems for you. After discussing it with Fred I thought that you could get . . . . Please let me know if that would work for you.

Jane

Email Transmission

The first transformation of the completed email file will be its encoding into a digital signal that can be sent over the Internet. If you were to look at email file

(continued)

(*continued*)

The Origin of the Keyboard

The most common modern-day keyboard layout on English-language computer and typewriter keyboards is known as QWERTY. It takes its name from the first six letters on the keyboard's top first row of letters. The QWERTY design was patented by Christopher Sholes in 1868 and sold to Remington in 1873, when it first appeared on typewriters.

Originally, the characters on the typewriters were arranged alphabetically, set on the end of a metal bar, which struck the paper when its key was pressed. However, once an operator had learned to type at speed, the bars attached to letters that lay close together on the keyboard became entangled with one another, forcing the typist to manually separate the type bars, and frequently caused blotting to the document. A business associate of Sholes, James Densmore, suggested splitting up keys for letters commonly used together to speed up typing by preventing common pairs of type bars from striking the platen at the same time and sticking together.

as a computer stores it, you would find that each byte contains not a letter but a number—an ASCII code corresponding to the character.

Beth pushes the Send button on the email program (email client), which starts the transmitting process via her mail server or mail transfer agent (MTA) that sends the message to Walt's mail server.

The encoded email message uses the Internet as the communication channel. To ensure a proper ''handshake'' between the two ends, SMTP (Simple Mail Transfer Protocol) is used by both. In other words, L&M.rr.com sends the message to the mail server with the domain name JCFooting.com using SMTP. The mail server, managed by Walt's ISP (Internet service provider), in turn puts the message into the inbox of user Walt.

Assuming that Walt uses Eudora, Microsoft Outlook, or similar email clients, he can retrieve the message, still encoded in ASCII decimal, using either POP (Post Office Protocol) or IMAP (Internet Message Access Protocol). This occurs when Walt presses the Get Mail button, or happens automatically according to preset intervals. The computer can be connected either to a cable network within the company or a wireless laptop.

In addition, if Walt owns a wireless BlackBerry with access to the Internet, he could have it programmed to receive the email there as well. Similar to office

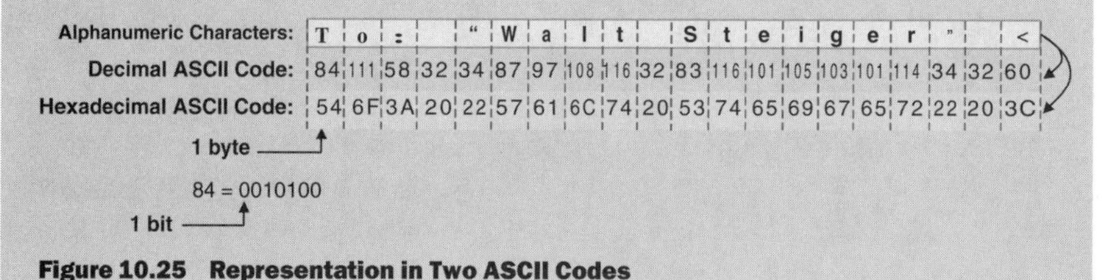

Figure 10.25 Representation in Two ASCII Codes

computers, BlackBerry uses POP3/IMAP to retrieve messages from the same box on Walt's mail server.

The built-in SMTP ensures the proper decoding of information, still a string of 0s and 1s, by separating headers, text, style, and so on. Eventually, Walt will look at—either on his office, laptop, or BlackBerry—an exact copy of the version sent by Jane.

Since the message does not constitute a definite instruction to act from a person with the appropriate authority, no direct action will occur.

These two functions mirror functions 5 and 4 (refer back to communication model in Figure 10.10), in that they model the processes of Walt:

1. Reading and understanding the email.
2. Deciding whether to do something.
3. Initiating a response if the answer to 2 is positive.

Phase D: Walt's Decision-Making Process

Gaining access to the site for assembling the cages would be fantastic for Walt, for many reasons. Although it is not on his critical path, and would only slightly affect his schedule, he would be able to save money and become extremely flexible when things change (which they always do). The offered space is very tight; but considering the circumstances, he will make it work. However, Walt does not trust the offer, which might cost him more later if he takes it. In fact, Walt's approach to dealing with general contractors and other subcontractors can be classified as playing the traditional zero-sum game. He knows that situations exist where everybody is able to win (a win-win situation), but not in construction, where the lowest-bid culture is so entrenched in everyone's thinking.

Reading Beth's e-mail (that young "gal" working for L&M—which has treated JC Footing fairly in the past), he ponders how he should respond, since it is not the usual "I trade you" offer. The thoughts in his head develop as follows.

> **Zero Sum Game**
>
> In a zero-sum game, one participant's gain is the other participant's loss. Adding the gains and the losses always ends with zero. Tennis is a zero-sum game: one loses and one wins.

Walt considers the decision situation as a multistep process that can be represented in a decision tree. The nodes are decision points by either Beth or Walt; the arrows show the decision. Playing a zero-sum game in his head Walt can't know if Beth is trying to get something from him, and if so, what. Such a game is referred to as a game with incomplete information, as compared to a game with complete information like chess, where both know the rules about how to win the game.

(*continued*)

(continued)

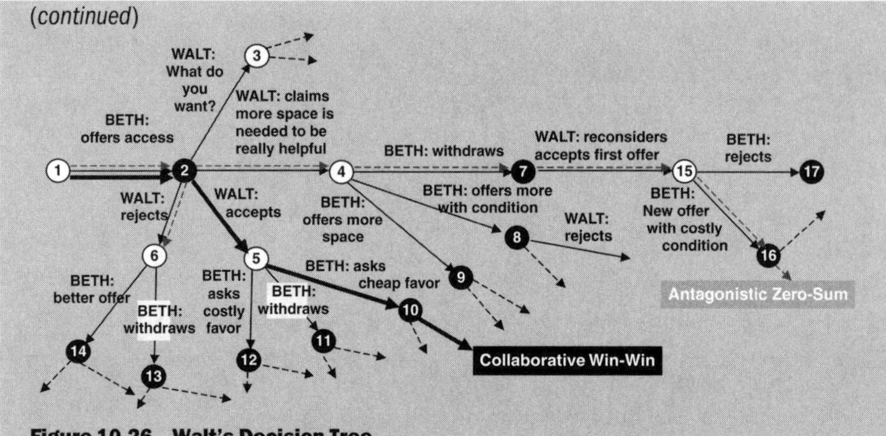

Figure 10.26 Walt's Decision Tree

As indicated, Walt has one collaborative path and three possibly antagonistic paths. Because he is under pressure on that job, his information about Beth's "political power" is sketchy; but because he has had good experiences working with J&M, he decides to accept the offer via an email back to Beth he sends from his BlackBerry.

CHAPTER REVIEW

Journaling Questions

1. A study found that weekly meetings with all subcontractors on a construction site is used mostly used to: (a) report progress, (b) inform about short-term plans, and (c) review safety standards. Since the travel cost for attending such meetings, in addition to the time it takes, represents a substantial effort. What kind of Web-based system other than teleconferencing might you design that would improve one the present system whereby users have to wait until Monday to share information. Address specifically the three main content areas of such meetings.
2. Face-to-face interaction is the preferred method for many communication needs. It has been proven most effective in creating trust, familiarity, and common spirit. On the other hand, such get-togethers can be costly and time-consuming, in particular when they involve global travel. What do you have to consider when putting together a new working group for designing a large project with international participants?

3. Engineering drawings belong to one of the oldest methods of communication. Why is it so special? How did the introduction of computers and CAD lead to costly problems in the communication with drawings? How do you predict the future of the engineering drawing technology/system will develop? What will be gained, and who will benefit most?

Traditional Homework

1. In Chapter 3, we studied the present method of rebar delivery as an example of a supply chain that would offer an opportunity for a win-win collaboration that would benefit the rebar supplier as well as the steel placing company. Although we made significant improvements, the paperwork that parallels that rebar supply channel is extensive, including lost tags and short shipments. RFID technology, active and passive, has some key capabilities that might help to totally automate the paperwork, without anybody having to key in data, print tags or bills of lading, and so on. Furthermore, such a system would ensure that the shipments are complete every time.
   a. Draw an information supply path for rebar that begins with the design engineer and ends with an individual bar being placed into the form at the proper place. Indicate the form or medium that holds the messages in each phase on the supply chain.
   b. Consider a company that is interested in implementing active or passive RFID technology. What would you tell management are the (dis-) advantages of each?
   c. Assume that the company decides on the active RFID. Design a system that would totally automate the data supply, store data about each rebar in a bundle as well as its laydown area on the formwork, ensure that a shipment is complete, and recycle the tags.
   d. Where would you install RFID readers, transponders, and beacons?
2. Consider a backhoe operator sitting in his cabin that includes a flat screen.
   a. Design a simple wireless communication system that would allow the operator access to the design drawing for the trench he has to build, and verify electronically that he is indeed at the correct location.
   b. Create a module that creates automatic as-built updates of the design drawing each time a new pipe segment is installed.

Open Ended Question

1. RFID tags combined with a real-time locating systems (RTLS) provide many opportunities to track not only materials, but also laborers and

equipment. One important beneficiary of such a system would be construction safety, since many accidents happen when a laborer and a piece of equipment collide, or a laborer falls through an opening or forgets to tie himself properly. Sketch two opportunities using RFID with RTLS to avoid dangerous accidents.

BIBLIOGRAPHY

Acar, E., I. Koc, Y. Sey, and D. Arditi. Use of Information and Communication Technologies by Small and Medium-Sized Enterprises (SMEs) in Building Construction. *Constr. Mgmt. Eco.*, vol. 23, September 2005.

Adams, S., S. G. Roch, and R. Ayman. Communication Medium and Member Familiarity: The Effects on Decision Time, Accuracy, and Satisfaction. *Small Group Res.*, vol. 36, no. 3, 2005.

Becker-Beck, U., M. Wintermantel, and A. Borg. Principles of Regulating Interaction in Teams Practicing Face-To-Face Communication versus Teams Practicing Computer-Mediated Communication. *Small Group Res.*, vol. 36, no. 4, 2005.

Bernold, L. E. Ubiquitous Communication to Link Islands of Information in Construction. *Proc. Earth & Space 2006*, ASCE, League City, TX, March 5–8, 2006.

Bernold, L. E., J. Loyd, and M. Vouk. Equipment Operator Training in the Age of Internet2. 19th Inter. Symposium Auto. Rob. Constr., NIST, Gaithersburg, MD, September 2002.

Camilli, A., C. E. Cugnasca, A. M. Saraiva, A. R. Hirakawa, and P. Correa. From Wireless Sensors to Field Mapping: Anatomy of an Application for Precision Agriculture. *Comp. Elect. Agric.*, vol. 58, no. 1, 2007.

Campbell, J., and G. Stasser. The Influence of Time and Task Demonstrability on Decision-Making in Computer-Mediated and Face-to-Face Groups. *Small Group Res.*, vol. 37, no. 3, 2006.

Cheng, E. W., H. Li, P. E. Love, and Z. Irani. Network Communication in the Construction Industry. *Intr. J. Corp. Comm.*, vol. 6, no. 2, 2001.

Cruz, W. Differences in Nonverbal Communication Styles between Cultures: The Latino-Anglo Perspective. *Leadership Mgmt. Eng.*, October 2001.

Darling A. L., and D. P. Dannels. Practicing EngineersTalk about the Importance of Talk: A Report on the Role of Oral Communication in the Workplace. *Com. Edu.*, vol. 52, no. 1, 2003.

Doucet, M. 7 Lessons on Working with Asia: Understanding Cultural Differences Can Help Engineers Adapt Behavior to Optimal Communication Across Cultures. *Control Eng.*, April 2006.

Du-Babcock, D. Language-Based Communication Zones and Professional Genre Competence in Business and Organizational Communication. A Cross-Cultural Case Approach. *J. Asian Pacific Com.*, vol. 17, no. 1, 2007.

Dunnewijk, T., and S. Hulten. A Brief History of Mobile Communication in Europe. *Telem. Informatics*, vol. 24, no. 3, 2007.

Dupouet, O., and M. Yildizoglu. Organizational Performance in Hierarchies and Communities of Practice. *J. Eco. Behavior Org.*, vol. 61, no. 4, 2006.

East, E. W., J. G. Kirby, and G. Perez. Improved Design Review through Web Collaboration. *J. Constr. Eng. Mgmt.*, vol. 20, no. 2, 2004.

Eng, F., H. Tay, and A. Roy. CyberCAD: A Collaborative Approach in 3D-CAD Technology in a Multimedia-Supported Environment. *Comp. in Industry*, vol. 52, no. 2, 2003.

Facilities Information Council. *National Building Information Modeling Standard (NBIMS)—Version 1.0.* National Institute of Building Sciences, Washington, DC, 2007.

Fischer, M., and J. Kunz. *The Scope an Role of Information Technology in Construction.* CIFE Technical Report #156, Center Integra. Facility Eng., Stanford University, Stanford, CA, February 2004.

Forbes, L. H., and S. A. Ahmed. Construction Integration and Innovation Through Lean Methods and E-Business Applications. *Constr. Res.*, 2003.

Forza, C., and F. Salvador. Information Flows for High-Performance Manufacturing. *Int. J. Prod. Econ.*, vol. 70, no. 1, 2001.

Galbraith, J., D. Downey, and A. Kates. How Networks Undergird the Lateral Capability of an Organization—Where the Work Gets Done. *J. Org. Excellence,* Spring 2002.

Gallivan, M. J., V. K. Spitler, and M. Koufaris. Does Information Technology Training Really Matter? A Social Information Processing Analysis of Coworkers' Influence on IT Usage in the Workplace. *J. Mgmt. Info. Syst.*, vol. 22, no. 1, 2005.

Ghanem, A. G., and Y. A. AbdelRazig. A Framework for Real-time Construction Project Progress Tracking. *Proc. Inter. Conf. Earth Space*, ASCE, March 5-8, Houston, TX, 2006.

Gibson, W. H., D. Megaw, M. S. Young, and E. Lowe. A Taxonomy of Human Communication Errors and Application to Railway Track Maintenance. *Cog. Tech. Work*, vol. 8, no. 1, 2006.

Gillard S., and J. Johansen. Project Management Communication: A Systems Approach. *J. Info. Sc*, vol. 30, no. 1, 2004.

Gilleard J., and J. D. Gilleard. Developing Cross-Cultural Communication Skills. *J. Prof. Issues Eng. Edu. Pract.*, vol. 128, no. 4, October 2002.

Gilsdorf, J. W. Organizational Rules on Communicating: How Employees Are—and Are Not—Learning the Ropes. *J. Bus. Comm.*, vol. 35, no. 2, April 1998.

Grosse, C. U. Managing Communication within Virtual Intercultural Teams. *Bus. Com. Qtrl.*, vol. 65, 2002.

Guimera, R., L. Danonb, A. Diaz-Guilera, F. Giralt, and A. Arenas. The Real Communication Network behind the Formal Chart: Community Structures in Organizations. *J. Econ. Behavior Org.*, vol. 61, no. 4, 2006.

Halfawy, M., and T. Froese. Building Integrated A/E/C Systems Using Smart Objects: Methodology and Implementation, *J. Comp. Civil Eng.*, vol. 19, no. 2, 2005.

Hanna, A. S., and J. Swanson. Risk Allocation by Law—Cumulative Impact of Change Orders. *J. Prof. Issues Eng. Educ. Practice*, vol. 133, no. 1, January 1, 2007.

Hernandez, M. A., and S. J. Stolfo. Real-World Data Is Dirty: Data Cleansing and the Merge/Purge Problem. *Data Mining and Knowledge Disc.*, vol. 2, no. 1, 1998.

Hinds, P., and S. Kiesler. Communication across Boundaries: Work, Structure, and Use of Communication Technologies in a Large Organization. *Org. Sci.*, vol. 6, no. 4, 1995.

Huang, C., M. Chen, and C. Wang. Credit Scoring with a Data Mining Approach Based on Support Vector Machines. *Expert Syst. Appl.*, vol. 33, no. 4, 2007.

Huang, X., and L. Bernold. CAD-Integrated Excavation and Pipe Laying. *J. Constr. Eng. and Mgmt.*, vol. 123, no. 3, 1997.

Jaselskis, E. J., and T. El-Misalami. Implementing Radio Frequency Identification in the Construction Process. *J. Constr. Eng. Mgmt.*, vol. 129, no. 6, 2003.

Johnson-Cramer, M. E., S. Parise, and R. L. Cross. Managing Change through Networks and Values, *Cal. Mgmt. Rev.*, vol. 49, no. 3, 2007.

Kalman, M. E., P. Monge, J. Fulk, and R. Heino. Motivations to Resolve Communication Dilemmas in Database-Mediated Collaboration. *Comm. Res.*, vol. 29, no. 2, 2002.

Kang, J. H., S. T. Anderson, and M. J. Clayton. Empirical Study on the Merit of Web-Based 4D Visualization in Collaborative Construction Planning and Scheduling. *J. Constr. Eng. Mgmt.*, vol. 133, no. 6, 2007.

Keller, M., R. J. Scherer, K. Menzel, T. Theling, D. Vanderhaeghen, and P. Loos. Support of Collaborative Business Process Networks in AEC. *ITconstr.*, vol. 11, special issue, 2006.

Kraemer, S., and P. Carayon. Human Errors and Violations in Computer and Information Security: The Viewpoint of Network Administrators and Security Specialists. *Appl. Ergo.*, vol. 38, no. 2, 2007.

Lee, J., and L. E. Bernold. Ubiquitous Agent-Based Communication in Construction. *J. Comp. Civil Eng.*, vol. 22, no. 1, 2008.

Liu, S., C. A. McMahon, M. J. Darlington, S. J. Culley, and P. J. Wild. A Computational Framework for Retrieval of Document Fragments Based on

Decomposition Schemes in Engineering Information Management. *Adv. Eng. Informatics*, vol. 20, no. 4, 2006.

Lowry, P. B., T. L. Roberts, N. C. Romano, and C. Hightower. The Impact of Group Size and Social Presence on Small-Group Communication: Does Computer-Mediated Communication Make a Difference? *Small Group Res.*, vol. 37, no. 6, 2006.

Manvi, S. S., and P. Venkataram. Applications of Agent Technology in Communications: A Review. *Comp. Comm.*, vol. 27, no. 15, 2004.

Maznevski, M. L., and K. M. Chudoba. Bridging Space over Time: Global Virtual Team Dynamics and Effectiveness. *Org. Sc.*, vol. 11, no. 5, 2000.

McElroy, J. C., K. P. Scheibe, and P. C. Morrow. Computer Technology as Object Language: Revisiting Office Design. *Comp. Human Behavior*, vol. 23, no. 5, 2007.

Mehrani, E., and A. Ayoub. Remote Health Monitoring of the First Smart Bridge in Florida. *Proc., Structures Congress*, ASCE, St. Louis, MI, May 18–20, 2006.

Moon, S., and L. E. Bernold. Graphic-Based Human-Machine Interface for Construction Manipulator Control. *J. Constr. Eng. and Mgmt.*, vol. 124, no. 4, 1998.

———. Operator-Interfaced Intelligent Path Planning for Robotic Bridge Paint Removal. *J. of Comp. Civil Eng.*, vol. 11, no. 2, ASCE, 1997.

Nuntasunti, S., and L. E. Bernold. Experimental Assessment of Wireless Construction Technologies, *J. Constr. Eng. Mgmt*, vol. 132, no. 9, 2006.

———. Wireless Site-Network for Construction: A Win-Win Strategy for GCs. *Proc. Constr. Res., Winds of Change*, March 19–21, Honolulu, HI, 2003.

Pansupap, V., and D. H. Walker. Innovation Diffusion at the Implementation Stage of a Construction Project: A Case Study of Information Communication Technology. *Constr. Mgmt. Eco.*, March 2006.

Pena-Mora, F., and G. H. Dwivedi. Multiple Device Collaborative and Real-Time Analysis System for Project Management in Civil Engineering. *J. Comp. Civil Eng.*, vol. 16, no. 1, 2002.

Pokorny, J. Database Architectures: Current Trends and Their Relationships to Environmental Data Management. *Environmental Model. Soft.*, vol. 21, no. 11, 2006.

Rob, M. A. Project Failures in Small Companies. *IEEE Software*, November/December, 2003.

Rukanova, B., K. Van Slooten, and R. A. Stegwee. Towards a Meta Model for Describing Communication: How to Address Interoperability on a Pragmatic Level. *Enterprise Info. Sys.*, vol. 6, 2006.

Salim, M., and L. E. Bernold. Process-Oriented Intelligent Planning for the Fabrication and Placement of Concrete Reinforcement. *Structural Eng. Review*, vol. 8, no. 2/3, 1996.

———. A Design-Integrated Process Planner of Rebar Placement. *J. Comp. in Civil Eng.*, vol. 9, no. 2, 1995.

Shelby, L. R. The Role of the Master Mason in Mediaeval English Building. *Speculum*, vol. 39, no. 3, 1964.

Shohet, I. M., and S. Frydman. Communication Patterns in Construction at Construction Manager Level. *J. Constr. Eng. Mgmt.*, vol. 129, no. 5, 2003.

Stoman, S. H. Effective Management Style. *J. Mgmt. Eng.*, January–February, 1999.

Street, C. T., and D. B. Meister. Small Business Growth and Internal Transparency: The Role of Information Systems. *MIS Quarterly*, vol. 28, no. 3, September 2004.

Subramanian, S. Vision: An Open Eye and Ear Approach to Managerial Communication. *J. Bus. Persp.*, vol. 10-l, no. 2-l, 2006.

Tatara, E., A. Cinar, and F. Teymour. Control of Complex Distributed Systems with Distributed Intelligent Agents. *J. Proc. Cont.*, vol. 17, no. 5, 2007.

Terry, P. Communication Breakdowns. *Pract. Period. Struct. Design Constr.*, vol. 1, no. 4, 1996.

Thomas S., R. L. Tucker, and W. R. Kelly. Critical Communication Variables. *J. Constr. Engrg. Mgmt.*, vol. 124, no. 1, 1998.

Turk, Z. Construction Informatics in European Research: Topics and Agendas. *J. Comp. Civil Eng.*, vol. 21, no. 3, 2007.

Van der Walt, P. W., and A. S. A. Du Toit. Developing a Scalable Information Architecture for an Enterprise-Wide Consolidated Information Management Platform. *New Info. Perspective*, vol. 59, no. 1, 2007.

Williams, T., L. Bernold, and H. Lu. Adoption Patterns of Advanced Information Technologies in the Construction Industries of the United States and Korea. *J. Constr. Eng. Mgmt.*, vol. 133, no. 10, 2007.

Winkler, S., R. Zimmermann, and F. Bodendorf. An Agent-Based Information Logistics Architecture for Process Management. *Proc. 2005 Inter. Conf. Comp. Intel. Model. Contr. Auto.*, Computer Society IEEE, Vienna, Austria, November 28–30, 2005.

Waldeck, J. H., D. R. Seibold, and A. J. Flanagin. Organizational Assimilation and Communication Technology Use. *Comm. Monographs*, vol. 71, no. 2, June 2004.

Worley, J. M., and T. L. Doolen. The Role of Communication and Management Support in a Lean Manufacturing Implementation. *Mgmt. Decision*, vol. 44, no. 2, 2006.

Yoon, S. H., M. Yoon, and J. Lee. On Selecting a Technology Evolution Path for Broadband Access Networks. *Technol. Forecasting Soc. Change*, vol. 72, no. 4, 2005.

Yu, A. T., Q. Shen, J. Kelly, and K. Hunter. Investigation of Critical Success Factors in Construction Project Briefing by Way of Content Analysis. *J. Constr. Eng. Mgmt.*, vol. 132, no. 11, 2006.

Zhu, Q. Topologies of Agents Interactions in Knowledge Intensive Multi-Agent Systems for Networked Information Services. *Adv. Eng. Informatics*, vol. 20, no. 1, 2006.

Zahn, L. Face-to-Face Communication in an Office Setting: The Effects of Position, Proximity, and Exposure. *Comm. Res.*, vol. 18, 1991.

Zipf, P. J. Technology-Enhanced Project Management. *J. Mgmt. Eng.*, January/February 2000.

Zwikael, O., and S. Globerson. From Critical Success Factors to Critical Success Processes. *Inter. J., Prod. Res.*, vol. 44, no. 17, 2006.

Chapter 11

Performance Management

Performance management encompasses the numerous topics we have addressed so far, and so is the ideal subject for the final chapter in this book. Performance management is about past performance, to be sure; more importantly, it is about the *process* of improvement, of individuals, teams, and departments, and of entire organizations. It covers measures of value-added processes, self-assessments, 360-degree assessments, as well as goals set to align with the mission of the company. As such, performance management is intended to determine the appropriate tool and method that will enhance the efficiency and effectiveness of an operation or a company.

The word "appropriate" implies that the manager is familiar with the entire range of techniques that he or she could utilize to move toward the desired goal. For that reason, it is appropriate to begin here with a historical review of the development of some of the key management tools that have emerged over time.

11.1 HISTORICAL RECAP OF KEY MANAGEMENT CONCEPTS

For all intents and purposes, the name Frederick Taylor stands for the first modern management techniques; as noted earlier in the book, he is regarded as the founder of Scientific Management, whose purpose was to improve industrial efficiency by maximizing the productivity of laborers through tight planning and control by management. Indeed, managers were ordered to plan and train their laborers to follow standard methods that had been tested, and thereby eliminate any reason for the laborers to stop working. Stopwatch methods were used to measure the productivity of laborers, who were paid more for higher rates. This dehumanization of work, turning laborers into mere automata, was also referred to as Taylorism.

Henri Fayol (July 1841–November 1925)

Henry Fayol was a French a mining engineer who researched geology, and later became the director of the French mining company Comambault, in 1888, saving it from bankruptcy. Contrary to Taylor's bottom-up approach, he took a top-down approach to optimizing business organizations.

(*continued*)

Fayol's five key management functions are still relevant today:

1. Forecasting and planning future
2. Organizing needed resources
3. Commanding/instructing
4. Balancing available resources
5. Controlling

Fayol's detailed understanding of control has been lost in translation but his overall approach was clear from the beginning. Instead of coercing laborers to follow the prescribed plan, he recommended the use of cybernetic control, where the manager establishes a performance feedback loop as a basis for adjusting the plans.

In France, a compatriot of Taylor's, Henri Fayol, took a more holistic view of the production process, by considering management responsible not only for planning but also for supporting or controlling the ongoing production process. According to Fayol, measurements provided the data needed to make sure that the resources were available as needed. He listed five key management functions, all in support of the worker. He understood control as the use of feedback data for the purpose of improving the operation, and not, as Taylor saw it, to coerce the laborers to follow time-optimized paths.

Frank and Lillian Gilbreth, whom we will discuss later in this chapter, developed a different approach to improve productivity. Their work focused on reducing the amount of energy a body has to exert to do a certain task. Referred to as *motion studies*, the logic behind their theory was that reducing the length and complexity of motion paths would result in less fatigue and fewer repetitive motion syndromes. Obviously, laborers who use less energy will work smoother and with less pain, make fewer mistakes, and become more productive while feeling better.

Frank Gilbreth Sr. (1868–1924) and Lillian Moller Gilbreth (1878–1972)

Parents of 12 children, Frank and Lillian Gilbreth were also collaborating scientists who believed that the workplace could be steadily improved by eliminating inefficient motions. They first focused on eliminating fatigue-causing motions at a workplace, but ended up investigating the entire workflow. They developed a new motion study technique, which he called *micromotion study*. The Gilbreths formulated a basic alphabet of work motions, naming them *therbligs*, consisting of 18 basic motions such as search and grasp. Frank Gilbreth also developed better cement mixers and the technology for driving concrete piles.

Recall from our earlier discussions of the Hawthorne project (1927–1932), which proved that there were other important factors that impacted the performance of workers who assembled telephones for Western Electric, Inc. That study was the beginning of the Human Relations Movement, which argued that worker motivation had to include the social aspects of human behavior, as well as interpersonal relations. As the Hawthorne project demonstrated, removing human relations from the way management interacts with laborers will have seriously negative effects on productivity.

Maslow, Herzberg, and Vroom are but a few of the names of researchers who studied the human aspects of work, their motivators and demotivators, and demonstrated the power of treating each person as an individual in need of recognition and attention by fellow workers. The desire of people to avoid boredom and narrowly focused work gave rise to the concept of the self-managed team, giving greater autonomy to entire groups to plan the way they carried out their tasks. The success of teamwork rejuvenated the concept of job enrichment, one of Herzberg's ideas from the 1960s. The center point of job enrichment is to give people an opportunity to use multiple skills and their brainpower to complete several sequenced work tasks, instead of a single one. Furthermore, a worker with a special skill could be involved in cross-departmental study teams, to develop a new project or product. All in all, the goal was to increase satisfaction and performance. However, it was learned that performance does not necessarily increase with higher levels of satisfaction.

The rapid success of the Japanese in manufacturing high-quality cars and electronics forced the American automotive industry to catch up to what W. Edward Deming had taught unsuccessfully in the US 20 years earlier: statistics in quality control. Taiichi Ohno of Toyota was a careful listener and believer, and learned that applying statistics reduced the number of defective cars that were sold to unhappy customers, and that the measurements also revealed problems in the manufacturing long before they became real problems. Consequently, machinery could be fixed proactively, reducing the risk of interruptions during operation. The combined result was fewer defects, lower downtime, and higher productivity and performance. Toyota's success led to its leadership role in the industry and to the use of the term *muda* to mean operational waste, lean manufacturing, just-in-time production, and Kaizen (continuous process improvement).

In 1980, the term Total Quality Management (TQM) started to appear as U.S. companies and the U.S. Armed Forces began to implement what had been so successfully tested in Japan. The globalization of manufacturing, exemplified by Boeing, which had begun to buy entire components from other countries, provided the incentive to establish quality standards that applied all over the world. This brought to the forefront the International Organization for Standardization (ISO).

W. Edward Deming (1900–1993)

As a mathematical physicist working for the U.S. Department of Agriculture, W. Edward Deming was greatly influenced by the engineer and statistician Walter Shewhart at Bell Laboratories, who developed the concept of statistical process control and the related control chart. Deming developed the sampling techniques that were used for the first time during the 1940 U.S. Census. He also adopted the idea of separating common-cause from special-cause variations of a process output. The special-cause variations are unexpected and indicate that something dramatic has happened (e.g., failure of a valve or gate).

The ISO defined TQM as " . . . a management approach for an organization, centered on quality, based on the participation of all its members and aiming at long-term success through customer satisfaction, and benefits to all members of the organization and to society." The most important impact of this effort on construction came in the form of the ISO 9000 family of standards for quality management systems. Many large owners required a construction company working for them to be ISO 9000 certified. That meant that such a company had to implement specific procedures into their operation, train its people, and document its adherence so it could be audited by an independent group. Certification to an ISO 9000/1 standard did not guarantee quality of end products and services; rather, it certified that formalized business processes were being applied.

The globalization of business, which saw both Europe and China catching up to the United States and Japan, and the steady rise in the price of oil, required further shifts in the way a company responded to changing conditions on the global market. Agility of the entire organization from top to bottom became of the essence. In addition, the quarterly report orientation of top management served as a red flag, in that it discouraged a long-term view needed for research, innovation, and cultural changes. The spectacular collapse of both GM and Chrysler will long serve as reminders of companies that were stuck in the past and protected by politics.

The new tool to circumvent the quarterly earnings approach to assessing companies was provided by Robert Kaplan and David Norton in the mid-1990s. They called their holistic approach to evaluating a company the Balanced Scorecard, an evaluation of the entire corporation from different angles, with the underlying rationale that management cannot directly influence quarterly outcomes, since they represent decisions that were made much earlier.

The Balance Scorecard evaluates the operational, marketing, and developmental activities, in addition to the financial performance for the shareholder. This tool is also used to address business responses to such environmental issues as climate change and greenhouse gas emissions.

11.2 FROM MEASURING TO MANAGING PERFORMANCE

What is performance, and how can it be measured? We learned in Chapter 1 that a performance assessment of individuals, groups, departments, or companies includes two clusters of measures: (1) efficiency (productivity), and (2) effectiveness. Whereas efficiency focuses on operational ratio (e.g., m^3 concrete placed/person hour), effectiveness comprises measures that focus on how closely long-term goals have been met, including the trend of improvement since last time. The amount of rework, the number of accidents, and owner satisfaction of completed projects all point toward the future, especially when their patterns over time are reviewed and revealed.

Treating a corporation as the subject of a performance analysis is fairly new, but this concept is already being expanded to include entire supply chains. This represents a shift from focusing solely on production costs, to include key indicators of how well a company is run beyond the monetary level—wages, safety, equipment obsolescence, emotional intelligence, scrap resources, and quality of work. Because companies today are more closely tied to supplier alliances or supply chains, it is being argued that performance be assessed all the way back to the supplier chain and forward to the end user.

That said, some researchers have pointed out that measuring past performance does nothing to steer the company in the direction of its mission. Measurement systems reveal only what happened in the past, with environmental conditions and resources that are irrelevant and may already be different today. Furthermore, the

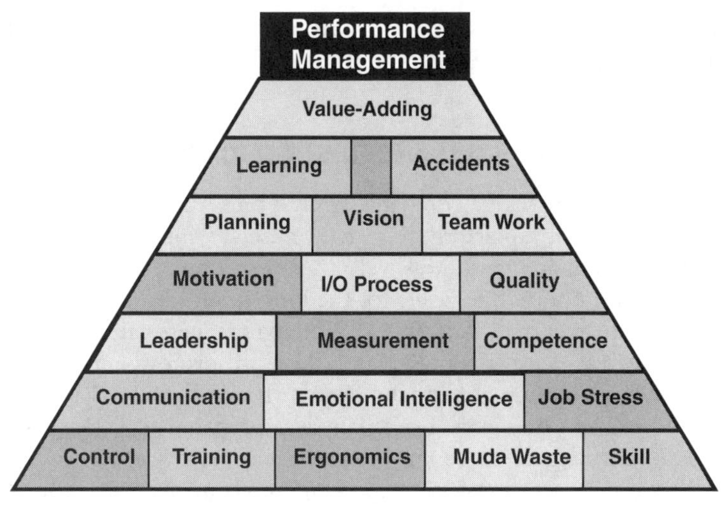

Figure 11.1 The Integrated Nature of Performance Management

measurements do not explain why performance is this way or that, and how things should be done differently; nor do they identify what to anticipate and address proactively. The collection and analysis of performance measures, updating of past performance goals, implementation of improvements, and communication of shared vision have since been subsumed under the new managerial function called performance management.

Figure 11.1 illustrates that performance management is an effort that reaches all aspects of running a company, from defining a long-term vision to promoting emotional intelligence for the managers to promoting skills training for the workers. The process starts with a clear mission statement, to give direction to the entire staff and provide measurement scales for performance within the entire organization.

It should go without saying that such an alignment of goals throughout the company is extremely important to its success. It has been found that the failures of many companies were frequently accompanied by a total lack of understanding by the workers of the company vision and mission.

One of the key stones in the pyramid is communication—vertically, horizontally, and transversely—including exchanges between employees and supervisors. Performance is seen as an ongoing process, rather than a one-time, snapshot assessment occurring every four months for departments, teams, and employees. Most importantly, the primary mission of performance management is *not* assessment, as it is merely a tool, but advancement toward agreed-upon goals.

One of the most important functions of performance management is to meld past performance measurements and goals into action plans for the future and, of course, match them with a motivational reward system. The questions are: What is the best path to take to reach the target? And, what actions are necessary to hit the target? The answers to these questions have to be tied to the establishment of meaningful performance measures.

11.3 A CORPORATION'S BALANCED SCORECARD

The traditional assessment of a company focused on profit, and for public companies of the past, was the quarterly report. Bonuses for upper-level management were linked to a company's profit as well, thus providing a disincentive for management to invest money strategically for the long term, since it reduces the profit during the next quarter and, with it, the money managers can take home in bonuses. In the early 1990s, Robert Kaplan, a Harvard professor, began publishing his findings after evaluating a company from a strategic point of view—meaning that he was measuring how well a company was prepared to be successful in the future. His approach came to be known as the Balanced Scorecard (BSC). It required an all-around view, measuring company performance using four main perspectives, shown in Figure 11.2.

The name Balanced Scorecard invokes two images: that of a balance and that of a scorecard used to keep a record by making marks or notches on a card or tally board. Balance is sought in terms of assessing the performance of a company by moving away from the traditional one-sided metrics of short-term financial success. By including a scorecard in the concept, the importance of performance

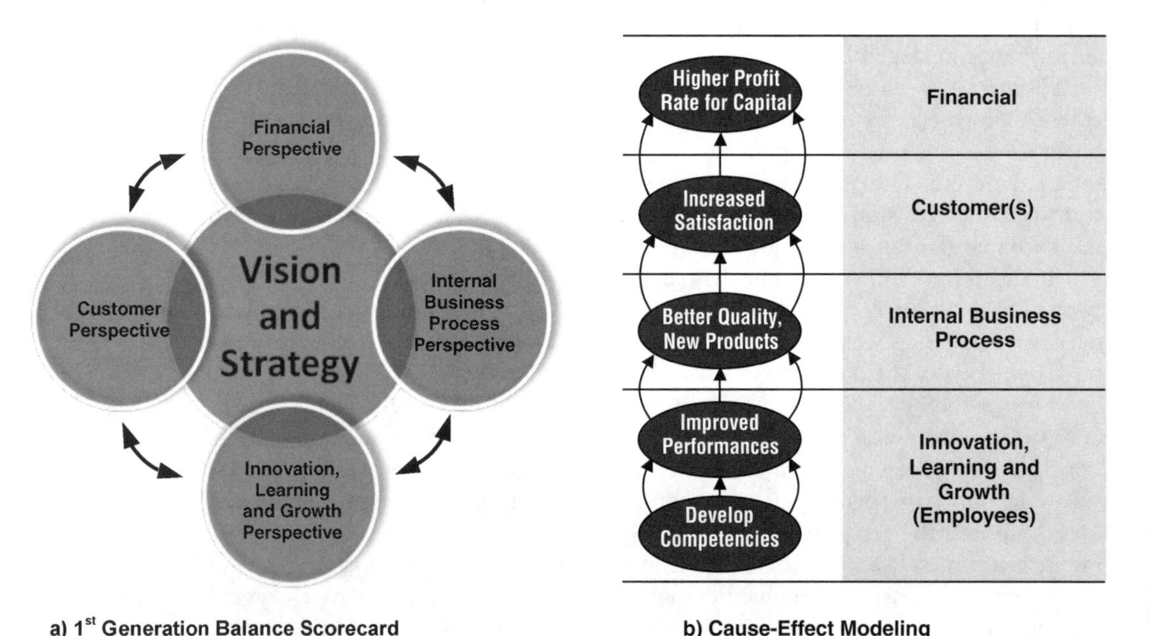

Figure 11.2 The Balanced Scorecard Concept by Kaplan and Norton

measurements of a department, division, or even an entire company can be highlighted. Of course, the scoring should be aligned with the four balancing perspectives shown in Figure 11.2. Selecting a perspective really means to "inspect" the company from a specific point of view or angle. One of the four questions being asked is: How does the company look from the perspective of a customer? Or, how would the customer evaluate the performance of company?

BSC is based on four perspectives that are all linked to the overarching visions and strategic goals set up by top management and, of course, to each other. In that, each perspective must have a long-term and strategic view. For example, the financial perspective needs to include financial objectives and measures that match the company mission and the expectations of its shareholders.

Some of the most common financial measures that are incorporated in the financial perspective are Economic Value Added (EVA), revenue growth, cost ratios, profit margins, cash flow, and net operating income.

The specific measures chosen to consider the customer's perspective depend on the products that the company offers. Generally, companies are interested to learn how much value is delivered to the customer. In the case of construction, issues would be: on-time record, amount of rework, number of accidents, responsiveness to changes, and meeting budget.

The third focal area is the internal operation of the company. How well does it produce the value that the customer desires and expects? Is it working on improving its operation on a short- and long-term basis? Is it using the innovative and creative capital of its employees? Issues are the performance of its supply-chain management, customer relations, environmental impact, equipment maintenance, quality rejects, and others.

The last perspective focuses on a company's human capital, in particular addressing the nurturing of skill, creative thinking, and innovation. Does the company offer education and training opportunities for its employees, helping them to grow within the company? Does it encourage and actively work on knowledge generation and sharing, to improve operation and customer value generation?

The second and third generations of the BSC highlighted the fact that its many objectives and outcomes are interrelated in a cause-and-effect relationship. It was discovered that drawing those interrelationships within a company was an excellent way to deconstruct the complexity while forcing management to focus on those that are most critical. At the outset, however, management needs to create a long- and short-term vision, on top of which the different groups within the company can define their strategies (i.e., operation, supply chain, marketing, finance) using the cause-and-effect logic. The relationships in the BSC are specific to the organization, based on beliefs and assumptions, as in a hypothesis. For that reason, relationships are best defined between measures, not between measurement areas.

The scorecards need to reflect the cause/effect maps, in that lower-level measures relate to higher-level ones. Similarly, the success of one measure in one perspective is rooted in the success of measures in others. This concept is depicted in Figure 11.2 (b), where improved proficiencies of laborers will reduce equipment downtime, improve operator efficiency, and raise product quality in the internal business process. The effect of these improvements will be to produce lower operating cost (finances) and higher customer satisfaction, due to superior quality.

Worked-Out Problem 11.1: Balanced Scorecard System for AspdinCement

The cement company AspdinCement, Inc. received a forecast that shows a slow but steady increase in the need for cement in its area of service. Its board of directors also foresees long-term economical benefits from pursuing an ecofriendly strategy for its cement production. However, it may have to improve some of the cement products, as well as its financial services to its customers, such as precast plants. It might even be necessary to add a new $120 million kiln system to provide greater capacity and flexibility in the use of alternative fuels. The board of directors is requesting a proposal on how it could use the Balanced Scorecard (BSC) to coordinate the different organizational departments more closely around its vision. The directors believe that AspdinCement would also transition more smoothly through the adjustment process that will result from its new vision by using BSC. The new vision statements, in brief, are as follows:

1. AspdinCement will become an ecofriendly company within five years.
2. AspdinCement will develop new uses of Portland pozzolana cement (PPC) to increase its demand and thus the use of waste material in the production of cement.
3. AspdinCement will not import large amounts of cement to meet regional demand.
4. AspdinCement will increase its return on investment by 5 percent.

Here are the seven main departments of the organization:

Joseph Aspdin (1778–1855)

Joseph Aspdin was a British mason-inventor who, in 1824, received a patent to produce hydraulic cement, which he called Portland cement. Although the cement was not very strong at first, the basic principles of its production are still in use today. A mixture of limestone and clay is heated to form a hardened mass, called *clinker*, to be ground into fine powder. When water is added, it turns into "mud" for a while, before it hardens. Because the cement had the same color as the stone on the Isle of Portland, Aspdin named his invention after it.

Cement Galore

There are several varieties of cement, based on compositions developed for specific end uses:

- Ordinary Portland cement (OPC): OPC has 95 percent clinker and 5 percent gypsum and other add-ons. It accounts for 70 percent of the total consumption.
- Portland pozzolana cement (PPC): PPC has 80 percent clinker, 15 percent fly ash/burnt clay/coal waste and 5 percent gypsum, and accounts for 18 percent of the total cement consumption.
- White cement: White cement is basically OPC, clinker using fuel oil with an iron oxide content below 0.4 percent, to ensure whiteness. A special cooling technique is used in its production. It is used to enhance aesthetic value in tiles and flooring. White cement is much more expensive than gray cement.
- Portland blast furnace slag cement (PBFSC): PBFSC consists of 45 percent clinker, 50 percent blast furnace slag, and 5 percent gypsum, and accounts for 10 percent of the total cement consumed. It has a heat of hydration even lower than PPC and is generally used in the construction of dams and similar massive structures.
- Rapid-hardening Portland cement: Rapid-hardening Portland cement is similar to OPC, except that it is ground much finer, so that on casting, the compressible strength increases rapidly.
- Waterproof cement: Waterproof Cement is similar to OPC, with a small portion of calcium stearate or nonsaponifibale oil to impart waterproofing properties.

1. Operation (limestone quarry, maintenance and repair, production)
2. Technical Services (engineering, R&D, quality control, etc.)
3. Marketing and Sales
4. Purchasing (extra raw material, fuel, equipment, etc.)
5. Human Resources Management
6. Logistics (storage facilities, trucks, optimization, etc.)
7. Finance (budgeting, cost control, accounting, cash flow, etc.)

Facts and Figures

Portland cement is made by grinding and mixing limestone and clay. The mixture is heated to 1400°C in a long, slowly rotating kiln, resulting in

(*continued*)

(*continued*)
golf-ball-size products called clinker. After the clinker is sufficiently cooled, it is ground into fine cement and, possibly, supplemented with add-ons such as fly ash or gypsum. The manufacturing of Portland cement creates about 5 percent of CO_2 emissions. Transporting cement via trucks over long distances is uneconomical. However, a substantial amount of cement is moved worldwide using special ships. The gap between growing demand and cement produced locally could be filled by supplemental importing, using ships and trucks. However, transportation costs cut into the profit of a cement producer that imports to keep the market share of cement sales in the region.

Concern about sustainability and ecofriendliness of the cement industry offers many opportunities to reduce the environmental impact while lowering production costs. One example is using scrap and nonhazardous wastes as fuel for the 1400°C kiln(s) to decrease the demands on local landfills and incinerators. The effect of this will be a smaller waste stream, less groundwater pollution, and lower fuel cost. There might be opportunities to collaborate with other industries to use their wastes and by-products. Such an effort might even synergize technologies to reduce polluting substances in the air, ground, or water.

A company strategy to use more ecofriendly opportunities could be developed by the innovative and creative engineers working for the company.

Model of Integrated Strategy

As a first step, we need to draw a map of integrated activities that will lead to the four visions established by the board of directors.

A proposal for a Balanced Scorecard, with an example entry for the first measure is outlined in Table 11.1.

Management of BSC

The BSC is a tool to achieve cohesion between different departments, as well as to monitor progress and performance of each entity. For this reason, each group involved should be asked to review their performance measures, and modify them if necessary. Furthermore, they need to develop a five-year progress plan, as the board of directors set a five-year goal. Subsequently, the planned values for the performance measures need to be chosen for each intermediate reporting period. A department may choose to set its own more detailed goals for its divisions.

After each quarter, the BSC should be filled out, posting the actual progress for each measure and assigning a score to account for nonmeasurable accomplishments. After the second and third quarters, a total review of the performance should be done in order to correct for unreasonable expectations. Possibly, changes in the plans need be made.

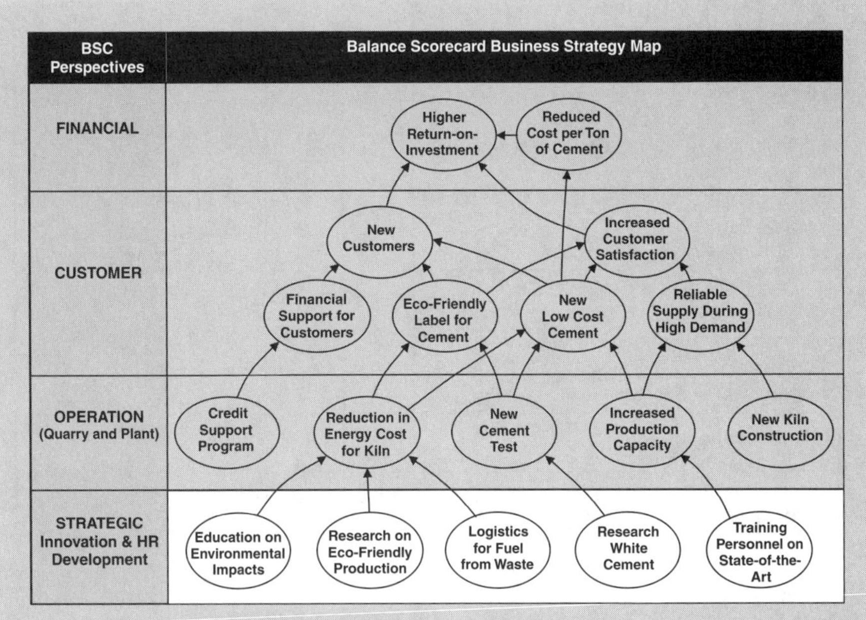

Figure 11.3 Map of Strategic Relationships of Initiatives

Table 11.1 Layout of Scorecard

Strategic Objectives with Performance Measures				
Date: September 4, 2009	Score	Actual	Target	Leader
Strategic Innovation and HR Development	79			
We educate the workforce on environmentally friendly cement production.				HR Management Department
Number of employees participating in seminars	90	35	39	
Postseminar evaluation score	100	89	90	
Number of entries into idea box for innovations of ecofriendly actions	25	3	12	
Number of rewards for excellent ideas	0	0	3	
We experiment with new ecofriendly production methods.				
Number of projects being researched actively				
Number of final reports with results				
Number of high-potential technologies				
Projected reduction in environmental impacts and production cost				

(continued)

(*continued*) *We develop new ecofriendly channels to fuel our kilns.* Sources of waste products in region Number of waste products successfully tested by research group Number of long-term agreements with waste sources Projected cost savings per ton of cement				
We research white cement to find ways to reduce production cost by 50%. Number of experiments conducted providing useful results Probability for success				
Operations (Quarry and Plant)				
We develop a financial support program for both our loyal and new customers. Total credit given to customers Number of new customers using credit Payback rate Cost of financing/ton cement sold				
We expand capacity when demand is high enough, by building a new kiln. Design status Demand projections Contractor prescreening Procurement of mechanical hardware Construction progress				
Testing new low-cost cement in production Assemble production line Quality testing by marketing department Cost/benefit analysis of new cement				
Reduction of energy cost of kiln Measure savings due to employee recommendation—rewards Savings due to new fuel sources				
Increased production capacity Number of new pieces of equipment installed and operators trained Return on investment (ROI) Capacity increase per investment				
Customers				
Reliable supply during high demand Accuracy of forecasts Storage capacities over time Number of unfilled orders				

Customer satisfaction Number of comprehensive surveys made by marketing group Satisfaction over time Dissatisfaction over time Effect of ecofriendly label Effect of low-cost white cement				
New Customers Number of new customers over time Reasons for becoming customer				
Financial				
Shareholder value Total shareholder return ROI of new investments of new initiative (long term) Production cost per ton of cement Total financial effect of ecofriendly initiative				

All participants in this project will be kept informed on the overall progress in order to encourage collaboration among departments, teams, and the individuals focusing on accomplishing the goals.

11.4 PERFORMANCE MANAGEMENT OF THE SUPPLY CHAIN

The first question we have to answer is, what does performance of the supply chain mean? As it is defined, it encompasses both efficiency and effectiveness of the supply chain, which in construction consists of suppliers of bulk and engineered materials, as well as rental companies. As we learned, efficiency relates to the productivity of production process, something that is directly applicable—unless we consider the delivery process as key. As a consequence, for managing the performance of the supply chain we have to focus on its effectiveness.

Now the questions at hand are: What constitutes an effective supply chain, and how can we measure it? A look at the I/O process model will help us find some answers. Supplies will, by design, provide the construction process with the necessary materials, prefabricated components, parts, equipment, tools, and so on. We also defined as the goal of process planning and control to minimize muda-waste, which has many sources, among them resource idleness, damages to output requiring rework, errors to be corrected, accidents causing damage and interruptions, nonvalue-added unnecessary work, equipment or tool breakdowns, and others. It is easy to recognize that these muda-waste examples are directly related to the quality of the supplies that are a major feed into the processes.

Who better than Edward Deming to "consult" about high-quality performance of the supply chain? We based the following list of key issues on his theory (which related to the manufacturing industry), and adapted it to the construction industry.

Factors to Ensure Optimal Supply for Construction Processes (from Deming's 11 Points)

1. Create consistency of purpose toward the improvement of product and service, and communicate this goal to all employees.
2. Adopt the new philosophy of quality throughout all levels in the organization.
3. No longer select suppliers based solely on price.
4. Improve processes, products, and services constantly; reduce waste.
5. Drive out fear of expressing ideas and concerns.

11.4.1 Defining Appropriate Performance Measures

Following the well-known rules put in place by Deming, we recognize that the key purpose of a performance management system is to strive for constant improvements in reducing muda in the future, not the past. Deming did not measure the quality of outputs in order to discard parts that did not meet a desired standard; rather, he plotted production charts to analyze the variations over time. Measuring the ever-changing differences relative to the target value provided critical information about how to improve the work. Complicating the effort, however, is that quality does not have a single standard measure; instead, each metric is highly specialized. And every so often, it is impossible to come up with a metric that can be used directly. For example, how can we measure the quality of a flat roof or a transmission system of a car?

Cybernetics

From the Greek *kybernētēs*, meaning "steersman," "pilot," or "rudder," cybernetics is the study of processes or systems able to set objectives, to act, to sense, and evaluate intelligently their functioning. It provides a model for studying business management and organizational learning, to make them more efficient.

The scientific background of measuring to control of systems is provided by *cybernetics*. A measurement system is, by nature, oriented to the past; therefore, it is critical that we create a system that:

1. Provides feedback data in real time.
2. Uses "feed-forward" predictors.
3. Searches data for trends

The obvious purpose is to manage the supply chain proactively, meaning eliminate the sources of imminent problems before they have time to manifest. The same principle was at work at Toyota, when the Japanese car company found that trends in quality data became indicators of trouble brewing on the production line. It was also found that *preventing* poor quality is cheaper than *fixing* poor quality. Equally important, repairing worn parts that caused quality problems, but that were not broken yet, eliminated breakdowns along the active production line. As a result, the efficiency of the process improved, due to less downtime, and the level of quality rose at the same time.

What are the causes of muda related to supply? This is impossible to answer with a specific list, because there are many construction supplies and with various risks to interrupt the process. For example, price and delivery time may be of varying importance under different conditions. Should a particular material item be needed for an activity on the critical path, delivery time has to receive special attention. Furthermore, the various owner requirements and contract specifications also make it impossible to define such a catalog. If the planned use of equipment/material is for a government-type project, the importance of the criteria will differ from those applied if the planned use is for a commercial project. Government-type projects emphasize submittal lead time, resubmittal time, cooperation in identifying deviations between the product, and the specifications and warranty. In contrast, commercial projects emphasize time reliability, expediting costs, and coordination with the engineer.

In other words, for the purpose of discussing the performance management of the supply chain, we have to pick measures knowing full well that each company will have to establish its own list. The following sections identify areas of concern from three main perspectives: (1) purchasing, (2) site operation, and (3) postproject repairs/warranties.

Perspective 1: Purchasing

The purchasing department of a mechanical contractor places heavy emphasis on criteria such as trust, problem resolution by the vendor, adherence to specifications, modifications to the product, and payment terms. Of less importance are submittal response time, lead time for delivery, company location/size, and discounts. Of course, the project management department in the same company would emphasize submittal time, time reliability, cooperation in identifying deviations between product and specification, and order tracking.

Table 11.2 presents an incomplete list of measures that could be used as the basis to assess comprehensive effectiveness.

Highlighted in the table are the first three measures to indicate their special nature in their capacity to serve as indicators and predictors early on. Obviously, trust with the purchasing department must have been earned on past projects. Whereas the financial situation of a company could have drastically changed since the last project, and so needs to be reevaluated in order to avoid the possibility of the supplier suddenly going bankrupt during the project. Another early-warning sign is the performance of the supplier during the bidding phase. Poor performance might be indicative of what could be expected after a contract is signed.

Perspective 2: Site Operation

The performance of a supplier during construction is tightly linked to the performance of the project overall. Six of the more critical criteria of performance include: (1) quality of delivered products; (2) timeliness, not only of product delivery but of key documents; (3) responsiveness to solving difficulties during the procurement process; (4) completeness of orders; (5) flexibility to changes; and (6) errors.

Table 11.2 Performance Measures Important for the Purchasing Department

	Measure	Description	Quantity	Quality
1.	Response time	Response time to questions or modifications during bidding phase alerts a contractor of potential problems.		
2.	Trust	Past dealings with the supplier, fulfilling unwritten agreements, standing by quoted prices, not trying to slip out of basic understandings.		
3.	Financial stability	Is the credit rating of the supplier high enough to pay its vendors?		
4.	Adherence to specifications	Accepts specifications required by the owner that may not all include the supplier's standard products.		
5.	Capacity	Is supplier willing to use its production capacity to ensure optimal flow?		
6.	Payment terms	Does the supplier stick with payment conditions, such as discounts for on-time payment?		
7.	Logistics	Is the supplier managing the logistics, proper bill of lading, protection, security against theft, and so on?		
8.	Bid price	How competitive are the supplier's prices?		

Again, supplier performance is not a one-time issue, but rather of ongoing importance throughout the project. Maintaining up-to-date pricing and catalog data at a contractor's office is critical to the contractor's ongoing operation. Prebid collaboration, including picking up drawings and specifications, pricing of material on short notice, and highlighting potential problems are other crucial services of a supplier.

Table 11.3 highlights the difficulty of creating a system that is mainly based on quantitative measures. Is it possible to translate the past or present performance of a supplier in logistics into a value between 0 and 100, the later meaning perfect? The table presents one approach, whereby performance measures are defined in terms of their value for the contractor. As shown, each measure is separated into three levels, each "earning" the supplier a predefined number of points (arbitrarily selected).

Of the 15 measures, none can be considered a predictor before the start of the project. However, some can be used as early-warning signs. For example, how smoothly the design approval process is handled or how complete the shipments are serve as indicators of how effective the supplier's design and fabrication shops are tracking projects. The effectiveness of the supplier's lead manager can be gauged early, using measures 4, 8, and 9, which indicate whether he or is organized and proactive, or operates reactively to problems.

Table 11.3 Example Performance Assessment Structure for a Supplier of Mechanical Systems

	Measure	Performance Levels with Earned Value Points 10 Points	5 Points	1 Point
1.	Timeliness of submittals	Provided within 2 weeks	Provided within 2 to 6 weeks	Between 6 and 8 weeks
2.	Lead time for release	2 to 8 weeks	8 to 12 weeks	12 to 14 weeks
3.	Response time to rejected submittals	Within 1 week	Within 3 weeks	3 to 4 weeks
4.	Cooperation in planning project	Provides information in timely manner	Provides information, after several requests	Does not provide complete information
5.	Design approval process	Excellent approval process	Submits sometimes 2 weeks late or with inadequate information	Consistently late approvals
6.	Responsiveness to problems	Meets and brings technical people promptly	Not prompt in responding to technical problems	Unwilling to provide technical expertise when needed
7.	Identifies deviations to specifications	Provides complete deviation list by specification	Provides incomplete list	Does not provide deviations
8.	Reporting of order status	Calls weekly to report status of orders	Reports only after prompting	Does not correspond regarding status
9.	Lead-time reliability	Lead times always met	Lead times met 40% of time	Lead times met between 40% and 60%
10.	Flexibility during installation	Has competent personnel (informed); can respond within 3 days	Has personnel (not always competent); can respond within 3 to 7 days	Does not have competent personnel; can respond only after 1 week
11.	Expediting costs	Does not charge for expediting unless unusual request	Charges 10% or less for expediting	Charges between 10% and 15%
12.	Product quality	0 to 2 quality defects, and only cosmetic	2 to 5 quality defects, of which at least 1 is operational	6 to10 operational and/or cosmetic defects

(*continued*)

Table 11.3 *(continued)*

	Measure	Performance Levels with Earned Value Points: 10 Points	5 Points	1 Point
13.	Timeliness/ competence during start-up	Responds within 1 week, with competent craftsman	Responds after 1 week, or person lacks competence	Responds after 10 to 15 days, or person is not competent
14.	Production of operation and maintenance manuals	Produces documents prior to equipment installation	Produces prior to project completion	Produces only after project completion
15.	Completeness of shipments	Shipments are always complete	Maximum 5% of shipments are incomplete	Between 5% and 10% of shipments are incomplete

Perspective 3: Postproject Repairs/Warranties

The performance after the completion of the physical facility is important, too. Warranty servicing and the quality of the operating and maintenance manuals are critical to a contractor, in that retained payment reduction by the owner, as well as project release, are based on good performance in these areas. In summary, the service of a supplier after completion is as important as the supplies.

Three measures might be at the forefront here:

1. Number of warranty calls that the supplier has to respond to
2. Responsiveness to warranty calls
3. Overall satisfaction of the owner and user with the supplied system

11.4.2 Framework for Managing Supply Performance

As we learned, the difference between performance measurement and performance management is the desire to influence the process, with the goal of making further improvement. Hence, the data from the measurement needs to become the basis for fault identification and preventive interventions. Since projects and subcontracts can last an entire year, there should be enough time for managing the performance of suppliers, as they eventually impact the performance of the entire project.

The core of this approach is the timely recording of supplier performance in an agent-based project website—and by "timely" we mean every time a measurement can be made that involves one of the measures established at the beginning. A software agent monitors and updates a performance chart before processing the new

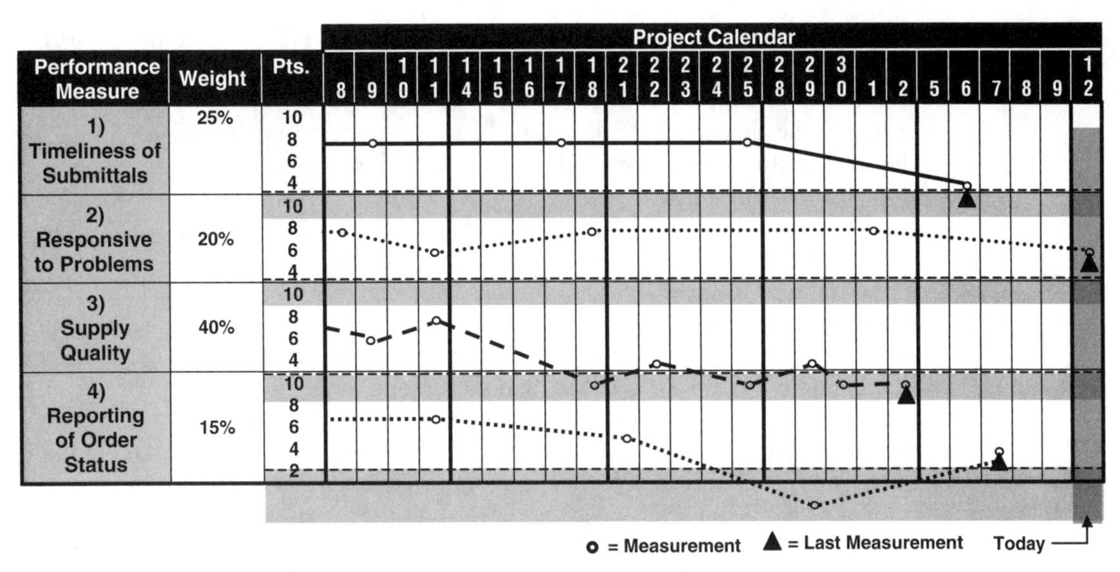

Figure 11.4 Example Performance Charts with Four Performance Measures

data. This chart is a time-based plot of the performance points earned by the supplier. The plot will readily highlight poor performance, as well as any changes that might indicate alterations inside the supply chain, such as a new project manager hired by a supplier. Figure 11.4 presents a four-measure example with a five-level scale: 0, 2, 4, 6, 8, and 10.

Each of the four performance measures is accompanied by a weight in percentage, shown in the second column. The weights identify their relative importance, and add up to 100 percent. Table 11.4 shows how weights are used to calculate

Table 11.4 Performance Analysis and Trends

Performance Measure	Weight	no. of Entries	Total Performance Points	Mean	Weighted Mean	Mean Previous Four Entries	Last Entry	Trend
1.	25%	4	10+10+10+6	9	0.25*9=2.25	9	6	Down
2.	20%	5	10+8+10+ 10+8	9.2	0.2*9.2=1.84	9.5	8	Down
3.Supply Quality	40%	8	8+10+4+6+ 4+6+4+4	5.75	0.4*5.75=2.3	4.5	4	Down
4.	15%	4	10+8+2+6	6.5	0.15*6.5=0.97	6.5	6	Down
	100%	21		7.6	7.36	← Overall weighted performance		

weighted performance values for each measure, and overall. The example performance chart covers the last five weeks, with each measurement entered at the date it was made. The plots reveal clearly the trend of each performance measure. For example, the timeliness of submittals had been perfect until the unusual case on the sixth of the month. We would have to wait to see whether that was an aberration or the beginning of a trend. At least, somebody should inquire what happened. Of greater concern, supply quality has been steadily deteriorating, to below average. This must be addressed, especially since quality has, at 40 percent, the highest weight assigned. Taken together with the very low performance in status reporting, we must assume that the supplier's fabrication shop is facing major problems. The performance chart certainly justifies a high-level meeting with the supplier to find out what needs to be done to stem the sliding performance.

Computing weighted means and trends results in the same observations as the graphical plots, but provides the quantitative information necessary to compare them against benchmark values. The overall weighted performance of the supplier is 7.36, while the unweighted mean is 7.6. This means that the supplier performed poorly at higher-weighted measures. Table 11.4 supports this fact, showing that for the performance measure number 3, supply quality, the supplier reaches only a weighted mean of 5.75, far below the other measures. What makes it worse is that it is most important to the contractor indicated by the weight of 40 percent, the highest. Finally, all the trends are set the down. For this example, the trend is based on the difference between the mean of the previous four entries and the last.

11.5 PERFORMANCE MANAGEMENT AT THE TASK LEVEL

At this point, we should remind ourselves that performance management includes both the measurement of efficiency and effectiveness, as well as the use the data to improve the present situation. Unlike Taylor, whose time studies focused solely on improving productivity, this theory includes other aspects of the work. For example, the information a worker receives might be so erroneous or incomplete that much of his or her work has to be redone; or the designed setup for the workplace might cause so much strain on the workers' backs that it will fatigue them quickly, cause pain, require frequent rest stops, and lead to increased risks of back injuries. As we learned in Chapter 6, high levels of fatigue and work stress raise the risk of accidents, and demotivate employees.

Two approaches to labor-intensive performance management at the task level are motion studies and preventive ergonomics.

11.5.1 Improvement through Motion Studies

In the early twentieth century, the path to improve the work of laborers was opened up by the couple Frank and Lillian Gilbreth, whose goal it was to make work easier and more straightforward and eliminate the risk of injuries. In other words, they looked beyond efficiency to the effectiveness of the work. The motion

Table 11.5 Four Steps of a Motion Study

Step	Goal	Possible Actions
1	Elimination or reduction of motions	Add simple mechanisms, tools, supports, or material layouts
2	Making motions to be executed simultaneously	Design improvements in the methods and tools that allow both hands to be used at the same time
3	Shortening of motion distances	Reduce walking, reaching, stretching, squatting, turning, and other movement
4	Making motions easier	Smooth the flow of work; reduce fatigue-producing motions such as bending, kneeling, pulling, lifting,and promote safety

studies they developed extended beyond the immediate workplace, to include how materials and parts were delivered. Frank Gilbreth, who was a building contractor himself, designed a climbing scaffold that could be raised as the masonry wall grew taller. The scaffold platform for the masons featured raised benches for the brick and mortar, eliminating the need to bend. Finally, the brick was delivered by laborers in such a fashion that it was easy for bricklayers to grab the bricks with one hand and apply the mortar with the other, without having to turn around.

Gilbreth also invented a forerunner of the "live action" movie to study the motion of subjects such as horses or athletes. He took advantage of the "etching" capability of a light beam on a light-sensitive film to capture the motion of a worker, using a small electrical light attached to his hands. Dubbed a "cycle--graph," the photograph presented white lines on a single time-exposed photograph created by the bright, moving light. A picture full of twisting lines represented an inefficient movement that needed to be simplified by applying the 18 therbligs, the basic motions that the Gilbreths had identified as underlying work.

It is well known that the repeated lifting of heavy loads, the climbing of steps, bending over, or squatting for long periods of time will fatigue the body to the point at which it needs to slow down, rest, or stretch. Let's study how the principles of motion studies can be used to assess the work performance of individuals.

The Gilbreths proposed four basic steps to pursue a motion study targeted at improving the work of an individual. These are shown in Table 11.5.

Work can be improved on an ongoing basis, especially in construction, as the conditions and the environment are constantly changing. Comparing laborers with different experience levels, it is easy to recognize that figuring out how to make someone's work easier and more efficient requires a systematic approach. The Gilbreth's proposed four-step approach provides a system for managing the performance of labor that can be easily integrated into worker training, teamwork, and recognition and reward of excellent work by the company. Worked-Out Problem 11.2 gives an example of how to conduct a motion study to improve the performance of bricklaying.

Worked-Out Problem: 11.2: Motion Study of Bricklaying

A building contractor, Elisabeth Cook, learned that Frank Gilbreth used to be a contractor who early on improved the performance of his masons. For one, he had invented an adjustable scaffold that permitted quick adjustment of the working platform so that the masons were at the most convenient level at all times. Elisabeth is challenging us to assess the work performance of her masonry crew that just started a new job. In order to save time, Fred, one of the two masons, and one helper started laying the bottom courses while the scaffold was installed on one side wall. The site visit presented in Figure 11.5 shows the sequence of how Fred laid brick at one end of the wall.

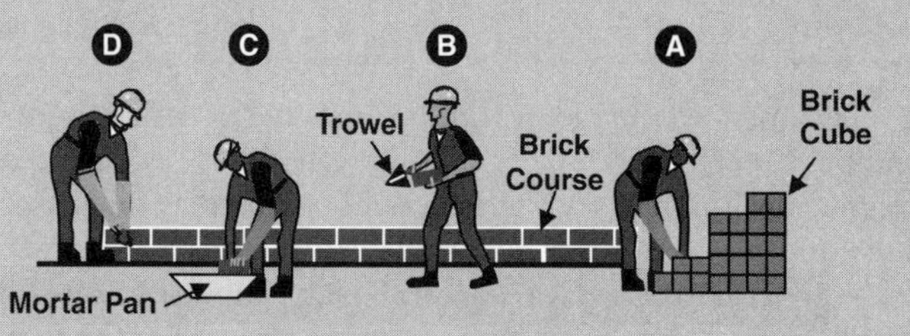

Figure 11.5 Fred's Setup to Lay the First Courses for the Brick Wall

Assumptions and Models

Laying bricks requires the mason to place and tap one brick at a time after laying a mortar bed on top of the previous course. The ends of the bricks need to be offset at each course, ending in a head joint filled with mortar. Bricks at the end of the wall need to be cut in half with a saw or a hammer. A cube of bricks is positioned at one end of the wall and serves as the brick supply. One helper, not visible in the figure, mixes the mortar and pours it into the mortar pan on the ground.

Task A: Picking next brick with left hand

Task B: Walking

Task C: Getting mortar with the trowel in the right hand and brick in left hand

Task D: Laying brick with both hands

The motions necessary to move from position to position can be defined as shown in Table 11.6.

Analysis of Motion Sequence

Frank Gilbreth used photographic methods to capture the sequence, time, and exact paths of the hands and body of a worker. One approach he then

Table 11.6 Basic Bricklaying Motions

Basic Motion	Fatigue Producing	Basic Motion	Fatigue Producing
1. Apply mortar for head joint		6. Lay new brick	
2. Apply mortar for bed joint		7. Reach and grasp new brick	
3. Bend	x	8. Tap brick and strike off mortar	
4. Get mortar on trowel		9. Turn body	
5. Hold brick		10. Walk	x

used was to develop a process chart, which he subsequently inspected for possible improvements. One important feature of a process chart for motion study is that the motions of the left and right hands are shown separately.

Figure 10.6 (a) presents the chart for Fred's situation. Let's apply Gilbreths' four phases to identify all the opportunities for improvement.

Phase 1: Eliminate motions

It is apparent that the fatigue-producing motions of *walking and bending* re appear several times and so should to be prime targets. Reviewing Fred's setup, it's clear that he created for himself a triangle with the brick, mortar, and the moving location, to lay a brick at its corners. Moving the bricks and mortar together will eliminate one edge of the triangle. There are two different reasons for bending: storing material on the ground, and laying brick for the lowest courses.

The first reason can be easily fixed by using simple elevated platforms; but not much can be done about the second, as it is part of the job. Its fatiguing impact will, however, be quickly reduced as the wall rises.

Phase 2: Simultaneous Motion

By moving bricks and mortar closer together, the mason will be able to grab a new brick with his left hand while getting mortar onto the trowel with his right hand.

Phase 3: Shorten Motion Distances

Because of the nature of assembling a long wall with small elements by hand, turning and walking can't be eliminated, but they can be shortened. Setting up multiple mortar pans, instead of one located centrally, adjacent to

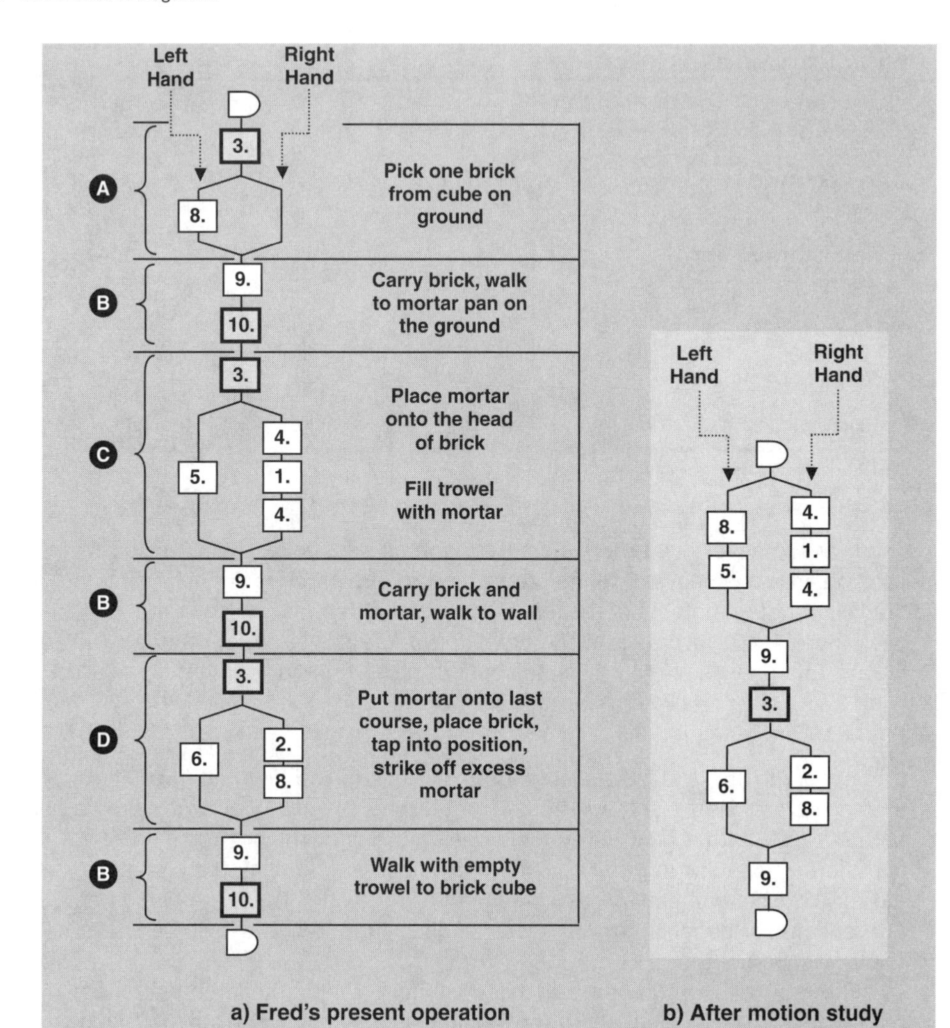

Figure 11.6 Process Charts of Existing and Modified Method

multiple brick cubes, and parallel and close to the wall, will minimize the need to walk back and forth. The 180 degrees of turning, however, is impossible to reduce.

Phase 4: Easier Motions

Two fatigue-producing repetitive motions have already been removed and minimized. The only other possibility is to use mortar pumps and hoses that would get rid of the mortar pan but introduce other tasks. We will not consider this option at this time.

Setup of New Operation

The final step in the motion study is to redesign the workplace, showing the mason's new motions and the physical relationships between the laborer, materials, tools, and brick wall (see Figure 11.7). This step will ensure that the masons can still operate comfortably within the redesigned environment.

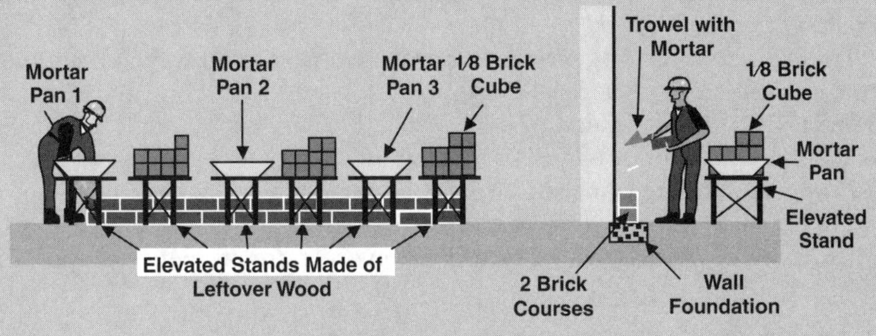

Figure 11.7 Sketch of Masonry Setup after Motion Study

Discussion

It is apparent that the new operation requires additional mortar pans and elevated stands. Also, it will mean slightly more work for the helper, in that a cube has to be broken up, distributed to three tables, and the mortar poured into three elevated pans. However, this should not incur an additional cost, because the pans are made of cheap reusable metal pieces, and the stands can be made collapsible, from leftover lumber.

The distribution of bricks and mortar is commonly done with the help of a clamp tool, wheelbarrow, or mechanical device. If the stands are in place early enough, the breaking up of the cubes can be done when one of the helpers is idle. The higher productivity of the mason measured in m^2 of wall per hour will more than justify the minimal additional costs. The key to improved performance will be the dramatically lower level of fatigue and the drastic reduction in stresses experienced by the mason to his vertebra.

Performance management does include training and employee involvement, to align the entire workforce behind the new performance goals. Is it easy to recognize that motion studies can be used to establish benchmarks for the company's entire workforce. Still, for this approach to succeed, several additional steps are necessary: (1) training the workforce, (2) brainstorming with the staff to find further improvements, and (3) setting up a reward and recognition system.

11.5.2 Improvement through Intervention Ergonomics

One key aspect of performance management is proactive problem avoidance, as demonstrated by the supply-chain application. One major area in construction that could benefit from the collection and analysis of performance data, followed by the implementation of improvements, is safety and health. Well-known sayings such as "Prevention is better than healing," "Prevention is cheaper than treatment," and "Knowledge is power" all apply here, for prevention depends fundamentally on the knowledge of a risk. Thus, the first step in any prevention plan is awareness of a potential risk.

To that end, NIOSH generated a list of "red flags" and turned them into a simple questionnaire that can be used to assess the risk level of a given task (DHHS (NIOSH) Publication 97-117). Table 11.7 presents a simple task analysis checklist, an extension of NIOSH's original list. It should serve as a reminder of those issues and items that need review before beginning a new operation—similar to the way a pilot prepares for takeoff. Note that every line that receives a check in the "No" column signals that a closer look is in order, for prevention purposes.

Table 11.7 Checklist for Preventing Health and Safety Problem

Task Analysis Checklist for Problem Avoidance	Yes	No
1. Does the primary task require:		
Bending or twisting of the back or trunk?		
Crouching?		
Bending or twisting of the wrist?		
Extending the arms?		
Raised elbows?		
Static extended muscle loading?		
Finger pinch grip?		
2. Are mechanical lifting devices used?		
3. Will the created noise level stay below the acceptable decibel		
4. Can the task be done with two hands?		
5. Can pushing or pulling forces be exercised using proper handles?		
6. Are supportive tools assigned for tasks with extensive kneeling?		
7. Are the materials to be moved:		

Able to be held without slipping?		
Free from sharp edges and corners?		
8. Are jigs, tables, stands, and vises used to elevate workplaces?		
9. Are hand gloves of the proper material used?		
10. Is the material to be carried or lifted less than 23 kg (51 lbs)?		
11. Are the laborers asked to wear personal protective equipment?		
12. For repetitive motions tasks, are the following measures required:		
Job rotation?		
Self-pacing?		
Sufficient pauses?		
Adjusting the job skill level of the worker?		
13. For tasks requiring > 200 kcal/hr, are rest periods planned?		
14. For operators of equipment, are vibration exposures mitigated?		
15. Do power hand-tool operators know about vibration exposure?		
16. Is the employee trained in:		
Proper healthy and safe work practices?		
How to make adjustments in order to mitigate hazards?		
Recognizing signs and symptoms of health problems?		

Prevention during Design

NIOSH launched a national initiative called Prevention through Design (PtD), to promote the inclusion of health and safety issues in the design process. The premise was that risk-avoidance activities frequently are least expensive when done during the phase where simple modifications will cost nothing but will dramatically reduce a hazard potential. Take, for example, the roofing activity, a major contributor to deaths through falls. If the design of a roof were to include simple anchorage points for lifelines, or supports for toe-boards, roofers would realize that that the designer, owner, and contractor care about their health and safety. Similarly, a steel structure might be designed to include connection points to anchor lifelines, horizontal or vertical, that hook to body harnesses.

As an example, consider that, today, a worker nailing the floor boards to the joists or beams underneath with a nail gun not only has to bend over constantly, but will feel the recoil action from the air hammer. Keep in mind that occupational back injuries result in more lost workdays than any other illness, second only to the common cold. Many of these problems can be avoided using the Task Analysis checklist shown in Table 11.6; yet many other hazards are systemic in nature, caused by the lack of ergonomic tools, difficulty in applying what is available in construction, and, finally, by the high cost of prevention.

Part of the preventive effort involves technological intervention, through modification of traditional methods, with to goal to prevent risks. Backaches, as common as they are, are rarely lethal; but statistics show that they cause an extreme amount of misery. These recurrent attacks of pain generally occur throughout a person's adult years, often interfering with the prime time of life. Back injuries are cumulative, too, meaning that the discs in the back erode slowly with overuse and may rip one day lifting an object that is not particularly heavy or cumbersome. Anybody who has experienced the agony of a herniated disc will attest to the fact that the ensuing pain cannot be controlled. In addition, the sufferer becomes totally incapable of doing anything requiring walking, or even sitting. An example of a technical intervention to eliminate the bending and recoil action of a nail gun is the Ergonomic Nailing Device (END). As Figure 11.8 shows, the basic mechanism links the space between the handle of the nail gun on the floor to the hand of an upright, standing person. The END is an attachment that serves as an extension device capable of handling any commercial pneumatic nail gun. The characteristics of the new mechanism are:

Figure 11.8 Ergonomic Nailing

1. Lightweight aluminum structure
2. Height and angle adjustment capability
3. Trigger operation at waist level
4. Adaptability to any pneumatic nail gun
5. Mobility with the use of multidirectional, height-adjustable casters
6. Safe operation by utilizing the safety mechanisms of the nail gun itself
7. Quick and efficient method of attaching and detaching the nail gun

The flexible extension of the END, as pictured in Figure 11.8, is made of lightweight aluminum tubing that is both height and angle adjustable. The trigger is located on the right (or left) handle grip and is operated with two fingers. A cable connects the finger trigger and the actual trigger-housing mechanism. This housing mechanism engages the trigger of the pneumatic nail gun, and allows only one nail to be shot at a time, and is only operational when the nail gun is in the correct position for nailing, thereby retaining all the safety features implemented by the manufacturer. The entire device is completely mobile in all directions, with the addition of two multidirectional, height-adjustable casters.

The wheeled gun carrier, with extensions for control, was a big hit with plant laborers who needed frequent breaks and complained of back pain. A simpler version was tested on the construction site. Figure 11.9 (b) shows how the carpenter is able to operate while standing upright yet still experiencing the recoil effect that was eliminated for the heavier gun.

Ergonomic intervention efforts to improve performance can focus on many different problem areas, among them: (1) layout of the job (e.g., brick supply), (2) horizontal or vertical movement of heavy material by hand (e.g., mortar), (3) exposure to fumes and dust, or (4) work in confined spaces.

CHAPTER REVIEW

Journaling Questions

1. Henri Fayol was a contemporary of Frederick Taylor, but their approaches to managing the performance of construction could not be more different. Discuss the differences and compare them with today's approach to performance management.
2. Edward Deming and Taiichi Ohno applied statistical quality control similar to the way performance management uses performance measurements. Explain the commonalities of the two approaches, in the way they were feeding into performance improvement efforts.
3. The Balanced Scorecard and the stock market impose vastly different and even contradictory pressures on the upper management of a company.

a) Casters and extension eliminates recoil and fatigue

b) Simple extension for light nail gun

Figure 11.9 Field Testing of Ergonomic Nailing Device: (a) Casters and Extension Eliminates Recoil and Fatigue; (b) Simple Extension for Light Nail Gun

Describe the differences between the two methods for assessing the value of a company. As a CEO of a large construction company trying to open a new market, requiring substantial investments, how would you deal with or use the two concepts?

4. Performance management of operations or suppliers during construction is difficult to do, as measurements can only be taken of what already has happened. Sketch out a system that would still provide valuable input to proactively minimize the occurrence of major problems in the future.

Traditional Homework

1. Worked-Out Problem 10.2 applied a motion study to improving the performance of bricklaying. The example showed Fred, the mason, starting at the bottom of a new wall. It is apparent that he was still required to bend in order to lay the first few layers of the wall. Sketch a scaffold system needed to reach the second floor, which would not only make the brick and mortar available at a convenient height but also would eliminate the need for Fred to bend during the placement of the brick.
2. Consider the performance chart in Figure 11.4 Assume that on the 13th (tomorrow) which is not yet shown on the calendar, each of the four

measures receives an 8 as a new entry. How would the trends change? Would you still recommend a big meeting with the supplier? Why or why not?

Open Ended Question

1. History tell us that construction people lack the desire to change, even if it might make their work easier. In a recent survey, construction foreman were asked what would they buy to make their job easier if they were given $75,000. The overwhelming answer was: a bigger pickup truck. Thus, it is not surprising that new but ergonomically optimized tools are not being bought by construction companies, which in turn discourages tool manufacturers from investing in their production.

   Develop a marketing strategy for a new tool that improves the performance of a construction worker. How would you convince a manufacturer to make it? How would you convince enough contractors to buy it? What would you recommend the contractors use in order to make sure that the new tool would be adopted by the employees?

BIBLIOGRAPHY

Bernold, L. E., and J. F. Treseler. Vendor Analysis for Best Buy in Construction. *J. Constr. Eng. Mgmt.*, vol. 117, no. 4, 1991.

Busi, M., and U.S. Bititci. Collaborative Performance Management: Present Gaps and Future Research. *Int. J. Prod. Perform. Mgmt.*, vol. 55, no. 1, 2006.

Chan, A. P. C., and A. P. L. Chan. Key Performance Indicators for Measuring Construction Success. *Int. J. Benchmarking*, vol. 11, no. 2, 2004.

Doolen, T. L., E. M. Van Aken, J. A. Farris, J. M. Worley, and J. Huwe. Kaizen Events and Organizational Performance: A Field Study. *Int. J. Prod. Perform. Mgmt.*, vol. 57, no. 8, 2008.

Evans, I. R. Impacts of Information Management on Business Performance. *Int. J. Benchmarking*, vol. 14, no. 4, 2007.

Fernandes, B. H., J. F. Mills, and M. T. Fleury. Resources That Drive Performance: An Empirical Investigation. *Int. J. Prod. Perform. Mgmt.*, vol. 54, no. 5/6, 2005.

Halachmi, A. Performance Measurement Is Only One Way of Managing Performance. *Int. J. Prod. Perform. Mgmt.*, vol. 54, no. 7, 2005.

Hervani, A. A, M. M. Helms, and J. Sarkis. Performance Measurement for Green Supply Chain Management. *Benchmarking: An International Journal*, vol. 12, no. 4, 2005.

Pheng, L. S., and P. Ke-Wei. A Framework for Implementing TQM in Construction. *TQM Magazine*, vol. 8, no. 5, 1996.

Radnor, Z. J., and D. Barnes. Historical Analysis of Performance Measurement and Management in Operations Management. *Int. J. Prod. Perform. Mgmt.*, vol. 56, no. 5/6, 2007.

Randall, W. S. Utilizing Cash-to-Cash to Benchmark Company Performance. *Int. J. Benchmarking*, vol. 16, no. 4, 2009.

Sharif, A. Benchmarking Performance Management Systems. *Int. J. Benchmarking*, vol. 9(1), 2002.

Shepherd, C., and H. Günter. Measuring Supply Chain Performance: Current Research and Future Directions. *Int. J. Prod. Perform. Mgmt.*, vol. 55, no. 3/4, 2006.

St-Pierre, J., and L. Raymond. Short-Term Effects of Benchmarking on the Manufacturing Practices and Performance of SMEs. *Int. J. Prod. Perform. Mgmt.*, vol. 53, no. 8, 2004.

Wiese D. S., and M. R. Buckley. The Evolution of the Performance Appraisal Process. *J. Mgmt. History*, vol. 4, no. 3, 1998.

Wren, C. R., and A. P. Pentland. Dynamic Models of Human Motion. *Proc. FG'98, IEEE*, April 14–16, Nara, Japan, 1998.

Appendix

An Overview of Modeling Elements in Simphony

In describing the modeling elements of the General Purpose Modeling Template in Simphony, we will use a simple structure as shown below. First we will give the element name and its symbol.

Element Name	**The modeling element name as it appears in the template, e.g., task**
Element symbol	The symbol used to describe the element. The shape for each element is unique.
Properties box of the element	The properties box contains all input parameters and out parameters pertaining to the element. The user can specify the values of various input variables to the modeling element and the simulator will display all outputs after the simulation.
Input to parameters:	The inputs that the user can specify for the element
Output parameters and statistics:	The output form of the simulation is displayed in the property box in this section. Statistics are also produced and displayed in the property box when applicable.
Description of the element	Description of how the element is intended to be used, what it does, etc.

The reader should note that due to the flexibility of Simphony, the element appearance and functionality in the version that he/she is using may have been customized and may differ from the descriptions here. Indeed, different versions of Simphony may have different modeling element looks and functionality. What we describe here are the basic modeling elements for Version 1.1.13.

A.1 DESCRIPTIONS OF THE ELEMENTS IN SIMPHONY VERSION 1.1.13

A.1.1 Comment Element

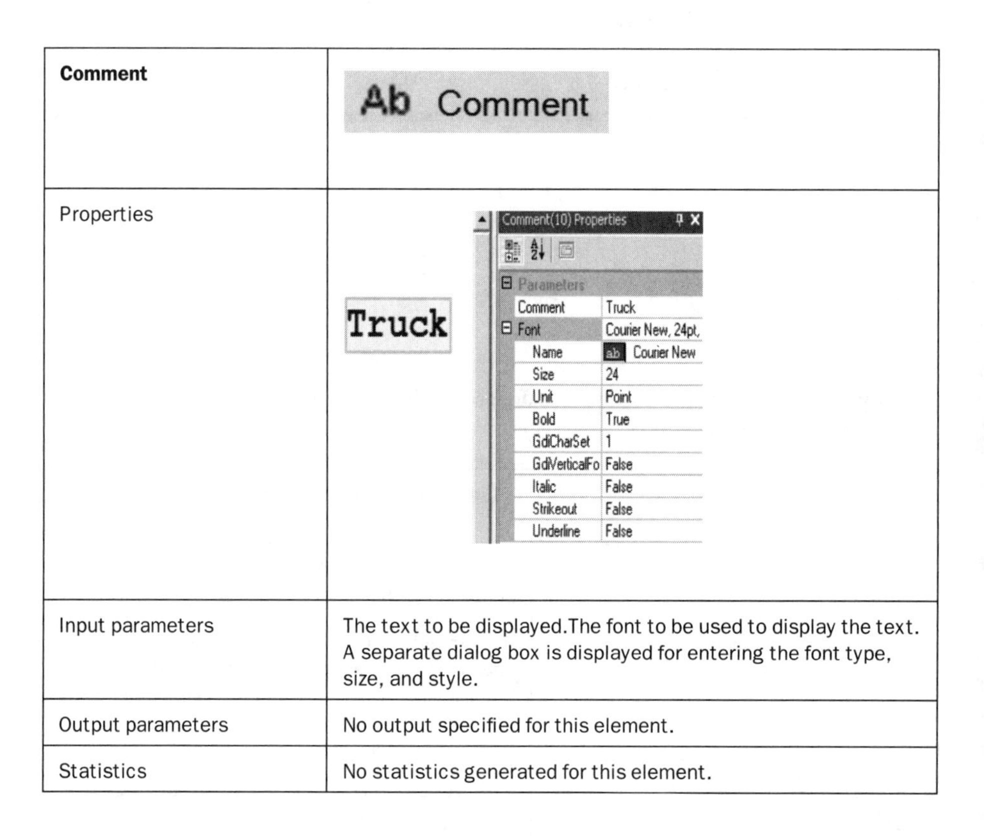

Comment	Ab Comment
Properties	Truck
Input parameters	The text to be displayed.The font to be used to display the text. A separate dialog box is displayed for entering the font type, size, and style.
Output parameters	No output specified for this element.
Statistics	No statistics generated for this element.

A.1.1.1 Description

The *Comment* element, as the name implies, is an element that has no function in the model other than documentation. It is best used to describe what the model is doing. The element properties box allows the user to add comments (as text). The user can also change the font of the text being displayed.

A.1.2 New Entity Element

New Entity	
Properties	NewEntity(2) Properties Parameters NA: 1 TBA: Constant(0) (DistType): Constant 1:Constant V: 0 TFA: 0
Input parameters	*NA (Number of Arrivals):* Total number of entities that would be produced by the element. *TBA (Time Between Arrivals):* The time interval between the creations of entities. The user can set this time to a constant or to a random value sampled from a statistical distribution. The user can also write VB code in the "Formula" option to specify a user-defined way for determining this variable. *TFA (Time of First Arrival):* The time at which the first entity will be created.
Output parameters	No output specified for this element.
Statistics	No statistics generated for this element.

A.1.2.1 Description

The *NewEntity* element, as the name implies, creates new entities and sends them to the next connected element. The number of entities created is specified in the parameter NA, or Number of Arrivals, in the properties box. The time between creations (arrivals) of entities is specified by the parameter TBA, or Time Between Arrivals. The time of the first creation is, by default, set to 0 but can be specified by the user using the Time of First Creation parameter, TFA, as shown in the properties box.

A.1.3 Task Element

Task	 Task2 Unconstrained Dur: Constant (1)
Properties	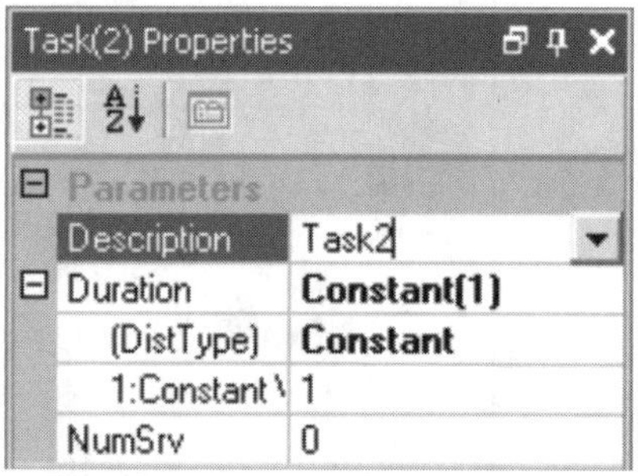 Task(2) Properties Parameters Description: Task2 Duration: Constant(1) (DistType): Constant 1:Constant: 1 NumSrv: 0
Input parameters	*Duration:* The time after which the entity will be transferred out from the element. The time is specified by the Duration Parameter by specifying the distribution type and the parameters of the distribution. The duration can be: *Deterministic*—Distribution type is constant and the value of the duration to be used is specified in the parameter box. *Probabilistic*—Distribution type is selected from available sampling distributions (e.g., Uniform, Triangular, Normal, Log Normal, Beta, Exponential). The parameters of the distributions in this case must be entered by the user in the appropriate parameter box. *User defined*—A formula writing in VB code by the user is specified. The duration is returned from the computations carried out in that code. *NumSrv (Number of Servers):* The number of available servers. If zero is entered, it means that task is ''Unconstrained,'' entities are processed without limitations.
Output parameters	Each entity transferred to the element is transferred out after the duration time specified in the input parameters.
Statistics	If the number of servers is specified, the element displays statistics on the utilization, queue length, and waiting time for the servers in the properties box. If the task is unconstrained, the displayed statistics will all have a value of 0.

A.1.3.1 Description

The *Task* element is used to represent an activity in the model. In general, an entity has to complete a given task either alone or with the assistance of a server or resource. The task simply delays the entity by the amount specified by the task duration before it releases it to the next element.

There are two types of tasks: constrained and unconstrained. A constrained task has a limited number of servers associated with it. When an entity passes through such a task it must check to see if the servers are available or tied up with other entities. The entities queue when the servers are all busy and then de-queue and commence the task once a server becomes available. An unconstrained task has no such limitations. Any entity passing through is processed immediately and never queued.

A.1.4 Counter Element

Counter	 Counter
Properties	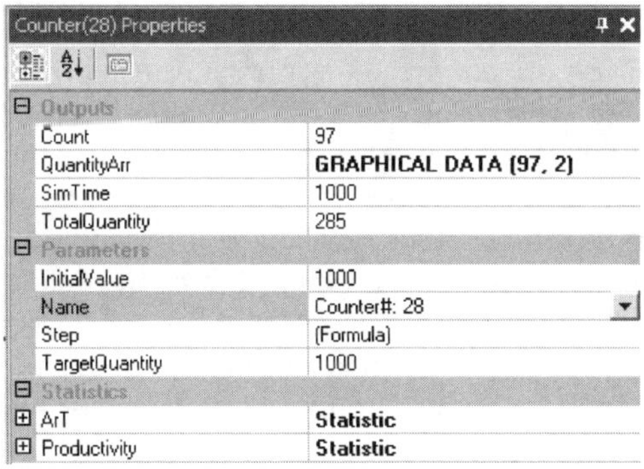 Counter(28) Properties **Outputs** Count: 97 QuantityArr: GRAPHICAL DATA (97, 2) SimTime: 1000 TotalQuantity: 285 **Parameters** InitialValue: 1000 Name: Counter#: 28 Step: (Formula) TargetQuantity: 1000 **Statistics** ArT: Statistic Productivity: Statistic
Input parameters	*Name*: Name of Counter element. *Initial Value*: This element allows the user to define an initial value and change it by a step function through the simulation. Whenever an entity arrives to this element, total quantity is reduced by a step value and its value is updated in the screen. *Step (multiplier)*: The user can specify step value (or function) to represent the amount with which the counter increments upon the arrival of an entity. *Target Quantity*: The target total amount of entities can also be specified. The first counter that reaches its target quantity will terminate the simulation. Its default value is 1,000.

Output parameters	*Count*: The number of passing entities. *Total Quantity*: The total accumulated amount, and the change of this amount with time in a graphical format.
Statistics	The element collects statistics on the number of served entities.

A.1.4.1 Description

The *Counter* element in Simphony mimics the Counter element from CYCLONE (Hapin 1976). It is a useful element whose main function is to count entities that pass through it. In productivity studies this is most useful as production is generally realized when a particular customer completes a given cycle.

The most basic form of a counter in Simphony has an initial value of 0, a step value of 1. This means that at the start of the simulation, the counter has nothing. As entities pass through it, it will record a production value of 1 for each entity that passes through. When a truck, for example, passes through a dump location, the counter would record 1 truck load produced.

The counter can also provide a means for stopping the simulation. If the Quantity parameter is reached during simulation for any of the counters, the model run will be considered complete. This provides an effective means of estimating the time required to produce a certain amount of product, for example. The basic form of a counter is shown in Figure A.1, Model A.

We can use the counter in a more advanced manner if desired. For example, we can specify the starting value of the counter at the beginning of the simulation with the InitalValue parameter. In this case, the counter starts accumulating the count from that initial value. In addition, we can specify a multiplier, or a step value, for the counter. This can be a real number. For example, in Model B, we have a counter where the start value is 20 truck loads. When trucks pass through we want to record 1.1 to reflect a 10 percent swell factor. We also want the simulation to stop once 80 truck loads have been completed. The reader should note that the time required is only 10 minutes (since all 100 entities are created at time 0 and all require 10 minutes at the task which is unconstrained). When the count exceeds 80 (in our counter it shows 80.5), the simulation stops. Focusing the cursor on the counter allows us to see the simulation total time as 10, the number of entities as 55, and the total recorded production equaling 80.5 units ($20 + 1.1 \times 55$).

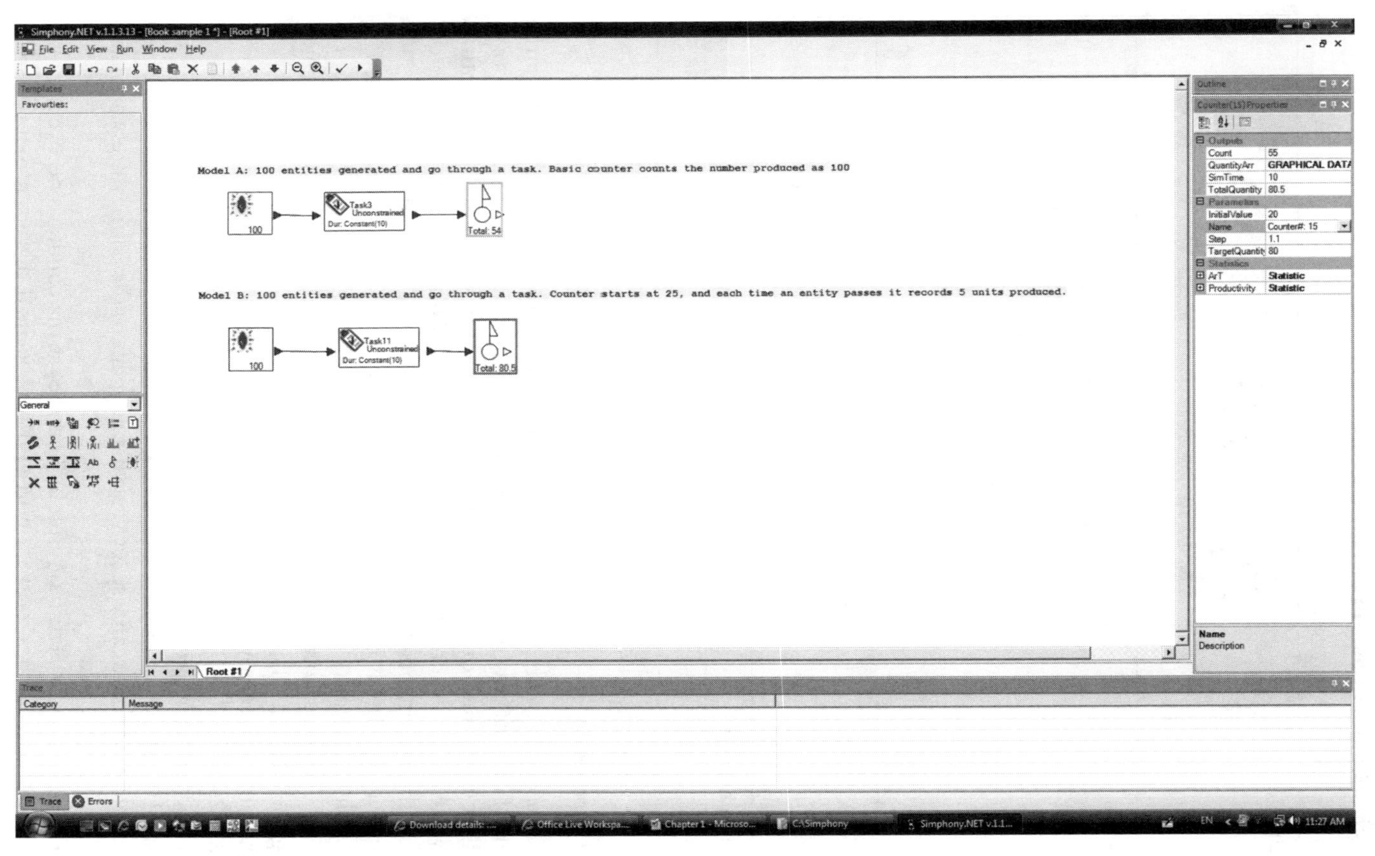

Figure A.1 Counter element—basic form

A.1.5 Consolidate/Generate Element

Consolidate	Consolidate
Properties	Consolidate(3) Properties Parameters Generate 1 Quantity 5
Input parameters	*Quantity*: The number of entities that need to be accumulated before the element releases. *Generate*: The number of entities that will be generated each time the quantity is reached.
Output parameters	No output specified for this element.
Statistics	No statistics generated for this element.

A.1.5.1 Description

The *Consolidate/Generate* element provides a means for the user to manage the number of entities flowing through the model. The element has two output ports. The **right-hand side output** connection point will transfer out the same entities that are transferred into the element. The **bottom output port** transfers out clones of these elements depending on **the setting of the element**.

The element holds incoming entities until the specified *quantity* is reached, at which point it outputs the specified *number of entities, as designated by the user under the Generate parameter*. The generated entities are clones of the last entity passing the element when the number to consolidate was reached.

To demonstrate how this element works, consider the simple model shown in Figure A.2:

The screen shot shows three submodels, Model A, B, and C. In all three we generate 10 entities and send them to a Consolidate node, then from there to a Counter element so we can visually see what happens to the entities after they leave the Consolidate element.

In Model A, the 10 entities result in 2 entities at the counter since we have the counter connected to the bottom port and we specified that each 5 entities will consolidate into 1. Note that the first entity arriving at the counter will have the

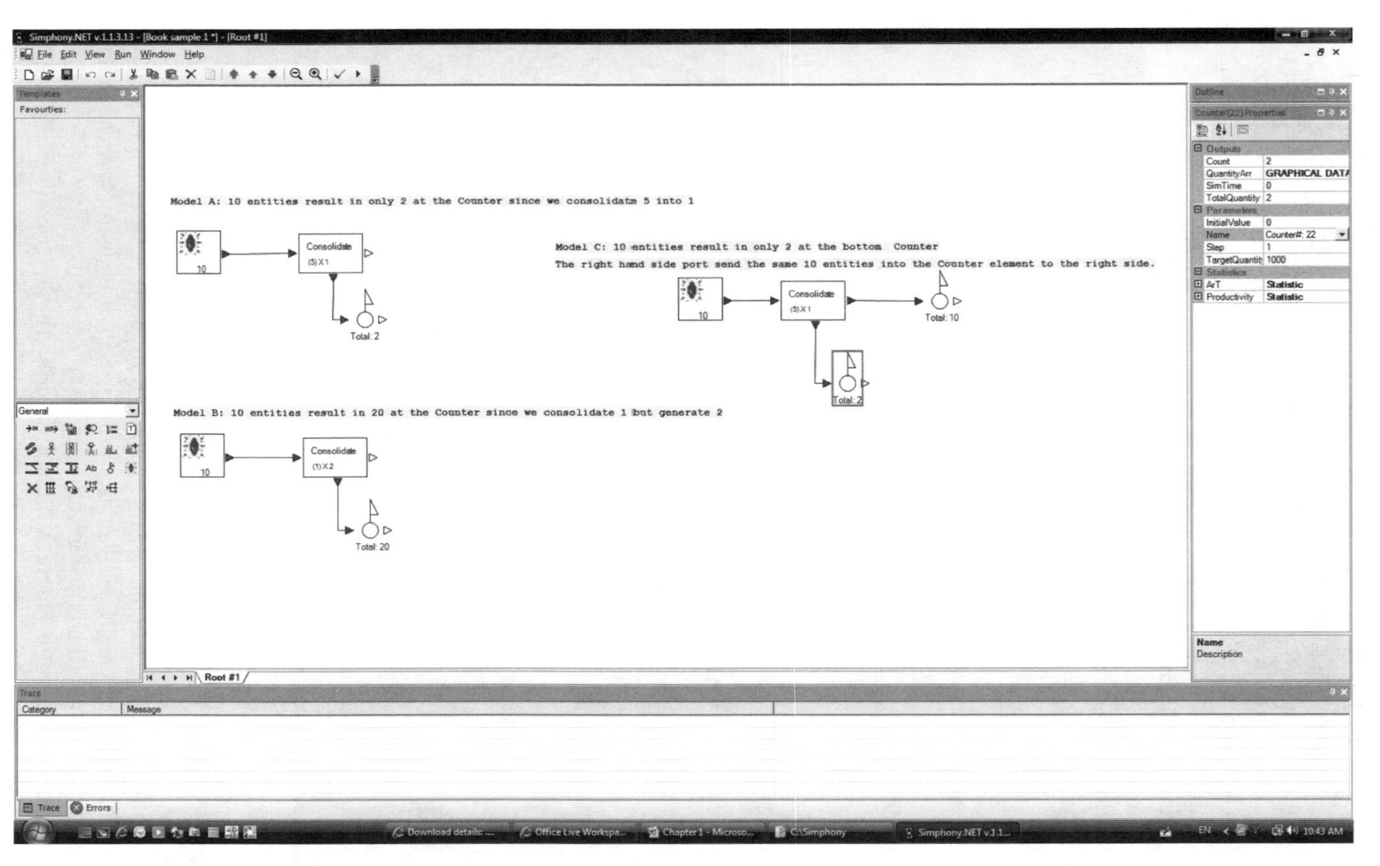

Figure A.2 Consolidate element in model

same attributes as the fifth entity that passed through the Consolidate element. The second will have the same properties as the tenth entity.

In Model B, 10 entities produce 20 entities at the counter, since we have the counter connected to the bottom port and we specified that for every 1 entity consolidated we generate 2. Each pair of generated entities will be identical. Therefore the 10 entities produce 10 similar ones upon leaving the Consolidate element.

In Model C, 10 entities produce 2 entities at the lower counter and 10 at the right counter. The right port of the Consolidate window allows the user to have the same entities that entered to leave intact. Therefore the same 10 entities produced 10 at the right counter while clones were consolidated into 5s thus producing 2 at the bottom counter.

A.1.6 Probability Branch Element

Probability Branch	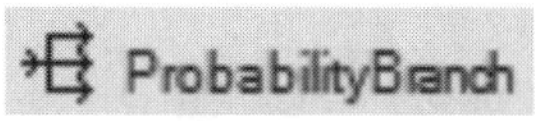 ProbabilityBranch
Properties	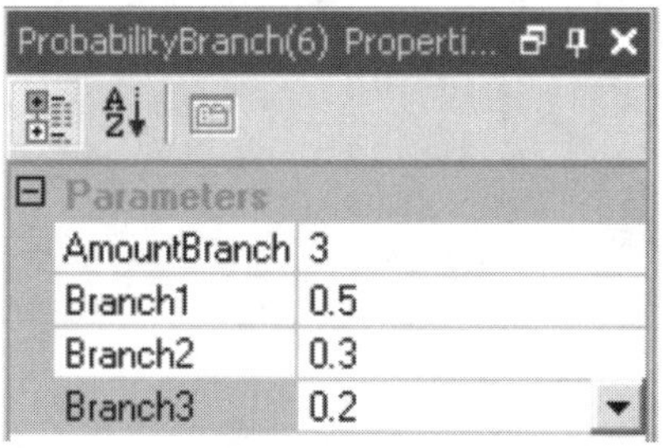 ProbabilityBranch(6) Properti... Parameters AmountBranch 3 Branch1 0.5 Branch2 0.3 Branch3 0.2
Input parameters	*Number of branches.* *The probability of each branch.* (The sum of all probabilities should equal 1.)
Output parameters	No output specified for this element.
Statistics	No statistics generated for this element.

A.1.6.1 Description

The *Probability Branch* element enables us to route entities using a probability value to different elements. First, the user must specify how many branches there will be and for each one the probability corresponding to it.

The example shown in Figure A.3 helps describe the functionality of this element. In the model, we desire to have 5 percent of the entities (0.05 probability)

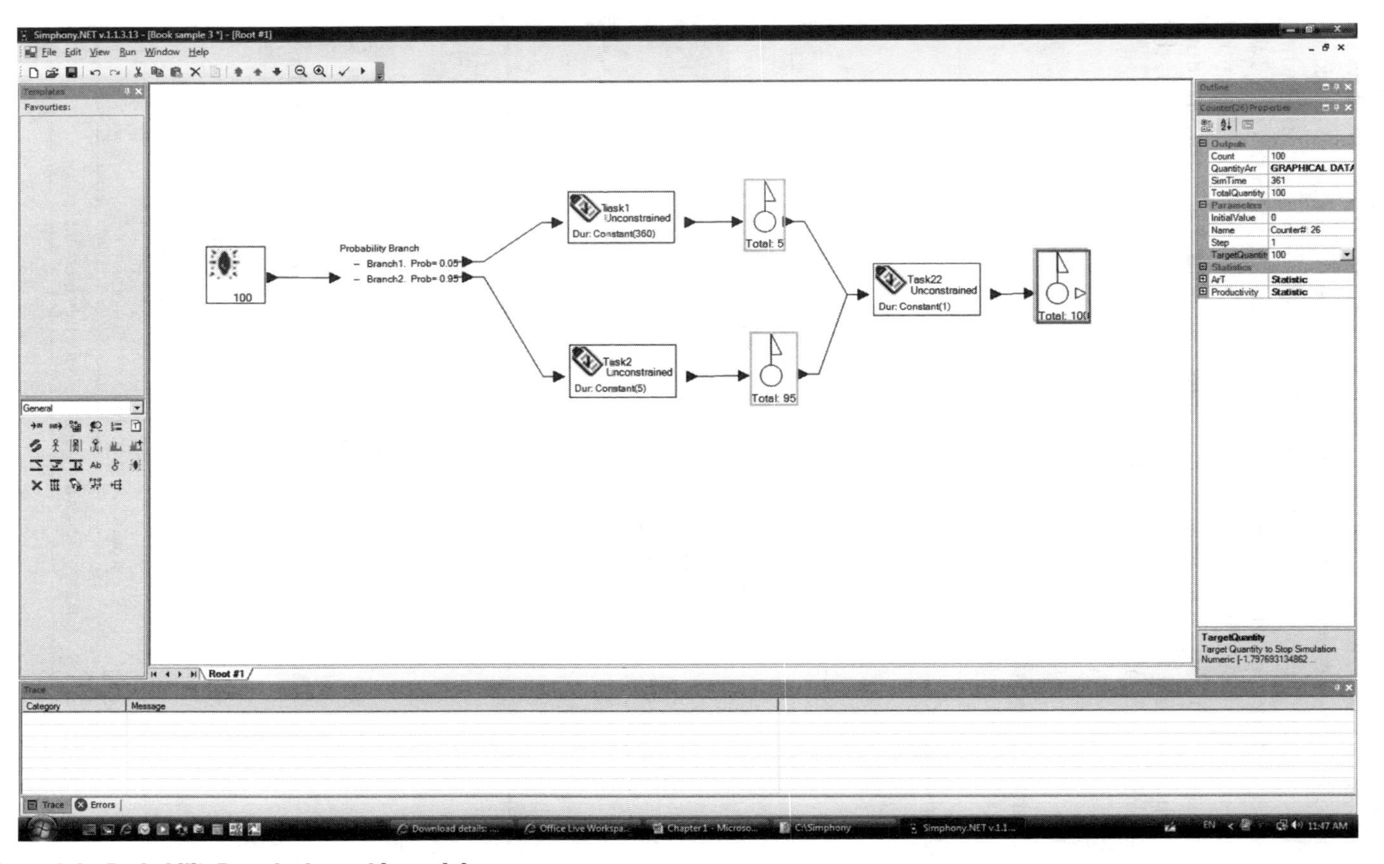

Figure A.3 ProbabilityBranch element in model

go into Task 1 (360 minutes service time), while 95 percent (0.95 probability) go into Task 2 (10 minutes service time). We inserted counters after the two tasks to show the count at the end of the simulation. The reader should note that while 100 entities were generated, 5 of them went through Task 1 and 95 through Task 2. What happens when a probability branch is encountered is simple: The entity arrives at the Probability Branching element and Simphony generates a random number (between 0.0 and 1.0). If the number is less than or equal to 0.05, the entity is routed through the upper branch as shown. If it is larger than 0.05, it is routed to the other one.

A.1.7 Conditional Branch Element

Conditional Branch	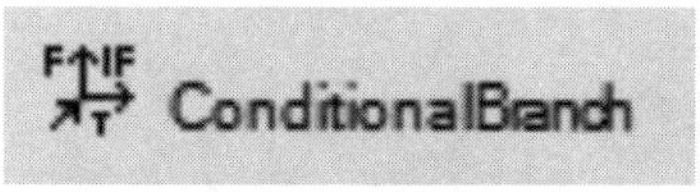 F IF T ConditionalBranch
Properties	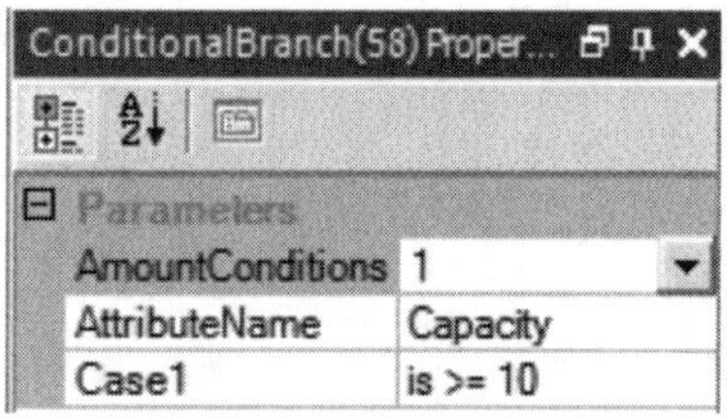 ConditionalBranch(58) Proper... Parameters AmountConditions: 1 AttributeName: Capacity Case1: is >= 10
Input parameters	*AmountConditions*: Determines number of desired predefined conditions for the output branches in the element. The number of output branches will be the number of conditions plus 1 (it counts for the "else" condition for the element). *AttributeName*: Specifies the name of the attribute of the incoming entity which holds the value to be examined by the Condition statement. *Case1 (Casei)*: Specifies the condition that needs to be met by the value of the incoming entity to send the entity through its corresponding port. Entered condition in this parameter is shown at the front of its associated output branch in the element graphics. The condition syntax is the same as the Select . . . Case condition in Visual Basic.
Output parameters	No outputs are specified for this element.
Statistics	No statistics are generated for this element.

A.1.7.1 Description

The *Conditional Branch* element routes the incoming entities into output branches that correspond in number to what the user specifies in the Amount Condition parameter plus 1. The Conditional Branch element has the same effect as the following statement:

```
If Case 1 is true
  Send the entity through top port branch
If Case 2 is true,
  Send the entity through next port branch
If Case i is true
  Send the entity through next port branch
Else If
  Send the entity through the bottom port branch.
End If
```

The Case i is simply the condition that needs to be met as specified in the Case i parameter. It can be one value (e.g., is = 4) or a list of values (e.g., value can be 1, 3, or 7 and allowed into this branch. The statement in the Case i will be 1, 3, 7). It can also be specified with an operator (e.g., is > 10).

The example shown in Figure A.4 illustrates the use of this element. We have four New Entity elements generating 1, 10, 2, and 20 entities, respectively. We assign these entities a type in the Attribute "T." The first has a Type 1, the second T = 2, the third T=4, and the last T = 7.

The conditional branching is accomplished at the Multi Branch element using two Case statements. In the first, we want the entities that have a value T = 1, 2, or 3 to go to the top counter. In the second, we want entities that have T = 4 to go to the middle 1. Everything else goes into the bottom counter. Notice the counter values shown at the end of the simulation. The top counter shows 11, since there is 1 Type 1, 10 Type 2s and 0 Type 3s. The count is 2 at the middle counter since there are only 2 entities of Type 4 and the bottom counter shows 20, since there are 20 Type 7 entities.

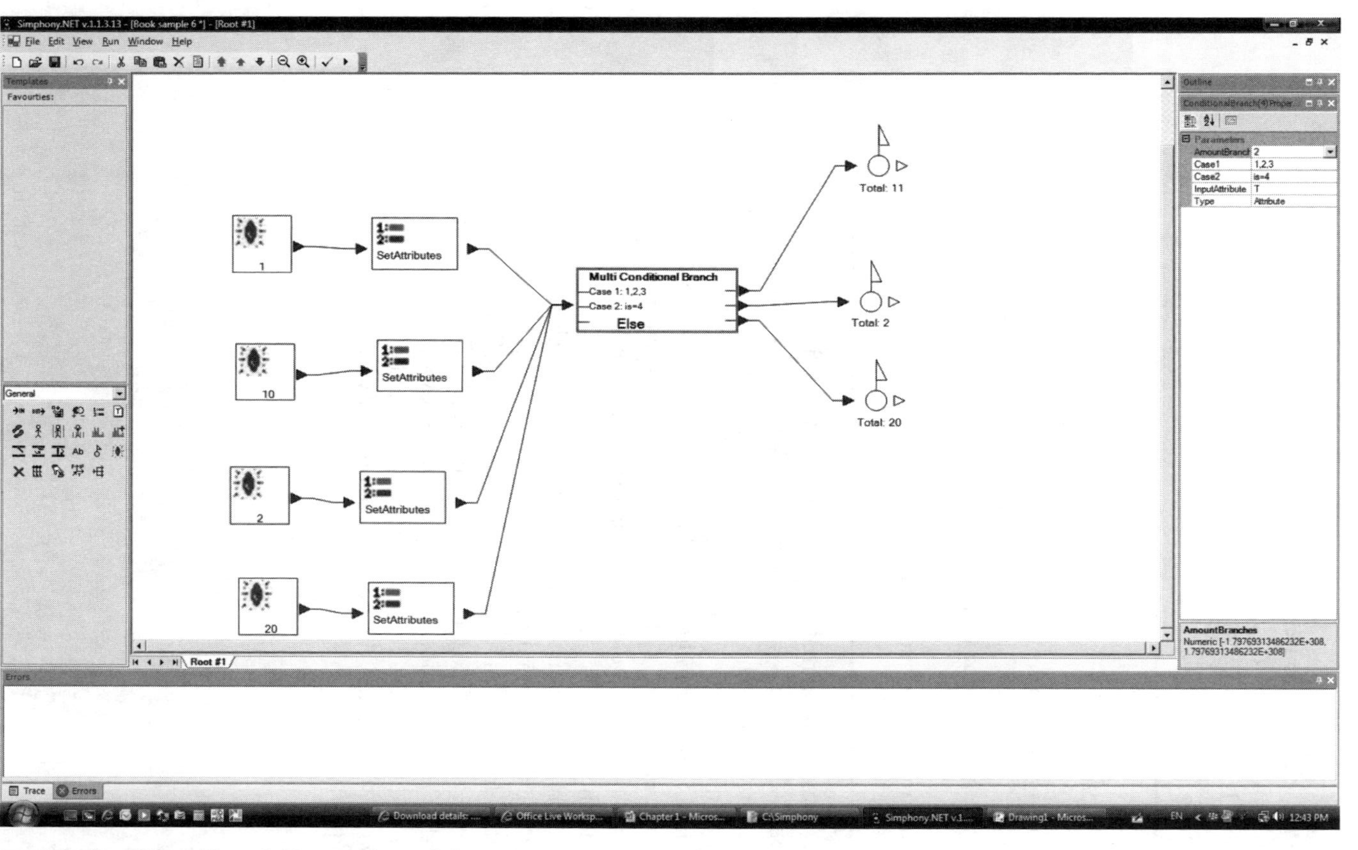

Figure A.4 ConditionalBranch element in model

A.1.8 Set Entity Attributes Element

Set Attribute	 SetAttribute
Properties	 SetAttribute(64) Properties Parameters Attr1Name: Capacity Attr1Val: 5 Attr2Name: StartTime Attr2Val: 0 Attr3Name: Hauling Time Attr3Val: 30 Attr4Name: Res Attr4Val: Loader NumofAttributes: 4
Input parameters	*Number of attributes*: Reflects the number of attributes the entity will hold in total. As a default, five attributes are set for each entity. For each attribute, there are two input parameters required: Attrib*ute # Name*: A text for describing the attribute. The same text should be used for referencing the attribute when retrieving its value at another element. *Attribute # Value*: The value to be assigned to the attribute. Only single values are allowed but a distribution can be used by using the link property.
Output parameters	No output specified for this element.
Statistics	No statistics generated for this element.

A.1.8.1 Description

The *Set Attribute* element allows us to specify various properties to distinguish between entities in the model. For example, when an entity arrives at this element we can give it an identity attribute to help us track it in the simulation. We can set the attributes by simply specifying the property name and its value. The value can be changed in the model.

Consider the example shown in Figure A.5. The top New Entity creates one entity and sends it through a Set Attribute element. In the element, we assign it an attribute name T and give it a value of 1. The lower New Entity creates 10 entities

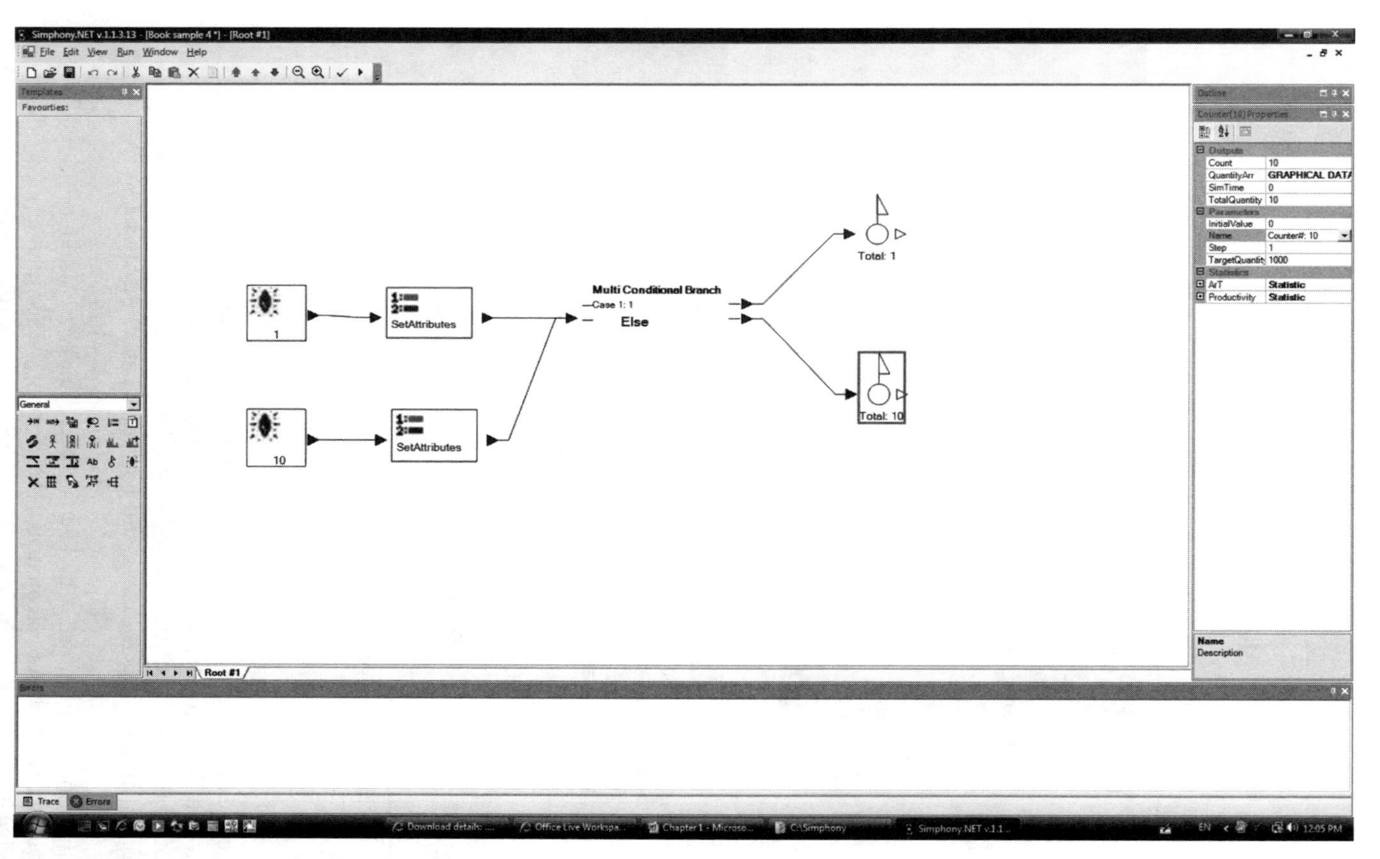

Figure A.5 ConditionalBranch element in model

and sends them through another Set Attribute element. They are assigned the value T = 2. The elements go through a conditional branch where, based on the value of T, they are routed to the top counter or lower counter. The reader should note that the total entities in the top counter is 1, as only one of this type was generated. The one at the bottom shows 10, since we generated 10 of this type.

A.1.9 Delete Entity Element

Delete Entity	DeleteEntity
Properties	DeleteEntity(117) Properties Outputs Count 0
Input parameters	No inputs specified for this element.
Output parameters	*Count:* Number of deleted entities.
Statistics	No statistics generated for this element.

A.1.9.1 Description

The *Delete Entity* element deletes any entities that pass through it. This helps optimize the utilization of system resources by freeing the memory occupied by unneeded entities. It also counts the number of deleted entities and shows the count as an output.

A.1.10 Statistics Elements

Collecting statistics on various parameters in the simulation model requires us to use two different elements in concert. First, we need to declare the statistic we wish to observe during simulation. For example, we may wish to collect statistics on the cycle time of a particular entity. We declare a statistic and call it Cycle Time. This would then allow us to observe cycle time values during the simulation and record them into this variable. Once the statistic is declared we can use a Collect Statistic element to collect relevant observations related to this statistic. The Statistic Declaration element does not receive or process entities. It only receives values from the Collect Statistic element.

A.1.11 Statistic Declaration

Statistic	Statistic
Properties	STATISTIC CycleTime
Input parameters	*Statistic's Name:* A description of the statistic to identify it when collecting values during simulation. *Intrinsic Statistic:* Yes/No parameter to identify whether to treat the statistic as intrinsic or nonintrinsic. *Do Full Tracking:* Yes/No parameter to switch full tracking of the statistic on/off. A statistic that is not fully tracked will not produce any graphs.
Output parameters	No outputs specified for this element.
Statistics	The element produces the mean, standard deviation, minimum, and maximum values for the collected observations. If the statistic is fully tracked, graphs also will be produced (indicated by the "Y" letter under the graph's columns). First StdDev stands for the standard deviation of observations made in the last run. Global StdDev stands for the standard deviation of observations across the total number of runs.

A.1.11.1 Description

The *Statistics* element is used to facilitate collection of statistics on interesting parameters in the model. This element allows us to declare the statistic we wish to track. The declared statistic can then be used in "collect" elements to add observations to it.

For each declared statistic, a description has to be provided to distinguish it from other statistic elements. The description provided for each statistic will be used in the graphical representation of the element in the model layout. In addition, it provides an ID for the element when selecting it through a statistics-collection element.

A statistic can be declared as intrinsic or nonintrinsic. For an intrinsic statistic, the times at which observations are collected affect the analysis of these observations while for nonintrinsic statistics the time at which an observation is made is ignored.

A full tracking of a statistic can also be specified, which will cause a histogram, time graph, and cumulative density function (CDF) to be generated for the statistic. If the full tracking of the statistic is not required, it is better to switch it off to save memory and processing time.

A.1.12 Statistics Collection Element

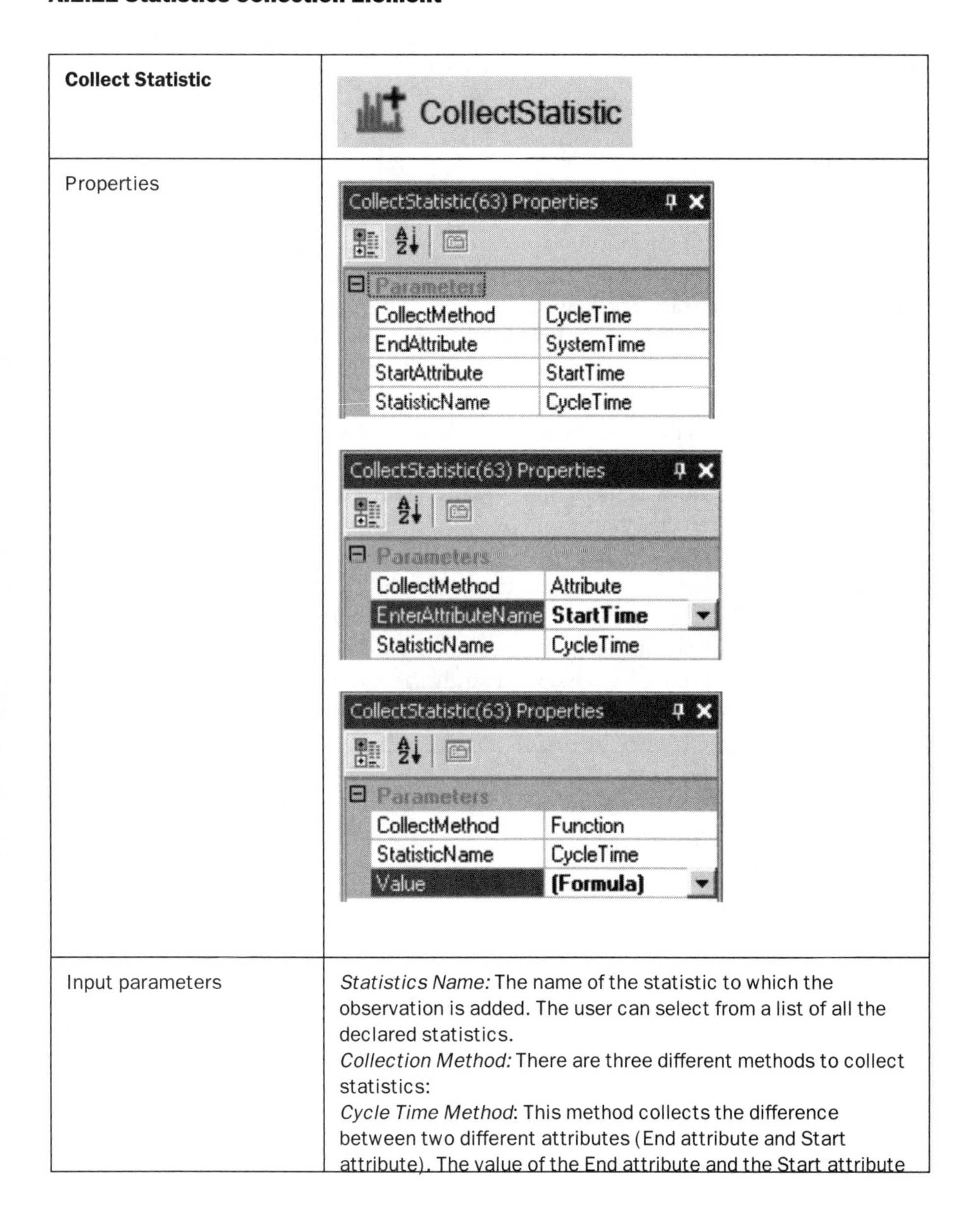

Collect Statistic	CollectStatistic
Properties	
Input parameters	*Statistics Name:* The name of the statistic to which the observation is added. The user can select from a list of all the declared statistics. *Collection Method:* There are three different methods to collect statistics: *Cycle Time Method*: This method collects the difference between two different attributes (End attribute and Start attribute). The value of the End attribute and the Start attribute

	can be the name of a predefined attribute and the simulation time of the system, which is SimEnvironment.SimTime. *Attribute Method:* This method collects the value of an entity's attribute. We only specify the name of the attribute for this parameter. *Function Method*: This method collects the value, which is written as a formula.
Output parameters	No outputs specified for this element.
Statistics	No statistics generated for this element.

A.1.12.1 Description

The *Statistics Collection* element adds observations to the declared statistic elements. Each Statistics Collection element can be assigned one statistic to which it adds observations. Whenever a statistic is declared, its description (name) is added to the list of available statistics in all

In the example shown in Figure A.6, we have 2 trucks starting at the New Entity element. We simply assign one attribute to house the time the truck starts a cycle in it. We called it T1. To set the value of the attribute we used a formula, as shown in Figure A.7. The value we keep in the attribute is "SimEnvironment.SimTime," which is the simulation time of the system at the time the entity passes through the Set Attribute element. The truck then has a 5 percent chance of breaking down and a 95 percent chance of continuing its cycle. When it breaks down, it takes 360 minutes to fix. Once done, it completes its backcycle and records a production of 1 truck load. We collect statistics for the truck in the Collect Statistics element shown. We create a statistic called S1 to contain observations of cycle time in it. To collect the cycle time, we compute the difference between the current simulation time (as the entity passes through the Collect Statistics element) and the time T1 when it started the cycle. We specify the following:

```
Collect Method: Cycle Time
End Attribute: System Time
Start Attribute: T1
Statistics Name: S1
```

When we run the simulation for 4,800 minutes, we can find the observations in the statistics element S1. By clicking on the element and checking its properties (see Figure A.6) we notice that the mean cycle time was 20.02 minutes, the minimum recorded cycle time was 20.1, and the maximum time was 370 minutes. There were 10 breakdowns recorded during the simulation out of 243 truck load cycles.

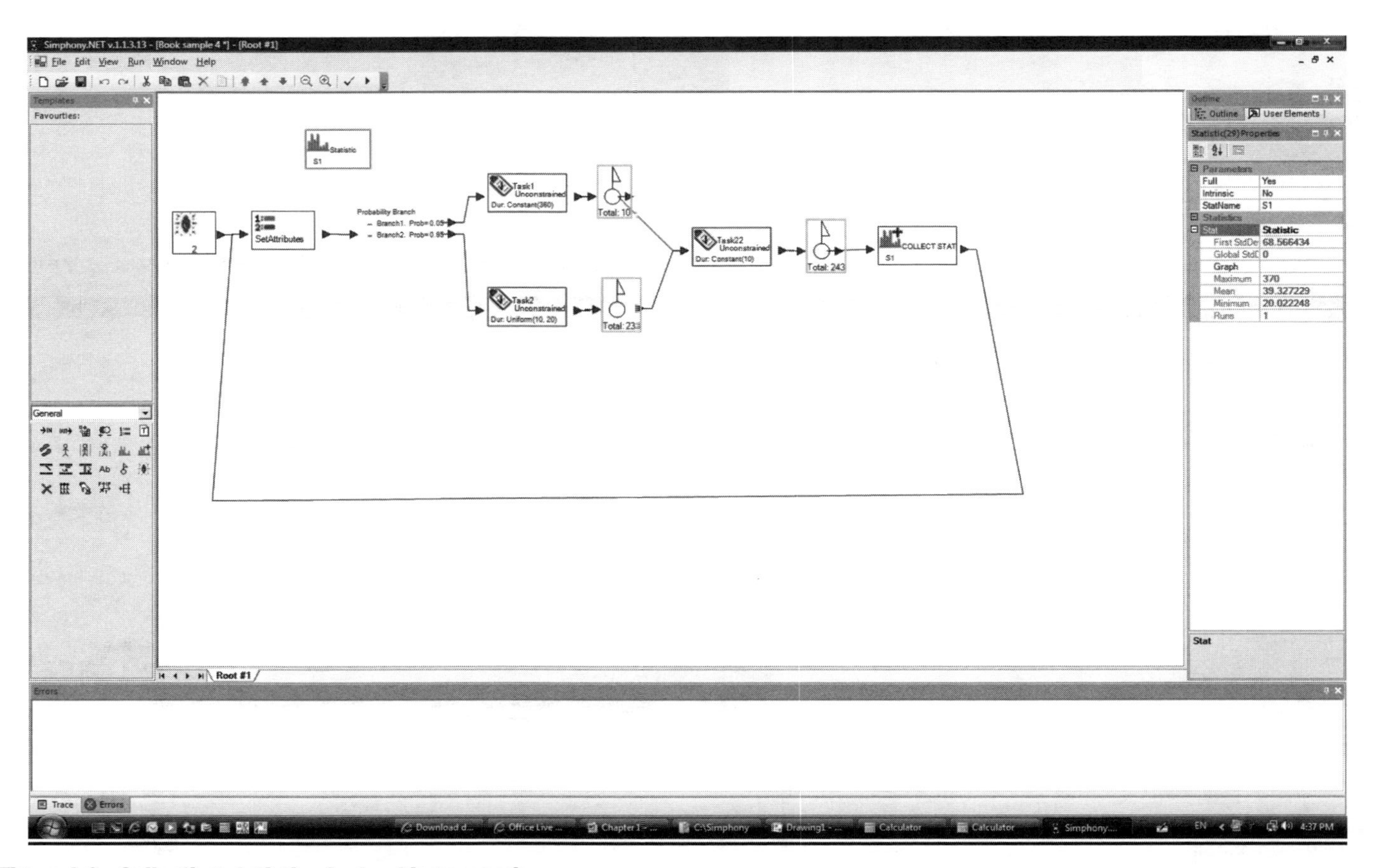

Figure A.6 Collecting statistics for trucking scenario

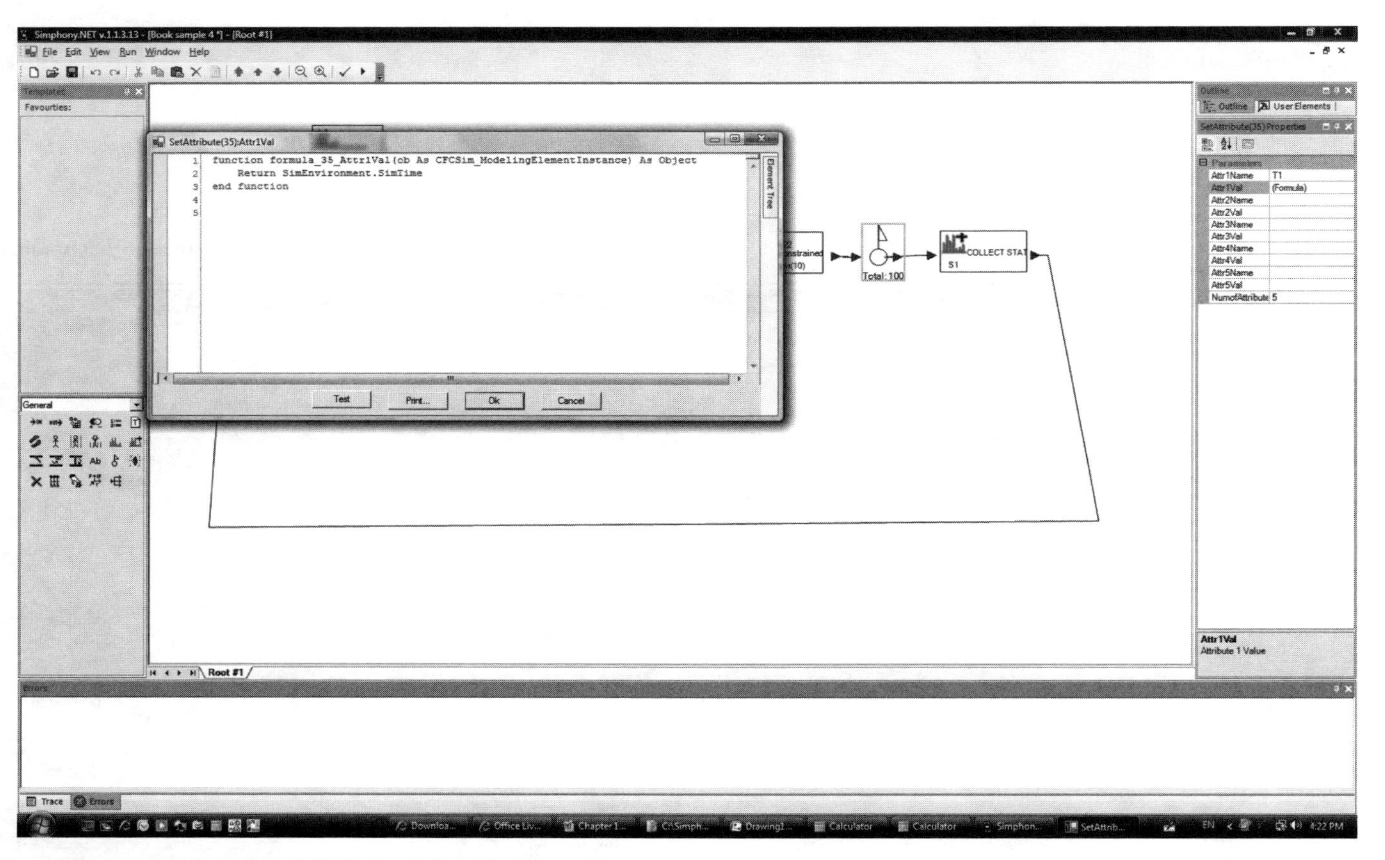

Figure A.7 Formula for setting attributes in trucking scenario

A.1.13 Valve Elements

The valve elements include the Valve, Close Valve, and Open Valve elements.

Valve	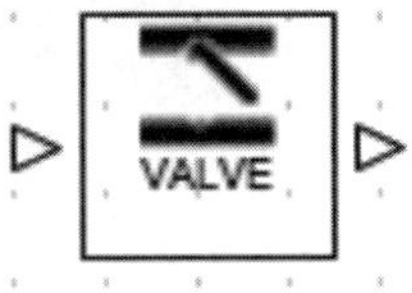 VALVE
Close Valve	 CLOSE
Open Valve	 OPEN
Properties (Valve)	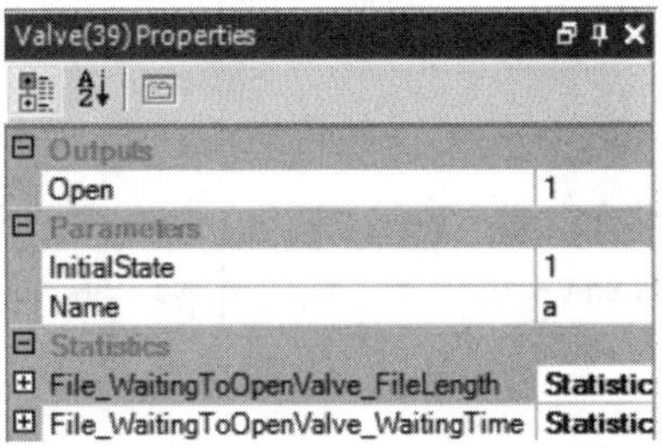 Valve(39) Properties Outputs Open 1 Parameters InitialState 1 Name a Statistics File_WaitingToOpenValve_FileLength Statistic File_WaitingToOpenValve_WaitingTime Statistic
Properties (Close Valve)	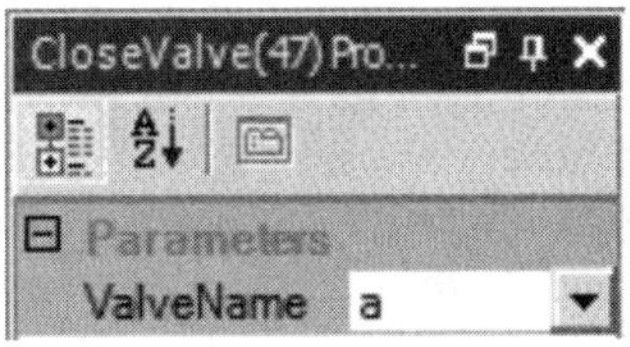 CloseValve(47) Pro... Parameters ValveName a

Properties (Open Valve)	OpenValve(48) Pro... Parameters ValveName a
Input parameters (Valve)	*Name*: The name of the valve, which will be shown below the element graphics. *Initial State*: Determines the initial state of the valve; 1 means open valve and 0 means close valve.
Input parameters (Close Valve)	*Valve Name*: The name of the related valve which is controlled by this Close Valve element. This name will be shown below the element graphics.
Input parameters (Open Valve)	*Valve Name*: The name of the related valve which is controlled by this Open Valve element. This name will be shown below the element graphics.
Output parameters (Valve)	*Open*: Returns the state of the valve during the simulation and final state of the valve after simulation; 1 means open valve and 0 means close valve.
Output parameters (Close Valve)	No outputs are specified for this element.
Output parameters (Open Valve)	No outputs are specified for this element.
Statistics (Valve)	The element collects statistics on Queue Length and Waiting Time for the entities trapped behind the valve.
Statistics (Close Valve)	No statistics are generated for this element.
Statistics (Open Valve)	No statistics are generated for this element.

A.1.13.1 Description

These three elements work together to control the flow of entities through the Valve element.

The *Valve* element is set in between the flows of the entities and may stop the flow or allow the entities to pass based on its current state; open or close.

The *Open Valve* and the *Close Valve* elements work as the controllers to the specified valve state. There is an attribute embedded in both elements which relates them to the specific valve by writing the valve name in it. When entities pass through these controllers, they turn the valve state to the state that they are meant to, i.e., open or closed. It is also possible to have several Open or Close elements to control the valve state in different places in the model.

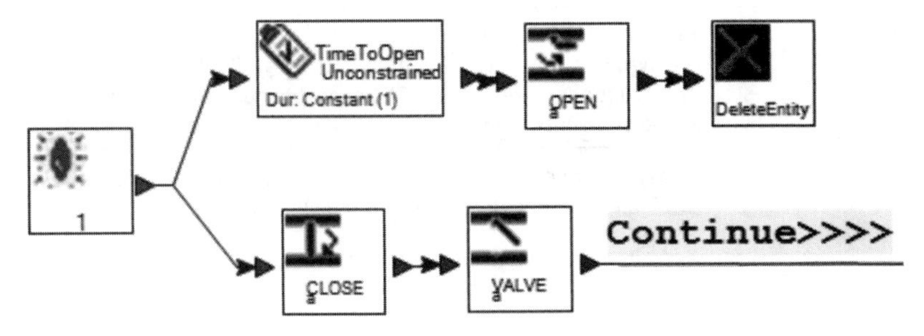

Figure A.8

In the submodel shown in Figure A.8 there are two lines—the first line is a line to control the entity flow at the second line. After entity creation, one entity goes to the first line and the other one goes to the second line. The entity in the second line is passing through Close Valve "a," closes the valve "a" immediately after, and a new entity is generated. The entity in the first line enters the Time To Open task which lasts 1 time unit, and after that passes through Open Valve "a" and opens the valve "a" after 1 time unit and continues its way to the other parts of the model.

A.1.14 Resource Element

Resource	Resource
Properties	Resource(7) Properties Outputs Current: 1 Parameters ResName: Crane Total: 1 Statistics Utilization: Statistic First StdDev: 30.788848 Global StdDev: 0 Graph Maximum: 100 Mean: 10.603974 Minimum: 0 Runs: 1

Input parameters	*Resource Name*: A text for describing the type of resource and distinguishing it from different resources available in the model. *Total Number of Resources*: The available number of resources at the start of the simulation.
Output parameters	*Current Number of Available Resources*: Shows the number of available resources at the end of the simulation.
Statistics	*Resource Utilization*: Shows the utilization of the resource as a percentage (busy time/total time).

A.1.14.1 Description

Handling resources in the common template involves a number of operations: declaring a waiting file, declaring a resource, capturing a resource, releasing it. To be able to use a resource in the *Capture* and *Release* elements, the resource must first be declared through the *Declare* element. Once a resource is declared, it can be captured or released by the *Capture Resource* or *Release Resource* element. This *Resource* element defines one type of resource in the model in addition to the total number of units of that resource.

A.1.15 Capture Element

Capture	 Capture
Properties	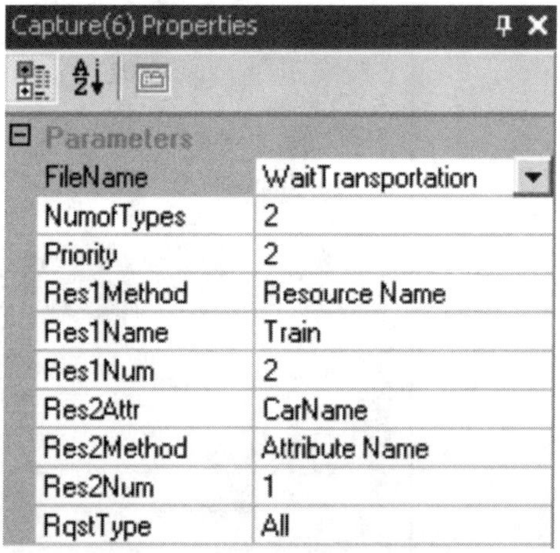 Capture(6) Properties Parameters FileName: WaitTransportation NumofTypes: 2 Priority: 2 Res1Method: Resource Name Res1Name: Train Res1Num: 2 Res2Attr: CarName Res2Method: Attribute Name Res2Num: 1 RqstType: All

Input parameters	*File Name*: Defines which file to use for entities waiting for a resource. *Number of Types:* Defines the number of types of resources that the entity wants to capture. *Capture Priority*: A number to specify the priority of the entity in capturing resources, among other entities in waiting files. The higher the number, the higher the request priority. *Res1Method:* There is a list box containing three different methods to specify which resources need to be captured. For each type of resource you need to specify the method. *Res1Name:* Specifies the name of the resource which is to be captured. For each type of resource you need to specify the name. *Res1Num*: Specifies how many units of the resource are to be requested. *Res2Attr*: If the resource method is Attribute name, it means that the user links the resource name to an entity's attribute, and the name of that attribute should be specified through this Input parameter. If the resource name is not linked, this parameter is ignored and does not show up. *Number of Resources to Capture*: Specifies if *ALL* the resources defined in the element are required or just *ANY* of them.
Output parameters	No outputs are specified for this element.
Statistics	No statistics are generated for this element.

A.1.15.1 Description

The *Capture* element is triggered by any entity that is transferred into it. Upon the arrival of an entity, the element adds it to the file defined in its Input parameters. A check for the entities in that file is then triggered at the File element. More than one element can be captured by the same Capture element.

Each Capture element can be assigned a capture priority number. This number specifies how the entities should be ranked in the waiting file. The higher the number is, the higher the priority of the entity in getting its requested resources.

A.1.16 Release Element

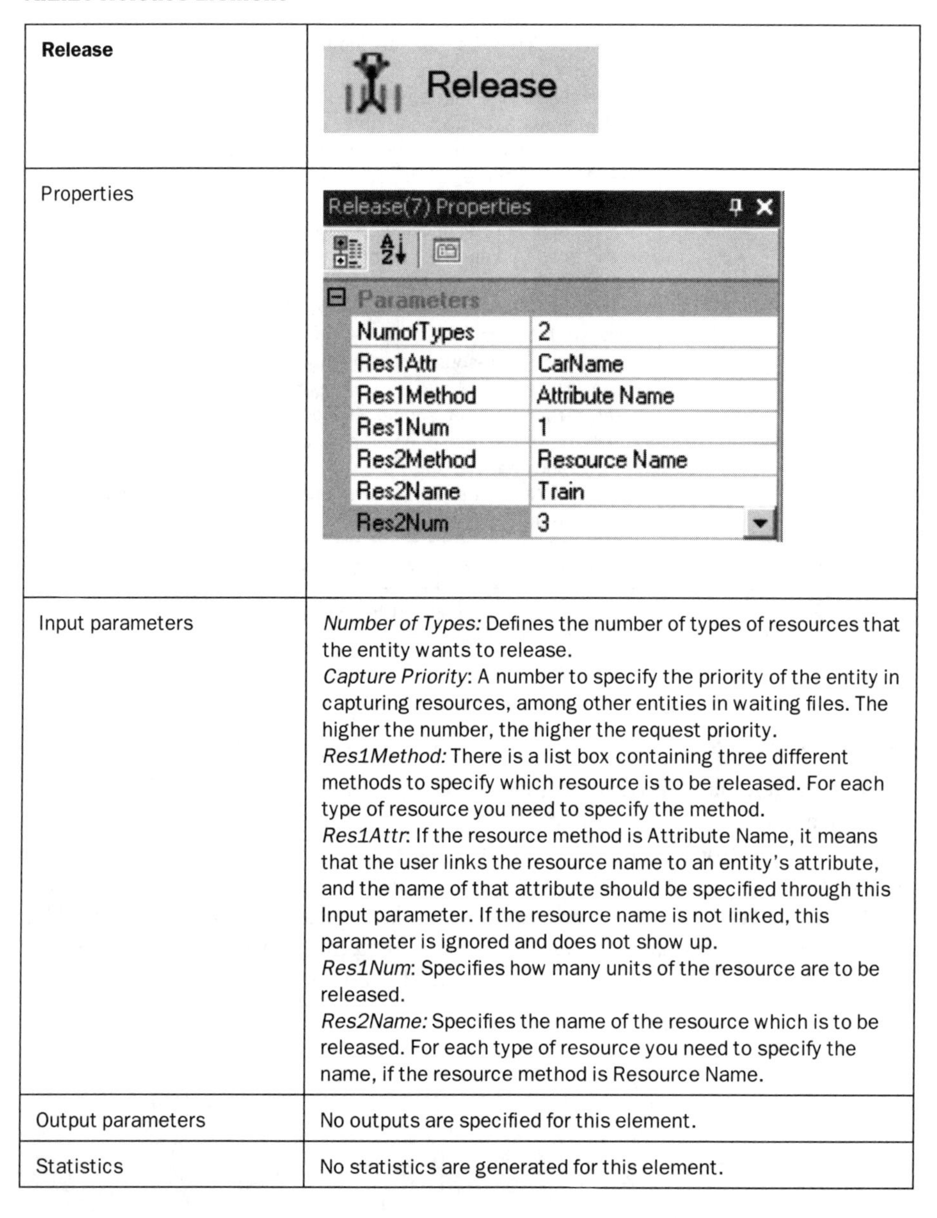

Release	Release
Properties	Release(7) Properties Parameters NumofTypes: 2 Res1Attr: CarName Res1Method: Attribute Name Res1Num: 1 Res2Method: Resource Name Res2Name: Train Res2Num: 3
Input parameters	*Number of Types:* Defines the number of types of resources that the entity wants to release. *Capture Priority*: A number to specify the priority of the entity in capturing resources, among other entities in waiting files. The higher the number, the higher the request priority. *Res1Method:* There is a list box containing three different methods to specify which resource is to be released. For each type of resource you need to specify the method. *Res1Attr*: If the resource method is Attribute Name, it means that the user links the resource name to an entity's attribute, and the name of that attribute should be specified through this Input parameter. If the resource name is not linked, this parameter is ignored and does not show up. *Res1Num*: Specifies how many units of the resource are to be released. *Res2Name:* Specifies the name of the resource which is to be released. For each type of resource you need to specify the name, if the resource method is Resource Name.
Output parameters	No outputs are specified for this element.
Statistics	No statistics are generated for this element.

A.1.16.1 Description

The *Release* element is triggered by any entity transferred into it. Upon the arrival of an entity, more than one element can be captured by the same Capture element.

A.1.17 Waiting File Element

Waiting File	 WaitingFile
Properties	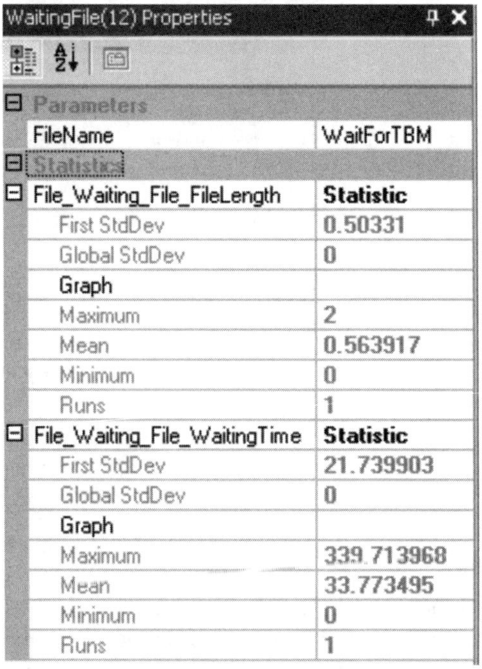 WaitingFile(12) Properties Parameters FileName: WaitForTBM Statistics File_Waiting_File_FileLength: Statistic First StdDev: 0.50331 Global StdDev: 0 Graph Maximum: 2 Mean: 0.563917 Minimum: 0 Runs: 1 File_Waiting_File_WaitingTime: Statistic First StdDev: 21.739903 Global StdDev: 0 Graph Maximum: 339.713968 Mean: 33.773495 Minimum: 0 Runs: 1
Input parameters	File Name: A text for describing the file and distinguishing it from other files in the model. This text is used as an ID for the file when selecting it in a Capture element. In addition, it is used in the graphical representation of the element.
Output parameters	No outputs are specified for this element.
Statistics	File Length: Shows statistics on file length during the simulation session. *Waiting Time*: Shows statistics on waiting time during the simulation session.

A.1.17.1 Description

The *Waiting File* element defines a waiting file for entities. The entities waiting in the file will be ranked according to the priority associated with each of them. The higher the number is, the higher the priority.

A.1.18 Trace Element

Trace	
Properties	
Input parameters	*Expression to Trace:* A text to be displayed as a trace message. *Trace Category:* A text representing a category for classifying trace messages.
Output parameters	No outputs specified for this element.
Statistics	No statistics generated for this element.

A.1.18.1 Description

The *Trace* element can be used to track the model by producing trace messages at user-specified locations. The user can define the message to be produced and define different categories under which different trace messages can be classified.

For example, let's assume that we want to keep track of an entity's attribute called Start Time (Figure A.9). If we set the trace category as Start Time Trace View while running, the simulation model will show the results for Start Time.

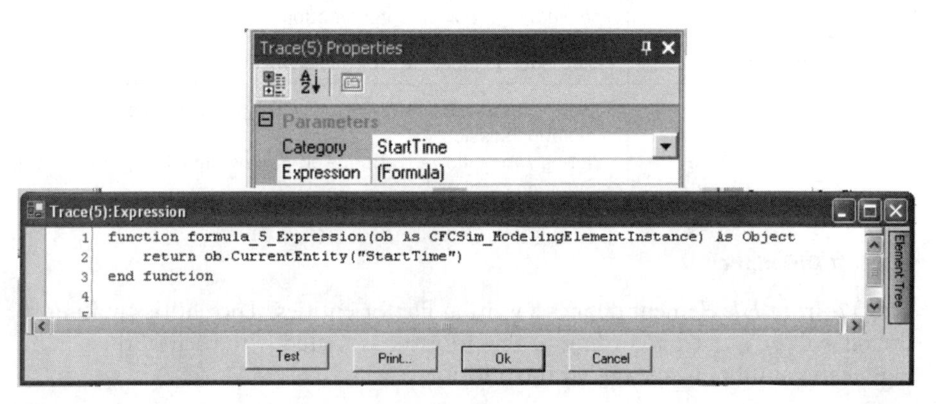

Figure A.9

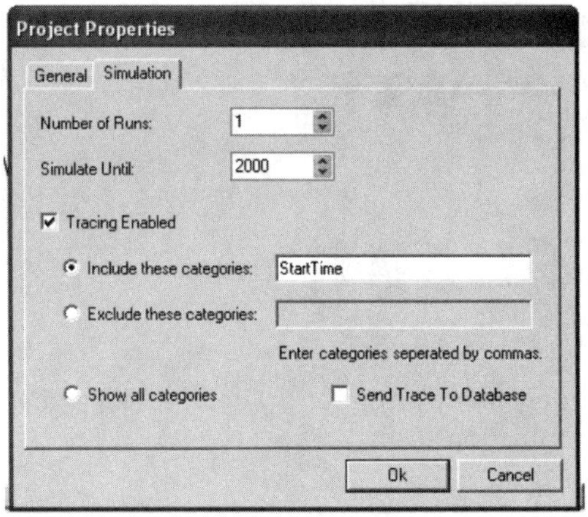

Project Properties

General | Simulation

Number of Runs: 1

Simulate Until: 2000

Tracing Enabled

Include these categories: StartTime

Exclude these categories:

Enter categories seperated by commas.

Show all categories

Send Trace To Database

Ok | Cancel

Figure A.10

In order to be able to see the tracing results in Trace View, the following procedure should be followed:

In File, Properties and Simulation, the radio button for "*Tracing Enabled*" should be activated and then one of the options of "*show all*" categories or "*include these categories*" should be selected.

For this particular example (Figure A.10), the second option was chosen and the Start Time category was included.

In the "View" menu, "Trace" should be chosen (Figure A.11). In Trace View the results can be seen. For repeating the process and seeing the results for further

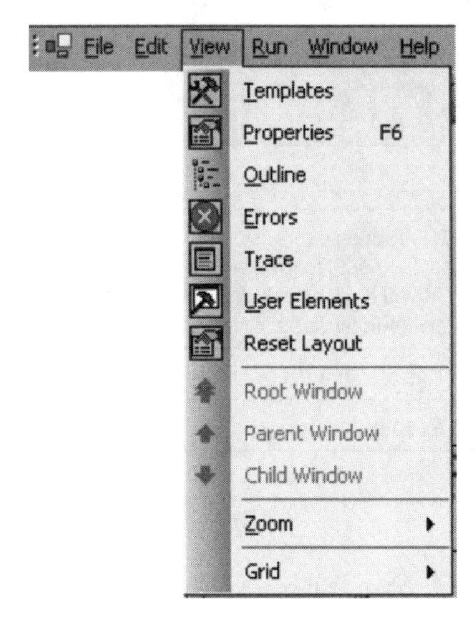

Figure A.11

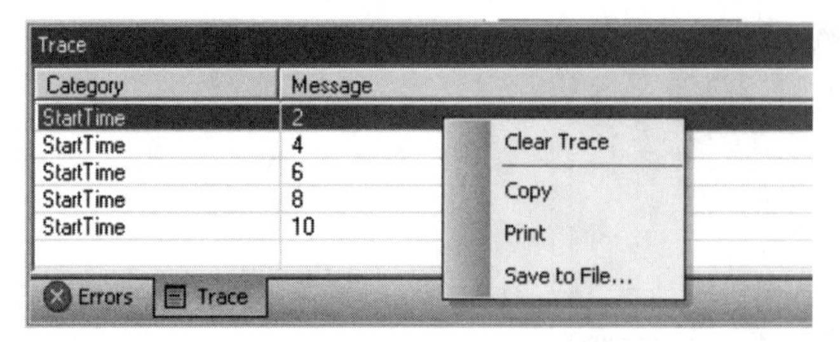

Figure A.12

runs, first the current results should be cleared (right-click in Trace View and choose Clear Trace) and then run the simulation again.

A.2 OTHER ELEMENTS (NOT USED IN THIS BOOK)

A.2.1 Execute Element

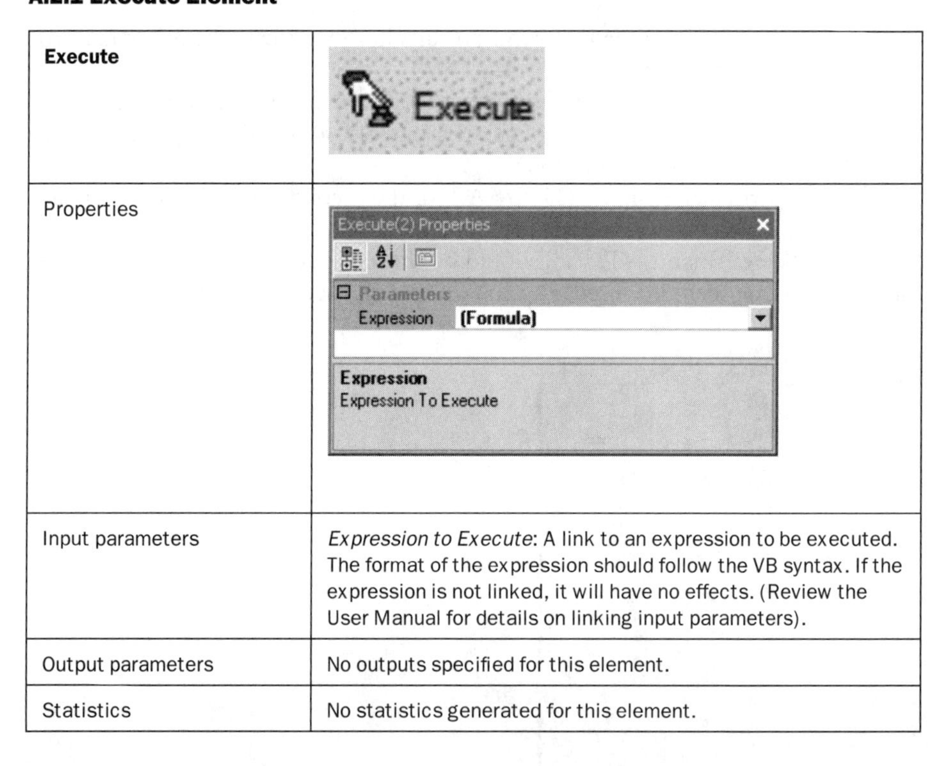

Execute	Execute
Properties	Execute(2) Properties Parameters Expression (Formula) Expression Expression To Execute
Input parameters	*Expression to Execute*: A link to an expression to be executed. The format of the expression should follow the VB syntax. If the expression is not linked, it will have no effects. (Review the User Manual for details on linking input parameters).
Output parameters	No outputs specified for this element.
Statistics	No statistics generated for this element.

A.2.1.1 Description

The *Execute* element is used to enforce the execution of an expression defined through the link property of the Input parameter of the element. Whenever an entity passes through the element, the expression linked to the element is executed. That expression may include formulas for modifying attributes of the current entity or of other elements in the model. The Execute element by itself does not have any effect on the passing entity or other elements. The only effects are produced through the executed expression.

A.2.2 Submodel Element

Submodel	SubModel
Properties	SubModel(17) Properties Parameters 1_Name: Excavation 2_Shape: Diamond 3_FrameColor: Blue 4_FillColor: Green
Input parameters	*1_Name*: The name of the submodel, which will be shown below the graphics in the model. *2_Shape*: Choose a shape to represent the submodel. *3_FrameColor*: Choose a color which is shown on the frame of the submodel shape. *4__FillColor*: Choose a color which is shown inside the frame of the submodel shape.
Output parameters	No outputs are specified for this element.
Statistics	No statistics are generated for this element.

A.2.2.1 Description

The *Submodel* element is used to build submodels inside the main model. The user can create other elements as children inside a *Submodel* element and link them to higher-level elements through the *Inport* and *Outport* elements. When an Inport or Outport is created inside a Submodel element, it will automatically create a corresponding input or output connection point to the Submodel element.

Entities flowing into the input connection points will be passed to the Inport element at the child level. When an entity comes to the Outport, it is passed to the output connection point of the Submodel element where it flows back to the higher-level model.

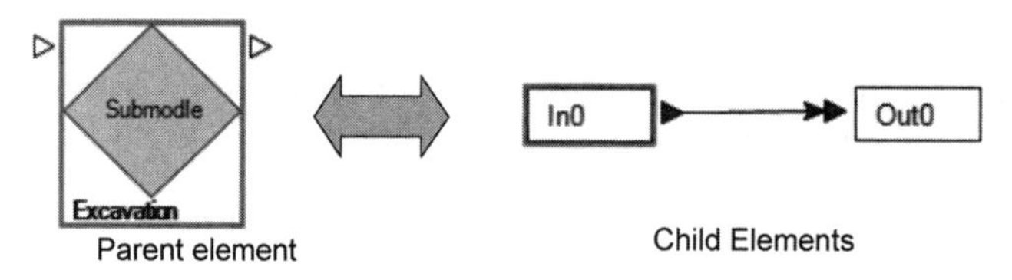

Figure A.13

A.2.3 Inport Element

Inport	InPort
Properties	This element is a connector used to link other elements, and has no properties of its own.
Input parameters	No inputs are specified for this element.
Output parameters	No outputs are specified for this element.
Statistics	No statistics are generated for this element.

A.2.3.1 Description

The *Inport* element represents the link between the elements at one hierarchical level and the elements in a lower level represented by a submodel inside a Submodel element. The creation of an Inport inside a Submodel element causes a simultaneous creation of an input connection point for the Submodel element. The entities transferred to this input connection point are directly routed to the Inport element (Figure A.13).

A.2.4 Outport Element

Outport	OutPort
Properties	This element is a connector used to link other elements, and has no properties of its own.
Input parameters	No inputs are specified for this element.
Output parameters	No outputs are specified for this element.
Statistics	No statistics are generated for this element.

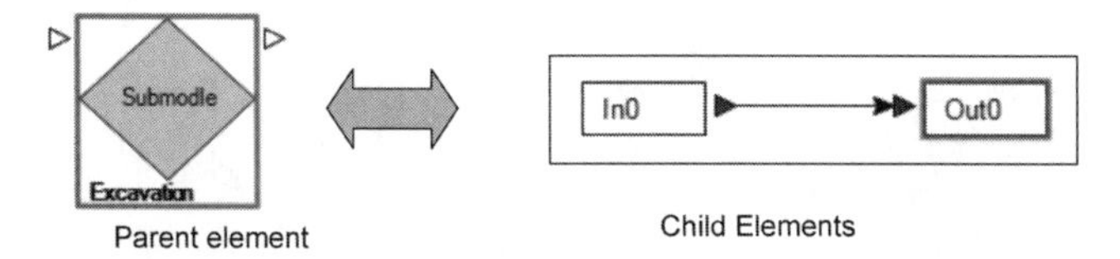

Figure A.14

A.2.4.1 Description

The *Outport* element represents the link between the elements in a lower level represented by a submodel inside a Submodel element and the elements at the higher hierarchical level.

The creation of an Outport inside a Submodel element causes a simultaneous creation of an output connection point for the Submodel element. The entities transferred to the Outport element are directly routed to the corresponding output connection point.

Glossary

14 Points—W. Edwards Deming's 14 management practices to help companies increase their quality and productivity.

Access point—Connects wireless communication devices to a wired network; also, relays data between wireless devices.

Accident prevention plan (APP)—A document that outlines occupational safety and health policies, responsibilities, and program requirements.

Accident—An unplanned event that results in injury, illness, death, property damage, work interruption, or other loss that has a negative effect on the work site.

Accountability—Being answerable for, but not necessarily personally charged with, doing specific work. Accountability cannot be delegated, but it can be shared.

Accuracy—In quality management, the degree of freedom from error or the degree of conformity to a standard. Accuracy is different from precision. For example, four significant-digit numbers are less precise than six-significant-digit numbers; however, a properly computed four-significant-digit number might be more accurate than an improperly computed six-significant-digit number.

Action—Conscious or reflective execution of a behavior strategy.

Activity—In project management, an element of work on a project that usually has an anticipated duration, estimated cost, and estimated resource requirements. An activity can be linked to a Work Package in the Work Breakdown Structure.

Activity hazard analysis (AHA)—A documented process by which the steps (procedures) required to accomplish a work activity are outlined, the actual or potential hazards of each step are identified, and measures for the elimination or control of those hazards are developed.

Adaptive control—The ability of a control system to change its own parameters in response to a measured change in operating conditions.

Alphabet—Standardized set of written symbols representing both consonant and vowel sounds.

Analogue signals—A current/voltage that is continuous but varies in amplitude. When calibrated, the changes represent information (e.g., changes in sensor output reflect changes in pressure).

Applet—A small application or utility program, usually written in the Java programming language, designed to do a very specific and limited task. Applets are most commonly used in handheld mobile devices.

ASCII (American Standard Code for Information Interchange)—A character encoding concept using decimal or hexadecimal system. ASCII codes are used to represent text in computers, communications equipment, and other devices that work with text. The character A = 65 in decimal and 45 in hexadecimal; Z = 90 and 5A.

Assembly—A group of subassemblies and/or parts that are put together and that constitute a major subdivision of the final product. An assembly may be an end item or a component of a higher-level assembly.

Assembly line—An assembly process in which equipment and work centers are laid out to follow the sequence in which raw materials and parts are assembled.

Awkward posture—Deviation from the natural or "neutral" position of a body part. A neutral position is one that puts minimal stress on the body part. Awkward postures typically include reaching above or behind or twisting, and can cause fatigue, pain, and musculoskeletal injury.

Back order—Product that is out of stock, on order, and scheduled to be shipped when it becomes available.

Backbone, Internet or wireless—Collection of major communications "pipelines" that transfer data from one major hub to another.

Balanced Scorecard (David Norton and Robert Kaplan)—A structured measurement system based on a combination of financial and nonfinancial measures of business performance. A list of financial and operational measurements used to evaluate organizational or supply chain performance. The dimensions of the Balanced Scorecard might include customer perspective, business process perspective, financial perspective, and innovation and learning perspectives. It formally connects overall objectives, strategies, and measurements. Each dimension has goals and measurements.

Bar code—A symbol consisting of a series of printed bars representing values. An optical character system of reading, scanning, and unit tracking by translating a series of printed bars into a numeric or alphanumeric identification codes. The most common example is the UPC used on retail packaging.

Benching—A method of protecting employees from cave-ins by cutting the sides of the excavation in the arrangement of one or more horizontal levels, usually with vertical or near-vertical walls between steps.

Benchmarking—The process of comparing performance against the practices of other leading companies for the purpose of improving performance.

Companies also benchmark internally by tracking and comparing current performance with past performance. Benchmarking seeks to improve any given business process by exploiting best practices rather than merely measuring the best performance.

Best practice—A specific process or group of processes that have been recognized as the best method for conducting an action. Best practices may vary by industry or geography, depending on the situation. Best practices methodology may be applied with respect to resources, activities, cost objects, or processes.

Bill of lading (BOL)—A transportation document that is the contract of carriage containing the terms and conditions agreed upon by the shipper and carrier.

Bill of materials (BOM)—A structured list of all the materials or parts and quantities needed to produce a particular finished product, assembly, subassembly, or manufactured part.

Bit—Short for "Binary digit," the numbering system used by computers, which consists only of 0 and 1. The binary number 1011 equals in the decimal system: $(1 \times 2 \wedge 3) + (0 \times 2 \wedge 2) + (1 \times 2 \wedge 1) + (1 \times 2 \wedge 0) = 8 + 0 + 2 + 1 = 11$.

BlackBerry—A wireless handheld device that supports push email, mobile telephone, text messaging, Internet faxing, Web browsing, and other wireless information services.

Body harness, full—Straps that are secured about a body in a manner that distributes the arresting forces over, at minimum, the thighs, waist, chest, shoulders, and pelvis, with provision for attaching a lanyard, lifeline, or deceleration device.

Bounded rationality—Concept introduced by Herbert Simon that challenges the assumptions that people make rational decisions. The many reasons include the lack of full information or unwillingness to explore all available options.

Byte—A packet of 8-bits with which 256 characters can be represented. Computers are built to operate with 1, 2, or 4 bytes (e.g., 32-bit computer). CDs use 16 bits per sample.

Capital—The resources, or money, available for investing in assets that produce output.

Carpal tunnel syndrome (CTS)—A condition caused by nerve compression where it passes through the wrist into the hand. Symptoms can include pain, tingling, or numbness in the hand, wrist, and/or arm. These symptoms are often most noticeable at night.

Carrier—A firm that transports goods or people via land, sea, or air.

Cartilage—Thick, white connective tissue attached to the surfaces of bones where they contact other bones, forming a low-friction cushion. Cartilage is structurally more rigid than a tendon.

Cause-and-effect diagram—In quality management, a structured process used to organize ideas into logical groupings. Used in brainstorming and problem-solving exercises. Also known as the Ishikawa or fishbone diagram.

Certificate of Compliance (COC)—A supplier's certification that the supplies or services in question meet specified requirements.

Certified supplier—A status awarded to a supplier that consistently meets predetermined quality, cost, delivery, financial, and count objectives. Incoming inspection may not be required.

Cervical vertebrae—Seven small irregular bones in the neck that support and allow head movement.

Change agent—An individual from within or outside an organization who facilitates change within the organization. May or may not be the initiator of the change effort.

Change order—A formal notification that a design must be modified in some way. This change can result from a revised quantity, date, or specification by the customer; an engineering change; a change in inventory requirement date; other.

Communication—Process by which information is exchanged between or among individuals and computer agents through a common system of symbols, signs, and behavior.

Communication channel—Medium used to convey information from a sender (or transmitter) to a receiver. Includes electronic media such as e-mail, intranet, Internet, teleconference; print media such as memos, bulletin boards, newsletters; and personal contact.

Communication, content—Information, knowledge, advice, commands, questions.

Communication disorder—Condition that inhibits perfect exchange of information because of defects in transmitting or receiving. Reasons include speech disorder or hearing problems.

Communication flow—Downward, upward, or lateral distribution of content, either formal or informal, internal or external.

Communication format—Includes written, oral, nonverbal, visual (graphs, maps, paintings), analogue, digital.

Communication message—An object in a transmittable format and containing content.

Communication network, digital—A collection of transmitters, receivers, or transceivers that communicate with each other. Digital networks may consist of one or more routers that transmit data to the correct user. A repeater may be necessary to amplify or re-create the signal when it is being transmitted over long distances.

Communication, noise—Unwanted disturbances that block, distort, or change the meaning of a message in both human and electronic communications.

Communication, nonverbal—Exchanging thoughts, opinions, or information using sign language, gestures, body language, touch, eye contact.

Communication, speech—Use of vocal apparatus and a linguistic structure to convey information.

Communication, theory—Based on Lasswell's maxim, this still-evolving theory intends to address: "Who says what to whom in what channel with what effect."

Communication, verbal—Use of speech or song.

Communication, visual—Use of sketches, diagrams, photos, paintings, or objects to communicate.

Communication, written—Use of letters or characters that serve as visible signs of ideas, words, or symbols in composing a letter, note, or notice.

Complete information—A term used in economics and game theory to describe an economic situation or game in which knowledge about other market participants or players is available to all participants. For example, every player knows the strategies available to other players.

Computer graphics—The science of discrete digital representation of objects in two and three dimensions.

Computer networks—A set of computing devices linked together via a cable or a wireless system to share resources.

Concept map—A technique for visualizing the relationships between concepts, articulated in linking phrases—for example, "gives rise to," "results in," "is required by," or "contributes to."

Confined space—A space that: (a)is large enough and so configured that a person can bodily enter and perform assigned work; and (b) has limited or restricted means of entry or exit, such that the entrant's ability to escape in an emergency would be hindered (e.g., tanks, vessels, silos, storage bins, hoppers, vaults, and pits are spaces that may have limited means of entry; doorways are not considered a limited means of entry or egress); and (c) is not designed for continuous worker occupancy.

Containerization—A shipment method by which commodities are placed in containers; and after initial loading, the commodities per se are not rehandled in shipment until they are unloaded at the destination.

Continuous process improvement (CPI)—An ongoing effort to expose and eliminate root causes of problems; small-step improvement, as opposed to big-step improvement.

Conveyor—A materials-handling device that moves freight from one area to another in a warehouse. Roller conveyors make sue of gravity, whereas belt conveyors use motors.

Cribbing—A system of timbers, arranged in a rectangular pattern, used to support and distribute the weight of equipment.

Critical lift—A nonroutine crane lift requiring detailed planning and additional or unusual safety precautions.

Critical success factor (CSF)—Those activities and/or processes that must be completed and/or controlled to enable a company to reach its goals.

Cycle time—The amount of time it takes to complete a sequence of tasks comprising a cycle.

Data—Numbers, words, images, and other formats that are further processed by a person or entered into computer, stored and processed there, and/or transmitted (output) to another person or computer.

Data, derivative—Data that can be extracted or inferred from other data.

Data, primary—Raw data that is being stored.

Data processing—Process that converts data into information or knowledge.

Data, secondary—Converse of primary data.

Data visualization—Use of interactive, visual representations of abstract data to reinforce understanding.

Decibel (dB)—A measure of sound pressure.

Decision making—A cognitive process leading to the selection of a course of action from among alternatives.

Decision support system—A computer-based system consisting of a database, modeling and analytical tools, and a user interface.

Decoding, of transmitted message—Transforming content from one format into another (e.g., reading a letter, listening to speech, examining a picture, watching hand signals).

Dialogue—A form of communication wherein both parties are involved in sending/receiving information, with feedback being encoded information, either verbal or nonverbal, sent back to the original sender.

Disorder—A medical condition in which some body function does not work as it should.

DMZ—In computer security, a demilitarized zone (DMZ), more appropriately known as demarcation zone or perimeter network, is a network area (a subnetwork) that sits between an organization's internal network and an external network, usually the Internet.

Downstream—Referring to the demand side of the supply chain. One or more companies or individuals who participate in the flow of goods and services moving from the manufacturer to the final user or consumer. Opposite of upstream.

Dunnage—The packing material used to protect a product from damage during transport.

Dust—Solid particles generated by handling, crushing, grinding, or detonation of organic or inorganic materials.

Dynamic Host Configuration Protocol (DHCP)—A server that automatically assigns IP addresses to network hosts. Each machine on the network must have a unique address. Rather than manually having to enter the IP address, DHCP tracks all of this automatically. Each device that is assigned to use the DHCP will request an IP address when it starts up.

Economic Order Quantity (EOQ)—An inventory model that determines how much to order by determining the amount that will meet customer service levels while minimizing total ordering and holding costs.

Electronic mail (e-mail)—A software system used to compose, send, store, and receive messages over electronic communication systems.

Empowerment—A condition whereby employees have the authority to make decisions and take action in their work areas without prior approval.

Encoding, message content—Transforming message content from one communication format into another, to be stored or transmitted to another unit.

Enterprise Resource Planning (ERP) system—A class of software for planning and managing enterprisewide the resources needed to construct, including planning, supply-chain management, time, and cost controls, and more. Often includes electronic commerce with suppliers. Examples of ERP systems are the application suites from SAP, Oracle, PeopleSoft, and others.

Equity theory (John Stacey Adams)—People value fair treatment whereby an individual perceives his or her rewards as equivalent to those of others, who provided an equal amount of inputs.

Ergonomics program—A systematic process, often spelled out in writing, for identifying, analyzing, and controlling ergonomic hazards at a particular workplace.

Ethernet—The most popular international standard technology for wired local area networks (LANs). It provides from 10 Mbps to 100 Mbps transmission speeds.

Expectancy theory (Victor Vroom)—People make conscious choices among alternatives to maximize personally valued returns for the minimum amount of input. The individual effort is increased if there is a clear relationship between effort and outcome, as evaluated by the company.

Expediting—To take extraordinary actions in acquiring engineered materials because of an increase in relative priority.

Extranet—A network that uses Internet protocols, network connectivity, and possibly the public telecommunication system to connect an organization with outside suppliers, vendors, partners, customers, or other businesses. An extranet can be viewed as part of a company's intranet that is extended to users outside the company

Fabricator—A manufacturer that turns the product of a raw materials supplier into a larger variety of products. For example, a fabricator may turn steel rods into nuts, bolts, and mats.

Fatigue—A condition that results when the body cannot provide enough energy for the muscles to perform a task.

Feedback—Performance appraisal of an individual or group, as part of career development. Some companies have implemented a 360-degree appraisal system whereby employees receive assessments from their manager, peers, subordinates, and customers; they also perform a self-assessment.

Firewall—A system of software and/or hardware that resides between two networks to prevent access by unauthorized users. The most common use of a firewall is to provide security between a local network and the Internet. Firewalls can be set to intercept, analyze, and stop a wide range of Internet intruders and hackers.

First in, first out (FIFO)—Warehouse term meaning that the first items stored are the first used. In accounting, FIFO is associated with the valuation of inventory, such that the latest purchases are reflected in book inventory.

Fishbone/cause-and-effect diagram—Also called an Ishikawa diagram, this graphic is useful to identify and organize/categorize possible causes of an effect or problem.

Fixed-order quantity—A lot-sizing technique in MRP or inventory management that will always cause planned or actual orders to be generated for a predetermined fixed quantity.

Flatbed—A type of truck trailer that consists of a floor and no enclosure. A flatbed may be used with "sideboards" or "tiedowns," which keep loose cargo from falling off.

Flatcar—A rail car without sides; used for hauling machinery.

Flowchart—A schematic representation of an algorithm or a process using symbols, arrows, and text.

Flow rack—Storage rack that utilizes shelves (metal) that are equipped with rollers or wheels. Such an arrangement allows product and materials to "flow" from the back of the rack to the front, making the product more easily accessible for small-quantity order-picking.

Free on board (FOB)—Contractual terms between a buyer and a seller that defines where title transfer takes place.

Freight bill—The carrier's invoice for transportation charges applicable to a freight shipment.

FTP (File Transfer Protocol)—Protocol used to transfer data from one computer to another over the Internet or through a network. FTP server software "listens" on the network for connection requests from a second computer running FTP client software that initiates a connection to the server. Once connected, the client can do a number of file manipulation operations such as uploading/downloading files, renaming, or deleting files on the server.

Fume—Very small suspended solid particles created by condensation from the gaseous state.

General Agreement on Tariffs and Trade (GATT)—The General Agreement on Tariffs and Trade started as an international trade organization in 1947, and has since been superseded by the World Trade Organization (WTO). GATT (the agreement) covers international trade in goods. An updated General Agreement is now the WTO agreement governing trade in goods. The 1986–1994 "Uruguay Round" of GATT member discussions gave birth to the WTO and created new rules for dealing with trade in services, relevant aspects of intellectual property, dispute settlement, and trade policy reviews.

Gestures—Form of nonverbal communication; a paralanguage that expresses itself through body movements, including winking, and the pitch, volume, and intonation of speech.

Graphic organizer—Visual representations of knowledge, concepts, or ideas: relieves learner boredom, enhances recall, provides motivation, creates interest, clarifies information, assists in organizing thoughts.

Grip force—Physical force applied by the hand when holding or gripping an object.

Guardrail system—A rail system erected along the open sides and ends of platforms. The rail system consists of a top rail and mid rail and their supports.

Hand-arm vibration—Vibration (generally from a hand tool) that travels through the hand and can advance to the arm and other areas of the body.

Hawthorne effect—Result from a study conducted at the Hawthorne Works of Western Electric Company between 1927 and 1932, which found that the act of showing people that management is concerned with their welfare usually results in better job performance. Studying and monitoring of activities is typically interpreted as concern and results in improved productivity.

Hazard—A dangerous condition, potential or inherent, that can bring about an interruption to, or interfere with, the expected orderly progress of an activity. A source of potential injury to person or to property.

Hazardous [physical] agent—Noise, nonionizing and ionizing radiation, and temperature exposure of durations and quantities capable of causing adverse health effects.

Hazardous environment—An environment whose atmosphere poses a risk of injury, incapacitation, illness, or death due to flammable or explosive hazards; hazardous substances or agents; oxygen concentrations below 19.5 percent or above 22 percent; or any other atmospheric condition recognized as IDLH.

Herniated disc—A condition where the soft inner part of an intervertebral disc pushes out through a tear in the disc.

Hotspot—Venue that provides Wi-Fi access to the Internet.

HTML (Hypertext Markup Language)—A language used in the creation of Web pages. HTML provides a means to describe the structure of text-based information in a document—by denoting certain text as headings, paragraphs, lists, and so on—and to supplement that text with interactive forms, embedded images, and other objects. HTML is written in the form of labels (known as tags), surrounded by less-than (<) and greater-than signs (>).

HTTP (Hypertext Transfer Protocol)—A communications protocol used to transfer or convey information on the World Wide Web. Its original purpose was to provide a way to publish and retrieve HTML hypertext pages. HTTP is a request/response protocol between clients and servers. The originating client, such as a Web browser, is referred to as the user agent. The destination server, which stores or creates resources, such as HTML files and images, is called the origin server.

Hub/switch—A multiport device used to connect client devices to a wired Ethernet network. Hubs can have numerous ports and transmit data at speeds ranging from 10 to 1000 Mbps per second to all the connected ports. A small wired hub may only connect 4 computers; a large hub can connect 48 or more.

Hygiene factors (Frederick Herzberg)—Job factors that can cause dissatisfaction if absent, but do not necessarily motivate employees if increased. In other words, these factors are important or notable only when they are lacking, such as from working conditions.

Ideogram—A graphic symbol that represents an idea.

IMAP (Internet Message Access Protocol)—With POP, the most prevalent Internet standard protocol for email retrieval from an email server that stores received messages in the recipient's inbox. Email clients can generally be configured to use either POP3 or IMAP4 to retrieve email, and in both cases use SMTP for sending.

Impulse noise—Noise is considered impulse when the variations in sound-pressure level involve peaks at intervals greater than one second.

Inflammation—A protective response of the body to infection and injury. Symptoms may include tissue swelling, redness, pain, and a feeling of warmth.

Infographics—Visual representations of information, data or knowledge.

Informatics—Study of the structure, behavior, and interactions of natural and artificial systems that store, process, and communicate information. Since computers, individuals, and organizations all process information, informatics has computational, cognitive, and social aspects.

Information—The result of processing, manipulating, and organizing data that can be sent as a message to be received and understood by a second agent or person. General definition of information (GDI): Consists of one or more data elements, presented following accepted rules (syntax) meaningful to the receiver.

Information entropy—A quantitative measure representing a signal pattern.

Information management—The collection and management of information from one or more sources, and distribution to one or more audiences who have a stake in that information or a right to that information.

Information processing—A process that uses sets of data and information as inputs and transforms it into another form (total of a bill plus taxes, regression curve, translation into another language).

Information system—A system, automated or manual, that comprises people, machines, and/or methods organized to collect, process, transmit, and disseminate data that represent user information.

In-line grip—A hand tool handle that is straight.

Integrated carrier—A company that offers a blend of transportation services such as land, sea, and air carriage, freight forwarding, and ground handling.

Internet—Originally called Arpanet, created by the United States government in conjunction with various colleges and universities for the purpose of sharing research data. There is no central server or owner of the Internet; every computer on the Internet is connected with every other computer.

Internet Protocol (IP)—Used to deliver data packets to their proper destination. Each packet contains both the originating and destination IP address. Each router or gateway that receives the packet will look at the destination address and determine how to pass it to the next device until it reaches the matching address.

Internet relay chat (IRC)—A form of real-time Internet chat or synchronous conferencing. It is mainly designed for group (many-to-many) communication in discussion forums called channels, but also allows one-to-one communication and data transfers via private message.

Interstate Commerce Commission (ICC)—An independent regulatory agency that implemented federal economic regulations controlling railroads, motor carriers, pipelines, domestic water carriers, domestic surface freight forwarders, and brokers.

Intranet—A computer network that uses Internet protocols and network connectivity to securely share part of an organization's information or operations with its employees. The same concepts and technologies of the Internet, such as clients and servers running on the Internet protocol suite, are used to build an intranet. HTTP and other Internet protocols are commonly used as well, such as FTP.

IP address—A number made of 4 bytes uniquely identifies each device on the Internet. In standard decimal numbers, each byte can be any number from 0 to 255. A standard IP address would look something like 192.168.45.28.

ISO 10303—Standard for the Exchange of Product (STEP) model data. The goal is to establish a common or neutral file exchange system that also allows the sharing of product databases and archiving.

ISO 15489—The international standard on records management.

ISP (Internet service provider)—A company that owns the servers, routers, communication lines, and other equipment necessary to establish a presence on the Internet. ISPs sell access to their equipment in the form of Internet service as dial-up, cable modem, DSL, or other types of connections.

Job enlargement—An effort to minimize the boredom and alienation of employees through an increase in job scope by adding duties and responsibilities.

Job enrichment (Frederick Herzberg)—An effort to motivate employees by giving them the opportunity to use a broad range of their abilities. Compare with *job enlargement.*

Joint—The area where two bones are attached to allow body movement. A joint is usually formed of ligaments and cartilage.

Just-in-Time (JIT)—An inventory control system that controls material flow to the point so that desired materials arrive just in time for use. Developed by the auto industry, it refers to shipping goods in smaller, more frequent lots.

Kaizen—The Japanese term for improvement; continuing improvement involving everyone—managers and workers. In manufacturing, kaizen relates to finding and eliminating waste in machinery, labor, or production methods.

Kanban—Japanese for "visible record," which, loosely translated, means card, billboard, or sign. Popularized by Toyota Corporation, it uses standard containers or lot sizes to deliver needed parts to the assembly line just in time for use.

Key performance indicator (KPI)—A measure of strategic importance to a company or department. For example, a supply-chain flexibility metric is supplier on-time delivery performance, which indicates the percentage of orders that are fulfilled satisfactorily.

Knowledge visualization—Aims to transfer insights, experiences, attitudes, values, expectations, perspectives, opinions, and predictions beyond the mere facts.

Lading—The cargo carried in a transportation vehicle.

Lagging—Timber planks, steel plates, or other structural members used for transferring loads and supporting soil or rock.

LAN (local area network)—A system of connecting PCs and other devices within the same physical proximity for sharing resources such as an Internet connections, printers, files, and drives. When Wi-Fi is used to connect the devices, the system is known as a wireless LAN or WLAN.

Language—A system of signals, such as voice sounds, intonations or pitch, gestures, or written symbols communicating thoughts or feelings.

Lanyard—A flexible line that is used to secure a safety belt or harness to a lifeline or directly to a point of anchorage.

Lead time—The total time that elapses between an order's placement and its receipt. It includes the time required for order transmittal, order processing, fabrication of engineered material or order preparation, and supply.

Lifeline—A line provided for direct or indirect attachment to a worker's body belt, body harness, lanyard, or deceleration device; may be horizontal or vertical in application.

Ligament—Strong ropelike fibers that connect one bone to another to form a joint.

Logistics—The process of planning, implementing, and controlling procedures for the efficient and effective transportation and storage of goods, including services, and related information from the point of origin to the point of consumption for the purpose of conforming to customer requirements.

MAC (Media Access Control) address—A unique hardware number that identifies each device on a network. A device can be a computer, printer, etc.

Mail transfer agent/mail server—A computer program or software agent that transfers electronic mail messages from one computer to another.

Material Safety Data Sheet (MSDS)—A sheet that provides information on substance identification; ingredients and hazards; physical data; fire and explosion data; reactivity data; health hazard information; spill, leak, and disposal procedures; and special precautions and comments.

Materials handling—The physical handling of products and materials between procurement and shipping.

Materials requirements planning (MRP)—A decision-making methodology used to determine the timing and quantities of materials to purchase.

Matrix organizational structure—An organizational structure in which two (or more) channels of command, budget responsibility, and performance measurement exist simultaneously. For example, engineering, purchasing, and project management of a construction company could be implemented simultaneously; that is, all managers have equal authority, and employees report to all three managers.

Mbps—(megabits per second) A measurement of data speed that is roughly equivalent to a million bits per second.

Measure—A number used to quantify a metric, showing the result of part of a process often resulting from a simple count.

Meeting—Two or more people coming together to discuss a predetermined topic (e.g., weekly, staff, board, off-site retreat)

Memory—The ability to store, retain, and subsequently recall data and information. Sensory memory of humans corresponds approximately to the initial 200 to 500 milliseconds after an item is perceived. Short-term memory allows a person to recall something from several seconds to as long as a minute. Long-term memory can store much larger quantities of information for potentially unlimited duration.

Message—An object of communication that encapsulates its content (e.g., information, question, directive, knowledge, greeting).

Metrics—Specific areas of measurement. A metric must be quantitative and must support benchmarking, and it must be based on broad, statistically valid data. Therefore, it must exist in a format for which published data exists within the enterprise or industry.

Mind map—A diagram used to represent ideas, issues, tasks, or other entities linked to and arranged radially around a central key word or idea. Aids in study, organization of information, problem solving, brainstorming, and decision making.

Motivators—Sources of motivation, intrinsic or extrinsic, that lead to actions that lead to pleasure or the avoidance of threats.

Musculoskeletal disorders (MSDs)—A group of conditions that involve the nerves, tendons, muscles, and supporting structures such as intervertebral discs. The various conditions can differ in severity from mild, infrequent symptoms to severe chronic and disabling disorders. Examples include carpal tunnel syndrome, tenosynovitis, tension neck syndrome, and low back pain.

Neutral body posture—The natural position of body parts; the best position to minimize stress. For example, when standing, the head should be aligned over the shoulders, the shoulders over the hips, the hips over the ankles, and the elbows at the side of the body.

NIOSH (National Institute for Occupational Safety and Health)—NIOSH, is part of the Centers for Disease Control and Prevention (CDC), under the Department of Health and Human Services, a federal government agency. Its mandate is to conduct and fund occupational safety and health research and training.

North American Free Trade Agreement (NAFTA)—Agreement implemented January 1, 1994, between Canada, the United States, and Mexico. It includes measures for the elimination of tariffs and nontariff barriers to trade, as well as many more specific provisions concerning the conduct of trade and investment that reduce the scope for government intervention in managing trade.

Ontology—The study of the reality or identity of an object.

Organizational communication—The transactional, symbolic process in which the activities of a social collective are coordinated to achieve individual and collective goals.

OSHA (Occupational Safety and Health Administration)—A federal government agency, part of the U.S. Department of Labor, whose mission is to help prevent workplace injuries and protect the health of workers. OSHA adopts and enforces workplace health and safety standards.

OSI Data Communications Model—The Open System Interconnection (OSI) model was defined by the International Standards Organization (ISO) and consists of seven layers, each performing a specific task.

Pattern recognition—Classifying and clustering patterns in observations based on learned relationships between characteristics of patterns and their meaning.

PDA (personal digital assistant)—Smaller than laptop computers but with many of the same computing and communication capabilities, PDAs range greatly in size, complexity, and functionality. PDAs can provide wireless connectivity via embedded Wi-Fi card radios, slide-in PC card radios, or compact flash Wi-Fi radios.

Peer-to-peer (P2P) network—Links computers with large bandwidth connections. Users believe that they all are connected to the same server. Sharing of large files containing audio, video, and data is done very efficiently.

Performance test—A test to determine, for example, the proper operation of a crane and the capability of the crane to safely lift loads within its performance rating. A performance test includes operational performance tests and load performance tests.

Pictogram—Early example of communication with drawings to tell a story.

POP (Post Office Protocol)—With IMAP, the most prevalent Internet standard protocol for email retrieval from an email server that stores received messages in the recipients inbox.

Power grip—A grasp by which the hand wraps completely around a handle. The handle runs parallel to the knuckles and protrudes on either side.

Problem solving—Considered the most complex of all intellectual functions, problem solving has been defined as higher-order cognitive process that requires the modulation and control of more routine or fundamental skills. It includes problem identification and definition.

Process—A series of time-based tasks linked to provide a specific output.

Process benchmarking—Benchmarking a process (e.g., steel erection) against organizations known to be the best in class in this process. Process benchmarking is usually conducted on firms outside of the organization's physical area.

Process improvement—Designs or activities that improve quality or reduce costs, often through the elimination of waste or nonvalue-added tasks.

Protocol—A set of rules or guidelines that enable communication, like the syntax of a language. If one party speaks French and one German, the communications will most likely fail.

Qualified person—In construction, someone with the skills and knowledge related to a specific operation and related equipment and has received safety and health training on the hazards involved.

Quality control (QC)—Management function that attempts to ensure that the constructed objects meet the product specifications

Radio—Wireless transmission of signals by modulating electromagnetic waves.

Receiver—Tool that converts a signal (e.g., radio) from a transmitter into a content understandable by the communication target.

Record—Information about business transactions retained because of its value. ISO 15489 defines records as "information created, received, and maintained as evidence and information by an organization or person, in pursuance of legal obligations or in the transaction of business." According to the International Committee on Archives (ICA) a record is "a specific piece of recorded information generated, collected or received in the initiation, conduct or completion of an activity and that comprises sufficient content, context and structure to provide proof or evidence of that activity."

Repeater (wireless mesh)—An electronic device that receives a weak or low-level signal—for example, from an wireless access point—and retransmits it at a higher level or higher power, so that the signal can cover longer distances without degradation.

Repetitive stress injury (RSI)—An injury caused by working in the same awkward position, or repeating the same stressful motions, over and over. This is one type of musculoskeletal disorder.

RFID (radio frequency identification)—An electronic identification technology that uses radio frequency signals to read identifying data contained in tags on equipment and merchandise. An alternative to bar codes.

Risk factor—An action and/or condition that may cause an injury or illness, or make it worse. Examples related to ergonomics include forceful exertion, awkward posture, and repetitive motion.

Routing—Forwarding of logically addressed data packets from their source networks toward their ultimate destination through intermediary nodes using hardware devices called routers.

Safety harness—A design of straps that is secured about the employee in a manner to distribute the arresting forces over, at least, the thighs, shoulders, and pelvis, with provisions for attachment to a lanyard, lifeline, or decelerating device.

Scorecard—A performance measurement tool used to capture a summary of the key performance indicators (KPIs)/metrics of a company. Metrics scorecards should be easy to read and usually have red, yellow, and green indicators to flag when the company is not meeting its targets for its metrics. Ideally, a dashboard/scorecard should be cross-functional in nature and include both financial and nonfinancial measures. In addition, scorecards should be reviewed regularly—at least on a monthly basis, and weekly in key functions such as manufacturing and distribution, where activities are critical to the success of a company. The scorecard philosophy can also be applied to external supply-chain partners such as suppliers, to ensure that suppliers' objectives and practices align.

Self-efficacy—Confidence that leads to belief that one is capable of performing in a manner needed to attain desired goals.

Semantic network—A graph to represent knowledge and expertise using

Semantic transform—To create meaning from chunks of words, phrases, sentences, or texts in general.

Server—A computer that provides resources or services to other computers and devices on a network. Types of servers can include print servers, Internet servers, mail servers, and DHCP servers. A server can also be combined with a hub or router.

Shoring—A support member that resists compressive forces imposed by a load.

Six Sigma Quality—A term used generally to indicate that a process is well controlled (e.g., tolerance limits are ±6 sigma; 3.4 defects per million events) from the centerline in a control chart. Six Sigma's goal is to define processes and manage those processes to obtain the lowest possible level of error—thus, it can be applied to virtually any process, not just manufacturing. The term is usually associated with Motorola, which named one of its key operational initiatives Six Sigma Quality.

SMTP (Simple Mail Transfer Protocol)—SMTP provides mechanisms for the transmission of mail from the sending user'' host to the receiving user's host. The MAIL command gives the forward-path or source route, while the reverse-path is a return route (which is used to return a message to the sender when an error occurs with a relayed message).

Staging—Laying out material on the construction site to ensure that all required materials are and will be available for use at time of construction.

Strain—An injury caused by a muscle, tendon, or ligament stretching beyond its natural capacity.

Stress—Demand (or "burden") on the human body caused by something outside of the body, such as a work task, the physical environment, work/rest schedules, and social relationships.

Supply chain—Includes material and informational interchanges as part of the logistical process stretching from acquisition of raw materials to delivery of finished project.

Symbol—Object, character, or other representation of an idea, concept, or other abstraction.

Tagout—A form of hazardous energy control procedure using the placement of a tagout device, in accordance with established procedures, on an energy-isolating device to indicate that the energy-isolating device and the system being controlled may not be operated until the tagout device is removed.

Taguchi method—A concept of offline quality control methods conducted at the product and process design stages in the product development cycle. This concept, expressed by Genichi Taguchi, encompasses three phases of product design: system design, parameter design, and tolerance design.

TCP (Transmission Control Protocol)—TCP takes the information and breaks down the larger data packages into smaller pieces for transmission using the IP addresses. TCP reassembles the packets at the destination end and performs error-checking to ensure all of the packets arrived properly and were reassembled in the correct sequence so that they can be understood.

TCP/IP—A suite of protocols making up the basic framework for communication on the Internet.

Telecommunication, analog—Continuous but changing signals travel along a medium such as copper wire or are carried over the air as radio waves. Includes traditional telephone, cellular phones, and television.

Telecommunication, digital—The transmission of signals in discrete pulses of 0s and 1s rather than continuously. A vast array of networks are used to connect the needed devices.

Teleconferencing—Live exchange of information among persons and machines remote from one another but linked by a telecommunications system, usually over the phone line.

Telegraph—A device for transmission of written messages without physical transport of letters over wire using the Morse code(s). Optical telegraphs included smoke signals and beacons.

Telephone—Telecommunications device to transmit and receive sound (most commonly, speech) across distance using analog (traditional) or digital signals.

Tendon—Tough ropelike material that connects the muscles to the bones. Tendons transfer forces and movements from the muscles to the bones. Tendons do not stretch, and excessive force or twisting may cause them to tear or fray like a rope.

Tension neck syndrome (TNS)—Fatigue, stiffness, tenderness, swelling, weakness, or pain in the neck or shoulder area, or headache radiating from the neck. It is caused by strain on various neck and shoulder muscles, often from long periods of looking upward. The trapezius muscle is particularly affected and may develop a "knot."

Trailer drop—When a driver drops off a full truck at the construction site and picks up an empty one.

Translation—The interpretation of the meaning of a text in one language (the "source text") and the production, in another language, of an equivalent that communicates the same meaning

Transmitting message—Converting message content into signals for transmission to receiver.

Upstream partners—The suppliers that provide goods and services to the construction company.

USB—A high-speed bidirectional serial connection between a PC used to transfer data between the computer, and peripherals such as digital cameras and memory cards. The USB 2.0 specification provides a data rate of up to 480 Mbps.

Verbal communication—See communication, verbal.

Videoconferencing/e-meeting—Use of audio, video, and computer technologies to bring people at different sites together for a meeting. This can be point-to-point or involve several sites (multipoint) with more than one person in large rooms at different sites. Includes the sharing of documents, computer-displayed information (CAD), and whiteboards.

Waste (muda)—Effects of poorly planned and controlled production processes. Includes time wasted, damaged materials, errors in production, accidents, etc.

Webcasting—A live media file distributed over the Internet using streaming media technology. Essentially, webcasting is broadcasting over the Internet.

Web conferencing—Live meetings or presentations over the Internet.

Web/Internet forum—A web application for holding discussions and posting user-generated content.

Webinar—A type of Web conference. The presenter speaks in a teleconference mode, pointing out information being presented on-screen, and the audience can respond over their own telephones.

Web page—A source of information that can be accessed through a Web browser. This information is usually in HTML or XHTML format, and may provide navigation to other Web pages via hypertext links. The Web server may restrict access only to a private network (e.g., a corporate intranet), or it may publish pages on the World Wide Web. Web pages are requested and served from Web servers using the Hypertext Transfer Protocol (HTTP).

Whole body vibration (WBV)—Working conditions that involve sitting, standing, or lying on a vibrating surface. Excessive exposure may contribute to back pain.

Wireless 802.11g—An IEEE standard for a wireless network that operates at 2.4 GHz Wi-Fi, with rates up to 54 Mbps.

Wireless bridge—A wireless device that connects multiple networks together.

Wireless, mesh network—A communications network with least two pathways to each node, forming a netlike organization. When each node is connected to every other node, the network is said to be fully meshed. When only some of the nodes are linked, switching is required to make all the connections, and the network is said to be partially meshed, or partially connected.

Wireless repeater—A device that extends the coverage of an existing access point by relaying its signal. A wireless repeater does not do intelligent routing performed by wireless bridges and routers.

Wireless RFID (Radio Frequency Identification)—An electronic identification technology that uses radio frequency signals to read identifying data contained in tags on equipment and merchandise. An alternative to bar codes.

Wireless router—A device that accepts connections from wireless devices to a network (Ethernet) and includes a network firewall for security, and provides local network addresses.

Wireless, Bluetooth—A 2.4 GHz frequency communication technology designed for short-range (30 feet), wireless communications among computing devices and mobile products, including PCs and laptop computers, personal digital assistants, printers, and mobile phones.

Wireless, CSMA/CA (carrier sense multiple access/collision avoidance)—The principal media access control strategy used in 802.11 networks to avoid data collisions. It is a "listen before talk" method of minimizing collisions. The network node checks to see if the transmission channel is clear before a data packet is sent.

WLAN (wireless local area network)—A data communications network. The term is used to distinguish between phone-based data networks and Wi-Fi networks. Phone networks are considered WANs and Wi-Fi networks are considered wireless local area networks.

WMAN (wireless metropolitan area network)—A wireless data network that is comparable to a cellphone network, in that users throughout a metropolitan area can freely access the Internet.

Work-in-process (WIP)—Parts and elements built to add to the constructed facility. Work in process generally includes all of the materials, labor, and overhead charged against the project.

Work-related musculoskeletal disorder (WMSD)—A musculoskeletal disorder caused or made worse by the work environment. WMSDs can cause severe symptoms such as pain, numbness, and tingling; reduced productivity; lost time from work; temporary or permanent disability; loss of motion; inability to perform job tasks; and an increase in workers' compensation costs.

WPAN (wireless personal area network)—A network that wirelessly connects personal devices centered within a radius of about 30 feet, such as an individual's workspace or room environment in a home. WPAN technologies include Bluetooth and others defined by the IEEE 802.15 standard. Device specifications include low data rates (250 kbps, 40 kbps, and 20 kbps), and multi-month to multi-year battery life a

Writing—Act of composing a text with words of a written language.

Index

Page numbers followed by t and f refer to tables and figures